THE URBAN AND REGIONAL PLANNING READER

The Urban and Regional Planning Reader draws together the very best of classic and contemporary writings to illuminate the planning of cities and metropolitan areas. Forty-seven generous selections include contributions from Lewis Mumford, Jane Jacobs, Ian McHarg, Paul Davidoff, Charles Harr, Susan Fainstein, and Charles J. Hoch through to Timothy Beatley, Jonathan Barnett, Alex Garvin, Tom Daniels, Andres Duany, and Barbara Faga. The variety and wide selection of readings offers one of the most innovative amalgamations of planning research and practice.

The Reader lays out the context, range of concerns, history, methods and key topics for 21st century urban and regional planning. Sections on the world of planning, history and theory, classic readings, practice and current issues include writings with a focus on the distribution of space and place, housing, transportation design, environment, community development, the effects of cultural diversity and information technology on land use and other topics. It displays the techniques used to direct and control growth, including zoning, master planning, public budgeting and citizen participation. It explores different types of plans distinguished by their scale and reference type. It references analytical and presentation techniques and outlines ethical issues confronting planners.

This *Urban and Regional Planning Reader* provides an essential resource for students of planning, drawing together important but widely dispersed writing, while the associated bibliography enables deeper investigations. The synthesis is also valuable for lecturers and researchers in the area and the pertinent editorial commentaries preceding each entry not only demonstrate its significance, but also outline the issue surrounding the topic.

Eugénie L. Birch FAICP, Lawrence C. Nussdorf Professor of Urban Research and Education, Department of City and Regional Planning, School of Design, University of Pennsylvania.

124

THE ROUTLEDGE URBAN READER SERIES

Series editors

Richard T. LeGates

Professor of Urban Studies, San Francisco State University

Frederic Stout

Lecturer in Urban Studies, Stanford University

The Routledge Urban Reader Series responds to the need for comprehensive coverage of the classic and essential texts that form the basis of intellectual work in the various academic disciplines and professional fields concerned with cities.

The readers focus on the key topics encountered by undergraduates, graduates and scholars in urban studies and allied fields. They discuss the contributions of major theoreticians and practitioners and other individuals, groups, and organizations that study the city or practice in a field that directly affects the city.

As well as drawing together the best of classic and contemporary writings on the city, each reader features extensive general, section and selection introductions prepared by the volume editors to place the selections in context, illustrate relations among topics, provide information on the author and point readers towards additional related bibliographic material.

Each reader contains:

- Approximately thirty-six *selections* divided into approximately six sections. Almost all of the selections will be previously published works that have appeared as journal articles or portions of books.
- A *general introduction* describing the nature and purpose of the reader.
- Two- to three-page *section introductions* for each section of the reader to place the readings in context.
- A one-page *selection introduction* for each selection describing the author, the intellectual background of the selection, competing views of the subject matter of the selection and bibliographic references to other readings by the same author and other readings related to the topic.
- A plate section with twelve to fifteen plates and illustrations at the beginning of each section.
- An index.

The types of readers and forthcoming titles are as follows:

THE CITY READER

The City Reader, fourth edition – an interdisciplinary urban reader aimed at urban studies, urban planning, urban geography and urban sociology courses – will be the anchor urban reader. Routledge published the first edition of *The City Reader* in 1996, a second edition in 2000 and a third edition in 2003. *The City Reader* has become one of the most widely used anthologies in urban studies, urban geography, urban sociology and urban planning courses in the world.

URBAN DISCIPLINARY READERS

The series will contain urban disciplinary readers organized around social science disciplines. The urban disciplinary readers will include both classic writings and recent, cutting-edge contributions to the respective disciplines. They will be lively, high-quality, competitively priced readers which faculty can adopt as course texts and which will also appeal to a wider audience.

TOPICAL URBAN ANTHOLOGIES

The urban series will also include topical urban readers intended both as primary and supplemental course texts and for the trade and professional market.

Books in the series so far:

INTERDISCIPLINARY ANCHOR TITLE

The City Reader, fourth edition
Richard T. LeGates and Frederic Stout (eds)

URBAN DISCIPLINARY READERS

The Urban Geography Reader
Nick Fyfe and Judith Kenny (eds)

The Urban Sociology Reader
Jan Lin and Christopher Mele (eds)

The Urban Politics Reader
Elizabeth Strom and John Mollenkopf (eds)

The Urban and Regional Planning Reader
Eugénie L. Birch (ed.)

TOPICAL URBAN READERS

The City Cultures Reader, second edition
Malcolm Miles, Tim Hall with Iain Borden (eds)

The Cybercities Reader
Stephen Graham (ed.)

The Sustainable Urban Development Reader
Stephen M. Wheeler and Timothy Beatley (eds)

The Global Cities Reader
Neil Brenner and Roger Keil (eds)

For further information on the Routledge Urban Reader series, please contact:

Andrew Mould
Routledge
2 Park Square
Milton Park
Abingdon
Oxfordshire OX14 4RN,
England
(andrew.mould@routledge.co.uk)

Richard T. LeGates
Urban Studies Program
San Francisco State University
1600 Holloway Avenue
San Francisco
California 94132
U.S.A.
(dlegates@sfsu.edu)

Frederic Stout
Urban Studies Program
Stanford University
Stanford
California 94305-6050
U.S.A.
(fstout@stanford.edu)

The Urban and Regional Planning Reader

Edited by

Eugénie L. Birch

Routledge
Taylor & Francis Group

LONDON AND NEW YORK

First published 2009 by Routledge
2 Park Square, Milton Park, Abingdon, Oxon, OX14 4RN

Simultaneously published in the USA and Canada
by Routledge
270 Madison Avenue, New York, NY 10016

Routledge is an imprint of the Taylor & Francis Group

Typeset in 9.5/12pt Amasis MT by Graphicraft Limited, Hong Kong
Printed and bound in Great Britain by Bell and Bain Ltd, Glasgow

British Library Cataloguing in Publication Data
A catalogue record for this book is available from the British Library

Library of Congress Cataloguing in Publication Data
The urban and regional planning reader/Eugénie L. Birch.
p. cm.
Includes bibliographical references and index.
1. City planning. 2. City planning—United States. 3. Regional planning.
4. Regional planning—United States. I. Birch, Eugénie Ladner.
HT166.U7125 2008
307.1′216—dc22

2008019153

ISBN 13: 978-0-415-31997-3 (pbk)
ISBN 13: 978-0-415-31998-0 (hbk)
ISBN 13: 978-0-203-62640-5 (ebk)

ISBN 10: 0-415-31997-8 (pbk)
ISBN 10: 0-415-31998-6 (hbk)
ISBN 10: 0-203-62640-0 (ebk)

To my wonderful family and, especially, Bob.

To the brilliant Penn city planning students and those in planning programs elsewhere who will shape the future world.

To my sensational Penn colleagues and those professors everywhere who are guiding the next generation of urban planners.

To the splendid cadre of urban planners in the United States and throughout the world whose work inspires us.

Contents

List of plates

List of contributors

Alan Altschuler Ruth and Frank Stanton Professor of Urban Policy and Planning, Graduate School of Design and Kennedy School of Government, Harvard University, Cambridge, MA; author, *Megaprojects: The Changing Politics of Urban Public Investments* (Washington, DC: Brookings Institution, 2003) with David Luberoff.

Miquel Barceló President, Barcelona@22 SA.

Jonathan Barnett FAIA, FAICP, Professor of Practice, Department of City and Regional Planning, School of Design, University of Pennsylvania, Philadelphia, PA; consultant, WRT; author, *Smart Growth in a Changing World* (Chicago: APA Planners Press, 2007).

Carol D. Barrett FAICP, Planning Manager, City of San Gabriel, California; author, *Everyday Ethics for Practicing Planners* (Chicago: APA Planners Press, 2001).

Timothy Beatley Theresa Heinz Professor, Department of Urban and Environmental Planning, School of Architecture, University of Virginia, Charlottesville, VA; author, *Native to Nowhere: Sustaining Home and Community in a Global Age* (Washington, DC: Island Press, 2004); *Green Urbanism, Learning from European Cities* (Washington, DC: Island Press, 2000).

Alain Bertaud Consultant and former Principal Urban Planner, Urban Development Division, World Bank, Washington, DC; author, *Analyzing Building Height Restrictions: Predicted Impacts, Welfare Costs, and a Case Study of Bangalore, India*, World Bank Policy Research Working Paper No. 3290 (April 22, 2004) with Jan K. Brueckner, and "Transit and Density: Atlanta, the United States and Western Europe" in Harry W. Richardson and Chang-Hee Christine Bae (eds) *Urban Sprawl in Western Europe and the United States* (Aldershot: Ashgate, 2004).

Eugénie L. Birch FAICP, Lawrence C. Nussdorf Professor of Urban Research and Education, Department of City and Regional Planning, School of Design, and Codirector, Penn Institute for Urban Research, University of Pennsylvania, Philadelphia; editor, *Rebuilding Urban Places after Disaster: Lessons from Katrina* (Philadelphia: University of Pennsylvania Press, 2006) and *Growing Greener Cities: Urban Sustainability in the Twenty-first Century* (Philadelphia: University of Pennsylvania Press, 2008) with Susan Wachter.

F. Stuart Chapin Jr FAICP, Professor of City and Regional Planning *emeritus*, Department of City and Regional Planning, University of North Carolina, Chapel Hill, NC, 1949–78; author, *Urban Land Use Planning*, first through fourth editions (Urbana, IL, and Chicago: University of Illinois Press, 1957–95).

Linda C. Dalton AICP, Vice-president of Planning and Enrollment Management, California State University, East Bay, Hayward, CA; author, *The Practice of Local Government Planning* (Washington, DC: International City/County Managers Association, 2000) with Charles J. Hoch and Frank S. So.

Katherine Daniels AICP, Senior Planner, New York Planning Federation; Adjunct Professor, SUNY Albany, Albany, NY; author, *The Environmental Planning Handbook* (Chicago: APA Planners Press, 2003) with Thomas L. Daniels.

Thomas L. Daniels Professor of City and Regional Planning, Department of City and Regional Planning, School of Design, University of Pennsylvania, Philadelphia, PA; author, *The Environmental Planning*

Handbook (Chicago: APA Planners Press, 2003) with Katherine Daniels; *When City and Country Collide* (Washington, DC: Island Press, 1998).

Paul Davidoff (1930–84) Professor of City and Regional Planning, Department of City and Regional Planning, Graduate School of Fine Arts, University of Pennsylvania, Philadelphia, 1958–65; Professor and Director, Master of Urban Planning program, Hunter College/CUNY, 1965–69; founding Executive Director, Suburban Action Institute (later Metropolitan Action Institute), 1969–84.

Dawn Dhavale Planner, Environmental and Development Branch, Department of Planning and Zoning, Fairfax County, VA; doctoral candidate, Virginia Tech.

Andrés Duany FAIA, Founding Principal, Duany Plater-Zyberk & Co., Miami, FL; author, *New Civic Art* (New York: Rizzoli, 2003) with Elizabeth Plater-Zyberk and Robert Alminana.

Ann-Margaret Esnard Associate Professor of Urban and Regional Planning and Director, Visual Planning Technology Lab, College of Architecture, Urban and Public Affairs, Florida Atlantic University, Fort Lauderdale, FL.

Barbara Faga FASLA, Executive Vice-president, EDAW; author, *Designing Public Consensus: The Civic Theater of Community Participation for Architects, Landscape Architects, Planners and Urban Designers* (Hoboken, NJ: Wiley, 2006).

Susan Fainstein Professor of Urban Planning, Department of Urban Planning and Design, Graduate School of Design, Harvard University, Cambridge, MA; author, *The City Builders: Property Development in New York and London, 1980–2000* (Lawrence, KS: University of Kansas Press, 2001); editor, *Gender and Planning* (New Brunswick, NJ: Rutgers University Press, 2005) with Lisa Servon.

Alexander Garvin AICP, founder and principal, Alexander Garvin & Associates, New York, NY; adjunct professor, Yale School of Architecture, Yale University, New Haven, CT; author, *The American City: What Works, What Doesn't* (New York: McGraw-Hill, 2002).

Judith Getzels Former Director of Research, American Planning Association; author, *The Practice of Local Government Planning*, second edition (Washington, DC: International City Management Association, 1988) with Frank S. So.

Oliver Gillham AIA, architect and planner; author, *The Limitless City: A Primer on the Urban Sprawl Debate* (Washington, DC: Island Press, 2002).

David R. Godschalk FAICP, Stephen Baxter Professor *emeritus*, Department of City and Regional Planning, University of North Carolina, Chapel Hill, NC; author, *Urban Land Use Planning*, fifth edition (Urbana, IL, and Chicago: University of Illinois Press, 2006) with Philip R. Berke, Edward J. Kaiser and Daniel Rodríguez.

Charles M. Haar Louis D. Brandeis Professor of Law *emeritus*, Harvard Law School, Harvard University, Cambridge, MA; author, *Mastering Boston Harbor: Courts, Dolphins, and Imperiled Waters* (Cambridge, MA: Harvard University Press, 2004); *Suburbs under Siege: Race, Space and Audacious Judges* (Princeton, NJ: Princeton University Press, 1998).

Gary Hack AICP, Dean *emeritus* and Paley Professor of City and Regional Planning, School of Design, University of Pennsylvania, Philadelphia; editor, *Practice of Local Planning* (Washington, DC: International City/County Management Association, 2008) with Eugénie L. Birch, Paul Sedway and Mitchell Silver; editor, *Global City Regions: Their Emerging Forms* (London: Taylor & Francis, 2001) with Roger Simmonds.

Dolores Hayden Professor of Architecture, Urbanism and American Studies, Yale University, New Haven, CT; author, *A Field Guide to Sprawl* (New York: Norton, 2004); *Building Suburbia, Green Fields and Urban Growth, 1820–2000* (New York: Pantheon, 2003).

Tony Hiss Author, *The View from Alger's Window: A Son's Memoir* (New York: Vintage Books, 2000); *A Region at Risk: The Third Regional Plan for the New York–New Jersey–Connecticut Metropolitan Area* (Washington, DC: Island Press, 1996) with Robert D. Yaro; *The Experience of Place* (New York: Vintage Books, 1991).

Charles J. Hoch Professor of Urban Planning, College of Urban Planning and Urban Affairs, University of Illinois, Chicago; author, *The Practice of Local Government Planning* (Washington, DC: International City/County Managers Association, 2000) with Linda C. Dalton and Frank S. So.

Lewis D. Hopkins FAICP, Professor of Urban and Regional Planning and Landscape Architecture; author, *Urban Development: The Logic of Making Plans* (Washington, DC: Island Press, 2001); editor, *Engaging the Future: Forecasts, Scenarios, Plans, and Projects* (Cambridge, MA: Lincoln Institute of Land Policy, 2007) with Marisa Zapata.

Jane Jacobs (1916–2006) Journalist; author, *The Death and Life of Great American Cities* (New York: Random House and Vintage Books, 1961); *The Economy of Cities* (New York: Random House, 1969); *Cities and the Wealth of Nations* (New York: Random House, 1984); *The Nature of Economies* (New York: Modern Library/Random House, 2000); and *Dark Age Ahead* (New York: Random House, 2004).

Edward J. Kaiser FAICP, Professor of City and Regional Planning *emeritus*, Department of City and Regional Planning, University of North Carolina, Chapel Hill, NC; author, *Urban Land Use Planning*, fifth edition (Urbana, IL, and Chicago: University of Illinois Press, 2006) with Philip R. Berke, David R. Godschalk, and Daniel Rodríguez.

Joel Kotkin Journalist; Irvine Senior Fellow, New America Foundation, Washington, DC; author, *The City: A Global History* (New York: Random House, Modern Library Edition, 2005) and *The New Geography: How the Digital Revolution is Reshaping the American Landscape* (New York: Random House, 2000).

Robert E. Lang Associate Professor, School of Planning and International Affairs, College of Architecture and Urban Studies, Virginia Polytechnic Institute and State University, Alexandria Center, Alexandria, Virginia; Senior Fellow, Brookings Institution; author, *Boomburgs: The Rise of America's Accidental Cities* (Washington, DC: Brookings Institution Press, 2007) with Jennifer LeFurgy.

Richard T. LeGates Professor of Urban Studies, San Francisco State University, author, *Think Globally, Act Regionally: GIS and Data Visualization for Social Science and Public Policy Research* (Redlands, CA: ESRI Press, 2005), editor, *The City Reader* (London and New York: Routledge, fourth edition 2006) with Frederic Stout.

Nancey Green Leigh FAICP, Professor of City and Regional Planning, College of Architecture, Georgia Institute of Technology, Atlanta Georgia; author, *Economic Revitalization: Cases and Strategies for City and Suburb* (Thousand Oaks, CA: Sage Publications, 2002) with Joan Fitzgerald.

David Luberoff Executive Director, Rappaport Institute for Greater Boston, Kennedy School, Harvard University, Cambridge, MA; author, *Mega-projects: The Changing Politics of Urban Public Investments* (Washington, DC: Brookings Institution, 2003) with Alan Altschuler.

William H. Lucy Lawrence Lewis Jr Professor, Department of Urban and Environmental Planning, School of Architecture, University of Virginia, Charolottesville, VA; author, *Tomorrow's Cities, Tomorrow's Suburbs* (Chicago: APA Planners Press, 2006) with David L. Phillips.

Kevin Lynch (1918–84) Professor of Urban Planning, Department of Urban Studies and Planning, School of Architecture and Planning, Massachusetts Institute of Technology, Cambridge, MA (1949–84); author, *Site Planning* (Cambridge: MIT Press, 1984) with Gary Hack, *A Theory of Good City Form* (Cambridge, MA: MIT Press, 1981), *Image of the City* (Cambridge, MA: MIT Press, 1960).

Richard A. Matthew Associate Professor of Planning, Policy and Design, School of Social Ecology, University of California, Irvine; Director, Center for Unconventional Security, University of California, Irvine; editor, *Land Mines and Human Security: War's Hidden Legacy* (Albany: SUNY Press, 2004) with Bryan L. McDonald and Ken Rutherford.

Bryan McDonald Assistant Director, Center for Unconventional Security Affairs, University of California, Irvine, CA.

Ian McHarg (1920–2001) Founding Chair and Professor of Landscape Architecture, Department of Landscape Architecture (1954–85), Graduate School of Fine Arts, University of Pennsylvania, Philadelphia; author, *Design with Nature* (New York: Ballantine Books, 1969).

Matthew J. Michel Attorney, Beery Elsner & Hammond, Portland, OR.

Jay H. Moor Chief of Strategic Planning, Office of the Executive Director, UN-Habitat.

Lewis Mumford (1895–1990) Author, *The City in History: Its Origins, its Transformations, and its Prospects* (New York: Harcourt Brace, 1961); *The Urban Prospect* (New York: Harcourt Brace, 1968); *Sketches from Life: Autobiography of Lewis Mumford* (New York: Dial Press, 1982); Professor of City and Regional Planning, Graduate School of Fine Arts, University of Pennsylvania (1951–61); author, "The Skyline" column, *New Yorker Magazine* (1931–63).

Dowell Myers Professor of Urban Planning, School of Policy, Planning and Development, University of Southern California, Los Angeles; author, *Immigrants and Boomers: Forging a New Social Contract for the Future of America* (New York: Russell Sage Foundation, 2007).

Michael Neuman Associate Professor, Department of Landscape Architecture and Urban Planning, College of Architecture, Texas A&M University, College Station, TX.

John R. Nolon Professor, Pace Law School, Pace University, White Plains, NY; Visiting Professor of Environmental Law, Yale School of Forestry, Yale University, New Haven, CT; editor, *Losing Ground: A Nation on Edge* (Washington, DC: Environmental Law Institute, 2007) with Daniel B. Rodriguez; author, *Land Use: Cases and Materials*, sixth edition (Minneapolis, MN: Thomson West, 2003) with Morton Gitelman, Patricia E. Salkin and Robert R. Wright.

Myron Orfield Associate Professor, Law School, University of Minnesota, Minneapolis, MN; author, *American Metropolitics: The New Suburban Reality* (Washington, DC: Brookings Institution Press, 2002).

Milton R. Ospina Urban and Regional Planning and Economic Development Solution Manager, ESRI, Redlands, CA; editor, *Measuring Up: The Business Case for GIS* (Redlands, CA: ESRI Press, 2004) with Christopher Thomas.

David L. Phillips Associate Professor, Department of Urban and Environmental Planning, School of Architecture, University of Virginia; author, *Tomorrow's Cities, Tomorrow's Suburbs* (Chicago: APA Planners Press, 2006) with William H. Lucy.

Elizabeth Plater-Zyberk FAIA, Founding Principal, Duany Plater-Zyberk & Co., Distinguished Professsor and Dean, School of Architecture, University of Miami, Coral Gables, FL; author, *New Civic Art* (New York: Rizzoli, 2003) with Andrés Duany and Robert Alminana.

Mohammad A. Qadeer Professor *emeritus*, School of Urban and Regional Planning, Queens University, Kingston, Ontario, Canada.

Catherine L. Ross Harry West Professor of Quality Growth and Regional Development; Director, Center for Quality Growth, Georgia Institute of Technology, Atlanta, GA; editor, *The Inner City: Urban Poverty and Economic Development in the Next Century* (New Brunswick, NJ: Transaction Publishers, 1997) with Thomas Boston.

Nancy Sappington Marketing writer, ESRI, Redlands, CA; editor, *ESRI Map Book*, Vol. 21 (Redlands, CA: ESRI Press, 2006).

Jeff Speck Architect, Canopy Development, Northampton, MA; former Director of Design, National Endowment for the Arts, and Director, Town Planning, Duany Plater-Zyberk & Co.; author, *Suburban Nation: The Rise of Sprawl and the Decline of the American Dream* (New York: North Point Press, 2000) with Andrés Duany and Elizabeth Plater-Zyberk.

Frank S. So FAICP, former Executive Director, American Planning Association; author, *The Practice of Local Government Planning* (Washington, DC: International City/County Managers Association, 2000) with Charles J. Hoch and Linda C. Dalton.

Frederic Stout Lecturer, Urban Studies Program, Stanford University, Palo Alto, CA; editor, *The City Reader* (London and New York: Routledge, fourth edition 2006) with Richard T. LeGates.

Edward J. Sullivan Attorney and partner, Garvey Schubert Barer, Portland, OR; author, "Year Zero: The Aftermath of Measure 37," 36 *Environmental Law* 131 (2006).

Nigel Taylor Principal Lecturer, Planning and Architecture, Faculty of the Built Environment, University of the West of England, Bristol, England.

Michael B. Teitz Professor of City and Regional Planning *emeritus*, College of Environmental Design, University of California, Berkeley, CA; Senior Fellow, Public Policy Institute of California, San Francisco.

Vukan R. Vuchic UPS Foundation Professor of Transportation Engineering, School of Engineering and Applied Science, University of Pennsylvania, Philadelphia; author, *Urban Transit Systems and Technology* (Hoboken, NJ: Wiley, 2007) and *Urban Transit Operations: Planning and Economics* (Hoboken, NJ: Wiley, 2005).

Rasna Warah Photographer and former Editor-in-chief, *UN-Habitat Debate.*

Robert D. Yaro President, Regional Plan Association of New York; Professor of Practice, Department of City and Regional Planning, School of Design, University of Pennsylvania, Philadelphia; author, *A Region at Risk: The Third Regional Plan for New York–New Jersey–Connecticut Metropolitan Area* (Washington, DC: Island Press, 1996) with Tony Hiss.

Acknowledgments

Many thanks to the large numbers of colleagues, associates and students who have offered help in assembling this reader and critiquing the editor's commentary. Andrew Mould, the ever-patient editor at Routledge has been a key supporter. His assistants Michael P. Jones and Jennifer Page have also helped keep track of important details. Richard LeGates, series editor has offered wise and steady guidance. Over the years, numerous students have read and reacted to the contents especially those in my introduction to city planning course at Penn and doctoral candidates (now graduated and teaching their own courses) who were teaching assistants including Peter Brown, Daniel Campo and Melissa Saunders. Several research assistants helped with the myriad details, and Ian Anderson, Emily Yuhas and Collins Makunda diligently pursued copyright data and images. Jane Margolies was a wonderful editor whose intelligent comments were much valued. David Gest offered organizational advice. Finally, several anonymous reviewers offered helpful observations that shaped this *Urban and Regional Planning Reader*.

Introduction

The *Urban and Regional Planning Reader* lays out the context, the range of concerns, history, theory, methods and key topics for 21st century urban and regional planning. It is intended to stimulate students to either pursue the field in more depth or to be informed as professionals in related fields or as citizens with regard to the planning issues whatever their future careers. The *Urban and Regional Planning Reader* contains forty-six selections chosen from the vast literature of a field that encompasses both a scholarly discipline and a profession, a dual identity that provokes lively debate and a constant quest for state-of-the-art knowledge among its practitioners.

The editor has assembled readings that define the field as it has emerged in the beginning of the 21st century – one that has the distribution of space and arrangement of place as its defining nature. While the editor will be sensitive to political, economic and social conditions in which urban and regional planning is practiced, she will focus closely on the processes and products that define the planners' contemporary expertise and niche. She anticipates that this *Reader* would serve well as either a primary or supplemental text for the hundreds of urban planning courses now offered worldwide. It would also partner well with the *City Reader* as there would be little duplication of selections. There is a substantial trade market for this reader among practicing planners and the general public interested in urban planning issues.

The authors of the majority of the selections are urban and regional planners who write from the perspective of academic or professional practitioners, seeking to inform and contribute to the field. Most are instructors in departments or programs of city and regional planning and thus have exposure to the field's research and pedagogical issues. While the text will refer to material from geography, sociology, political science, and other disciplines and professional fields, the emphasis on writings on planning by planners will distinguish this reader from other offerings.

This *Reader* has seven parts. The first, "The World of Urban and Regional Planning," outlines the environment in which urban and regional planning operates. The second, "History and Theory of Urban and Regional Planning," reviews the history of the profession and its theoretical underpinnings. The third, "Classic in Urban and Regional Planning," offers the views of selected pioneers in the field. The fourth, "The Plan: Its Origins and Contemporary Uses," surveys how planners developed the comprehensive plan over time and how they use it today. The fifth, "Planning Practice and Methods," explains just what planners do and how they do it. The sixth, "Key Topics in Urban and Regional Planning," highlights subfields in urban and regional planning, ranging from transportation to urban revitalization to environmental planning. The seventh, "Emerging Issue in Urban and Regional Planning," highlights topics that will become increasingly important in the decades to come.

A note about the artwork: the cover is a rendering from the award-winning, year-long Penn Praxis planning program focused on Philadelphia's industrial waterfront. The section divider images portray the physical growth of a typical northeastern city from 1680 to 2000. Andrew Whittemore, a PhD student at UCLA is the artist. In collaboration with Sam Bass Warner, he will use these drawings as the centerpiece of the forthcoming book *The City Grows: An Illustrated History of the American Metropolis* forthcoming from the University of Pennsylvania Press's City in the 21st Century Series (www.upenn.edu/pennpress/series/C21.html).

PART ONE

The World of Urban and Regional Planning

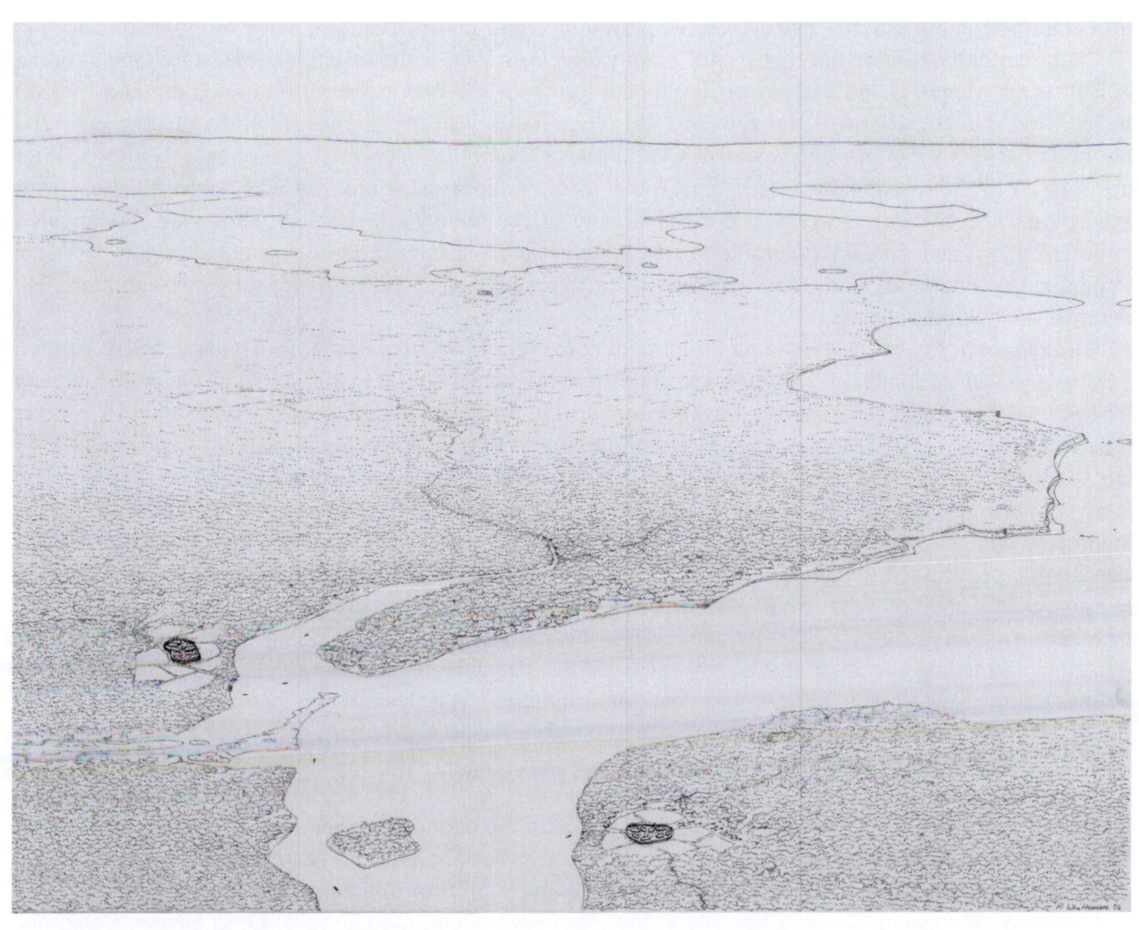

Growth of a typical Northeastern city: 1600s. (Drawing by Andrew Whittemore)

INTRODUCTION TO PART ONE

"Honey, I think the world is flat," whispered Pulitzer Prize winner Thomas L. Friedman to his wife. He was immersed in a research project on globalization, or worldwide economic integration, that would become the basis of his book *The World is Flat: A Brief History of the Twenty-first Century* (New York: Picador/Farrar Straus & Giroux), originally published in 2005 and revised twice since then. After visiting Bangalore, India, and other fast-growing cites, he had concluded that modern technology was leveling the playing field, allowing more people in distant places to participate in global markets. Friedman's assessment is a good way to think about the world in which urban and regional planners now operate. While he focuses on markets, competition, industry, and trade, his findings reveal a lot about how people live and work, thus informing the metrics – population, employment – that planners use to envision communities.

Friedman had confided his "flat world" observation via a long-distance phone call to his wife at home in Bethesda, Maryland. He had visited Bangalore to take a first-hand look at the area's reputed "Silicon Valley." He was impressed – and shocked. When embarking on the journey, Friedman had mused that his quest was the modern equivalent of Christopher Columbus's fifteenth-century search for a shorter trade route to the wealthy markets of East India. Like the Italian navigator, he was seeking riches – not the spices and silks of yesteryear, but the bounty of the twenty-first century: "software, brainpower, complex algorithms, knowledge workers, call centers, transmissions protocols, breakthroughs in optical engineering" (Friedman, p. 4). And in India, China, Singapore, and Tokyo he found them, and more. He discovered "Globalization 3.0."

Globalization 3.0, he proposed, is the natural outcome of two previous eras. The first, "Globalization 1.0," began with Columbus's 1492 voyage and lasted to about 1800. It featured mighty imperialist nations trading beyond their boundaries because they had the naval power or religious conviction to impose their will. The second, "Globalization 2.0," extended from 1800 to 2000, when multinational corporations led the way in developing global markets for goods and services by taking advantage of technological innovations in transportation and communications. Globalization 3.0, starting in 2000 and built on the immediate past, began with massive investment in worldwide digital communications infrastructure (broadband hookups, undersea fiber optic cable), the proliferation of inexpensive computers, and the explosion in communications software. It has strengthened worldwide economic growth by enabling individuals to collaborate in ways never before possible. They invented new means of doing business that involved outsourcing (moving a function that a firm was performing in-house to another location), offshoring (moving a factory to a place with less expensive costs of production), supply-chaining (purchasing products and goods from all over the world), insourcing (taking on new jobs, complementary to a firm's core business), and widespread use of wireless connections, instant messaging, and teleconferencing.

These Globalization 3.0 innovations have had enormous impacts on the physical arrangements among and within cities all over the world. For example, much of the consumer goods manufacturing has moved from the United States and Western Europe to Asia and Eastern Europe. New and old industrial towns and districts, transportation routes, and port facilities have felt the effects of this change.

Consider Shenzhen, China, a key city in the Pearl River Delta megalopolis; Long Beach, California, a major U.S. port; and Detroit, Michigan, a former industrial center. Thirty years ago, Shenzhen had a population of less than 25,000. Today it numbers 9 million. It supports more than 3 million jobs, primarily in manufacturing, and is the world's fourth busiest container ship port. This rapid urbanization has put pressure on the city to provide services and amenities. Air pollution, crowded housing, and traffic congestion are common. Shenzhen ships a substantial portion of its production to the United States, a massive consumer goods market, via Long Beach, California, the second largest U.S. container port after Los Angeles. Thirty years ago, this city had only 360,000 inhabitants, but as a result of growth fueled by shipping, it now has a population of 500,000. Goods distribution in Long Beach and its surroundings has spawned badly congested highways and rails, sprawling warehouses and storage facilities. Detroit, in contrast, which was the center of the U.S. automobile industry thirty years ago, has seen its population shrink from 1.5 million in 1970 to about 920,000 today, while jobs dropped from over 600,000 to around 380,000. The city has one of the nation's highest rates of abandoned housing and concentrated poverty. Each of these places has a unique profile related to its population and economy. Each has its own set of planning problems.

In the Globalization 3.0 era, people pursue the promise of employment and prosperity by moving across national boundaries, from rural areas to cities, from small cities to large cities, and from cities to suburbs. These population shifts affect all income levels. In the global South, for example, highly paid managerial workers settle in enclaves where they have services comparable to those of the global North. People with jobs that require less skill, or who are involved in the informal economy, gravitate to slums at the periphery of cities that in Africa, Asia and Latin America have sprung up rapidly. The housing is flimsy, constructed from scraps of metal or cardboard, largely without water and sewers. As households prosper, they upgrade their dwellings piecemeal, but cities are hard pressed to provide basic services. Meanwhile, in the global North, centers of communications, trade, and finance such as Chicago, Los Angeles, and New York have benefited enormously from Globalization 3.0. But prosperity can bring about the loss of affordable housing, traffic congestion, and air pollution, presenting planning problems far different from those of cities in the global South or even less prosperous areas of the United States.

The readings in Part One, "The World of Urban and Regional Planning," lay out the dimensions of the environment for planning in the twenty-first century. Jay H. Moor and Rasna Warah provide an overview of global urbanization in "State of the World's Cities." They outline urban conditions in the global South and global North, highlighting each area's urban problems. "Megaforces shaping the Future of the Nation's Cities," an excerpt from the 2000 U.S. Department of Housing and Urban Development's *State of the Cities* report, outlines key social and economic variables yielding today's metropolitan conditions. Notably, even though the HUD researchers report on decade-old data, the report's general themes – highlighting the important role of cities as engines of economic development, demonstrating selective city revitalization and outlining the disparities among U.S. cities – persist. Regrettably, HUD has failed to comply with the congressional mandate to produce an annual *State of the Cities* report. The most recent one dates from 2000.

Catherine Ross and Nancey Green Leigh's "Planning, Urban Revitalization and the Inner City: An Exploration of Structural Racism" argues that poor, minority communities have disproportionate numbers of brownfields (usually abandoned industrial sites), experience high rates of poverty, and lack residential choice owing to exclusionary suburban zoning. The culprit: thirty years of misguided urban policy. Mohammad A. Qadeer's "Pluralistic Planning for Multicultural Cities" takes the reader to Canada to show how urban planning can and should be sensitive to the values and belief systems of varied racial and ethnic groups. Although Qadeer focuses on immigrant communities in Toronto, his findings are applicable in any place that has welcomed newcomers.

Turning to the suburbs, two authors depict contemporary settlement patterns in the United States. Joel Kotkin's "Suburbia, Homeland of the American Future" asserts that the now dominant suburbs deserve more specialized attention from urban planners. Myron Orfield's "The New Suburban Reality" categorizes and maps suburbs to show their heterogeneous character and location in metropolitan areas. But, he observes, some types of suburbs have common issues. Primary among them is that their need to supply municipal services outstrips their ability to raise tax revenues to pay for them. This condition alone makes them natural allies. Unite, he advises. Form a powerful voting bloc to command state and national attention, because the problems at hand do not have municipal solutions.

Together, these six readings profile the conditions – from global to local – that shape the world in which urban and regional planners work. The environment is dynamic. The rates and levels of urbanization vary from place to place. The scale of urban problems is dramatically different worldwide. Understanding the character and nature of a place and the forces that are shaping it is the first task of planners. From this foundation they can begin to project ideas for enhancing local assets while simultaneously mitigating or reducing liabilities to yield beautiful, sustainable communities. That is planning.

1

The State of the World's Cities

Global Outlook: International Urban Research Monitor (2002)

Jay H. Moor and Rasna Warah

Editor's introduction

In 2006, for the first time in history, more people lived in cities than not. Around the world, cities in the twenty-first century face unprecedented challenges to meet the needs of this multibillion-person population, from the provision of basic services such as potable water and sewer systems to strong economic bases that employ city residents and raise their income and prosperity. Planners must understand the complex urban issues in both developed and developing countries, as well as the global trends that shape them. (See Plate 1.1.)

Data from the United Nations aid them in that task. The United Nations monitors world urbanization, and has focused attention on metropolitan growth and development and related issues through its Human Settlements Programme (UN-Habitat) based in Nairobi, Kenya. The United Nations established Habitat in 1978 following a conference known as Habitat I, in Vancouver, Canada, when only one-third of the world's population lived in urban areas. In 1996, leaders from more than 170 countries attended Habitat II in Istanbul, Turkey, overwhelmingly approving an urban agenda governed by three goals: ameliorating problems created by urban growth, promoting social and economic sustainability in urban areas, and alleviating urban poverty.

In 2000, the United Nations fortified the UN-Habitat urban agenda when 189 countries signed on to the Millennium Development Goals (MDGs), eight action items aimed at improving the quality of life of the world's citizens. Each of the MDGs has one or more numerical targets to be achieved by 2015. The MDGs are:

1 *Eradicate extreme poverty and hunger.* Examples of the numerical targets are: halve the proportion of people whose incomes are less than $1 a day, and halve the proportion of people who suffer from hunger.
2 *Achieve universal primary education.* Target: ensure that children everywhere, boys and girls alike, will be able to complete a full course of primary education.
3 *Promote gender equality and empower women.* Target: eliminate gender disparity in all levels of education.
4 *Reduce child mortality.* Target: reduce by two-thirds the under-five mortality rate.
5 *Improve maternal health.* Target: reduce by three-quarters the maternal mortality ratio.
6 *Combat HIV/AIDS, malaria, and other diseases.* Target: halt the spread of HIV/AIDS.
7 *Ensure environmental sustainability.* Target: halve the population without access to safe drinking water and basic sanitation.
8 *Develop a global partnership for development.* Target: address the special needs of the least developed countries, landlocked countries, and small, developing island states.

Particularly relevant to urban and regional planners is Goal 7, which aims to integrate environmental resource protection with national development policies. If the targets of this goal are met, an additional 350 million people will have clean water, 650 million will have basic sanitation, and 100 million slum dwellers will have significantly improved living conditions.

In a midterm assessment, the United Nations reported in *The Millennium Development Goals Report* (New York, 2007) that the results were uneven, showing substantial progress in some important areas but gains below expectations in others. Most places had met targets for extreme poverty rates, primary school enrollment, child mortality, and some diseases. For example, the rate of extreme poverty had fallen from one-third to one-fifth of the world's population. Primary school enrollment had increased from 80 percent to 88 percent in the developing world. Widespread immunization programs caused a decline of 60 percent in fatalities from childhood measles. However, only one of the eight regional groups is expected to meet all the MDGs, and sub-Saharan Africa is falling considerably behind. Obstacles to progress are: lack of jobs; fast, uncontrolled urbanization; water shortages; and disease, especially HIV. Among the most challenging issues are: maternal health, childhood obesity, basic sanitation, and the AIDS epidemic.

Since 2001 UN-Habitat has sponsored a biennial conference, the World Urban Forum, to examine urbanization and its impacts. The meeting held in Vancouver, Canada, in June 2006 and attended by 10,000 people representing more than 100 countries, reasserted that the twenty-first century's major challenge is population growth, especially in cities in developing countries. In light of annual increases averaging 70 million – equal to seven new "megacities" every year – conference participants emphasized the necessity of planning in managing growth and preventing future slums.

The UN-Habitat has a budget of approximately $300 million allocated to its three divisions: Shelter and Sustainable Human Settlements, Monitoring and Research, and Regional and Technical Cooperation. One element of its agenda has been to develop indicators to monitor urban trends related to population growth, economic conditions, housing, infrastructure, education, and other factors. As not all countries have the financial or administrative capacity to maintain regular censuses, UN-Habitat has set up a technical assistance program to assist in creating accurate and timely data collection systems. Furthermore, the organization works with countries to come up with other indicators to supplement basic data sets.

To undertake data collection, UN-Habitat established a Global Urban Observatory that began gathering information on a three-year cycle in 1993. It has provided a summary in its *State of the World's Cities* report, issued in 2001 and 2004. (It also has online databases – urban indicators, human settlements statistics, and city profiles – at www.unhabitat.org/programmes/guo/guo_databases.asp.) In these reports UN-Habitat provides global and regional demographic overviews of urban places, highlighting worldwide trends in metropolitan areas, focusing on the presence of poverty and such associated issues as shelter, environmental degradation, and governance difficulties.

Varying definitions of the term "urban" add complexity to global comparisons. For example, in an analysis of 228 countries, the United Nations reported that 14 percent either had no definition or included the whole nation as "urban," while 46 percent relied on political boundaries, and another 44 percent defined cities as places with a population ranging from 200 to 50,000.

The authors of this selection are Jay Moore, chief of Strategic Planning, Office of the Executive Director, UN-Habitat, and coordinator, Global Urban Observatory, and Rasna Warah, photographer and former editor-in-chief of *UN-Habitat Debate*, the quarterly publication of the UN Centre for Human Settlements, which can be accessed at www.unhabitat.org.

For background on issues of poverty, see Columbia University economist Jeffrey D. Sachs's *The End of Poverty: Economic Possibilities for our Time* (New York: Penguin Press, 2005), which presents national profiles. Sachs has received worldwide publicity for his work on developing countries and has been one of the intellectual forces behind the Millennium Development Goals. For more about the Millennium Development Goals, see UN Millennium Project 2005, *Investing in Development: A Practical Plan to Achieve the Millennium Development Goals* (Washington, DC: Communications Development, 2005), and visit the UN Millennium Development Goals website (www.un.org/millenniumgoals). Also see UN Human Settlements Programme, *The Challenge of Slums: Global Report on Human Settlements 2003* (London and Sterling, VA: Earthscan, 2003).

To focus on urban issues as defined by the United Nations, see Patricia L. McCarney, *Our Future: Sustainable Cities – Turning Ideas into Action, Background Paper, World Urban Forum III* (Nairobi: UN-Habitat, 2006). Marco Keiner, Martina Koll-Schretzenmayr, and Willy A. Schmid (eds) consider global warming and other twenty-first-century issues in *Managing Urban Futures: Sustainability and Urban Growth in Developing Countries*

(London: Ashgate, 2005), with essays covering planning and development in Asia, South America, and Africa. Focusing on Asia, former Philippine government official Aprodicio A. Laquian's *Beyond Metropolis: The Planning and Governance of Asia's Mega-urban Regions* (Baltimore, MD: Johns Hopkins University Press, 2005) documents the metropolitan growth of twelve major cities, including Shanghai, Beijing, Delhi, Tokyo, Osaka, Jakarta, and Singapore. This work is especially important since Asia will account for most future urban population growth and today includes six of the ten most populous cities in the world. Finally, for a fast, provocative read, see *Planet of Slums* (London and New York: Verso Books, 2006) by MacArthur Fellow Mike Davis, who provides his explanation for the growth of slums and the failure of contemporary housing programs.

A WORLD OF CITIES

With just under half of its population living in cities, the world is already urbanized. When measured in knowledge, attitude, aspiration, commercial sense, technology, travel and access to information, even most rural societies are, to one extent or another, woven into a global network of cities.

Globalization seriously escalated during the industrial revolution of the late eighteenth century. Since then, the steam engine, the telephone, the elevator, and now the internet and cheap air transport have conveyed people, goods, and information both horizontally and vertically at an unprecedented volume and velocity. The focal point of these activities has invariably been the city, a place of deals and decisions, takeoffs and landings – a place less concerned with the rhythms of nature, where everything can be bought or sold, especially one's ideas and labor.

In today's globalized world, cities no longer stand apart as islands. They are the nexus of commerce, gateways to the world in one direction and focus of their own hinterlands in the other. Tied together in a vast web of communication and transport, cities are concentrations of energy in a global force field. In a real sense, the world is completely urbanized, as this force field has the power to connect all places and all people into a productive, constantly adapting unity.

Three billion people – nearly every other person on earth – live in cities. Today the planet hosts nineteen cities with 10 million or more people; twenty-two cities with 5 million to 10 million people; 370 cities with 1 million to 5 million people; and 433 cities with 0.5 million to 1 million people. By 2030 over 60 percent of the world's population (4.9 billion out of 8.1 billion people) will be urban.

Cities in developed countries are rapidly losing their place on the list of the world's largest cities. Between 1980 and 2000 Lagos, Dhaka, Tianjin, Hyderabad and Lahore, among others, joined the list of the thirty largest cities in the world. By 2010 Lagos is projected to become the third largest city in the world, after Tokyo and Mumbai. By 2010 Milan, Essen and London will no longer be listed among the world's thirty largest cities, and New York, Osaka and Paris will have slipped farther down that list.

The current worldwide rate of urbanization (that is, the percentage, per year, that the urban share of the total population is expanding) is about 0.8 percent, varying between 1.6 percent for all African countries to about 0.3 percent for all highly industrialized countries. At the same time, urbanization of poverty is a growing phenomenon; it is estimated that between one-quarter and one-third of all urban households in the world live in absolute poverty.

Starting with the 2001 edition, *The State of the World's Cities* takes the reader through Africa, the Arab States, Asia and the Pacific, the highly industrialized countries, Latin America and the Caribbean, and countries with economies in transition to understand better how shelter, society, environment, economy, and, above all, systems of governance can contribute to urban vibrancy and viability in a globalizing world.

Africa

By 2020 sub-Saharan Africa's urban population will approach 440 million, or 46 percent of its projected total of 952 million. Today, Africa's urban areas account for 34 percent of the total population of

611 million and are credited with 60 percent of the Gross Domestic Product (GDP). Municipalities, however, capture only a small percentage of GDP – on average less than US$15 per capita per year – in revenue, creating disparity between the requirements for municipal governance and available resources.

Global economic processes have stalled in sub-Saharan Africa, with severe consequences for its urban areas. Africa is the only region in the world without a true newly industrializing economy. The failure to industrialize can partly be explained by external factors, but a variety of domestic factors must also be taken into account, including economic policies, the effects of personal rule, historical legacy, the role of the state and low levels of literacy. Structural adjustment, which has created shortages of imported materials, reduced investment, retrenched the public sector and led to declining effective demand, has badly affected urban-based manufacturing. Large-scale manufacturing, which created an impressive volume of jobs in the Asian and Latin American regions, has generated only a small number of employment opportunities in urban Africa and, consequently, the informal sector continues to remain the largest source of employment in the region.

Nonetheless, there is a forward movement. In recent years, national governments across Africa are increasingly adopting decentralization as one of their primary strategies for development. Africa has also spawned an "associative" sector built on local solidarity movements. Several countries in Africa have revised their constitutions and passed legislation that supports the participation of excluded and disadvantaged groups, especially women.

The Arab States

The Arab States' urban population is projected to be 260 million, or 66 percent of its estimated total of 395 million, by 2020. Today urban areas account for 56 percent of the total population of 270 million. Municipalities capture just over US$45 per capita in revenue per year.

Overall, the Arab States comprise a great diversity of socio-economic and human settlement profiles and characteristics: from least developed through developing to oil rich countries; conflict and post-conflict situations; from very open economies to economic isolation; and from highly urbanized to predominantly rural. The region's considerable internal diversity is reflected in the conditions in its cities and has resulted in widely varying domestic needs and priorities: rehabilitation and reconstruction (Iraq, Lebanon, Palestine and Somalia); poverty alleviation (Egypt, Jordan, Syria, Morocco and Yemen); urban management and housing needs (Egypt, Jordan and Algeria); and capacity building (Gulf countries).

Asia and the Pacific

In twenty years the urban population in the Asia and Pacific region is expected to be 1,970 million or 46 percent of its projected total of 4,298 million. Urban areas today account for 35 percent of the total population of 3,515 million. On average, municipalities secure about US$150 per capita in revenue per year.

Urbanization in Asia and the Pacific raises red flags, particularly because an increasing number of the region's poor live in urban areas. The size and urgency of the problem require different ways of managing cities and their related infrastructure and service requirements. Recently, macro-economic and financial crises have cast doubt on conventional approaches and concepts. The financial crises of the 1990s affected millions of lives and aggravated social vulnerabilities. In East and Southeast Asia, the social consequences of the financial crisis have been felt more in cities, reflected in increased poverty brought about by cutbacks in both public and private employment as well as in public expenditures for health and education.

However, the increased pace of urbanization and its linkages to economic globalization have reinvigorated interest in the process of governance and its links to economic growth. Decentralization and local autonomy are gaining more momentum, and with this, the interest in building the capacity of local governments is growing. While several Asian countries have adopted decentralization policies, excessive controls are still exercised in others on the functional, financial and administrative responsibilities of local government. As a result, there is a mismatch between the functional powers of local governments and the financial resources available to them.

Highly industrialized countries

The urban population in the highly industrialized countries is projected to be 547 million, or 84 percent of their combined population of 649 million, by 2020. Today, urban areas account for 80 percent of the total population of 597 million. On average, municipalities obtain about US$2,900 per capita in revenue per year.

In 2020, there will be five urban agglomerations larger than 5 million inhabitants in Europe: Paris, Moscow, London, Essen/Ruhrgebiet and St Petersburg. Most urban populations in Europe live in small or medium-sized towns; half the urban population lives in small towns of 10,000 to 50,000 people, and a quarter in medium-sized towns of 50,000 to 250,000. Only 25 percent of the urban population lives in cities with more than 250,000 people.

Cities in highly industrialized countries no longer deal with the effects of urbanization, but rather with a combination of other demographic issues and the impacts of global trends: increasing internationalization of metropolitan regions; changes in the distribution of responsibilities between the public and private sectors; a generally stronger role for a few major cities within each country; aging populations and the related problems of access to health care and pensions; international migration; and the highly detrimental impacts of social and economic polarization.

For the past two decades, the highly industrialized countries committed themselves to economic policies aimed at encouraging macro-economic stabilization, structural adjustment and the globalization of production and distribution. Although these policies have in general been effective in promoting short-term economic growth, low inflation and lower current account imbalances, negative longer-term societal implications are now emerging as major political and socio-economic dilemmas. Growing political disenchantment arising from widening income gaps, declining political participation and widespread social exclusion are manifesting themselves in cities across North America and Europe alike. In the United States, for instance, racial disparity is one of the most important issues facing cities.

Latin America and the Caribbean

The region comprising Latin America and the Caribbean is the most urbanized in the developing world. With 75 percent, or 391 million, of its people living in cities, it has an urban–rural ratio similar to that of the highly industrialized countries. However, unlike highly industrialized countries, urban areas are not geographically scattered in terms of physical size or populations. Urbanization patterns, with the exception of Brazil, typically involve one very large city that accounts for much of the country's urban population. The urban population will approach 539 million, or 81 percent of its projected total population of 665 million, by 2020. On average, municipalities capture less than US$90 per capita in revenue per year.

The biggest challenge facing Latin America and the Caribbean is the growing urbanization of poverty. Despite general economic growth, the number of low-income people in the region rose from 44 million to 220 million between 1970 and 2000. More than one-third of the households living in poverty are headed by women. Much poverty is concentrated in urban areas; a massive 40 percent of the population of Mexico City, for instance, and a third of São Paulo's population is at or below the poverty line.

Economies in transition

The urban population in countries with economies in transition will approach 420 million, or 78 percent of its projected total population of 541 million, by 2020. Today, urban areas account for 70 percent, or 382 million, of the total population of 543 million. On average, municipalities capture about US$275 per capita in revenue per year.

There are sharp differences among the various countries with economies in transition, notably in criminality, corruption and levels of democracy. Some, particularly in Central Europe, have clearly started to adjust to the market economy promoted by the West. Although the *laissez-faire* model was assumed by many to be the solution, experience during the past decade indicates that this does not necessarily hold true for all.

It is now recognized that sustainable urban development in these countries will depend on the creation and maintenance of efficient land and property markets; the development of more and better housing finance options; a greater emphasis on municipal finance and institution building; strengthening of urban utility systems; a growing interest in the preservation of cultural assets and heritage; and the responsiveness to emergencies such as earthquakes and flooding.

POLICY ISSUES

Urban shelter

More than one billion of the world's urban residents live in inadequate housing, mostly in the sprawling slums and squatter settlements of developing countries. A significant trend during the past decade has been the growing awareness of the relationships between human rights and sustainable development. In the field of shelter, this has led to a decline in human rights abuses, such as mass forced evictions. Negotiation and participation are increasingly being employed to secure the urban poor their right to shelter.

The right to adequate housing is recognized by more than 70 percent of the world's countries. Almost all countries in Asia and the Pacific promote housing rights in their legislation, and the Arab States provide the greatest protection against eviction.

Extending urban citizenship to low-income people, through the granting of secure tenure, for example, is one of the most far-reaching decisions that can be taken in promoting a sustainable shelter strategy.

Urban society

Despite the potential of cities to improve living standards, the benefits of urbanization are not shared equally. All too often, cities are still divided into haves and have-nots – the established and the marginalized – and offering different life opportunities for women and men. In developing countries, the statistics are particularly disturbing: over 50 percent of the population of Mumbai and New Delhi lives in slums, while in Lagos and Nairobi over 60 percent of households remain unconnected to water.

Exclusion, as a result of physical, social or economic barriers, prevents many groups from participating fully in urban life and services. Failure of local authorities to integrate such groups in their decision making is often a function of inertia, along with bureaucratic and unresponsive forms of governance. Participatory governance is a prerequisite to social cohesion and inclusion. It involves strengthening civil society and supporting local populations to engage in, and benefit from, urban opportunities.

Urban environment

Apart from its effect on health and well-being, environmental degradation and pollution continue to constrain development and growth of cities. Ill health and premature death not only cause suffering, they also impose heavy costs on the economy. Ultimately, the human population cannot escape the detrimental effects of unsustainable consumption and environmental degradation.

An increasing number of countries now recognize the key principles of environmental management. One is that the environment is not an end in itself – not something to be 'protected' from development – but that it is a resource to be carefully managed on a sustainable basis. Secondly, urban development necessarily depends on the natural resource base available to the city – which in turn has an impact on the state of those resources. It is, therefore, crucial to improve understanding of the two-way relationship between the environment and development.

Urban economy

Large cities typically produce a significant share of the Gross Domestic Product (GDP) of their countries. For instance, Bangkok, with only 12 percent of the national population, contributes 38 percent towards Thailand's GDP, more than the GDP of any of the agriculture-based economies in Africa. Yet, despite the significant economic role played by cities, they often receive less than they might warrant for their contribution to the national economy, negatively influencing their productive potential. Key constraints to urban productivity include infrastructure deficiencies, inappropriate regulatory frameworks for urban land and housing markets, weak municipal institutions, and inadequate financial services for urban development. Strategies to combat these constraints include:

1 strengthening the management of urban infrastructure by improving the level and composition of investments, reinforcing institutional capacity for operations and maintenance, and seeking opportunities for greater private sector involvement;
2 improving city-wide regulatory frameworks to increase land and housing market efficiency and to enhance private sector provision of shelter and infrastructure;
3 improving the technical and financial capacity of municipal institutions through more effective application of resources and division of responsibility between central and local governments; and
4 strengthening financial services for urban development, ranging from micro-finance to municipal bond markets.

Urban economies in the developing world are largely driven by the informal sector. Informal sector employment makes up 37 percent of the total employment in developing countries as a whole, and is as high as 65 percent in Africa. Cities can support the informal sector through the provision and maintenance of infrastructure, including adequate supply to electric power, water, transport and telecommunications networks. By relaxing rules and regulations, the informal sector can be expected to contribute further to the creation of jobs and, perhaps, eventually be integrated into the formal economy.

Urban governance

Effective governance is increasingly dependent on people assuming their responsibilities as citizens and participating in decision making and implementation. Citizens are learning to forge new alliances that strengthen their voice and make their concerns felt in legislation. Governments at all levels increasingly recognize the value of communication, consultation, and negotiation for joint decisions and joint implementation of policies that meet the changing needs of society.

However, reform of governance institutions and strengthening of local citizenship are necessary prerequisites for effective governance. Some of the emerging issues and priorities that need to be addressed include: redefining the roles and responsibilities of central and local governments; promoting city-wide development strategies; regulating equitable financial transfer between all levels of government; encouraging transparency through free flows of information; effective civic engagement and participation; and strengthening leadership while promoting ethical conduct in the governance of cities.

The key finding

A country's global success and its overall human development rest on the shoulders of its cities. For the good of all citizens, city and state must become political partners rather than competitors. If accommodation requires new political arrangements, institutional structures or constitutional amendments, it is never too early or too late to begin making changes. The nature of those changes can best be determined through empirical observation, analysis of information and political dialogue on both the condition of a nation's cities and a national strategy to improve that condition.

Each country should, indeed, prepare and publish, periodically, a report on the state of its cities, focusing especially on national and local policies and how they affect cities and their citizens. Gathering of evidence by putting in place proper information systems and diagnostic tools is a practical first step. Good information will provide the common platform for dialogue among stakeholders – an essential part of the process – as they approach a vision for the future and set priorities for conservation and change.

The main goal is to make the structures of governance more responsive to individuals, families, and communities so that both national and local authorities can better serve civil society, each through separate but complementary instruments.

2

Megaforces Shaping the Future of the Nation's Cities

From *The State of the Cities* (2000)

U.S. Department of Housing and Urban Development

Editor's introduction

That the United States was a definitively urban nation was revealed in 1970 when the U.S. Census Bureau reported that 74 percent of Americans were living in urban places, a figure that increased to 79 percent in the ensuing decades. The Census defines an urban place as a jurisdiction with a population of 2,500 or more. Between 1970 and the present, the nation's urban population grew 22 percent, and starting in 1970 more urbanites lived in the suburbs than in central cities.

A better way to understand the extent of U.S. urbanization is to look at the population in medium to large cities, defined here as places having 25,000 or more inhabitants. Presently, 130 million Americans (or 43 percent) live in the nation's 1,302 cities of 25,000 or more, and 27–8 percent of Americans are residents of cities having populations of 100,000 or more. About a third live in the 100 most populous places, and 12 percent live in the top twenty-five. While the percentage of city dwellers has remained stable for the past thirty years, individual city performance varies. With the exception of cities of 500,000 to a million, all city categories have experienced population growth rates of 25 percent or higher, with the 100,000 to 250,000 group seeing a 92 percent increase.

Three important trends contribute to this growth. First, *cities 25,000 and above increased in number* between 1970 and 2004 (up 61 percent). The highest numerical growth rate (82 percent) occurred in medium-sized cities (100,000 to 250,000). For example, in 1970, 100 cities fell into this category; by 2004, 182 did. Smaller cities (50,000 to 99,000) had the second highest growth rate (70 percent). Cities of 1 million or more increased 50 percent. Today, the United States has nine cities of a million or more, up from six in 1970. The newest million-inhabitant cities are in the South and West (Dallas, Phoenix, San Antonio, and San Diego), while the Midwestern city of Detroit fell below 1 million in this period. One group, cities of half million to a million, had a much lower growth rate (15 percent), due to its mix of cities. For example, in 2004 the group included Milwaukee and Jacksonville, each on different population trajectories: Milwaukee's population dropped 19 percent (from 717,099 in 1970 to 578,887 by 2004), while Jacksonville's increased 48 percent (from 528,865 to 783,623).

Second, *more cities have larger populations*. In 1970, only 153 cities had populations of 100,000 or more; by 2004, the number rose to 254. And in the list of the 100 most populous cities in 1970, the 100th city had 138,764 residents (Evansville, Indiana); in 2000, it held 194,973 (Glendale, California), a 41 percent increase; and in 2004, the cutoff was 198,550 (San Bernardino, California), a 43 percent rise since 1970.

Third, *rapidly growing cities have greatly expanded their capacity* to absorb more people or raise more revenues by extending their boundaries and/or by amending or building out their zoning. For example, the land areas of the twenty-five most populous cities increased 29 percent between 1970 and 2000. This spatial expansion occurred while these cities' populations grew at a lower rate, 8 percent. Emblematic is San Antonio, whose area grew 122 percent while its population rose only 75 percent. In the case of amending or building out zoning, increases in residential space account for the expanded capacity of cities to absorb more population. In 1970, the twenty-five most populous cities had approximately 14.2 billion square feet of residential area; by 2000 residential square footage increased 83 percent to 26.1 billion sq. ft.

The increases in land area and related population and household growth have had dramatic effects on the regional distribution of large-city populations since 1970. Among the top 25 cities in 2000, the Northeast dominated, having 34 percent of the total (down from 41 percent in 1970); the West came in second with 27 percent of the total population (up from 21 percent in 1970); the South was third with 22 percent (up from 16 percent); while the Midwest slipped to fourth place (18 percent of the total, down from 23 percent in 1970). These changes in city population do not parallel changes in regional growth: between 1970 and 2000 the South and the Midwest maintained their one and two ranks, while the West replaced the Northeast at third.

Although the United States, unlike other countries, has no stated national urban policy, it does have several agencies that deal with urban affairs. These include the U.S. Department of Housing and Urban Development, founded in 1965 and charged with overseeing homeownership, community development, and affordable housing. This agency has an Office of Policy Development and Research (PD&R) that maintains current information and conducts research on housing and urban development. It is responsible for the *State of the Cities* reports mandated by Congress. (Three of the four *State of the Cities* reports are on the HUD website: www.huduser.org/publications/polleg/tsoc.html.)

This selection provides an assessment of urban conditions for the nation's cities and comes from the *State of the Cities 2000*. It draws on the *State of the Cities* Data Systems (SOCDS). SOCDS is a U.S. contribution to the United Nations' 1996 Habitat II Conference, held in Istanbul. In conjunction with this effort, HUD issued four *State of the Cities* reports between 1997 and 2000. Although HUD has not published a *State of the Cities* report since 2000, PD&R has maintained and enlarged SOCDS as a web-based searchable database covering more than 1,200 cities and 17,000 suburban places. The SOCDS (socds.huduser.org) contains seven categories of information including the U.S. Census (1970–2000), employment and labor force (1990–2005), business patterns (1998–2001), FBI crime data (1997–2002), building permits (1997–2005), urban public finance (1992 and 1997), and relevant information from the Comprehensive Housing Affordability data for those cities that collect that information.

Other organizations collect and disseminate data and reports on cities and metropolitan areas. Like the HUD studies, they usually have an advocacy approach, attempting to bring attention to the sponsor's concerns. For example, the U.S. Conference of Mayors publishes an annual report, entitled *The Role of Metro Areas in the U.S. Economy*, which in 2006 showed that the nation's metro areas are responsible for 86 percent of the nation's gross domestic product. This study and many updates are on the Conference of Mayors' website (www.usmayors.org). Another source of topical urban reports is the Brookings Institution's Metropolitan Policy Program (see www.brookings.edu/metro/publications.htm).

The State of the Cities 2000 Report is part of an annual series in which HUD reports the most recent data on indicators of the social and economic viability of America's cities. . . . This year's State of the Cities report identifies four megaforces that are shaping the future of the Nation's cities and presents findings showing their impact.

The first is the new high-tech, global economy which has been a driver of recent economic expansion in the United States. New technologies in information and telecommunications – coupled with greater productivity – have produced record economic gains along with new opportunities and risks for the Nation's cities and suburbs.

A second is the new demography that is shaping cities. Major demographic shifts are under way that will have significant economic, social and political implications for both cities and suburbs. The

Nation is rapidly becoming more ethnically diverse, and at the same time our elderly population is growing dramatically.

A third is the new housing challenge that is presenting new threats to housing affordability. With the strong economy have come higher rents and housing prices, in some markets impacting all income groups in both cities and suburbs.

Finally, the fourth is the powerful major trend of continued decentralization – the continuing shift of jobs and people to the metropolitan edge – that is threatening the stability of existing communities and the development of new livable, sustainable communities.

These four megaforces frame the challenges for a twenty-first century urban policy agenda. *The State of the Cities 2000* presents the impact of these megaforces in four major findings for America's cities. These findings utilize new data from HUD's 2000 State of the Cities database (SOCDB) which tracks employment, population and other demographic trends in more than 300 metropolitan areas.

[. . .]

FINDING # 1 THE NEW ECONOMY

Most of America's cities are participating in the New Economy, with high-tech growth driving a new wave of economic prosperity – but at the same time creating both winners and losers. New HUD data find that high-tech employment is growing faster in the suburbs than in cities but that the proportion of new jobs that are high-tech is larger in cities than suburbs.

Cities are sharing in the unprecedented expansion of the new economy

The most recent data show that cities are enjoying new vigor in job growth, drawing closer to suburban growth rates. The number of private sector jobs in central cities increased dramatically, growing by 8.5 percent between 1972 and 1997. During this period, nearly 2.3 million private sector jobs were created in cities.

Business growth in cities is accelerating, and wage growth in cities surpasses that of their surrounding suburbs. From 1992 to 1994 businesses grew by just 0.7 percent in cities, but from 1994 to 1997 they grew by 3.7 percent – five times the previous rate. Overall, however, business growth in suburbs is still twice that of cities.

At the same time, wage growth in cities outpaced that of suburbs. Since 1992, central city wages have grown by 4.8 percent – faster than the suburban rate of 4.3 percent – and the current average wage in cities in now 10.5 percent higher than the average wage in suburbs.

Overall cities had a larger percentage point decline in unemployment rates than suburbs. Since 1992 jobless rates in central cities have fallen by 3.7 percentage points, to 4.8 percent. Suburbs experienced a smaller decline, of 3.2 percentage points, to 3.4 percent in 1999.

Incomes are steadily increasing in cities, and poverty has declined. The economic boom raised urban household income in 1998 to their highest levels since 1990. While all types of households throughout the country realized substantial gains in income, household income grew faster in cities (3.5 percent) than in suburbs (2.3 percent) between 1997 and 1998.

A new digital divide in high-tech jobs is emerging between cities and suburbs

High-tech growth is a substantial contributor to recent economic gains in cities. High-tech jobs account for 27 percent of new employment in cities. The high-tech job growth rate is three times that of overall job growth in central cities. From 1992 to 1997, there was a 27 percent increase in high-tech job growth in cities compared with a 8.5 percent overall job growth.

A new survey by the U.S. Conference of Mayors illustrates the breadth and depth of high-tech expansion in our cities. More than 80 percent of cities reported significant or moderate growth in high-tech jobs.

The South and the West lead the country in central city high-tech job growth. All regions saw high-tech job gains, but central cities in the South saw high-tech jobs grow the most, by 34 percent – followed by 27.2 percent in the West, 21 percent in the Midwest, followed by 19.5 in the Northeast.

But there is a new digital divide in high-tech jobs between cities and suburbs. High-tech job growth in suburbs is 30 percent faster than that of cities. Despite the positive gains in high-tech job growth in central cities, suburbs continue to outpace central cities. Most central cities are gaining high-tech jobs, but high-tech jobs in suburbs are, on average, growing 30 percent faster.

Fewer cities remain "doubly burdened"

Despite the overall dramatic record of job gains, one in eight cities are still "doubly burdened" according to HUD's index of distress. Doubly burdened cities face high unemployment and significant population loss or high poverty rates. This represents a modest improvement over last year, when one in seven cities were in this category. There are sixty-seven cities that have an unemployment rate 50 percent higher than the U.S. rate and either have lost more than 5 percent of their population since 1980 *or* have a poverty rate of 20 percent or higher. Of these cities, thirty-nine have unemployment rates at least double the national average.

Despite declines, unemployment and poverty still impact cities more than suburbs. Unemployment rates in central cities are still about one-third higher than the jobless rate in suburbs. Unemployment among minority youth remains unacceptably high at 22 percent in cities. The national poverty rate declined from 13.7 percent in 1996 to 12.7 percent in 1998. Encouragingly, the poverty rate also decreased in central cities during this period, from 19.6 percent to 18.5 percent – but remains twice the rate of poverty in suburbs.

Finding # 2 The new demography

The new demography is multigenerational, multiracial and multiethnic. While an increasing share of residents of both cities and suburbs are getting older, a disproportionate number of the elderly poor live in cities. At the same time, cities and suburbs are becoming more racially and ethnically diverse.

Overall population is on the rise, with metropolitan growth continuing at a faster pace in suburbs than in central cities. The 2000 estimated population of 275 million is projected to rise to 350 million by 2030. This projected 75 million more people, half of which will be new immigrants and their children, will drive economic expansion by providing both the demand for goods and services and the labor force to fill that demand. How best to meet these needs while protecting our dwindling open space and environment will pose difficult choices.

Cities are aging

In 2030, the elderly population will reach 70 million, doubling the current number of elderly Americans. These seniors will comprise 20 percent of the overall U.S. population. Many will age in place and remain in the cities or suburbs they have called home for decades. Central cities will continue to house disproportionate numbers of the Nation's seniors who live below or near the poverty line. As these populations of elderly age in place, they will pose special challenges for communities.

Suburbs and cities are becoming more racially and ethnically diverse

Diversity itself is changing as the traditional divide between blacks and whites blurs into a multiracial, multiethnic society. Cities – historically home to the Nation's newcomers as well as most of its minorities – remain the most diverse. But suburbs are becoming much more heterogeneous as well. Between 1980 and 1998, for example, the minority share of the population in central cities rose from 34.8 percent to 47 percent. In suburbs during the same period, the proportion of minorities nearly doubled from 13.4 percent to 21.7 percent. The proportion of Hispanics rose from 5.3 percent to 9.6 percent in suburbs. The percentage of African-American suburbanites expanded as well, from 6.1 to 7.6 percent.

Immigrants are fueling the new diversity in both suburbs and cities. Immigrants are more likely to live in central cities but are increasingly moving to the suburbs – a distinctly new phenomenon. They have transformed many traditionally ethnic neighborhoods in our major urban center from homogeneous enclaves to truly multicultural, multiethnic places. In the process, they have reversed the population decline of many cities and at the same time are blurring the ethnic and racial lines between cities and suburbs.

Finding # 3 The new housing challenge

As increases in the cost of housing surpass the rate of inflation, economic good times are paradoxically creating a housing crisis for many Americans. The economic growth that is pushing up employment and homeownership in most of the Nation's cities is also driving increases in rent more than one and a half times faster than inflation – and creating staggering jumps in home prices as well.

Homeownership has reached all-time highs in both central cities and suburbs

Between 1992 and 1999, over 8.7 million households became homeowners as the national homeownership rate reached 66.8 percent in 1999 – and rose even higher in the first quarter of 2000 to an all-time high of 67.1 percent. In 1999, homeownership in cities broke the 50 percent barrier for the first time – 50.4 percent in 1999 and 51.2 percent in the first quarter of 2000. All racial and ethnic groups have shared in this homeownership boom. As of the first quarter of this year, 45.7 percent of Hispanics and 47.8 percent of non-Hispanic African Americans, and 54.2 percent of other non-Hispanic minorities are now homeowners.

Nevertheless, important – and unacceptable – homeownership gaps still remain. The homeownership rate in central cities trails substantially behind the suburban rate of 73.6, and gaps between minority rates and the 73.4 percent ownership rate of whites remain unacceptably large. In addition, as homeownership has grown, a new problem has arisen, predatory lending, which occurs when lenders, often operating outside of the Federal regulatory structure, are able to engage in lending abuses such as charging excessive up-front fees, high interest rates, and prepayment penalties. Such practices contribute to skyrocketing foreclosures in the subprime mortgage markets, especially in minority and low-income communities.

The strong economy paradox

Paradoxically, the economic growth that is increasing employment and homeownership in most of the Nation's cities also is driving up rents and housing prices for many Americans.

Over the 1997–1999 period, house prices rose at more than twice the rate of general inflation, and rent increases exceeded inflation in all three years. For most of the goods and services that Americans routinely pay for – the items that go into the Consumer Price Index (CPI) – inflation has been very low throughout the economic expansion, but not so for the cost of housing. Over the last three years, the CPI rose 6.1 percent (just over 2 percent per year). During the same period, rents rose by 9.9 percent and house prices by 16 percent.

The hot high-tech markets are among the highest-cost housing markets. Among the top ten metropolitan areas that HUD identifies as the hottest high-tech markets, house prices rose more than 18 percent in seven of the ten areas from the end of 1995 to the end of 1999, and by more than 27 percent in three of the ten areas. During the same period, rents increased by more than 20 percent in such high-tech markets as Denver and San Francisco.

Housing affordability is both a central city and a suburban problem. In the late 1980s, both rents and house price increases in central cities lagged behind suburbs. By the late 1990s, however, this pattern changed. Central city house prices appreciated at a rate close to that of suburbs – and rent increases in central cities have been even greater than those of suburbs. In fact, since 1991 rents have risen faster in central cities than in suburbs.

[. . .]

Finding # 4 The new forces of decentralization

The New Economy's advances in information technology, coupled with rising income, population growth and infrastructure spending patterns, continue to drive residential and business development to the fringe. A new HUD analysis shows accelerating growth in land consumption, which threatens to undermine the quality of life in both cities and suburbs.

Improved information and communication technologies are encouraging the spread of jobs and people to the urban edge. But cities continue to have the inherent advantages of agglomeration – face to face contact, accessibility, and an already built-up, amenity-rich infrastructure, which have always been critical to economic growth and are valuable in the New Economy as well.

Cities' share of metropolitan jobs continues to decline. With a robust economy and cheap, open land on the urban fringe, businesses and housing are moving out to the periphery of metropolitan areas. In 1997, 57 percent of metropolitan-area jobs were located in suburbs up from 55 percent in 1992.

Population growth in suburbs relative to their central cities accelerated in the 1990s compared with the 1980s. Between 1990 and 1998, suburban population grew by 11.9 percent, compared with 4.7 percent for central cities. Central cities now house only 38 percent of the U.S. metro population compared with 45 percent in the 1970s.

At the same time land is being consumed at twice the rate of population growth. Land use grew in the 1990s at approximately two times the rate of the 1950s. Between 1994 and 1997, land consumption in the U.S. grew by 2 percent – but population grew by just 1 percent annually. In all, an average of 2.3 million acres of land are being consumed annually, with a substantial portion for residential development on lots of more than one acre in fringe suburbs or smaller cities.

Consequences for quality of life in cities and suburbs

Rapid growth in land use has potentially negative effects on the environment, transportation, and infrastructure in both cities and suburbs. Significant unintended costs for all parts of the metropolitan area – cities and suburbs alike – accompany the rush to the periphery.

- *Environmental quality.* As land is developed, water and air quality are degraded. Water pollution results from increases in impervious surfaces. Parking lots, for instance, generate almost 16 times more runoff than a meadow for comparable land areas. Air quality is harmed by automobile emissions from increased driving and decentralized development. Despite cleaner, more efficient cars and stricter regulation of emissions of industrial pollutants, air quality in many metropolitan areas is worsening and raising concerns about public health.
- *Transportation.* Many suburban residents are experiencing longer commutes and increasing traffic congestion. As metropolitan areas stretch out, Americans are driving more and spending an increasing portion of their productive time in daily commutes. The number of vehicle miles traveled (VMTs) increased sixfold between 1950 and 1993. As a result, household expenditures on transportation are up in many cities – less so in communities with strong public transit systems. In fact, congestion and gridlock are contributing to a resurgence in transit ridership, which in 1999 increased by 4.5 percent – twice the rate of increase of motor vehicle travel.
- *Infrastructure.* New development at the fringe requires investment in new infrastructure while existing infrastructure in cities is underused. Decentralized and low-density development on the fringe does not capitalize on existing infrastructure capacity that is already present in central cities and suburbs. In effect, citizens are paying twice – both to maintain existing infrastructure, and also to build new infrastructure to support new suburban growth.

The solution – livable communities at the core and the edge

The creation of livable communities requires reinvestment in the cities, Smart Growth practices and regional connections that encourage cooperation among all communities.

- Improving public safety and education are keys to livability in our cities. . . .
- Local land use/transportation management and planning play important roles in metropolitan development patterns. . . .
- Smart Growth in the suburbs . . . is a cooperative way to rationalize growth . . . [and] includes brownfields redevelopment, infill

housing investments and new business growth to take advantage of the untapped markets of our inner cities and older suburbs. . . .

- The answer to achieving livable communities lies in regional cooperation. Cities and suburbs are beginning to envision a new template based on regional cooperation and joining forces to address issues that cross local jurisdictional boundaries – transportation, environmental protection, housing affordability, education, concentrated poverty, and economic development. The bottom line, local leaders are learning, is that cities need suburbs and suburbs need cities to prosper in the New Economy.

3

Planning, Urban Revitalization and Inner City: An Exploration of Structural Racism

Journal of Planning Literature (2000)

Catherine L. Ross and Nancey Green Leigh

Editor's introduction

Urban and regional planners are involved in issues of race and discrimination in many ways. In the postwar period, they helped craft the public policies shaping metropolitan America – including slum clearance/urban renewal efforts and construction of the interstate highway system – that had such a profound effect not only on central cities but on suburban development as well. Selections in this *Reader* detail these phenomena (see, for example, "Megaforces Shaping the Future of the Nation's Cities"; "Suburbia: Homeland of the American Future"; "The New Suburban Typology"; "Reflections and Research on the U.S. Experience"; and "Sprawl and the Tyranny of Easy Development Decisions").

Many postwar programs, as has been well documented, had a deleterious effect on minority communities, which, due to segregation and discrimination, were often low-income. Residents of physically deteriorated neighborhoods chosen for demolition under citywide urban revitalization policies suffered displacement in the early years of slum clearance. Minorities were also denied access to new outlying housing because of outright prejudice that prevented their receiving equal treatment in the real estate market. Sociologist Herbert Gans's *The Urban Villagers: Group and Class in the Life of Italian-Americans* (Glencoe, IL: Free Press, 1962), an account of the destruction in the 1950s of a 7,000 inhabitant Italian-American community in Boston to make way for middle-income housing and hospital expansion, alerted planners and public officials to the necessity of understanding social as well as physical conditions when revitalizing urban neighborhoods. Similarly, Anthony Downs's *Opening Up the Suburbs: An Urban Strategy for America* (New Haven, CT: Yale University Press, 1973) argued that since cities and their suburbs were mutually dependent, a metropolitan area would be economically healthy only with the elimination of segregated residential districts.

Equally well documented is the work of planners as leaders in exposing institutional racism and fighting for environmental justice and meaningful citizen participation in the planning arena. They called for fair housing and anti-mortgage discrimination legislation, acted as advocates for underrepresented groups, and promoted equity and social justice for the disadvantaged.

Today, planners face enormous and complex tasks as they attempt to ameliorate past development decisions contributing to current regional income and social imbalances. For example, suburban land use regulations dating from the early twentieth century – such as those mandating large lot sizes – have resulted in hard-to-change settlement patterns and have, in many instances, excluded inexpensive housing. Unwanted, nuisance, or polluting facilities in politically weak urban neighborhoods defy change as such facilities are expensive to close down or relocate. Brownfields in such neighborhoods are also problematic because of the high costs of cleanup. Metropolitan transportation routes built in the 1950s and later often do not adequately link inner-city workers with the suburbs that have the majority of the nation's jobs.

The authors of this selection, which focuses on planners' work in cities, are faculty members in the School of Planning, College of Architecture, Georgia Institute of Technology, and they write authoritatively from their own professional experience. Catherine Ross, Harry West Professor of Quality Growth and Regional Development, and Director, Center for Quality Growth, is former Executive Director of the Georgia Regional Transportation Authority. She has written extensively on transportation and minority issues and is co-author of *The Inner City: Urban Poverty and Economic Development in the Next Century* (New Brunswick, NJ: Transaction Publishers, 1997) with Thomas Boston. Nancey Green Leigh, a professor, is co-author of *Economic Revitalization: Cases and Strategies for City and Suburb* (Thousand Oaks, CA: Sage Publications 2002) with Joan Fitzgerald and author of *Stemming Middle-class Decline: The Challenges to Economic Development Planning* (New Brunswick, NJ: Center for Urban Policy Research, 1994). She has consulted widely on brownfields and vacant land issues.

The literature surrounding race and urban policy is extensive, with contributions from the social sciences, law, and urban and regional planning. Kenneth Jackson's *Crabgrass Frontier: The Suburbanization of the United States* (New York: Oxford University Press, 1985) documents discriminatory practices in mortgage lending known as redlining. William Julius Wilson's *The Truly Disadvantaged: The Inner City, the Underclass, and Public Policy* (Chicago: University of Chicago Press, 1987) and *When Work Disappears: The World of the New Urban Poor* (New York: Vintage Press, 1997) draw bleak portraits of the social and economic isolation caused by urban poverty. Xavier N. De Souza Briggs's edited *The Geography of Opportunity: Race and Housing Choice in Metropolitan America* (Washington, DC: Brookings Institution Press, 2005) and Christopher Bonastia's *Knocking on the Door: The Federal Government's Attempt to Desegregate the Suburbs* (Princeton, NJ: Princeton University Press, 2006) bring the discussion of race and residential patterns into the twenty-first century.

Two authors offer contrasting views of local politics revolving around urban renewal and desegregation in the postwar period. June Manning Thomas's *Redevelopment and Race: Planning a Finer City in Postwar Detroit* (Baltimore, MD: Johns Hopkins University Press, 1997) criticizes planning decisions as ill conceived and discriminatory. In her book, Thomas relates how the seventy-eight-acre Mies van der Rohe-designed Lafayette Park, an integrated middle-income district adjacent to the downtown, displaced a poor African-American community in the 1950s and 1960s, and also traces other unfortunate planning efforts in declining Detroit. Thomas H. O'Connor's *Building a New Boston: Politics and Urban Renewal, 1950–1970* (Boston, MA: Northeastern University Press, 1995), on the other hand, argues that, on the whole, Boston's revitalization programs had long-term benefits for the city, especially in laying the foundation for the reinvention and refurbishment of the central business district and many neighborhoods. Notably, he touches on a key requirement for urban revitalization policy, the need for a steady revenue stream (usually real estate taxes) to support city services in the face of depopulation, a theme that awaits further research.

The issues of race and inner-city revitalization are woven together almost inextricably, presenting a complex and seemingly intractable problem for urban and regional planners, scholars, policy makers, activists, and citizens. In this article we present an overview of the dilemma from a city and regional planning perspective. . . . [W]e develop a description of the relevant planning climate. We then discuss four of the most important forces of structural racism that confront inner cities. We close with a discussion of those approaches that have shown some promise and with suggestions for potential new approaches for inner-city revitalization.

MACRO PLANNING AND POLICY ENVIRONMENT

Perhaps the most obvious component of the inner-city planning environment is federal policy for urban areas. . . . [F]ederal policy [is categorized] according to decades . . . [starting] with the New Deal and, . . . the 1950s, when programs centered on urban redevelopment and urban renewal. These programs aimed to transform urban blight into productive areas that could house more attractive, usually private, economic uses. Although the programs often did slow the physical deterioration of the neighborhoods in question, they also hastened social decay of the community.

The 1960s saw a shift in focus from physical planning to social issues with particular attention to the poverty that seemed to be pandemic in the inner city. This decade was marked by Lyndon Johnson's War on Poverty, which spawned many programs designed to empower inner-city residents, such as Head Start and Legal Services. In 1965, the government created the U.S. Department of Housing and Urban Development (HUD), which was intended to coordinate urban policies and programs for increased effectiveness. The Model Cities program was HUD's first major undertaking.

During the 1970s, cities saw a decrease in the amount of large-scale federal bureaucracy aimed at urban policies and programs. Nixon's "new federalism" consolidated several grant programs into a single Community Development Block Grant (CDBG), which dedicated federal monies to assist low-income residents. These grants were to be administered at the local level by urban municipalities, however, rather than by the federal government. Nixon also began to pursue a more conservative market philosophy by transferring housing assistance from supply subsidies to demand subsidies as a means of providing financial support directly to low-income residents. During his term, President Carter implemented two programs aimed at providing direct neighborhood benefits (Neighborhood Self-help Development Program and Urban Development Action Grants) and effected two laws directed at neighborhood reinvestment, the Home Mortgage Disclosure Act . . . the Community Reinvestment Act of 1977. . . .

During the 1980s, devolution of responsibilities from the federal to the state and local level often drastically reduced the level of federal intervention in housing and community development. The Neighborhood Development Demonstration program "provided matching grants to community groups to expand their development activities by facilitating public/private partnerships." Another major program from that era was the Low-income Tax Credit, by which tax credits were granted for corporate investment in low-income housing.

The 1990s brought competition for cities to win substantial grant awards under the Empowerment Zone (EZ) Program. The first awards, made in 1994, were for more than $3 billion and went to six cities: New York; Philadelphia-Camden; Baltimore; Chicago; Detroit, MI; and Atlanta, GA. The EZ Program focuses on the most distressed neighborhoods. It works through existing channels to involve the citizens of these areas in their community's revitalization; that is, the residents decide how the monies and tax incentives from the grant are to be spent. This focus on "empowering" the residents of distressed areas is a hallmark of the EZ approach and signals a change in federal policy toward community-based initiatives.

[. . .]

Neither the EZ Program nor the other major federally instigated community development move of the 1990s, the Affordable Housing Act of 1990 . . . have proven widely or sufficiently successful at fulfilling their goals. . . .

STRUCTURAL FORCES OF RACISM

Planning's misappropriated tool – zoning

Zoning policy has played a major role in creating the plight of today's inner city; in its various forms, it has affected the pattern of residential segregation that makes revitalization such a difficult undertaking. As jobs move from the inner city to the suburbs, affordable housing within a reasonable distance of those jobs becomes more and more difficult for low- and middle-income families to obtain. Because these families cannot afford expensive homes in the suburbs, they are confined to the inner city. In this way, racial or exclusionary zoning has mapped the metropolitan cities into clear, segregated areas.

Originally, city planners used zoning to protect residential areas from industrial expansion. As early as 1908, state governments allowed cities to regulate the physical details of any buildings within city limits. The U.S. Supreme Court upheld that zoning power in its rulings on *Hadacheck v. Sebastian* (1915) and *Village of Euclid v. Ambler*

Realty Corporation (1926). But during this time, municipalities also blatantly used racial zoning as a tool to exclude undesirable groups from entering their communities and to prevent the spread of slums into upscale neighborhoods. In 1917, however, the Supreme Court declared racial zoning unconstitutional in its ruling in the case of *Buchanan v. Warley* (1970). Therefore, municipalities became creative, developing more subtle ways to legally exclude undesired groups from their neighborhoods.

Although racial zoning is illegal today, special interest groups have gradually developed zoning ordinances that make moving into the upscale suburbs all but impossible for low- to middle-income residents, who are disproportionately racial minorities. These types of zoning ordinances have been labeled exclusionary, and they serve as de facto racial zoning. Between the 1960s and 1970s, exclusionary zoning practices were challenged in the Supreme Court. Three important concepts or definitional issues emerged to shape the Supreme Court's decisions: [who is a family member, who can sue regarding land use and how to determine if planning or zoning is racially discriminatory]. . . .

In the 1960s, the definition of family became a primary issue because of an increase in non-traditional living arrangements. Some of these nontraditional families include alternate families (cohabitation of unmarried members of the opposite sex) and extended families (which applied mostly to African-American families). Communities that wished to protect themselves from the non-traditional family often enacted zoning ordinances allowing only the more traditional nuclear family. Ultimately, the Supreme Court upheld the right of extended families to live together, thus removing an implicit source of discrimination against African-American families, but rejected the right for alternate families to exist in certain neighborhoods.

When considering potentially discriminatory zoning ordinances, the U.S. Supreme Court must also decide the question of who has the right to bring a lawsuit against a specific community pursuing a questionable land use policy. The U.S. Supreme Court refused to hear the 1975 case of *Warth v. Seldin* because it determined that the plaintiffs did not have standing or "a sufficiently personal stake in the outcome of a case to warrant its consideration by the Court." The plaintiffs in the *Warth* case were a minority group who wished to sue the town of Penfield, New York, because its zoning ordinances prevented the construction of multi-family housing, thereby restricting low-income residents from moving into its community. Even though the evidence supporting the *Warth* case was overwhelmingly strong, the U.S. Supreme Court denied standing to the litigants. The U.S. Supreme Court concluded that even though the minority group shared the same attitudes common to persons who may have been excluded from residence in Penfield, they did not have standing because they had not been personally injured by these ordinances. This decision, and the one described next, illustrates the insidious pattern that structural racism can take.

In the case of *Metropolitan Housing Development Corporation v. Village of Arlington Heights* (1977), the distinction between the harmful effect of zoning ordinances and the intent to discriminate was made by the U.S. Supreme Court. In this case, an all-white community refused to rezone to allow low- and moderate-income housing. The result was continued segregation in the community. Because no evidence showed that discrimination was the motive behind Arlington Height's refusal to rezone, the community had acted legitimately. However, the U.S. Supreme Court did find that under Title VIII, racially exclusionary zoning could be proven on the basis of disparate impact.

Although state court cases have abolished certain aspects of exclusionary zoning, U.S. Supreme Court decisions like the ones described here reveal the essence of zoning practice nationwide. The rulings on these three court cases have had the effect of greatly limiting the ability of minority groups to challenge exclusionary zoning on the basis of human rights issues. . . .

Contemporary exclusionary zoning measures take many different forms. Communities desiring to keep their neighborhoods "pure" employ different zoning ordinances that are not easily recognized as exclusionary. Expulsive zoning is an exclusionary mechanism that uses black neighborhoods as a dumping site for landfills or unwanted residential uses such as drug rehabilitation units or prisons. Some zoning ordinances explicitly prohibit lower-income homework such as beauticians and barbers but allow middle- to upper-income homework such as accounting or insurance practices. Other zoning ordinances that restrict the provision of child care or elder care and prohibit inappropriate uses of property are easily validated through nuisance and noise laws. The range of these exclusionary measures makes the survival of low-income families in a community very difficult.

The advent in the 1980s of growth management policies, which rely largely on stricter building codes and zoning laws, has been a major response to the attendant problems of urban sprawl. But policies to restrict growth have serious, negative consequences for blacks, according to [urban economist] Anthony Downs. Such policies are another exclusionary force. "Communities practicing growth management have become isolated from the cities around which they cluster, and they have become exclusionary. More important, these policies have not solved these problems; they have merely dumped them on others." Downs goes on to observe, "the most dangerous result of growth management policies is that they help perpetuate the concentration of very poor households in depressed neighborhoods in big cities and older suburbs."

Municipalities and states have begun to include the provision for low-income housing in their statutes. Approximately twenty states require an affordable housing component in comprehensive plans. Typically, the basis for inclusion of low-income housing is made through the requirement to provide land use opportunities development such as multifamily housing or small-lot zoning. These opportunities represent a "fair share" problem in which determination of a fair share is defined by the local jurisdiction or state. For example, both California and Oregon have statutes requiring a low-income housing component. California requires that local jurisdictions provide for a low-income housing component in their master plans to meet what has been determined to be their share of the regional housing need for affordable housing.

[. . .]

The Brownfield dilemma

[. . .]

A major land use issue whose resolution is critical to the revitalization of inner cities and minority communities is that of the environmental remediation and economic redevelopment of brownfields. As opposed to greenfields, or undeveloped land, a brownfield is an abandoned, idle, or underused property where redevelopment or expansion is limited by actual or perceived environmental contamination. The barriers to development that brownfields pose for our oldest urban areas may relegate substantial portions to permanent wasteland if present environmental policies and economic development practices continue. With their continuation, employment and economic development prospects are diminished for inner-city residents.

From an economic development and planning perspective, two factors are key for explaining failed efforts to revitalize inner cities during the latter half of this century. First, industrial restructuring has eroded the economic base of inner cities, thereby eliminating manufacturing jobs and leaving vacant industrial properties. Second, metropolitan regions' overly decentralized land use control, development practices, and public service funding have facilitated the growth sectors' (typically given some sort of services label) location outside the inner city. The resultant growing inequality between the urban poor and wealthy is well documented.

The inability to resolve the issue of widening inequality is further exacerbated by efforts in recent years to resolve another of our nation's critical issues: remediation and control of contaminated land. National environmental legislation has profoundly altered the rules of the urban redevelopment game. As a result of this legislation, we have under way a process of "environmental redlining," or "brownlining" (i.e., identifying areas to be excluded from redevelopment consideration), that significantly diminishes efforts to improve the economic status and quality of life of populations residing in areas containing contaminated land. For much of the inner-city minority community, this means victimization twice in this century by redlining practices that have actually been reinforced by federal policies.

Although industrial activity is most recognized as a source of environmental contamination, it is important to recognize that certain commercial activities – notably gas stations, auto repair shops, and dry cleaners – are also highly correlated with environmental contamination. Historically, gas stations and dry-cleaning establishments were neighborhood businesses, so most urban blocks contain at least one.

State and federal regulations intended to prevent or remediate environmental contamination have resulted in large-scale market failure for brownfield redevelopment projects. Knowledge and suspicion of a site's contamination typically result in it no longer being considered for redevelopment and can even taint prospects for contiguous sites. Developers shun such sites because of the liability that owners of a contaminated site assume by law, whether or not they were the actual contaminators, and because of the costs – which may not be fully calculable at the outset – incurred to clean up the site for reuse.

[. . .]

Brownfield properties may be unattractive for development for numerous reasons, including inadequate infrastructure, blighted neighborhoods, high crime rates, or lack of an adequately trained workforce. Perceived or actual contamination provides one more reason for developers to invest in the suburbs because the liability for cleanup falls on all potentially responsible parties. Potential buyers often shy away and instead seek uncontaminated land in suburban areas. What remains are vacant sites scattered throughout the urban center of industrial cities.

The inadvertent but serious consequences for brownfield redevelopment of environmental remediation requirements are beginning to be addressed. Some chipping away of the overall market failure is occurring as private developers and lenders become more knowledgeable about contamination and the risks and costs associated with redeveloping such properties. Where extraordinary returns may be expected, these properties will be redeveloped. Extraordinary returns will, of course, most likely be generated by properties that are well located, have particularly desirable structures on them, or are relatively large sites with the possibility of storing contaminants on one portion designated for non-intensive use. In other words, the private sector can be expected to "cream off" the best of the contaminated properties for redevelopment. However, the brownfields in most of the inner city's poorest neighborhoods will simply fall out of the market. Those sites put such neighborhoods at higher risks of hazardous material exposure, vagrancy and criminal behavior, and deflation of surrounding property values.

Within the public sector, program innovations by some states have occurred in response to the need to resolve the impasse created by national environmental legislation. To work, these programs must be able to effectively supersede federal legislation. At the local level, city governments . . . are beginning to evaluate their problem properties and develop means by which to dispose of, or redevelop, them. By what criteria will they choose properties for resolution? Clearly, if the properties number in the hundreds . . . not all can be remediated and redeveloped.

Serious consideration must be given to the criteria by which the number of brownfield properties is reduced to avoid promulgating yet another force for widening urban inequality.

[. . .]

Will fiscally squeezed city governments feel pressured to use only economic criteria in selecting brownfield properties to receive the limited remediation dollars that have recently been made available? If so, we risk undervaluing the costs of allowing brownfield sites to remain in poor neighborhoods on the basis of foregone tax revenue, redevelopment opportunities, profits, and residents' participation in the economy. Although the layer of contaminated properties below those with which the private sector would not deal is treated, the "bottom of the barrel" will still remain untouched, thereby contributing to a widening of the gap between the worst-off neighborhoods and the rest of the city.

The stigma of crime

Critically underrecognized crime, or the perception of crime, serves as a rationalizing force working against motivating a stronger political and metropolitan-wide effort to revitalize the inner city. Furthermore, efforts to "get tough on crime" concentrate societal focus on incarcerating criminals – who are disproportionately racial minorities from inner cities – rather than alleviating the economic and social conditions that contribute to their criminal development.

Although regional solutions are necessary to solve the problems of the inner city, the suburban portions of metropolitan regions, citing their concerns that it is a dangerous place, often turn their backs on the inner city. Interestingly enough, although the incidence of the kinds of crime most associated with inner cities is declining, concerns about crime have been increasing. The Federal Bureau of Investigation (FBI) reports that serious reported crime declined 9.4 percent between 1991 and 1996, and violent crime declined 12 percent. Yet, Gallup surveys that asked people what they considered to be the most important problems facing the nation found that those citing crime grew from only 1 percent in 1990 to 52 percent in 1994. . . .

This is not to deny that crime is a genuine problem in the inner city, but it is far more a problem for residents of the inner city than for the suburbanites who travel there. Harold Rose, the first and only African-American president of the American Association of Geographers, pioneered in studies of black violence and criminal behavior. "Black on black" violence and gang activity are, arguably, the nation's greatest crime and violence problems.

[. . .]

Rose correlates the rising levels of victimization and homicide in ghetto communities across the nation with the deterioration of the manufacturing economy and intensification of labor market problems for inner-city residents. . . .

[T]he media saddle the inner city with the stigma of crime. . . . [T]he U.S. mass media's private ownership, high centralization, and elite domination pose particular problems for poor, local, and nonwhite communities that lack the economic resources needed to effectively control or get access to the media. The media increasingly focus on creating mass audiences and, therefore, have little incentive to represent the interests of select communities such as poor, inner-city neighborhoods. Media outlets for information exchange within minority communities have decreased markedly. . . . Notable for our focus here is [the] discussion of the way the media frame social issues and prime public reaction to them. Television, the most common source of information in the United States, tends to frame issues episodically, as opposed to thematically. . . . To elaborate, episodic framing shows issues in specific or concrete terms (homeless person, unemployed worker, airline bombing) with a focus on high visual appeal, whereas thematic framing places an issue in a broader context such as an in-depth report consisting primarily of a "talking head."

[. . .]

Thwarted mass transportation systems

The provision of transportation services to inner-city residents is vitally important to improving their opportunities for participating in the economy. Continuing suburbanization of employment opportunities means that inner-city residents must have a means of traveling to the suburbs for work. . . . In the Atlanta city region [for example], the two highest growth counties of the metropolitan region, Gwinett and Cobb, cannot get enough workers to fill many of their retail and other service sector positions. As a consequence, jobs such as those in fast-food establishments, which normally pay at or close to a minimum wage, are being bid up to attract suburban teen and retired workers. Both of these counties, whose population is approximately 90 percent white, have refused to allow the Metropolitan Atlanta Rapid Transit System (MARTA) to be extended to them.

PROMISING APPROACHES TO INNER-CITY REVITALIZATION

For inner cities, the latter half of the twentieth century has been a prolonged period of disinvestment and outmigration of population and economic activity. That this tide of disinvestment and abandonment continues to this day does not result from inattention to the urban plight. We have reviewed a long history of redevelopment efforts undertaken by the public and nonprofit sectors at the local, state, and national levels, as well as by "good corporate citizens" across a wide spectrum of economic activity. At the end of this century, however, structural racism's subtle and not-so-subtle impacts continue to diminish African-American and other racial minorities' opportunities to become full-fledged members of the American society and economy and to have their fair share of the evolving American Dream.

Although much has been made of the failure of a half-century's efforts to revitalize the inner city, very little has been made of the failure to acknowledge how continued racism has defeated these revitalization efforts. At the very least, we should take from the past the understanding that large-scale, lasting success in inner-city revitalization must evolve from an approach that explicitly recognizes and seeks to eliminate racism. From this common understanding, we can renew past efforts that have shown some promise and develop new approaches. [We highlight some here:]

- We have seen past success in the financing of inner-city housing, which can mean providing low-cost loan advances to pay for the construction of affordable, multifamily housing or ensuring the availability of loans to inner-city residents for the purchase or renovation of single-family homes.

[. . .]

- The movement to develop housing in inner-city areas has its roots in the development of community-based organizations and coalitions of low-income housing supporters, social service agencies, housing authorities, neighborhood activists, tenants, community development organizations, and others committed to improving local housing and economic opportunity for residents of inner-city areas and some suburban areas. The formation of coalitions to improve housing in cities has, therefore, been one important avenue for community reinvestment.

[. . .]

- The successful revitalization of the inner city also requires the development of human resources or capital. Although the difficulty of the human resource problem in inner cities explains why it is often overlooked, human resources remain the foundation that must ultimately support American cities and their economies. The need to improve the state of human resources in our cities can be addressed through the joint efforts of social service agencies, school officials, churches, the business community, and other concerned groups. . . .
- At the same time, ensuring that the inner city is part not just of the regional economy but of the global economy as well will be increasingly important. The global economy has come about, in large part, as a result of advances in telecommunications and information technology. To be a part of this economy, an area must have the necessary information technology infrastructure, such as fiber optic networks. . . . [I]nadequate investment in information technology infrastructure for inner-city neighborhoods could prove to be another barrier to economic development. Back offices and call centers could be located just as well in the inner city as they are in exurban areas if present racial biases are eliminated. If inner-city schoolchildren are not given the same access to computers as suburban children, we have yet another – and potentially very serious – gap in human capital in the making. . . .

[. . .]

- [T]he core element of revitalizing inner cities is the restoration of human potential, which means creating communities where individuals have the opportunity and capacity to play an important role in determining their future. It also means integrating inner-city communities into the economic, political, and cultural life of surrounding metropolitan areas. Such integration can be accomplished only through a comprehensive approach that diffuses progress throughout the entire community rather than concentrating it among a privileged few. Individual profitability and the welfare of the community must be balanced. . . .

[. . .]

As we observed at the beginning of this final section of our article, there has been little acknowledgment of how continued racism has contributed to the defeat of a half-century's efforts to revitalize inner cities. The planning field is not alone in its culpability for failed revitalization efforts, but the misappropriation of its tools has, perhaps more than in other fields, made it a facilitator of social exclusion and economic isolation. Today, however, planners are at the forefront of efforts to create . . . the . . . city [that] promotes social justice by seeking a more equal distribution of resources among different racial and socioeconomic groups and between the inner city and the rest of the metropolitan area.

4

Pluralistic Planning for Multicultural Cities

Journal of the American Planning Association (1997)

Mohammad A. Qadeer

Editor's introduction

Since urban planners operate in a plural and multicultural world, "pluralism" and "multiculturalism" are important concepts to explore and understand, especially as they influence the future-oriented community-building work that is the basis of the profession. The first step in such an exploration is to define the terms, the next is to be aware of the differing interpretations of the definitions, and the last is to see how to integrate this awareness into plans and implementation strategies.

Dictionary definitions of the terms are quite straightforward. According to the *Oxford English Dictionary*, pluralism is "the presence or tolerance of a diversity of ethnic or cultural groups within a society or state; (the advocacy of) tolerance or acceptance of the coexistence of differing views, values, cultures, etc." Multiculturalism, according to the same source, is "the policy or process whereby the distinct identities of the cultural groups within such a society are maintained or supported." These ideas became important worldwide after the mid twentieth century as various societal changes challenged the dominance of Western ideals. Among them were the fall of colonialism, the beginnings of globalization, the emergence of large-scale labor force migrations, the rise of civil rights, and other social and political struggles for independence.

In the United States, the seminal study of urban multiculturalism was Nathan Glazer and Daniel Patrick Moynihan's *Beyond the Melting Pot: The Negroes, Puerto Ricans, Jews, Italians, and Irish of New York City* (Cambridge, MA: MIT Press, 1963). The book began as a *New York Post* series on the city's five major ethnic groups but soon became a research project encouraged by Martin Meyerson, the Director of the Joint Center for Urban Studies of Harvard and MIT, the original publisher of the book. Glazer, a sociologist and the book's senior author, recruited Moynihan, then assistant to the U.S. Secretary of Labor, to work on it. Moynihan wrote the chapter on the Irish and contributed to the conclusion. Synthesizing their own intensive fieldwork, they dispelled conventional melting-pot wisdom with their argument that ethnic groups were a "new social form," well defined and strong. These ethnic groups showed no signs of weakening, despite the absence of ongoing immigration – few new immigrants had entered the country over the past forty years – and decades of "Americanization," or assimilation, forced, in large part, through the public school system. The groups had distinct and robust cultural characteristics, evident in their voting patterns and residential and employment choices. The Irish, for example, lived on the Upper East Side and the West Side's Hell's Kitchen and dominated the civil service, especially the police force.

In 1965, shortly after the publication of *Beyond the Melting Pot*, Congress loosened the country's restrictive immigration laws dramatically. In the succeeding decades, New York and other "gateways" accommodated hundreds of thousands of newcomers. By 2000 many cities would see their foreign-born populations climb to double-digit percentages – 37 percent in Miami, 35 percent in Los Angeles, and 28 percent in New

York City. Other places in the world – especially in Western Europe, Canada, and Australia – experienced similar demographic changes.

This influx of immigrants stimulated public discourse on multiculturalism. Some condemned it, worrying that it would lead to the formation of disconnected communities subscribing to "particularism," the preservation of individual cultures at the expense of common, shared values and purpose.

With continuing immigration – New York City, for example, was gaining 100,000 immigrants a year in the 1990s – many cities and the nation as a whole became increasingly diverse. The idea of any city being uniform, or of there being one single type of American, vanished, replaced by the "gorgeous mosaic" evoked by New York City's mayor, David Dinkins, the first African-American to hold the post. Support for multiculturalism gained momentum as ethnic and other race-, gender-, and nationality-based groups became more visible through sheer numbers, self-identification, and/or political activism. Local, state, and national elected officials recognized and responded to their needs.

However, resistance to multiculturalism also grew, stimulating additional rounds of theorizing. The terrorist attacks of September 2001 fueled suspicions of certain ethnic groups; nativists made the issue of immigration fodder for national and local election campaigns.

Political scientists and, later, planners, explored ways of resolving the tensions between diversity and unity inherent in multiculturalism. Princeton University scholars Amy Gutmann (*Color Conscious: The Political Morality of Race*, Princeton, NJ: Princeton University Press, 1996) and Kwame Anthony Appiah (*Cosmopolitanism: Ethics in a World of Strangers*, New York: Norton, 2006) advocated creating opportunities for respectful deliberation and compromise. Canadian planner Leonie Sandercock (*Towards Cosmopolis: Planning for Multicultural Cities*, New York: Wiley, 1997, and *Cosmopolis II: Mongrel Cities of the Twenty-first Century*, London and New York: Continuum, 2004) proposed making the invisible visible – telling the untold stories of women, minorities, and others; engaging in dialogue; appreciating local knowledge. Sandercock's work applies "postmodern" planning theory to argue in favor of fostering communication – not rational decision making – to plan a multicultural city.

Discussions of multiculturalism first appeared in education policy circles but soon spread to other areas, including land use, service provision, zoning, and open space design and usage. On the professional level, the American Planning Association authorized five new divisions: Planning and Women Division (in 1979), Planning and the Black Community (in 1980), the Gay and Lesbian Planning Division (1998), Indigenous Planning Division (2004), and Latinos and Planning (2005). The APA also took up the topic of multiculturalism in newsletters and other official publications – this selection appeared in the *Journal of the American Planning Association*, and Kristin E. Sosnicki-Campbell's "Multiculturalism as a Tool for Economic and Community Revitalization" was published in the American Institute of Certified Planners' *Practicing Planner* (1: 4, 2003).

Germane to urban planning and urban development is the policy dilemma about how to treat ethnic enclaves. On the positive side, cultural agglomeration strengthens social ties and ethnic identification/pride; provides a critical mass for business opportunities related to the group's demand for specialized products and foods; serves as a bloc for political organization and/or representation; and fosters the development of supportive community facilities such as places of worship and social clubs. But there can also be a down side to ethnic enclaves: Their residents, especially children, may not learn the ways or language of the mainstream population, thus limiting opportunities for employment and economic mobility. Mohammad A. Qadeer, the Canadian author of this selection, discusses this issue in an article he co-authored with Sandeep Kumar, "Ethnic Enclaves and Social Cohesion," *Canadian Journal of Urban Research* (15: 2, 2006).

Canada, in fact, has been a leader in integrating multiculturalism and pluralism in national life. With 27 percent of its population foreign-born, Canada has been addressing these issues for more than four decades and was the first country in the world to adopt multiculturalism as national policy. It has worked multicultural protections into its Charter of Rights and, in 1988, passed a Multiculturalism Act that mandates the preservation and enhancement of diverse cultures, lodging oversight in the national Minister of Canadian Heritage. This selection shows how sensitivity towards multiculturalism is applied in urban planning.

Mohammad A. Qadeer, professor emeritus, School of Urban and Regional Planning, Queens University, has written extensively on urban communities in Toronto. His essay "Ethnic Segregation in a Multicultural City" was published in David P. Varady (ed.), *Desegregating the City: Ghettos, Enclaves, and Inequality* (Alb NY: SUNY Press, 2005). In addition, Qadeer has developed planning recommendations for multicultur environments. The object, he argues, is to promote social inclusion and integration while protecting individual cultures. This involves assuring equitable provision of services tailored to the neighborhood, allowing mixed land uses and a variety of housing types, using bilingual signage, and recognizing cultural and religious requirements and specialized commercial practices as a basis for zoning variances or adjustments to site plan regulations. *Vive la difference!*

A vast literature on ethnicity and multiculturalism has emerged. Geographers, for example, are documenting the new urban landscapes resulting from the nation's high rates of immigration. David H. Kaplan's *Landscapes of the Ethnic Economy* (New York: Rowman & Littlefield, 2006) analyzes how immigrant commercial activities have reshaped urban areas. Heather A. Smith and Owen J. Furuseth, *Latinos in the New South: Transformations of Place* (Burlington, VT: Ashgate, 2006), and Daniel D. Arreola (ed.), *Hispanic Spaces, Latino Places: Community and Cultural Diversity in Contemporary America* (Austin, TX: University of Texas Press, 2004), focus on the Latino population's effect on urban design. For a general assessment of ethnicity, its definitions and expression, see Wilbur Zelinsky, *The Enigma of Ethnicity: Another American Dilemma* (Iowa City, IA: University of Iowa Press, 2001). For more reading directly related to planning and design, see Leonie Sandercock's work cited earlier and also her edited volume *Making the Invisible Visible: A Multicultural Planning History* (Berkeley, CA: University of California Press, 1998), in addition to Michael A. Burayidi (ed.), *Urban Planning in a Multicultural Society* (Westport, CT: Praeger, 2000) and James Donald, *Imagining the Modern City* (Minneapolis, MN: University of Minnesota Press, 1999).

. . . A tree can be the source of neighborhood battles, as shown in Toronto by the "nature meets culture" headlines in *The Globe and Mail.* Italians and Portuguese like to keep trees short, allowing a better view of the neighbors. Ango-Saxons want trees to be tall and leafy, blocking any views from and to neighborhood houses. The Chinese believe trees in front of a home bring bad luck. As if these different preferences were not enough, the city has strict bylaws that prohibit cutting down trees, but allow pollarding (trimming trees into a high bush), which is favored by Europeans. This example illustrates how multiculturalism permeates even small details of urban life. It also embodies the planning issues that arise from the cultural diversity of local populations, such as the uniformity of policies and standards, differences in citizens' wants, equity in accommodating the needs of divergent groups, and public versus private interests in the spatial expressions of cultural values. . . .

. . . Multiculturalism as a public philosophy acknowledges racial and cultural differences in a society and encourages their sustenance and expression as constituent elements of a national social order. This philosophy envisages the society as a mosaic of beliefs, practices and customs, not as a melting pot assimilating different racial and cultural groups. . . . Multiculturalism has two defining principles: (1) the right to practice and preserve heritage, collectively as well as individually; that is, not only are Chinese or Ukrainians free to speak their languages at home, but they also have the right to form associations, organize communities, and practice their customs and religions as a group; and (2) equality of rights and freedoms under the law for all individuals and communities. Both these principles have direct bearing on urban planning.

Increasingly, the effectiveness of urban planning is assessed by its responsiveness to citizens' needs and goals. Given that interests and preferences differ by social class, race, gender, and cultural background, the responsiveness of urban planning depends on its ability to accommodate citizens' divergent social and cultural needs and to treat individuals and groups equitably in meeting those needs. To fulfill these requirements, the first step is to eliminate overt discrimination on the one hand, and cultural biases in the use of land, the housing market, and the provision of urban services on the other.

In Canada, the Charter of Rights and Freedoms and the post-1960s social consensus against racism have largely eliminated overt forms of discrimination, particularly of the type "Blacks or Indians need not apply." However, the cultural values embedded in historical practices, public policies, administrative procedures, and regulatory standards are another matter. Planning policies and standards presumably are based on universalist criteria. Often they are backed by historic practices and established professional conventions. Yet they originate from social patterns and cultural values of the dominant communities, namely, in Canada, the English or the French.

Toronto's tree bylaw, described earlier, is a case in point. A seemingly neutral regulation about tree maintenance is in fact an embodiment of English/European preferences. Should this bylaw be amended to accommodate the preferences of diverse communities, particularly now that trees have become icons of a new environmentalism? How should competing values be balanced? This is the challenge of dealing with cultural biases embedded in historical practices.

. . . In practice, intercommunity equity in fulfilling social needs is often pursued through political bargaining and administrative procedures. This is particularly the case with planning policies and regulations. At present, it is enough to "flag" the cultural predispositions of planning policies and standards as the area of contentions arising from multiculturalism.

A second aspect of urban planning with a bearing on multicultural communities is its reliance on property development as its instrument. Critics of urban planning have long held its property-centered approach to be a factor that limits realization of its promise to promote people's welfare and equity, and efficiency. Multiculturalism requires that planning instruments be both sensitive to and responsive to the social needs of particular communities and therefore calls all the more for people-centered approaches. Any cleavage between social objectives and institutional instruments is further sharpened by multiculturalism; the appropriateness of instruments can be called into question on the grounds of their relevance to culturally diverse communities.

Probably the most striking impact of multiculturalism on urban planning comes from the presence of ethnic neighborhoods and ethnic business enclaves. The emergence of ethnic residential and business districts brings up an enduring concern of urban planning, namely, striking a balance between social segregation and integration.

Urban planning has long espoused social integration as a guiding value, aiming to promote communities mixed across races and classes. Yet, a central process shaping urban structure is the spatial clustering of activities and social groups based on the advantages of agglomeration economies as well as on community sentiments. Such clusters usually occur as a result of the convergence of a multitude of individual choices and are seldom preplanned. They can take forms as different as a theater district or a Ukrainian village.

When the real estate market creates an ethnic concentration, pressure builds for the provision of appropriate services and regulations. In most such situations, the ethnic community's needs have to be reconciled with the requirements of other residents of an area and also with citywide objectives. A balancing act of public policy, making tradeoffs among different objectives and values, is called for.

The present public attitude toward ethnic concentrations reflects the notion that residential or business concentrations arising from individual choices freely exercised without prejudice to others should be sustained, whereas socially or racially homogeneous neighborhoods formed through discriminatory practices and explicit or implicit exclusionary policies should be recognized as prejudicial to the public interest. This distinction is based on structural factors: those external to individuals (discriminatory practices and policies) versus those internal to individuals (motivation and preference, free choice). Social segregation at the block or neighborhood level that arises from dispositional factors is voluntary as long as there are no barriers to others who may want to live there. The acceptability of social segregation or integration in a city is determined by the process of its formulation and by its function, not by the density of concentration.

To sum up the foregoing discussion, multiculturalism affects urban planning in two ways. The first is that it holds planning policies and standards up to the light of social values and public goals. Are policies equitable both procedurally and substantively in satisfying the needs of diverse individuals and groups? Is there a cultural bias in their universalist criteria? How can the competing interests of mainstream and of minority communities be balanced?

Second, multiculturalism recognizes the legitimacy of ethnic neighborhoods and enclaves. The emergence of these spatial concentrations affects urban planning by precipitating questions about the internal structure of the city as a whole, and challenging social policies to balance the advantages of neighborhood homogeneity with the public goals of openness and equal access by all.

A fundamental effect of multiculturalism is to call for pluralistic planning approaches and to question unitary conceptions of public interest and the ideology of "master plans." Davidoff's idea of pluralism as a planning approach comes close to accommodating multiculturalism. His concept has roots in equity and a commitment to open bargaining among competing interests that make it particularly relevant. However, multiculturalism, along with feminism, expands the definition of the interests to be accommodated beyond race and class, and thus extends the meaning of pluralism. In pluralistic planning, performance measures for policies and standards aim for the equal satisfaction of the needs and preferences of diverse groups. These are the expected directions of change in planning as influenced by multiculturalism. The following sections examine what actually is happening in multicultural Canada.

MULTICULTURALISM IN CANADA

Canada has been a multicultural country from its beginning. In the mosaic of aboriginal cultures and languages there, multiculturalism extends back to antiquity. The European settlement, though led by the two founding communities of the British and the French, also included Germans, Russians, Chinese, and Ukrainians, among others. Thus, Canada had been a multicultural society long before that was acknowledged in federal policy. Canada has long regarded itself as a mosaic of cultures, in contrast to the United States, which is described by the metaphor of a "melting pot." This distinction, albeit overdrawn, has shaped Canadian attitudes toward multiculturalism. . . .

HOW MULTICULTURAL IS CANADA?

The cultural characteristics of the Canadian population are measured by three indicators: ethnic origin, status as an immigrant, and language(s) spoken at home. Indicators yield different, though overlapping, images of Canadian society, but all reveal a wide range of ethnic diversity. . . .

On all three indicators – ethnic origins, immigrant population percentages, and home language – ethnic diversity in Canada is both wide-ranging and increasing in recent times. Cities are the loci of this multiculturalism. Multiculturalism in Canadian cities is not altogether new. . . .

MULTICULTURALISM: OLD AND NEW

. . . Canadian cities have long been strikingly multicultural on account of continual immigration from abroad. Historically, immigrant ghettoes and ethnic enclaves have been the emblems of urban Canada's multiculturalism. Winnipeg in the 1920s, for example, was composed of a number of subcommunities distinguished by ethnicity, religion, and class. Toronto's immigrant quarters were distinctly different from the main city even in the nineteenth century. The question then is: If multiculturalism is a long tradition of Canadian cities, what is new about it that affects the planning system?

Contemporary ethnic studies distinguish between "old" and "new" multiculturalism. The old multiculturalism or cultural mosaic was a private affair of immigrants, expected to dissolve or be diluted with their assimilation. It conferred no public rights. The old mosaic was confined to working-class, immigrant districts, typical of Burgess's zone of transitions, in the heart of a city.

The "new" mosaic is acknowledged by public ideology and official policy. It is based on the post-World War II notions of human rights and the equality of citizens. Though still largely driven by immigration, the new mosaic is also sustained by the circulation across national borders of corporations, labor, and information. Tourists and sojourning executives and professionals contribute to the new multiculturalism. The post-industrial economy and the rise of internationalism and globalization have turned cultural diversity into an economic asset. At the same time, they have also accelerated the diffusion of one culture into another. . . .

These changes mean that the new multiculturalism is not limited to the poor and to downtown core areas. It has spread to suburbs, creating bourgeois ethnic enclaves. It has spawned new spatial and architectural forms. The result is that today's Canadian metropolitan areas are having to consider the rights of ethnics to organize their social lives in accordance with their preferences and values. How such entitlements alter the planning system is observable in both its planning process and its policies about neighborhood development, housing, and public services. . . .

MULTICULTURALISM, SOCIAL DIVERSITY, AND THE PLANNING PROCESS

The cultural and racial diversity of citizens bears on the planning process in three ways. First, it affects the rational-technical component; race and culture have become significant analytical categories for assessing public needs and analyzing social conditions. Delineating neighborhoods by ethnocultural criteria, mapping catchment areas for community services along sociolinguistic lines, and analyzing housing conditions by race and ethnicity are examples of planning methodologies that are emerging with the acknowledgment of multiculturalism. They constitute a paradigm shift in the methods of defining a local community, as seen in the analytical work of planning departments.

Second, planners must now be sensitive to the needs of individuals (and groups) in new ways, largely in how they listen to clients and how they interpret and apply regulations to them, particularly those in minority communities. What treatment do persons of color, unusual names, thick accents, or non-English or non-French ancestry get from planning departments? Are there systematic biases in planning procedures and outlook that put minority communities at a disadvantage? The Royal (British) Town Planning Institute found that ethnic and racial minorities suffered high refusal rates for development permissions. No similar study for Canada is known. Yet it is the case that ethnic communities often complain about getting short shrift from planners.

Third, the scope and procedures of citizen involvement in the planning process have to be modified to accommodate multicultural policies. It is on this score that the planning process has shown the greatest responsiveness to the diversity of citizens.

Vancouver and Toronto have led in facilitating the participation of ethnic communities in the making of official plans. Vancouver's city plan process set up "citizens' circles" in neighborhoods and invited ethnic community associations, as well as individuals, to identify needs and articulate their goals. Separate info-lines were set up in four nonofficial languages (Chinese, Punjabi, Spanish, Vietnamese) to give out information and receive comments. Similarly, Metropolitan Toronto and the City of Toronto, separately, set up elaborate consultative procedures that facilitate the participation of ethnic communities. The City of Ottawa distributes notices and information about planning proposals in the heritage languages of an area's residents and regularly consults with ethnic community organizations. In metropolitan centers, then, the planning process is beginning to accommodate multiculturalism, at least in seeking the opinions and input of minorities. The practice has not yet filtered down to small cities or the rest of the country, where ethnic communities are not yet a significant political force.

The next step in pluralistic practices of planning is to include minorities on decision-making bodies. Toronto's proposed official plan urges the city council to include "ethnocultural and racial communities" in city committees and working groups. Also important is the appointment to planning departments of professionals from minority communities. The diversity of planners' backgrounds ensures appreciation of cultural and racial differences. In the same vein, representation of minorities among elected and nominated executives and local and provincial levels is a necessary condition for bringing a multicultural perspective to public decision-making bodies. On decision-making bodies, the progress towards accommodating multiculturalism is slower than that in the participatory aspects of the planning process.

A planning process is a means for developing policies and programs to fulfill diverse needs and goals. Ultimately, it is the relevance and appropriateness of the policies and programs themselves that determine how well planning accommodates diversity. The next section therefore turns to planning issues that arise at the neighborhood level, where multiculturalism is most evident.

NEIGHBORHOOD PLANNING: STABILITY AND CHANGE

Historically, ethnic residential concentrations have been features of Canadian cities. Toronto, Montreal, and Vancouver, of course, had Chinese, French, Jewish, and Italian districts in the late nineteenth century. Prairie cities had sizable concentrations of Eastern Europeans and Natives, and even a small town, Fort William, Ontario, had a Finnish colony. Clearly, residential segregation by ethnicity and class has been an historical feature of Canadian cities. . . .

The revised immigration law (1968) as well as the Charter of Rights and Freedoms have transformed Canadian cities. They have eliminated overt discrimination. In addition, immigrants now are by and large better educated, often technicians and professionals or, lately, investors and entrepreneurs. Large proportions of recent immigrants are visible minorities. Most settle in metropolitan areas and head directly to the suburbs. They are forming ethnic concentrations in new parts of cities. Furthermore, the urban structure itself is changing. Cities have spilled out to form multifocal metropolises. Edge cities have emerged. These new urban forms are socially more diverse. . . .

The social mix characteristic of the new multiculturalism is also reflected in the fashionable districts of central cities. Here clubs, bars, boutiques, gift shops, and restaurants offering a variety of ethnic goods intermingle, serving youth, yuppies, and tourists. Even new waterfront developments – Harbour front, Granville Island, and Old Montreal – are “hip” places by virtue of the cultural diversity of their commercial establishments. These new expressions of multiculturalism mix ethnicities to create cosmopolitan districts.

Thus, multiculturalism now serves two functions. First, it fulfills the social needs of ethnic communities, and, second, it weaves diversity of both activities and built forms into urban and regional structures. These functions bear directly on neighborhood planning, which must mediate between the competing interests of stability and change.

Socially, neighborhoods change continually; residents leave or die, and new households move in. Yet neighborhoods are considered stable as long as the change is incremental. With the movement of a new ethnic group, a different social class, or new activities into a neighborhood, however, the rate of change accelerates to the point that it reorganizes a local community, disturbs the social patterns, and necessitates the realignment of neighborhood services. Multiculturalism comes to permeate an area through such a process of change. Although the process is largely market-based, the planning system is called upon to intervene as different groups of residents respond to marked change in a neighborhood.

Old residents come calling on the planning institutions to protect their interests. The new arrivals seek fair treatment and accommodation of their needs and preferences. Zoning challenges, public hearings, and school board meetings become battlegrounds for these competing groups. Neighborly spats or personal biases often are pursued through planning hearings and appeals. The planning system becomes an arena not only for contesting ethnic interests, but for more personal conflicts as well. It may or may not have the mandate, resources, or instruments to resolve social conflicts and fulfill the demands of divergent groups. The system’s effectiveness lies in separating public policy issues from interest group agendas. If it succeeds in balancing competing interests fairly, it will have fulfilled its mandate. . . .

CASE HISTORIES OF PLANNING RESPONSES TO MULTICULTURALISM

[. . .]

Metropolitan Toronto: Kingsview Park (the “Dixon” case)

The “Dixon” case . . . is an example of cultural and racial tensions arising entirely from changes in the tenure mix of the housing market and the demographic mix of the community. Here the central question is how to reconstitute a local community after dramatic demographic changes.

A television documentary and numerous newspaper articles have made the "Dixon Complex" notorious as an example of a neighborhood gone awry. Kingsview Park is a high-rise condominium complex of six buildings containing 1,794 apartments, built in 1971 along the four-lane Dixon Road in Etobicoke, a suburban municipality of Metropolitan Toronto. Kingsview Park was first inhabited by young families and empty-nesters who were mostly owner-occupants; about half had migrated from Europe shortly after World War II.

Within a few years of its development, the social composition of this complex began to change. As young owner-occupants moved out, the vacancies were filled by recent immigrants from Southern Europe and Asia. The proportion of renters rose steadily, to about 12 percent in 1977.

Toronto's property boom of the 1980s attracted speculator-investors, absentee owners who rented out apartments while waiting for capital gains. By 1989–91, 25 percent of the apartments were renter-occupied. Then, because of high vacancies and downturn in the property market, occupancy gradually filtered down in the rental market to the Somalis who had come in the most recent wave of immigration.

By 1994, poor Somalis occupied 59 percent of the rental units and were 35 percent of Kingsview Park's population. This change in the complex's tenure mix and social composition precipitated ethnic tensions and cultural conflicts between white owner-occupants and the Condominium Board, and black tenants. A series of confrontations between Somali youths and the management and security personnel culminated in a riot that spilled on to Dixon Road and blocked traffic on July 30, 1993. This breakdown of the local community prompted public intervention.

The Mayor and the Municipal Council of Etobicoke responded to the "riot" by forming the Dixon Road Community Response Team and the Task Force on Residential Overcrowding. The Planning Department was represented on these bodies. The Task Force recommended a review of security and management in the buildings, race and cultural sensitivity training for security personnel, and formation of residents' committees, along with other measures. The planning issue was framed as a matter of housing occupancy: overcrowding in rented apartments became the focal point of the Task Force. The issue targeted mostly Somali households, which tend to be large and have long been viewed as harboring friends and relatives.

Conventional urban planning had no defined role in this profound cultural clash; yet there were continual calls from one or the other parties for the Planning Department to get involved despite the fact that on these issues it had little authority. The need was for social planning and community organization. Instead, the proposed solutions to the "Dixon Complex" include regulations to discourage overcrowding and a recommendation to amend provincial legislation to allow municipal inspectors the authority to enter private dwellings without warrants. In short, the planning response to this social and cultural clash was simply to tune up property standards and create the statutory instruments to enforce them.

[This] . . . case illustrate[s] public issues precipitated by the diversification of forms, functions, and populations in neighborhoods. The Canadian planning system usually muddles through such changes. It favors neighborhood stability, but promotes development. It has neither the authority nor the tools to intervene forcefully in real estate markets propelled by individual choices. And, indeed, multiculturalism calls into question current notions about stability and social mix in neighborhoods. It adds another reason for reexamining the social assumptions of urban plans, already being questioned on other grounds. . . . Social harmony and balance, rather than stability alone, should be the guiding principle for multicultural neighborhoods – this is the lesson from the Dixon case.

The overarching implication of the Canadian experience is that the role of the planning system in managing social change in neighborhoods is limited. In particular, planning regulations cannot and should not be based on the characteristics of clients, though policies aiming to satisfy people's needs do have to apply performance measures. It follows, then, that the planning system has not forged any systematic approach to sustain and promote cultural diversity of neighborhoods.

ETHNIC BUSINESS ENCLAVES

Multiculturalism also manifests itself in the commercial structure of cities, in the form of ethnic business enclaves, whether as sectoral specialization such as Koreans dominating the flower trade or Punjabis managing the taxi service, or spatial enclaves like the concentration of restaurants, food stores, and dress boutiques in Chinatown or the Greek village. Both are distinct economic formations in which ethnic solidarity and cultural norms undergird the transactions among establishments as well as among owners, managers, and workers.

In Canadian cities, local businesses of ethnic neighborhoods have been the basis of their cultural economies. Often they have extended their markets beyond their surrounding neighborhoods. In fact, not all immigrant-owned businesses target ethnic populations; many serve mainstream markets. An historical example of ethnic entrepreneurs in the mainstream economy is seen in the Chinese eateries that Chinese cooks who had been laid off by the Canadian Pacific Railway started in little villages of the Prairie provinces in the late nineteenth century. To this day, they are the main restaurants/hotels in small places – commercial manifestations of the old multiculturalism.

Since the 1970s, a new and vigorous form of ethnic business enclaves has emerged in Canadian metropolitan areas. Toronto's revitalized Chinatown and its Chinese suburban malls and office clusters, its Indian bazaars and Greek villages, and its Punjabi malls are obvious examples. Similar clusters have emerged in Vancouver and Montreal, and on a small scale in Calgary and Winnipeg. These enclave commercial districts serve metropolitan markets, creating a "niche" in the regional economy. They have been created through private initiatives; they were not planned or even anticipated by planning authorities.

City planning has often come to acknowledge the enclaves once they have been formed. Its reactive role consists of sustaining and consolidating them, frequently in the face of some local opposition that is expressed as concern about density and traffic, but that has an ethnic and racial subtext. At best, the planning system takes a circumscribed view of its jurisdiction in such situations. Focusing on development issues and regulatory requirements, it uses them as the means for mediating among competing interests and groups. The planning system faces a dilemma: it cannot plan or zone by the characteristics of persons, yet it has to acknowledge the cultural bases of an enclave's economy.

OTTAWA: SOMERSET HEIGHTS

Ottawa's Somerset Heights area, the city's hub of multicultural commercial establishments, illustrates the planning system's response to the formation of an ethnic business enclave. . . . Ottawa, the capital of Canada and center of a regional municipality, has been a bicultural city, though with the French language and culture not always obvious. In the past twenty-five years, however, the city has undergone a social and cultural transformation. Its Francophone population has gained prominence, and recently Italians, East Europeans, and Asians have become a presence.

Somerset Street has evolved into the center for Ottawa's ethnic businesses. Between Bay and Rochester Streets, it has become the spine of Ottawa's Chinatown or, more accurately, of a multicultural neighborhood. Its modest homes, which had declined in the postwar period, have been revived by immigrants: Italians and Chinese at first, and more recently, Arabs, South Asians, and Africans. Since 1971, dim sum halls, Chinese restaurants, groceries, and gift shops have made the area into a Chinatown. In the late 1980s, Vietnamese businesses started to appear, and now Caribbean and Middle Eastern stores are also springing up. The area has become a multicultural business enclave with a predominance of East Asian stores. . . .

Over a thirty-year period, 1961–91, area businesses increased from sixty to 101, mostly ethnic establishments. The Asian (Chinese) businesses dominated the street after 1971, steadily increasing from about 8 percent of ethnic businesses to 81 percent in twenty years. With the commercial transformation, the Asian population in the area has increased and a gradual gentrification of the neighborhood has occurred.

Ottawa's City Planning Department responded to the neighborhood change in the following manner.

- 1964 Ottawa City's comprehensive zoning bylaw designated Somerset Heights as a General Commercial Zone, permitting commercial uses for the ground floor and dwellings above.
- 1977 The Dalhousie Neighbourhood Plan designated the area for mixed land use – i.e. ground floor commercial uses and residences above; pedestrian orientation emphasized.
- 1984 The city's new zoning bylaw maintained the area as a General Commercial Zone.
- 1988 Somerset Street West Planning Study recommended "Multicultural Village" as the area's theme, and encouraged ethnic expression and the establishment of a Business Improvement Area (BIA).
- 1989 The city's Community Improvement Plan projected streetscape improvements and other capital works.
- 1992 A new official plan designated the areas as Neighbourhood Linear Commercial, recognizing the area's distinctive character and allowing mixed uses and pedestrian orientation.

Up to 1986, the Planning Department ignored the cultural dimension of the neighborhood change and maintained the fiction of uniformity. Only with the Business Improvement Area (BIA) designation, and consequent capital works, including the development of a Chinese community center, and signage and streetscape in 1988, were the cultural characteristics of the area officially acknowledged. Still, the official plan policies have not given much weight to ethnic commercial activities. They continue to treat the area as Neighbourhood Linear Commercial. Multicultural is evident in the facilities and services developed, but not in the city's zoning and land use policies. The planning response has been reactive and restrained.

The case of the Scarborough shopping malls also illustrates city planning's reactive role in such developments. Once the malls were being developed, Scarborough's city planning department duly accommodated them, first by resolving some contentious issues and then by upholding developers' rights in the face of neighborhood opposition. Yet north of Scarborough, in Markham, Chinese shopping malls have become a lightning rod for ethnic confrontation. An intemperate remark by the deputy mayor about the "excess" of Chinese malls driving away old residents has precipitated fierce public arguments about the ethnic composition of the community.

The policy issues that ethnic enclaves bring to the fore include the need for multilingual signage to ensure equal access, culturally sensitive services, and design guidelines for built forms and aesthetics that are both diverse and harmonious. Another issue that arises is the vulnerability of businesses to changing tastes and demands. The demographics of ethnic business enclaves tend to shift among groups, so planning policies should recognize the prospect of change. Supporting an enclave's special cultural character while keeping it open to all is the key challenge for multicultural planning in commercial areas. Canadian metropolitan areas have usually accommodated business enclaves through exceptions and variances of their statutory plans, as numerous thriving enclaves attest. These enclaves have, however, added a new element to the urban commercial hierarchy, necessitating reconsideration of planning policies and standards.

HOUSING CHOICES

Multiculturalism gives rise to two issues in the housing market. First is the question of discrimination against ethnic minorities in access to housing. Apart from visible physiognomic difference and class, language, cultural origin, or religion could also be the basis for restrictive practices in the market. Post-1960s legislation, including the Landlord–Tenants Act, has combined with the influence on the market of demand by ethnic minorities to eliminate overt forms of discrimination. There are cases of refusal to rent to Africans or East Indians, but they are sporadic and subject to legal recourse. The legacy of systematic bias against immigrants and visible minorities in the mortgage market, local codes, and housing standards is being exposed and eroded.

The second housing issue, more directly related to multiculturalism, is the matter of housing choices for ethnic minorities. Although their requirements do not differ from mainstream norms for housing type, location, and utilities, their preferences for dwelling size, layout, and neighborhood differ significantly. Most immigrants want single-family homes, but ethnic groups differ even among themselves in their choices of interior designs and community facilities.

The stereotypes are that Portuguese prefer two kitchens, Italians like large lots for gardens, and the Chinese notion of Feng-shui predisposes them against houses on cul-de-sacs. Neighborhoods in proximity to schools with language classes and to appropriate religious associations and cultural institutions are also preferred. Immigrants of modest income may want the option of dividing a home to create a rental apartment; their affluent counterparts prefer large homes, for prestige as well as for prospects of gathering together the extended family. For ethnic groups, these culturally based preferences, differing by economic class, are measures of housing satisfaction. They often fall within the purview of building, safety, and public health codes; and of planning standards, particularly for occupancy densities, household use, and the definition of a family. Often these do not accommodate ethnic minorities' preferences. Case studies reported earlier, particularly the . . . Dixon case . . . illustrate the divergence between the norms presumed to be universal and the minority groups' choices for home and family. Ironically, the very participatory procedures meant to give citizens a voice in planning provide the convenient means for some local groups to resist the accommodation of others' divergent needs and tastes. Public hearings on planning regulations have often been turned into the tools of NIMBYism and ethnoracism.

The overall effect of multiculturalism is to reveal the cultural biases embedded in the so-called universal standards. It follows that rethinking the bases of housing standards and planning policies is necessary, not only to accommodate divergent preferences, but also to integrate diverse architectural and functional elements into coherent local and regional idioms. The convenience and satisfaction of both old and new residents have to be ensured through balanced development. This is the agenda of planning from the multicultural perspective.

INSTITUTIONS AND SERVICES

Not only spatial and architectural forms, but also social institutions and community services, are affected by multiculturalism. Ethnic communities contribute new ceremonials, sports, practices, and organizations to the public life of a locality. Most of these institutions and services are provided privately, but public responsibility lies in facilitating their development and in a fair distribution of public funds for social programs. Multiculturalism requires ensuring equal opportunities to develop mosques and temples as well as churches; baseball diamonds and soccer fields; Highland dance troupes and Caribbean carnivals; and language classes as well as ballet schools. Canada's planning system is fitfully accommodating such diversity of institutions and services. How this institutional learning is evolving is illustrated by the case of Islamic mosques.

Recent waves of Muslim immigrants from the Middle East, Eastern Europe, and South Asia have necessitated the building of mosques and Islamic centers. The building of a mosque or Islamic center for a congregation typically moves through three stages. In the beginning, someone's living room serves as the gathering places for weekly prayers, which leads to renting a hall or buying an unused church for congregational gatherings, and finally to the stage of building a new mosque. At each of these three stages, there are planning issues. From the changes in the use of an existing building to the building of a new mosque, at numerous points public approvals and planning permissions are required. Yet many existing zoning and site plan regulations for religious uses are based only on the requirements of churches; therefore, mosques have to be developed through variances to planning standards and policies. In one case, a minaret had to be designated as a "clock tower," though once built its function and integrity were so obvious that the local council happily dispensed with the installation of a mock clock. Each proposal creates a battleground reverberating with ethnoracial overtones, as proponents and opponents contend over parking standards, lot coverage, etc.

Yet precedents and experiences that embody the multicultural perspective are accumulating; one by one, thirty mosques have been established in the Toronto area. The planning system obviously is accommodating mosques, gurdwaras and temples, but systematic policies and standards that embrace such options have not appeared. Planners have been more open to such developments than have their political bosses.

A similar process is under way with the provision of other social and cultural services. For example, varying burial customs require revisions in the regulatory standards for cemeteries and operations of crematoriums. Other examples of services for which demands are mounting are multilingual kindergartens, heritage language courses, soccer or cricket fields, banquet halls and social clubs. Many of these institutions and services require physical development, and others may necessitate changes in the policies for public grants. A meals-on-wheels service, for example, may have to provide ethnic foods to serve a non-English/French population appropriately. Even a service such as family counseling has cultural dimensions. Toronto, for example, has six family and women's centers specializing in counseling South Asian women. Similar developments are taking places wherever there are strong pressures of demand and community initiatives. From Caribbean parades to Iranian film festivals and Yiddish poetry readings, Canadian cities support a rich and diverse artistic and cultural life. Little by little, cultural diversity is being acknowledged by precedents being established in the planning system. Toronto, Vancouver, and Montreal are leading in the development of culturally sensitive modes of planning. Metro Toronto's "Social Development Strategy" recognized diversity and commits to nondiscrimination in the sharing of the city's resources and opportunities. Translating these goals into policies, standards, and practices is the challenge that remains to be met.

[. . .]

5

Suburbia: Homeland of the American Future

The Next American City (2006)

Joel Kotkin

Editor's introduction

In the United States in the last half of the twentieth century suburbs surpassed central cities in household and employment growth. By 2000, 60 percent of the nation's population lived in suburbs, according to the U.S. Census, a survey of population undertaken every ten years to determine states' congressional representation and collect nationally consistent statistics for government and general public use. Beyond counting the number of people, the census reports other information, including race, household numbers and forms, employment and labor force participation, income, housing data, commuting times, and age. It provides these data for many geographies, including blocks, cities, counties, states, regions, and the nation as well as urbanized areas, urban clusters, and special metropolitan areas as defined by the Office of Management and Budget and the U.S. Census Bureau.

The Bureau continuously reviews and revises its classifications so that the data it collects are useful and relevant. For instance, since 1930 the Bureau has used a set of definitions that distinguish urban and rural areas based mainly on political boundaries.

In 1949 the Bureau created a new kind of classification to reflect gradations of urbanization and rising suburban development. The new classifications, integrated in the 1950 census, were: Standard Metropolitan Area (SMA), defined as a place containing a central city of 50,000 or more, its county and adjacent counties integrated socially and economically as measured by commuting; and Urbanized Areas (UAs), defined as contiguous, densely settled (1,000 people per square mile) blocks that together encompass a population of at least 50,000. The intent of having two definitions for a metropolitan area was to differentiate between one defined by political boundaries (MSA), and one that showed urban development at its fringe (UA).

Over time the Bureau revised the name but not the definition of the SMA: in 1959 it became the Standard Metropolitan Statistical Area (SMSA); in 1983 the Metropolitan Statistical Area (MSA or Metro), the name it now has. (Urban planners tend to use the acronym, not the whole name, when referring to one of the categories in their writings and discussions. A typical statement might be: "MSAs in the South and West have had stronger population growth than those in the North and Midwest." Listeners would know that the initials stood for Metropolitan Statistical Areas.)

In the 1980s, the Bureau made additional revisions to capture two phenomena: (1) level of urbanization and (2) variance in the form of local government. In the first case, to acknowledge that large, contiguous MSAs were measurably different from those that stood alone, the Bureau created the Consolidated Metropolitan Statistical Area (CMSA) for adjoining MSAs with at least one central city of a million or more. Within the CMSA it also designated the Primary Metropolitan Stastistical Area (PMSA) to indicate that a MSA had economic linkages with its neighboring one. Thus, in 1990 the Detroit–Ann Arbor–Flint CMSA was composed of the Detroit PMSA, Ann Arbor PMSA, and Flint PMSA. In the second case, to recognize the unique governmental organization of New England, the Bureau delineated New England Country Metropolitan Areas (NEMCAs). Renamed in the 1990s as New England City and Town Areas (NECTAs), this special name mirrored the MSAs. All (MSA, PMSA, CMSA, and NECTA) fall under the general Metropolitan Areas designation.

By the late 1990s the Bureau realized that urbanization had taken yet a different course not captured by its current metropolitan definitions. To remedy the problem, it renamed the Metropolitan Areas as Core Based Statistical Areas (CBSA) and included under its rubric: the MSA, PMSA, NECTA, and UA. It also added three new designations: the Metropolitan Division, the Micropolitan Statistical Area (Micro), and the Urban Cluster. A Metropolitan Division is a place whose principal city has a population of 2.5 million or more and is surrounded by contiguous counties that are integrated socially and economically as measured by commuting. A Micro is a place containing a principal city of 10,000 to 49,000, its county and adjacent counties integrated socially and economically as measured by commuting. An Urban Cluster – which supplements the Urbanized Area category – is a densely settled area with a population of 2,500 to 49,999. The Combined Statistical Area (CSA) replaces the CMSA, and can include either two or more Metros, or one Metro and one Micro, or two or more Micros, or mulitiple Metros and Micros that meet the social and economic integration measure.

As population changes, the census adds or deletes places from these categories. In December 2005, the United States and Puerto Rico had 951 MSAs and Micros (369 MSAs and 582 Micros), 29 Metropolitan Divisions, and 124 Combined Statistical Areas. The federal government regularly publishes updated lists for these categories, for Urbanized Areas, and for Urban Clusters in the *Federal Register* and on its website (www.census.gov).

The Bureau collects data for this array of metropolitan areas and then public officials, academic researchers, and the press use them for specific purposes. For example, the Department of Housing and Urban Development uses household figures from the Metro data to determine the area median income (AMI) on which to establish rents in multifamily buildings that use Low-income Housing Tax Credits (a form of subsidy). Journalists who report on cities and suburbs make use of Metro data. Thus, an article may refer to Philadelphia as having a population of more than 4 million when the city's actual count is 1.5 million – its suburbs supply the 2.5 million difference. A cartographer might represent a city as an Urbanized Area or a Metro when drawing a map.

In using and interpreting urban data it is important to understand the metrics and definitions. Keep in mind that the building block of the MSA is the county, but counties across the country vary in size, with those in the west being much larger than those in the east. If all the U.S. Metros appear on a color-coded map, those places with large-size counties will appear to be more urbanized than those with small-size counties even though the large-county Metros are in reality not so urbanized and are much less heavily integrated with the principal city than their small-county counterparts. This might lead the uninformed observer to draw incorrect conclusions.

Author Joel Kotkin makes ample use of census data in this selection, which challenges many assumptions held by urban and regional planners. His essay highlights the importance of the suburb as a dominant feature of American life, offers explanations for its growth, and provides suggestions for its improvement.

Kotkin, an Irvine Senior Fellow at the New America Foundation, is a journalist. He writes the monthly "Grass Roots Business" column in *The New York Times* and contributes to *The Los Angeles Times*, *The Washington Post*, *The New Republic*, *The Weekly Standard*, *The American Enterprise*, and *The Wall Street Journal.* He has served as west coast editor for *Inc. Magazine* and as a television commentator. His several books include *The City: A Global History* (New York: Random House, Modern Library edition, 2005) and *The New Geography: How the Digital Revolution is Reshaping the American Landscape* (New York: Random House, 2000).

For more on metropolitan census definitions, see William H. Frey *et al.*, "Tracking American Trends into the Twenty-first Century: A Field Guide to the New Metropolitan and Micropolitan Definitions," and Robert E. Lang and Dawn Dhavale, "Micropolitan America: A Brand-new Geography" in Alan Berube, Bruce Katz and Robert E. Lang (eds), *Redefining Urban and Suburban America: Evidence from the Census 2000*, Vol. III (Washington, DC: Brookings Institution Press, 2006) and Anthony Flint, *This Land: The Battle over Sprawl and the Future of America* (Baltimore, MD: Johns Hopkins University Press, 2006). For contrasting views of modern metropolitan growth, see Abhijeet Chavan, Chrisitan Peralta and Christopher Steins, *Planetizen's Contemporary Debates in Urban Planning* (Washington, DC: Island Press, 2007). To stay up to date on these matters, go to www.planetizen.com.

For the better part of a half-century, America's leading urbanists, planners, and architects have railed against the growth of suburbia. Variously, the suburbs have been labeled as racist, ugly, wasteful, or just plain boring. Despite the criticism, Americans have continued to vote with their feet for suburban or exurban landscapes. These Americans now include not only whites, but also a growing proportion of recent immigrants, Asians, Latinos, and African-Americans. And it's not just people who are moving – suburbia is also snagging the lion's share of new economic growth and jobs.

The "action" in America's development is thus likely to remain heavily concentrated in suburbs and exurbs. Most projections show that the continued increase in the U.S. population and the projected 50 percent increase in space devoted to the built environment by 2030 will largely take place in the sprawling cities of the South and West, areas dominated by low-density, automobile-dependent development of residential, commercial, and industrial space.

For developers, builders, planners, and public officials, the key challenge will be to accommodate this growth in a way that both preserves the advantages of relatively low-density suburban living and addresses legitimate concerns about the environment and about family, cultural, and spiritual life.

In the most extreme cases, suburbia has been linked intimately both to global warming and America's involvement with the Middle East. James Kunstler takes an apocalyptic approach, warning that suburban places "are liable to dry up and blow away" due to the rising energy prices. "Let the gloating begin," Kunstler says, predicting a general catastrophe will impact the suburbs, and urges people to leave these places as soon as possible. Kunstler sees suburbia and other aspects of contemporary American life much the way an early Christian might have viewed classical Rome: "I begin to come to the conclusion that we Americans are these days a wicked people who deserve to be punished." The dismal collapse of suburbia serves this purpose for Kunstler and others who detest the places most Americans live.

The death of suburbs and the resurgence of traditional cities have been predicted before, most notably during the 1970s energy crisis and during the dot-com boom of the 1990s. Yet, despite blips in urban growth, the longer-term pattern could not be more clear. Since 1950, more than 90 percent of all the growth in U.S. metropolitan areas has been in the suburbs. During the 1970s – a period of radically higher energy prices – the suburbanization trend actually intensified as central cities went through one of their most sustained periods of population decline.

SUBURBIA AND ITS CRITICS

Some critics of suburbia hold the wistful conviction that our lost urban past may be recovered through the imposition of planning – for many of these people, something like the Portland, Oregon, model writ large. Others have responded in flat-out denial, interpreting phenomena, such as the shrinking percentage of people living in the traditional single-family unit, the rise of a hip "creative class," or the growth of aging empty-nesters, as reversing the suburban tide.

SUBURBIA'S RELENTLESS RISE

Suburbs may have started as places for living, but their ascendancy has come in large part by becoming places of work. By 2000, roughly three out of five jobs in American metropolitan areas were located in suburbs. More than twice as many people in the United States commuted from suburb to suburb, where the job growth was concentrated, than from suburb to city.

Studies have shown this preference for suburbs extends to a wide range of firms. Only 11 percent of the nation's largest companies were headquartered in the suburbs in 1969; a quarter-century later, roughly half had migrated to the periphery. The pattern of 1990s suburban job growth appears to have expanded further since the beginning of the new millennium. By 2000, in the 100 largest metropolitan areas, only 22 percent of people worked within three miles of the city center; another study focusing on areas with high levels of sprawl (such as Chicago, Atlanta, and Detroit) showed that more than 60 percent of all regional employment now occurs more than ten miles from the core.

Perhaps most importantly, suburbs have gradually become the preferred location for the burgeoning science and information-based industries, the biggest growth sectors of the modern economy. Since World War II, high-tech firms have migrated to the suburbs for many reasons, including space for large, campus-like office parks, less crime, lower taxes, and most critically, the access to educated workers. Areas like the Santa Clara Valley in Northern California, northeastern New Jersey, and the suburban ring around Boston, have provided ideal locations for aerospace, computer, and information industries.

These economic trends have reordered the fundamental relationship between the urban core and its hinterland. For much of the twentieth century, the city remained the prime focus of business activity, and so location close to the core was advantageous for proximity to other business activities. Today, central city living appeals more for lifestyle reasons than economic ones.

THE OVERSTATED URBAN RENAISSANCE

Although the downtown residential "boom" in some areas is heartening, it is important to keep it in perspective: the overwhelming demographic evidence suggests it pales in comparison to growth on the fringes. In a Brookings Institution study of forty-five metropolitan areas spread across the country, University of Pennsylvania professor Eugenie Birch calculated that total growth in housing units between 1970 and 2000 was about 9 percent, or approximately 35,000 units; in contrast, construction of suburban housing units almost doubled to 13 million units during the same period. In the late 1990s, a period in which some core cities enjoyed their first population gains in decades, many more people headed out to the suburbs than went the other way. This pattern held even among the twenty-five-to-thirty-four age group, considered the prime market for urban living.

And for many urban centers – including such relatively attractive places as Chicago, Minneapolis, San Francisco, and Boston – population actually declined in the first half of this decade. Outmigration has accelerated in other cities, and, in some, growth has slowed considerably from pre-2000 levels. Indeed, according to a 2001 report from the Brookings Institution and the Fannie Mae Foundation, the total projected growth for all major downtowns through 2010 is less than was the growth in the sprawling Riverside–San Bernardino metro area east of Los Angeles in 2004 alone.

Nearly every major region of the U.S. in this sense is undergoing suburbanization, even if the downtown is growing. In Houston, for example, there has been much talk about a downtown housing surge. But the entire inner ring of the city – which extends well beyond the central core – accounted for barely 6 percent of new units; the vast majority of the growth took place in the region's far-flung suburban areas.

THE "UNIVERSAL ASPIRATION"

The biggest reason for these patterns is not the "conspiracy" of big oil and freeway builders oft cited by enviro-activists and more radical New Urbanists, but rather what the Los Angeles urbanist Edgardo Contini called the "universal aspiration" – not only in America but in almost all rich countries – to own a piece of land, where families may live in relative comfort and privacy.

One clear indication of the vitality of the "universal aspiration" has been the surprising resilience of the single-family home market. Even in the last recession, the number of single-family homes increased. To the surprise of many forecasters and the U.S. Census Bureau itself, instead of dropping with the aging of the baby boomers, single-family home construction surged to levels not seen since the 1970s and 12 percent above those of the 1980s. Not only were more homes built in the late 1990s than expected, but the size of single-family homes actually grew, with the median expanding from 1,605 sq. ft in 1985 to over 2,100 sq. ft in 2001. Analysts such as Al Ehrbar suggest that demographers at the time had not only failed to see that many boomers would buy houses later in life, but they also underestimated the total prime homebuyer market – twenty-five-to-thirty-four-year-olds – by more than 4 million.

THE "OTHER" SUBURBANITES

Perhaps the most critical shift in sustaining the single-family home, however, has to do with immigrants. More than any demographic group, they are shaping the American future: by 2015, nearly one in three children in America will be an immigrant or a child of an immigrant. Immigrants' desire for homeownership is often overwhelming and has led a majority of them to seek homes in the suburbs. Seventy percent of minorities in the immigrant-rich Los Angeles–Long Beach area, for instance, live in suburbs. Fast-growing, sprawling markets such as Fort Lauderdale, Riverside–San Bernardino, Las Vegas, Atlanta, and Orlando have particularly high rates of immigrant homeownership. The new pattern of immigration also can be seen in places like greater Washington, DC, in economic and demographic terms the most dynamic region along the eastern seaboard. A recent Brookings study noted that 87 percent of the DC area's foreign migrants live in the suburbs.

Though suburbs are often seen as all-white, middle-class bastions, many suburbs, such as Fort Bend County, Texas, and small cities like Walnut, in the San Gabriel Valley, just east of Los Angeles, have among the most diverse populations in the nation: these places are majority–minority, though no single minority by itself composes over half the population. "If a multiethnic society is working out in America," suggests demographer James Allen, "it will be worked out in places like Walnut. The future of America is in the suburbs."

Immigrants aren't the only populations changing the nature of suburbia. Evidence from the 2000 census also revealed that singles, non-traditional families, and empty-nesters – widely reported to be moving from suburbs back to the inner city – grew far more rapidly in the suburbs than in the cities. In fact, largely due to the growth of singles and aging parents, there are now more non-families in the suburbs than traditional families.

Of these groups, the "empty nester" baby boomer may prove the most important to shaping suburbia. The baby boom generation far outnumbers its successor, Generation X, by roughly 76 million to 41 million. Due largely to this group, by 2030 more than one in five Americans will be over the age of sixty-five. Where these people – demographer Bill Frey calls them "downshifting boomers" – end up will prove critical in shaping future patterns of new residential and commercial development.

Roughly three-quarters of these "downshifters" appear to be sticking pretty close to the suburbs where most of them have settled, according to Sandi Rosenbloom, a professor of urban planning and gerontology at the University of Arizona. Those that do migrate, her studies suggest, tend to head further out into the suburban periphery, not back towards the old downtown. "Everybody in this business wants to talk about the odd person who moves downtown, but it's basically a 'man bites dog' story," Rosenbloom observes. "Most people retire in place. When they move, they don't move downtown; they move to the fringes."

The reasons they are staying put vary. Some have job commitments or need to stay close to their children or grandchildren; roughly 40 percent, according to one survey, expect their kids to move back in with them at some point. These people are also used to a suburban lifestyle and its amenities. For the most part, they are not acculturated to the density, congestion, and noise of inner-city life. "They don't want to move to Florida, and they want to stay close to the kids," suggests Jeff Lee, CEO of a prominent Washington, DC, real estate, architecture, and planning firm. "What they are looking for is a funky suburban development – funky but safe."

LOOKING AHEAD: SUBURBIA TO 2030 AND BEYOND

Suburbanization – and even ever greater sprawl – must be accepted as the future. Attempts to stomp out or control outward movement, as Portland tried, have not only failed but have driven settlement even further out beyond the areas of control.

By way of proof, during the 1990s over 80 percent of all population growth in greater Portland took place in the suburbs; since 2000 it has been closer to 90 percent. Mass transit, the other linchpin of the Portland "Smart Growth" legend, may also be less of a triumph than reported. Between 1986 and 2001, according to the most recent Texas Transportation Institute study, greater Portland has seen the biggest jump in congestion of any of the nation's seventy-five largest metro areas. More people rode the rails, but many more, it appears, have also decided to drive alone, perhaps in large part because they are living and often working further out.

The real issue is not so much how to prevent suburban growth, but how to make it more humane and capable of accommodating an increasingly diverse population. One key solution might lie in the growth of telecommuting, which could allow more suburbanites to work close to or at home. Already 20 million people work part-time or full-time from their residences. Some new suburban developments, such as Ladera Ranch in southern Orange County, have adapted their floor plans to serve the mixed uses of residence and business – by incorporating separate entrances for business clients, for instance. Suburban historian Tom Martinson believes the Ladera plan will "be in the history books in twenty years" because it anticipates "an incredible change in the way we live and work."

At the same time, the increasing decentralization of economic activity may spur the development of ever more self-sufficient "suburban villages." We can see this model emerging in new communities such as Valencia, California, or the Woodlands, outside Houston, which have developed their own successful town centers complete with thriving cultural and religious establishments. Scores of older suburbs have also used new commercial and cultural amenities to revitalize old town centers, such as Naperville, Illinois; Fullerton, California; and Bethesda, Maryland. Viewed from a long-term perspective, these places may represent not so much a rejection of city life, but a redefinition of what urban life is about – and where and how it takes place.

In this sense, we need to look at current suburbia not as a finished product, but something beginning to evolve from its Deadwood phase. During this evolution, our ancient sense of the city still has much to teach the suburbs, notably about the need for community, identity, the creation of "sacred space," and a closer relation between workplace and home life. Of course, the emerging suburbs won't be able to duplicate the forms of our great historical cities, but they will borrow from them as new public spaces are built and a sense of civic identity is established. In this, the suburbs can carry something of cities-past in their substance as they contribute to a new chapter in urban history, one that we today can play a role in forging.

The New Suburban Typology

from *American Metropolitics: The New Suburban Reality* (2002)

Myron Orfield

Editor's introduction

This selection – based on a massive study of twenty-five U.S. regions containing nearly half the nation's population – dispels many myths about the suburbs, primarily the one that holds that these outlying areas are monolithic. While scholars have argued for many decades that they are not, this study definitively demonstrates that suburbs are composed of multiple, politically independent localities defined by socioeconomic characteristics that are the result of their growth trajectories. Employing spatial analysis to measure demographic trends (including population size, race, and income), development patterns, and local fiscal portraits (revenues and expenditures), it identifies six types of communities. Maps display them in a mosaic across individual regions.

By graphically showing the types and locations of varied communities within a region, this selection's author, Myron Orfield, has made use of geographic information systems now available in the field. (See Plate 1.9.) And his descriptive work lays the foundation for thinking in new ways about metropolitan issues. For example, it allows assessment of common concerns related to fiscal problems across political boundaries. It suggests areas for potential regional alliances that promote mutual interests previously not identified.

That Orfield should explore regional issues is not surprising since he comes from the Twin Cities, one of the few places in the United States that has a regional government. Its Metropolitan Council (Met Council), created by state legislation in 1967 and amended several times through the 1990s, has responsibility for major regional infrastructure: waste disposal, open space, and public transportation, including aviation.

In its depiction of the dynamics of metropolitan growth, Orfield's research demonstrates the lack of regional social and economic balance throughout the study areas. By implication, it subtly tests longstanding sociological theories and measurements, including ideas of invasion and succession (the replacement of one population by another); tipping point (the point at which the movement of African-Americans or other minorities into a white area produces "white flight"); the use of the dissimilarity index (the percentage of people in a group that would have to move to achieve their random distribution in a given territory); the level of pernicious effects, such as poor school performance, of living in conditions of concentrated poverty; economic ideas such as the Tiebout model (individual households choose residential locations based on their perception of such public goods as schools and safety, or, more colloquially, "people vote with their feet"); and legal/political concerns (the equity and morality of concentrated poverty).

Myron Orfield is Associate Professor of Law and the Executive Director of the Institute on Race and Poverty at the University of Minnesota. He served six terms as a Minnesota state legislator, five in the House and one in the Senate, where he honed his interests in land use, fair housing, and school and local government finance. His first book, *Metropolitics: A Regional Agenda for Community and Stability* (Washington, DC: Brookings Institution Press, 1997), focused on the Minneapolis–St Paul region. *American Metropolitics: The New Suburban*

Reality (Washington, DC: Brookings Institution Press, 2002) applied the methodology developed for *Metropolitics* to more regions. More recently, he has been delving into school desegregation (see "Choice, Equal Protection, and Metropolitan Integration: The Hope of the Minneapolis Desegregation Settlement" in *Law and Equality*, 24: 2, 2006) and mobility (see "Access to Growing Job Centers in the Twin Cities Metropolitan Area," *CURA Reporter*, 36: 1, 2006).

To explore theories of race and metropolitan development, select a range of studies. In *American Apartheid: Segregation and the Making of the Underclass* (Cambridge, MA: Harvard University Press, 1998) Douglas S. Massey and Nancy A. Denton condemn current racial patterns as resulting from purposeful public policy that has aggravated African-American poverty rates, and they call for aggressive desegregation policies, including stepped-up prosecution of discriminatory actions by realtors and others who limit Black residential choice. Ingrid Gould Ellen's *Sharing America's Neighborhoods: The Prospects for Stable Racial Integration* (Cambridge, MA: Harvard University Press, 2001) presents a less harsh view, delineating progress in integrating neighborhoods. Ellen tracks the fragility of integrated neighborhoods, arguing that they resegregate not because of racial hatred but because of fear of declining quality of life or property values. She also outlines government strategies to address these issues. In "Continued Racial Residential Segregation in Detroit: 'Chocolate City, Vanilla Suburbs' Revisited" in the *Journal of Housing Research* (4: 1, 1993), Reynolds Farley, Charlotte Steeh, Tara Jackson, Maria Krysan, and Keith Reeves argue that Blacks in their sample freely chose to live in a predominantly Black neighborhood.

For a discussion of the contributions of the Tiebout model, see William A. Fischel (ed.), *The Tiebout Model at Fifty* (Cambridge, MA: Lincoln Institute of Land Policy, 2006). Of particular interest is Lee Anne Fennell's essay "Exclusion's Attraction: Land Use Controls in Tieboutian Perspective" and the accompanying commentary by Robert C. Ellickson.

Finally, the Metropolitan Council's *2030 Regional Development Framework* (adopted in 2004) shows how a regional body is coping with an anticipated population increase of 1 million, added to the 350,000 rise in the 1990s, by going beyond its traditional focus on infrastructure investment and environmental protection. The plan, which is on the Met Council website (metrocouncil.org), aims to guide development, promoting higher-density, mixed-use, transit-oriented settlements.

THE MYTH OF THE SUBURBAN MONOLITH

The suburbs contain more than half of the U.S. population, an even higher percentage of voters, and an overwhelming majority of elites. The perceived power of the supposed suburban monolith shapes American domestic policy and politics but, in truth, this power is fragmented. To judge from some eighteen commissioned regional reports, different types of suburbs are emerging in U.S. metropolitan regions.

Many suburbs and older satellite cities are beginning to experience rapid social changes, particularly in their school systems, but they lack the local resources to deal with these changes. Some places are more troubled than the central cities they surround. Another large and important group of fast-growing communities lacks adequate local resources for schools and infrastructure. Finally, a smaller, very affluent group of cities enjoys all the benefits of a regional economy without having to pay the costs. Ironically, this group often appears to be least happy with the status quo of America's suburban development patterns.

[. . .]

CLUSTER ANALYSIS OF SUBURBS

[. . .]

Six types of suburban communities [have emerged] from the analysis. Three types of communities under significant fiscal or social stress are designated *at-risk communities*. One group, made up of fast-growing communities with only modest fiscal resources at their disposal, is designated *bedroom-developing communities*. Finally, two groups of very affluent communities under little stress from either low tax capacity or high costs are labeled *affluent job centers.*

[. . .]

One of the most important characteristics that distinguish one community type from another is tax capacity. Four of the six groups – the three at-risk categories and the bedroom-developing communities – each command relatively low levels of local resources; their tax capacities vary from 66 percent to 90 percent of their regional averages. Only in the bedroom-developing communities were capacities keeping up with regional averages in terms of change in tax capacity. The at-risk communities were falling further behind.

What leads to these kinds of disparities in local resources? Unbalanced regional development patterns – unequal distribution of high-end housing and commercial–industrial development – are the likely source. This can be seen very clearly from the distribution of office space across the different types of communities. In a modern service economy, office development is often a key to local fiscal well-being. It is one of the few sources of tax base that more than "pays its way."

CB Richard Ellis, a large real estate firm, provided parcel-level, office location data for sixteen of the twenty-five largest regions in the study. Matching these data with the community types shows that office space, representing jobs and fiscal capacity, is very unequally distributed across metropolitan regions. . . . Not surprisingly, the central cities contain a large portion of regional office space: 51 percent of the total in the sixteen metropolitan areas with data, a percentage that far exceeds their share of households (24 percent). However, the distribution in suburban areas is highly skewed. The three at-risk categories and the bedroom-developing communities together contain 34 percent of office space, which is less than half their share of the region's households, while the proportion of office space in the two affluent job center categories is more than twice their share of population. These two job center categories also contain disproportionate shares of the "best" office space: roughly 70 percent of their office space is rated "A" or "B," compared with just 52 percent for central cities and 55 percent to 60 percent for all the other categories combined. In short, households in the affluent job center communities benefit disproportionately from the fiscal dividends generated by office development.

AT-RISK COMMUNITIES

The myth of urban deterioration and suburban prosperity suggests that social and economic decline stops neatly at the borders of central cities. Nothing could be farther from the truth. Once poverty and social instability permeate communities just outside the central city and begin to grow in older satellite cities, decline accelerates and intensifies. . . . At-risk suburbs-communities . . . have high social needs but relatively limited, and often declining, local resources. These communities include older suburbs, satellite cities and newer, lower-density communities with relatively high poverty rates. They have only about two-thirds of the fiscal capacity of the central cities, and their fiscal capacity is growing more slowly. Many at-risk communities lack the central cities' strong business district, vitality, resources, high-end housing, parks, cultural attractions, amenities, and public infrastructure (for example, police and social service agencies experienced in coping with social stress). As a result, these communities often become poor *faster* and lose local business activity even more rapidly than the cities they surround. . . . There are three kinds of at-risk municipalities: segregated communities, older communities, and low-density communities.

At-risk segregated communities

[. . .]

[V]irtually every region in the nation with a significant black or Latino population has a number of suburbs in this category. . . . The at-risk segregated communities – often experiencing dramatic social and fiscal change – are some of metropolitan America's worst places to live. Poor and segregated, they have a fraction of the resources of the central cities they surround. In 1994, the taxes on a $100,000 house in the at-risk segregated suburb of Maywood, Illinois, were $4,672. This level of taxation would support local school spending of $3,350 per pupil. In Kenilworth, an affluent suburb to the north, the taxes would be $2,688, yet this lower rate, applied to the whole tax base, would support almost three times the level of spending per pupil. . . .

At-risk older communities

The second group of at-risk communities . . . are much less racially diverse than at-risk segregated communities, they are often at the beginning stages of rapid racial change. These places often stand cheek by jowl with the at-risk segregated suburbs. There is a heavily defended racial line between many of them. This line, however, can quickly recede in a process of flight and resegregation.

[. . .]

The group comprises mostly older, inner-ring suburbs and small, outlying cities that have been swallowed up by metropolitan growth. Examples of the first type include Conshohocken, Lansdowne, and Cheltenham in the Philadelphia metropolitan area; Brooklyn Park, Burnsville, and Bloomington in the Twin Cities; and Brookline and Everett in Boston. The second type includes Media and Gloucester City in the Philadelphia area; Anoka and Center City in the Twin Cities metropolis; and Ann Arbor and Dearborn in the Detroit area.

The at-risk segregated and older communities have many common concerns. Both groups have slow population growth or (decline), relatively meager local resources, and struggling commercial districts. Their main street corridors and commercial districts cannot attract new, big businesses that could easily build on greenfield sites. . . . [B]oth of these groups stand to gain from state-level school equity programs that redistribute tax resources from other communities in order to reduce their local tax rates and increase their local spending. They are also likely to benefit from state and regional land use planning to curb urban sprawl and spread affordable housing to newer communities. Furthermore, both are likely to benefit from transportation policies that support rebuilding existing infrastructure rather than building new highways and that provide more flexible resources for transit. Despite these commonalities, segregated and older at-risk suburbs have not formed a cohesive political whole, probably because they are often divided on the issue of race.

Many at-risk segregated communities resemble segregated or deeply poor urban neighborhoods. Fear about what is happening to their neighbors often leads those in at-risk older communities to try to distance themselves from people in at-risk segregated communities. The at-risk older communities often struggle with lawsuits alleging housing discrimination against black and Latino homeowners or renters attempting to move from the at-risk segregated communities. This does not help relations on either side.

[. . .]

At-risk low-density communities

The final group of at-risk communities comprises 1,104 relatively low-density localities with low tax capacities that are growing more slowly than their regions and with higher-than-average poverty and population growth rates. . . . These communities . . . are typically located in the metropolitan areas' outer portions. Many of them are exurbs that still contain some pockets of rural poverty, but they are changing as they become integrated into the metropolitan area. . . . As they make the transition from rural to suburban, these localities must cope with the costs of greater-than-average population growth, relatively high poverty, and older-than-average housing stock. . . .

The Twin Cities, Atlanta, and San Francisco each have a band of at-risk low-density communities on the outermost fringes of the metropolitan area. In Atlanta and San Francisco, the band is in the unincorporated areas in the outermost counties. In the fully incorporated Twin Cities metropolitan area, it consists of a nearly continuous belt of semirural townships in the north and west.

BEDROOM-DEVELOPING COMMUNITIES

The fourth classification of suburbs comprises what many would regard as the prototypical suburb. The population – mostly white – is growing more quickly in the suburbs in this group than in any other. Density is low, housing is new, and tax capacity is just below average and growing at an average rate. . . . Of the six [types of suburbs] . . . this one had the most school-age children per household.

Though not experiencing the social stress of the at-risk communities, these localities must manage the costs of a high rate of population growth with only average (or below-average) local resources. How well they manage new growth will largely determine whether they experience fiscal stress. If new development is at a cost-effective density and coincides with the provision of adequate infrastructure, they may benefit from growth. If growth is scattered, low in density, and poorly coordinated with infrastructure provision, they may face significant fiscal stress as a result of excessive infrastructure costs. This group had an average 1990 median household income of $40,000 and an average median home value of $121,009 – about the same as the central city average.

SCHOOLS AND INFRASTRUCTURE IN AT-RISK LOW-DENSITY, AND BEDROOM-DEVELOPING SUBURBS

Both the at-risk low-density and the bedroom-developing suburbs share fiscal pressures arising from school and infrastructure finance. In all the large regions, the student-to-household ratio in at-risk low-density and bedroom-developing suburbs is much higher than the regional average. . . . Since schools account for the largest share of local expenditure, these sizable differences in the ratio of children per household are very important fiscally. Because of this ratio and their (at best) average tax base, these two kinds of suburbs have the lowest per-pupil spending in metropolitan America. In states without a significant aid system for equalizing the resources available to school districts, such as Illinois, Pennsylvania, and Ohio, these districts have extremely low spending, high school taxes, or both.

Developmental infrastructure such as roads and sewerage can also present large challenges for the at-risk and bedroom-developing suburbs.

[. . .]

Residents of these communities are also skeptical that their local elected officials will handle growth appropriately. Consequently, there has been an explosion in local ballot initiatives in the bedroom-developing suburbs. An analysis conducted in 2000 of a national database of ballot initiatives, separating the initiatives that could be identified with a specific city rather than a larger, more diverse county, suggests that bedroom-developing communities often drive local ballot initiatives. In Los Angeles, this group of suburbs put forward three of the ten ballot initiatives proposed that year; in the Bay Area, eight of sixteen; in Chicago, eleven of fourteen; in Philadelphia, ten of fourteen; and in New York, twenty-three of forty-one.

AFFLUENT JOB CENTERS

Many suburbs have moved well beyond their traditional role as bedroom communities for large cities and are now major players in their regional economy. "Edge cities" – suburban communities with more than 5 million sq. ft of office space and more jobs than bedrooms – are the major winners in the decentralization of U.S. metropolitan areas. *They* reap the benefits of extraordinary tax bases, while paying only a fraction of the costs of the central cities that once monopolized the office market. Although congestion and rapid loss of open space are among the costs they cannot escape, most of these communities have developed in ways (sometimes market-driven, sometimes planned) that enable them to evade the social costs associated with poverty. Edge cities in this study include the Route 128 corridor in Boston, the Schaumburg area west of O'Hare International Airport in Chicago, the Perimeter Center north of Atlanta, the Irvine area south of Los Angeles, several areas in New York (Stamford–Greenwich, Great Neck–Lake Success, and the Garden City area), and Pleasanton–Dublin in San Francisco, to name just a few.

The municipalities in these edge cities represent a significant proportion of the last two clusters in our analysis: "affluent job centers" and "very affluent job centers." . . . For example, along Route 128 in Massachusetts, from south to north, are Westwood, Dover, Needham, Wellesley, Newton, Weston, Waltham, Lincoln, Lexington, Concord, and Bedford. Of those localities, only Waltham is not in one of the affluent job center clusters. Among the others, 1998 tax capacities ranged from 172 percent of the regional average (Bedford) to 383 percent (Weston), and those capacities were growing faster than the regional average in all but one

(Bedford). Poverty among elementary school-age children ranged from just 4 percent of the regional average (Dover) to 22 percent of the average (Newton).

[. . .]

What is most striking about these kinds of suburbs is their enormous concentration of office space. . . . Concentration of commercial capacity is their distinguishing feature. Although many residents of other municipalities benefit from the employment opportunities and commercial activities in these areas, only a few can enjoy the tax (and resulting public service) benefits.

As might be expected, the political and business leaders in these affluent job centers work hard to maintain the quality of life of the community, and, of all types of suburbs, they are the ones that have revolted most successfully against growth and sprawl. Their residents are highly educated and quite willing to stand up for themselves. Some of the highest-profile local antigrowth controversies have taken place in America's wealthiest suburban areas: the anti-Disney theme park battle in Loudoun County, Virginia; the adoption of growth boundaries in Ventura County in 1998; and the growth moratoriums and slow-growth regulations in San Francisco's South Bay and eastern Contra Costa County. Affluent Bellevue and Redmond, Washington (home of Microsoft Corporation), contain powerful pockets of resistance to sprawl. Residents of Silicon Forest (in the western suburbs of Portland) have worked hard to hold Portland's urban growth boundary constant, constraining further growth in their direction. The lovely Saint Croix River Valley of the Twin Cities is home to strong opponents of growth – recent polls put growth above education and taxes as a key election issue. Antigrowth ballot measures were disproportionately found on the ballots in the affluent job centers of Los Angeles (four of ten of the region's initiatives), Chicago (three of fourteen), New York (sixteen of forty-one), and Cleveland (seven of twenty-one), although most antigrowth initiatives were on ballots in the bedroom-developing suburbs.

These places might seem to have it all: affluent people, a high tax base, an average load of children, and very low poverty. However, the mass of jobs and commercial activity also has its downside. First, because many workers cannot afford the local housing, these beehives of local activity generally have bad traffic congestion. Second, because land becomes so valuable it is often difficult to maintain open space.

DISTRIBUTION OF COMMUNITY TYPES WITHIN METROPOLITAN AREAS

. . . In the full sample, 68 percent of the population lived in a community experiencing fiscal stress of some sort: 28 percent were in central cities; 14 percent in the two high-density at-risk groups; and 26 percent in the low-density at-risk group. Localities in some kind of fiscal stress housed 88 percent of the people living in poverty in the sample metropolitan areas, but they controlled only 56 percent of local tax capacity. Although they received greater-than-average amounts of state aid, their share of total fiscal capacity (59 percent) was still well below their shares of total population and extremely poor people.

Another quarter of the population in the twenty-five metropolitan areas was in the bedroom-developing suburbs. Those localities controlled roughly average shares of tax capacity, state aid, and total capacity, but they captured 56 percent of total population growth in the metropolitan areas between 1993 and 1998. That growth and its management are important issues not only for the municipalities themselves but also for their metropolitan areas as a whole.

Just 7 percent of the population in the twenty-five metropolitan areas lived in communities in the two affluent job center categories. Those localities controlled 13 percent of total local fiscal capacity, but they were home to only 2 percent of people living in poverty. Although they captured more than their proportionate share of growth, they were not the hothouses of growth that the bedroom-developing suburbs were between 1993 and 1998. They thus enjoy a disproportionate share of the benefits of regional growth patterns with very few of the costs.

The six metropolitan areas in the comparison group have distinctly different mixes of community types. In the regions in which large shares of the population are in unincorporated areas (Atlanta, Chicago, Denver, and San Francisco), low-density communities are the most common type of at-risk community. In regions that are fully incorporated (Minneapolis–St Paul and New York), at-risk communities are much more likely to be segregated or older. Atlanta and Chicago have the smallest number of residents – only 51 percent of the regional population – in at-risk communities and central cities. In Atlanta, that reflects the fact that nearly 70 percent of the region's total population lives in unincorporated areas, most of which are classified as bedroom-developing communities. In the other four regions, the proportion of the population in communities in the at-risk categories and central cities ranges from 62 percent to 85 percent.

There is much less variation, however, in the extent to which poor populations tend to cluster in the at-risk communities and central cities. The share of the total poor population in at-risk communities ranges from 83 percent (Minneapolis–St Paul) to 89 percent (San Francisco). Similarly, the percentage of the poor population living in communities in the two affluent job center categories was very low everywhere, ranging from just 3 percent in Denver to 10 percent in Chicago.

... [T]he geographic pattern of community types in the six metropolitan areas [shows that] the fully incorporated metropolitan areas (Minneapolis–St Paul and New York) have the most distinctive patterns. Each has a distinct ring of at-risk suburbs surrounding the central cities; a mix of affluent job center and bedroom-developing suburbs in the second ring from the center; and a mix of at-risk, low-density, and bedroom-developing communities in the outermost parts of the regions. Atlanta shows a similar pattern, but large tracts of unincorporated areas predominate. Fulton County shows great variety, containing Atlanta, a set of at-risk segregated suburbs south of Atlanta, and a relatively affluent unincorporated area. By contrast, the entire county of DeKalb, just east of Atlanta, is in the at-risk, segregated category. Another set of at-risk, segregated municipalities (including Marietta and Smyrna) lie to the northwest of Atlanta. The outermost counties tend to fall into the at-risk low-density category; the remainder, in the bedroom-developing category.

... San Francisco ... is dominated by large expanses of at-risk low-density unincorporated areas. The inner suburbs include a group of at-risk segregated municipalities northwest and southeast of Oakland (including Hayward and Richmond) and farther south in the Menlo Park area. The more affluent suburbs, including places like Palo Alto, Portola Valley, Woodside, Santa Clara, and Milpitas, form a ring to the south.

Overall, inner suburbs tend to fall into the at-risk segregated or at-risk older categories. The middle ring of suburbs usually contains a mix of bedroom-developing and affluent job centers. Most outermost areas contain a mix of at-risk low-density and bedroom-developing suburbs.

PART TWO

History and Theory of Urban and Regional Planning

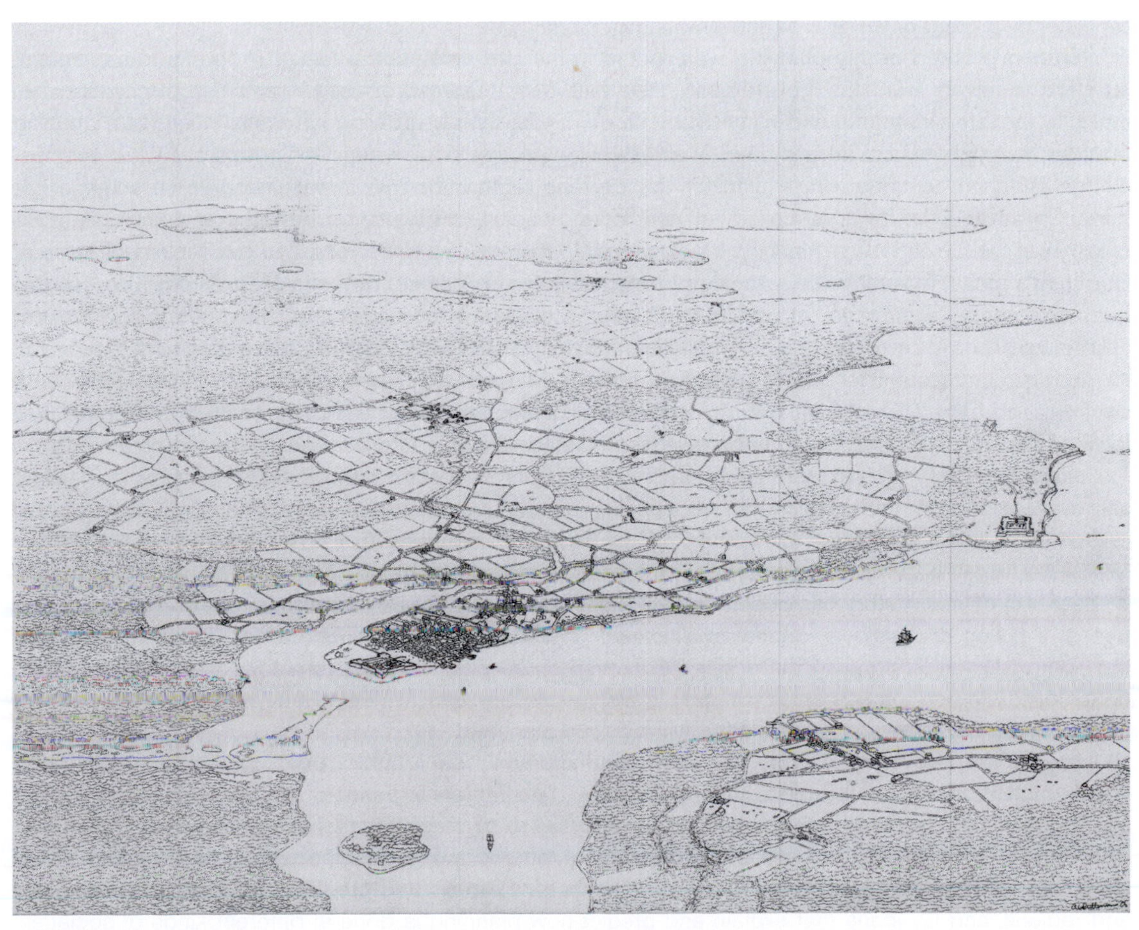

Growth of a typical Northeastern city: 1680. (Drawing by Andrew Whittemore)

INTRODUCTION TO PART TWO

"Make no little plans. They have no magic to stir men's blood and probably will not themselves be realized. Make big plans; aim high in hope and work, remembering that a noble, logical diagram once recorded will never die." If you had an undergraduate or graduate degree in urban planning you would immediately recognize this pronouncement by architect and urban planner Daniel Burnham. The familiar quotation might call up any number of images, from the Flatiron Building in New York, which Burnham designed, to the gleaming white structures of the 1893 Columbian World's Exposition, which he masterminded. As the author of the country's first comprehensive plan (Chicago, 1909) and the MacMillan Plan for Washington, DC, Burnham was the epitome of the visionary city planner of his day. He left his imprint on America's great cities and on the practice of planning.

Burnham's brand of city planning was rooted in the late nineteenth-century Progressive movement, an effort to reform municipal government in the face of widespread abuses in all areas of city management. In the late nineteenth and early twentieth centuries, rapidly growing industrial cities were choking on their own congestion. In response, businessmen's groups like Chicago's Commercial Club commissioned Burnham to undertake a plan for the city and region. Clearly, it was a top-down scheme but it was in advance of its time, called for new open space and circulation systems and stimulated the creation of the Chicago City Planning Commission. In the ensuing years, Chicago did implement many of Burnham's ideas, including the extension of North Michigan Avenue, now the Magnificent Mile. So this quotation is also a reminder of continuity – what is started in an earlier era continues in the present and future. Today, Chicago's planning tradition rooted in Burnham's 1909 plan endures, evidenced in its current commitment to urban greening efforts – tree planting, green roofs, Millennium Park – its confrontation of today's regional mobility issues through the creation of the Chicago Metropolitan Agency for Planning (CMAP) in 2005 to undertake coordinated land use and transportation planning.

Finally, the very words in Burnham's quotation bring into question the role of the planner in the past and present. Would modern planners design such stirring statements in addressing community groups, city councils, or other clients? Or would they think that they could never be so histrionic publicly – their role is not to tell people what to do but listen to them and let them lead?

Each era of U.S. history has posed its own planning concerns, and studying history serves many purposes. It builds professional identity by illuminating past accomplishments and key figures. It provides the context for understanding the origin and course of the planning decisions that affect contemporary conditions. It helps today's students learn from past techniques to see what worked and what didn't and to frame questions about how they will practice planning.

In addition to providing a sense of a profession and analytical tools for practice, history has an additional function: to act as a bridge to planning theory. Theory offers a means of predicting behavior, generalizing knowledge gained from observation for application to other cases. History and theory are close relatives. Theorists build their work by gathering data, often drawing on information about the past. They are concerned with two streams of thought: (1) ideas that explain and predict the physical form of cities and regions, and (2) ideas that explain and predict how planning is done in different kinds of societies. Seemingly separate, these themes intersect frequently.

Part Two, "History and Theory of Urban and Regional Planning," encompasses all of these ideas. It aims to address six basic questions, articulated succinctly in a useful collection, *Readings in Planning Theory* (Oxford: Blackwell, 1996, second edition 2003), edited by Scott Campbell, University of Michigan, and Susan S. Fainstein, Harvard University. They are: "What are the historical roots of planning?" "What is the justification for planning?" "What values do planners hold?" "What are the constraints on planners?" "What do planners do?"

In answering these questions, Part Two offers seven representative readings in history and theory. Richard T. LeGates and Frederic Stout's "Modernism and Early Urban Planning" tracks the transatlantic roots of contemporary planning from the nineteenth century through World War II. Michael B. Teitz, in "Reflections and Research on the U.S. Experience: The Profession of City Planning: Changes, Images and Challenges, 1950–2000," shows how land use, environmental, and economic development planning rose to the forefront of the field. Eugénie L. Birch traces how U.S. planners transmitted their ideas, values, and techniques after deciding to become a distinct profession with a specialized body of knowledge in the early twentieth century. In "*Ramapo* Plus Thirty: The Changing Role of the Plan in Land Use Regulation" Edward J. Sullivan and Matthew J. Michel illuminate the critical court case that strengthened the legal structure that supports planning in the United States. Dolores Hayden's "The Shapes of Suburbia" shows how the many different types of suburbs have evolved over time.

In the last two selections, Nigel Taylor and Susan Fainstein offer different takes on the development of planning theory. Taylor, in "Anglo-American Town Planning Theory since 1945: Three Significant Developments but no Paradigm Shifts," sees the field as a single, evolving intellectual stream. Fainstein, in "New Directions in Planning Theory," argues that there are substantial differences among three reigning groups of theorists – the communicative planners, the New Urbanists and the Just City supporters. Does this disagreement matter? Yes, because each view yields a different approach to planning. What is the answer? Readers must decide for themselves.

7

Modernism and Early Urban Planning

From *Early Urban Planning, 1870–1940* (1998)

Richard T. LeGates and Frederic Stout

Editor's introduction

A number of ideas influenced thinking about urban settlement patterns after the industrial revolution. Advances in technology with regard to energy and fabrication – steam replaced river power, machines took over hand labor – led to the development of large-scale factories in central locations close to markets and transportation hubs. Factory employment attracted myriads of workers, many of them desperate displaced rural dwellers or immigrants who moved to cities, settling near factories in inexpensive housing built by real-estate speculators.

Favorable locations near ports, natural resources, and markets caused population explosions in Northeastern and Midwestern cities. In the five decades following 1860, New York's population increased from 479,000 to 5 million, making it one of the most populous cities in the world. Philadelphia's population tripled to 1.5 million by 1910, and fast-growing Chicago expanded from under 112,000 at mid century to 2.1 million. In the United States, industrialization and urbanization peaked in the fifty-year period between 1860 and 1910, when the national population grew 194 percent, from 31 million to 91 million, and the urban population rose a colossal 650 percent. By 1910 urban residents were 46 percent of the total, up from 20 percent in 1860.

In the absence of strict development regulations, dense urban neighborhoods quickly turned into slums characterized by overcrowded housing, low-level or nonexistent municipal services (water and sanitation), and scarce recreational space. Poor health, poverty, and crime followed. Photographer Jacob Riis captured these conditions in *How the Other Half Lives* (New York: Scribner's, 1890).

By the end of the nineteenth century, crowded cities and neighborhoods became the focus of reform. Physicians, settlement-house workers, and clergy joined to call for better public health services, shelter, parks, and playgrounds. They later promoted more comprehensive, or systematic, changes, singling out congestion or overcrowding as the root of urban problems. Laying the foundation for later city planning analysis, social worker Paul Kellogg conducted the first extensive scientific study of living conditions in America, *The Pittsburgh Survey*, published in six volumes by the Russell Sage Foundation between 1909 and 1914.

Armed with growing skills in analysis and design, these professionals came up with solutions that often featured slum clearance and decentralization (see Plates 1.2–1.6). For the public, historian Lewis Mumford narrated a documentary movie titled *The City* that contrasted center-city slum housing with newly constructed suburban dwellings. Screened at the 1939 New York World's Fair to large audiences, it spread the word about the benefits of urban and regional planning. It was a short step from some of these approaches to the development of the basic professional tools of urban and regional planning.

Early on, urban planners developed techniques to deal with land use and traffic to manage growth. Like most Americans, they celebrated the advent of the automobile, devising ways to accommodate it that ranged from the fantastic to the practical (see Plates 1.7–1.8). Some planners called for leaving the city, while others advocated staying. Some recommended single-family housing, while others designed dense in-town apartment complexes. The Depression dampened planners' collective exuberance but put them in new arenas: they participated in projects that built low-cost housing and infrastructure in cities. They were involved in regional planning efforts, like the Tennessee Valley Authority, that transformed vast areas, and in federal mortgage programs where they established minimum standards for the location and type of government-insured dwellings.

The upshot of these experiences was that planners developed a broader scope. They were no longer restricted to working on city plans, zoning, and traffic patterns, as they had been prior to the Depression; they added new activities, including dealing with affordable housing and modeling transportation systems. They began operating at a number of scales – neighborhood, city, region, state, and nation. To reflect this change, they renamed their professional organization, removing the word "city" and becoming the American Institute of Planners. At the same time, they developed graduate programs to teach new professionals and began to accumulate a body of scholarly literature, based on their new activities in practice.

This selection outlines those beginning stages of the field's development, tracing its seminal efforts in Progressive Era reforms through the height of American urbanism in the 1920s, when urban residents became the majority, and beyond to the New Deal era. It also follows British planning, highlighting the Anglo-American interchanges of knowledge and theory.

The authors, Richard LeGates, Professor of Urban Studies at San Francisco State University, and Frederic Stout, Lecturer in Urban Studies, Stanford University, are co-editors of the popular *City Reader* fourth edition (London and New York: Routledge, 2006) and of the entire Urban Reader series of which this *Reader* is a part. In addition, they have edited a collection of important planning documents, *Early Urban Planning, 1870–1940* (London: Routledge, 1997), which is in many university library reference collections. LeGates has also published *Think Globally, Act Regionally: GIS and Data Visualization for Social Science and Public Policy Research* (Redlands, CA: ESRI Press, 2005).

For more on the history of urban and regional planning, see Sir Peter Hall's *Cities of Tomorrow: An Intellectual History of Urban Planning and Design in the Twentieth Century*, third edition (Oxford: Blackwell, 2004). Hall's *Cities in Civilization: Culture, Innovation and Urban Order* (London: Weidenfeld & Nicolson, 1998) provides detailed background, covering some of the same turf as Lewis Mumford's classic study *The City in History: Its Origins, its Transformations, and its Prospects* (New York: Harcourt Brace, 1961). Also see Stephen V. Ward, *Planning the Twentieth Century City: The Advanced Capitalist World* (Chichester: Wiley, 2002), a thorough review of planning traditions in Europe and Asia.

A number of edited volumes offer topical essays in the history of urban and regional planning. These include Robert Fishman's *The American Planning Tradition: Culture and Policy* (Baltimore, MD: Johns Hopkins University Press, 2000), Mary Corbin Sies and Christopher Silver's *Planning the Twentieth Century American City* (Baltimore, MD: Johns Hopkins University Press, 1996), and Donald A. Krueckeberg's *Introduction to Planning History in the United States* (New Brunswick, NJ: Center for Urban Policy Research, 1983).

Finally, many monographs on different eras of planning history have emerged over the years. Jon Peterson's *The Birth of City Planning in the United States, 1840–1917* (Baltimore, MD: Johns Hopkins University Press, 2003) is an authoritative study of the profession's roots. John Reps's *The Making of Urban America: A History of City Planning in the United States* (Princeton, NJ: Princeton University Press, 1965) is a classic account of the nation's town site planning. William H. Wilson's *The City Beautiful Movement* (Baltimore, MD: Johns Hopkins University Press, 1989) summarizes early twentieth-century planning in Washington, Kansas City, and Harrisburg. And Greg Hise's *Magnetic Los Angeles: Planning the Twentieth Century Metropolis* (Baltimore, MD: Johns Hopkins University Press, 1997) covers the 1920s onward, focusing on the application of the advanced planning techniques of the times to the city's development.

City building has preoccupied kings and cardinals, mayors and burghers, for thousands of years. But it was only in the modern period that urban planning became an accepted profession and a well defined field of study. Although the origins of modernism lie in the Renaissance rediscovery of classical learning and the Enlightenment attempt to impose a rational order on both external nature and the social nature of mankind, what we call modern urban planning – no less than modern art, or modern politics, or modern family life – was born of the fundamental transformations of economic, social, and political life that propelled and grew out of the Industrial Revolution.

NINETEENTH-CENTURY URBAN CRISIS AND REFORM

[. . .]

The parks movement

One of the first responses to the horrors and social dislocations of industrial urbanism was . . . the parks movement. . . . In 1844, the city of Liverpool engaged the gardener Joseph Paxton to lay out Birkenhead Park as the first urban garden, complete with recreation areas for sports, open to the general public. In London, Victoria Park was opened in 1845, and in 1872 it was enlarged to encompass more than 200 acres of gardens, walkways, ponds, meadows, and woods. And in Paris, the extensive demolition and redesign project of Baron Haussmann also involved extensive landscape design elements . . . such as the Bois de Boulogne and the Bois de Vincennes and new parks such as Montsoris and Buttes-Chaumon within the reach of the city's burgeoning population.

In America, the man who transformed landscape gardening . . . to a vehicle of democratic social reform was Frederick Law Olmsted (1822–1903). In 1811, the City of New York published a plan for the eventual development of the whole of Manhattan Island. . . . Almost immediately, the commissioners' plan was opposed by public-spirited community leaders, and over a period of three decades the opposition coalesced into an organized popular movement to build a great Central Park for the citizens of New York. Eventually Olmsted and his partner, the young British architect Calvert Vaux, won the design competition and together began work in 1857.

As it neared completion in 1863, Central Park was recognized as a masterpiece, and it remains today one of the most successful examples of the enhancement of urban space by the intervention of artfully designed nature. Composed primarily in the naturalistic English landscape tradition, Central Park also contains formal gardens and elegant esplanades. . . . The Central Park design . . . pioneered the use of a multilevel transportation network that separated pedestrian traffic from carriages and that permitted crosstown traffic to transverse the park unobtrusively. In addition, it functioned as an integral part of the great Croton Reservoir system that provided fresh water to the whole of Manhattan.

. . . [F]or Olmsted, parks and good urban design . . . meant the nurturing and preservation of social morality. The park was to be a vehicle to control vice and provide healthy outlets for the city's poor and working-class populations. . . . Offering active recreational opportunities (baseball) and passive entertainment (concerts), the urban park would be an alternative to the grog shop.

. . . The parks movement sought to provide the congested city with "lungs". . . . [T]he enormous increase of urban populations in the nineteenth century and the misery entailed by the Industrial Revolution greatly compounded urban health problems. . . . In Europe and America, cholera and yellow fever epidemics periodically killed large numbers of people, leaping across class boundaries to spread terror to middle- and upper-class sections of cities as well as the slums. The new generation of city dwellers brought up breathing air heavy with coal dust, drinking polluted water, living in lightless, airless tenements, and subjected to long hours of physical labor without concern for occupational health and safety, were increasingly found to be unfit for military service or any form of work.

[. . .]

Ebenezer Howard and the Garden City ideal

. . . The key to improved health was an urban plan that eliminated congestion and kept the open countryside close at hand. . . . But of all that . . . sought to reintegrate the urban and the rural, the Garden City plan of Ebenezer Howard (1850–1928) is by far the most important both as a unified vision addressing the full range of urban development issues and as an initiator and formulator of modern urban planning as a profession and a body of theory.

Howard's plan was first published in 1898 as *To-morrow: A Peaceful Path to Real Reform* (republished, in 1902 and subsequently, as *Garden Cities of To-morrow*). . . . In Howard's original vision, the Garden City would consist of 6,000 acres – a town of 1,000 acres surrounded by a permanent greenbelt of 5,000 acres – supporting a population of 32,000. All land would be collectively owned, with startup loans retired over time from yearly municipal revenues. Eventually, Howard argued, the municipality should capture the increment in land values achieved from buying land at its agricultural value and creating value by successfully building the Garden City. The city itself would feature a complete array of municipal services and amenities: parks, public gardens, tree-lined boulevards, hospitals and asylums, and an enclosed, centrally located Crystal Palace-style emporium. And although the Garden City would be connected to a larger system of "social cities" by rail lines and canals, it would be economically self-sufficient, with its own factories and workshops, not a bedroom suburb for commuters or a satellite to an existing urban center.

The Garden City movement begins

Ebenezer Howard attracted a cohort of dedicated followers, including Raymond Unwin and Patrick Geddes in Britain and Lewis Mumford, Henry Wright, and Clarence Stein in America. And within a few years of the publication of *To-morrow* he had attracted a number of financial backers as well. . . .

Working together, Howard's backers bought up some 3,800 acres in Hertfordshire, not far from London, and began to build Letchworth, the world's first Garden City. Success, however, came at a price. In order to make Letchworth a sound investment opportunity, Howard was forced to abandon some of the more radical elements of his original plan, including publicly owned land rented with 1,000 year leases and housing provided by cooperatives. The greenbelt was kept as a planning element, but greatly reduced in size and function.

[. . .]

Howard hoped that the Town Planning Law, passed by Parliament in 1909, would spur the construction of dozens of new garden cities, but it was not until after World War I, in 1919, that land for Britain's second Garden City, Welwyn, was purchased and construction begun on a bland neo-Georgian plan by the architect Louis de Soissons. But if the movement began slowly, it had nonetheless begun.

[. . .]

Urban aestheticism and the City Beautiful movement

In the 1890s, concern with the "adornment" of cities, with "civic design," "municipal art," and "the city beautiful" supplanted parks and public health as the dominant concern in city planning. This trend was strongly influenced by L'Enfant's plan for Washington, DC, and the work of Haussmann in Paris, particularly his grand public buildings and boulevards lined with neoclassical and neo-Baroque apartment buildings. But another branch of the aestheticist movement was equally strong: the neomedievalism that was exemplified in England by the art history of John Ruskin, the romantic Utopian fantasies of William Morris, and the Arts and Crafts movement. On the Continent, this tendency reached its peak with the brilliant writings and designs of Camillo Sitte. In 1889, Sitte published *Der Stadte-Bau nach seinen kunstlerischen Grundsatzen* ("City Planning according to Artistic Principles") in which he carried out a systematic spatial analysis of existing historic cities. Sitte paid special attention to buildings as parts of a larger compositional arrangement and to the way streets flowed into squares and plazas to form a pleasing, interconnected whole.

Daniel Burnham of Chicago

Late in his career, Frederick Law Olmsted accepted a commission to lay out the grounds of a new World's Fair to be held in Chicago in 1893. Called the Columbian Exposition in commemoration of the four-hundredth anniversary of the European discovery of America, ... the design of the fairgrounds and pavilions, extravagant with neoclassical splendor and Beaux-Arts pomposity, attracted some of the most talented designers, architects, and planners in the United States: in addition to Olmsted, men like Louis Sullivan, Dankmar Adler, Charles McKim, and, most especially, Daniel Burnham (1846–1912) [who oversaw the plan for the Fair].

The "White City" on the city's lakefront was both an outstanding achievement in integrated design and a clear expression of the new capitalist order's sense of self. ... The principal elements of City Beautiful design, and its allied Civic Art movement, were strong axial arrangements, magnificent boulevards, and impressive public buildings. The culmination of Burnham's career, his crowning achievement as a designer of cities, was the Chicago Plan of 1909. The plan featured an elaborate system of public parks and lagoons, an imposing civic center, harbor improvements, diagonal streets, a stately yacht harbor, and transit and open space connections throughout the metropolitan area. ... [A]spects of City Beautiful planning deserve more respect, especially the emphasis on city planning as a comprehensive and unified process. Burnham is famous for the motto "Make no little plans, for they have no magic to stir men's blood." But it was not just bigness that characterized Burnham's grand conceptions, but a truly visionary sense of the city as a metropolitan whole.

Planning comes of age

One radical reformer who played a brief, but influential, role in the creation of modern city planning was Benjamin Clark Marsh, author of an extraordinarily superficial and opinionated book titled *An Introduction to City Planning* (1909). ... [He] saw population congestion as a prime national evil. ... Noting that City Beautiful projects have little effect on the daily lives of working-class people, Marsh argued that all public improvements should be scrutinized with a view to the benefits they will confer upon those most in need. ...

Marsh's most important contribution to American city planning was organizing the first (U.S.) National Conference on City Planning, which took place in Washington, DC, in May 1909. Unfortunately, Marsh's confrontational personal style helped splinter the participants in this important first national meeting. The professional planners did not care to be associated with [him]. ... When they organized a second and subsequent annual conferences they dropped all mention of the problems of housing and population congestion from the conference title and the conference agenda.

Progressivism and the city efficient

During the years before and after World War I, United States politics witnessed the rise of a new, activist philosophy of government. In part, Progressivism, as the movement was called, grew out of "good government" reformers who battled the corruption and managerial inefficiency of big-city, immigrant-based political machines. Often frankly representing middle-class social and economic interests, the Progressives sought to apply the best scientific thinking – including new social scientific theories of education, welfare, and social work – to the management of America's cities. With the rise of the new social and political ethos, City Beautiful concerns about urban aesthetics during the early urban planning period gave way to an emphasis on making the modern city function efficiently. George B. Ford's address to the fifth National City Planning Conference in 1913, titled "The City Scientific," and Nelson P. Lewis's *Planning the Modern City* (1916) are exemplary of "city scientific" thinking. The City Scientific movement shared the boundless modernist faith in the power of rational, scientific decision making to determine the one best solution to any urban problem and the applicability of uniform standards across cities.

Other writers concerned with modern scientific efficiency focused attention on the theory of planning as a social activity. Among the most important is Frederick Law Olmsted, Jr, the son of the great park planner. In an address titled "A City Planning Program" to the fifth National Conference on Planning in Chicago in 1913 Olmsted laid out a remarkably subtle and visionary city planning program. ... Olmsted distinguished between real plans which actually express the collective will of the community and nominal paper plans. He envisaged a “city plan office” fulfilling three main functions: a custodian of ideas, an interpreter to make the plan consistent, and an amender of the plan.

Edward Bassett and the master plan

As American city planning matured, city planning departments began to develop general, comprehensive plans. . . . But the rationale for these plans and what they should contain was not well articulated until Edward Bassett sought to define what a general plan should contain and its relationship to the processes of city government.

Bassett’s answer about what the master plan should be, articulated in *The Master Plan* (1935), was a general, flexible document, adopted by the local planning commission, but deliberately not adopted by the local legislative body. The plan would consist of both map and text. The text would be organized in relation to a small number of plan elements, and the plan need not be consistent with existing zoning. Bassett felt that the plan should have a certain visionary quality but that it should emphasize physical land use planning. . . . He distinguished between a plan, which he felt should be easy to change, and an official map of streets, highways, plazas, and parks, which would be much more permanent.

. . . Bassett’s concept of the master plan as the core document of city planning agencies was both the culmination of the process that led to the professionalization of planning and the starting point of all future planning theory and practice.

NEW TOWNS AND REGIONALISM

While much early city planning thought focused on a single city as the unit of analysis, some of the new urban theorists and professional practitioners of city planning were concerned with entire regions. . . .

The contribution of Patrick Geddes

The construction of Letchworth and Welwyn, along with . . . important advocacy and implementation work . . . , established Garden City planning in the mainstream of British urban planning practice. But other important tributaries of planning thought were soon to join with the Garden City movement. . . . Much of this new thought came from a brilliant, eccentric Scot named Patrick Geddes (1854–1932).

Geddes outlined his theories of urban development and planning in *City Development: A Study of Parks, Gardens, and Culture Institutes* (1904) and *Cities in Evolution: An Introduction to the Town Planning Movement and to the Study of Civics* (1915). . . . [He] fully developed the regional vision that was implicit in Howard’s system of “social cities” and brought the abstraction down to earth. Before any changes could be made to a city or its neighborhoods, a survey would place the city within the environmental context of its region’s surrounding ecosystems. . . .

New Towns for America

Of the many disciples that Geddes attracted, perhaps none was more brilliant and influential than Lewis Mumford (1895–1990). . . . Mumford carried the ideas of both Howard and Geddes into his own philosophy of urban development and helped to popularize those ideas in America. He saw that the power of transportation and communication technologies . . . could actually permit decentralization of population and industry throughout regions. At the center of a small, but extremely influential group of intellectuals called the Regional Plan Association of America (RPAA) – a group that included Clarence Stein, Henry Wright, and Benton MacKaye – Mumford developed a powerful vision of regional planning that would turn existing urban development away from megapolitan sprawl toward a clearer, more humane pattern of small cities that would fit harmoniously within the greater New York region. The RPAA’s first important project was Sunnyside Gardens in Queens, New York, where Mumford and his family lived for six years, begun in 1924. This was followed by the even more ambitious Radburn, New Jersey, project begun in 1928.

In the Radburn Plan, Stein and Wright proposed a city using “superblocks” in place of the characteristic narrow rectangular blocks, roads for different uses (service lanes, secondary collector roads, main roads, and parkways), a complete separation of pedestrian and automobile traffic, and houses turned away from the street to face a series of parks forming the backbone of the community. . . .

Architect Clarence Perry (1872–1944) took the ideas of human scale development further and thought deeply about how to design neighborhoods that would function well in the automobile age. His thoughts are summarized in "The Neighborhood Unit" (1931). Perry envisioned the school as the centerpiece of the neighborhood, performing a role in the community well beyond educating primary school children, and argued that the neighborhood should have sufficient population to support one elementary school. Perry gave a good deal of attention to the relationship between the neighborhood and streets. He suggested that neighborhoods should be bounded on all sides by arterial streets for through traffic, but internal street systems should be almost exclusively for use by the residents. The use of culs-de-sac and careful separation of streets from pedestrian ways would harmonize transportation with living space.

The plan for New York and environs

Concerned with the nature of development in the New York region, the Russell Sage Foundation invested in a number of city planning and housing programs and in 1922 funded a monumental nine-year study of the New York region. As chair of the Russell Sage Foundation's Committee on the Regional Plan of New York and Environs, Charles Dyer Norton advocated a monumental planning effort and convinced other members of the Foundation board to fund this effort on a massive scale. Norton had been active in the Chicago Commercial Club and became deeply involved in Burnham's 1909 Plan for the City of Chicago. The person he chose to head this regional planning effort is one of the most influential of twentieth-century planners: Thomas Adams (1871–1941).

As a young man, Adams had worked with Ebenezer Howard and Raymond Unwin on Letchworth and other Garden City projects. He served as the first chair of the Garden City Association and became the first president of the British Town Planning Institute in 1904. From 1913 to 1921, he was the town planning advisor to the Governor of Canada and in this capacity developed the first real regional plan in North America for the area around Niagara Falls. Adams assumed the position of general director of plans and surveys for the New York regional planning effort in 1923.

In one of the most celebrated conflicts in American planning history, Lewis Mumford and other members of the RPAA attacked the completed regional plan. After poring over each sentence in the massive *Plan for the New York Region*, Mumford declared that he found little of value in the whole exercise. He dismissed the plan as an essentially conservative document which dodged hard choices, accepted continuation of the status quo as inevitable, and failed in its goal of providing a real vision of regional development. . . . Mumford argued [that] it really was a prescription for more congestion, . . . it really called for more chaotic land subdivision . . . [and] it actually would permit more, not less, overcrowded land development.

PROPHETS OF HIGH MODERNISM

Utopian modernism

As urban planning became professionalized and regularized, both in Europe and America, a new urban utopianism emerged to reinvigorate the movement at the level of theory. While the professionals planned for today, new dreamers planned for the city of tomorrow.

Le Corbusier and the international style

Charles Edouard Jeanneret, who reinvented himself as Le Corbusier, was the prophet of a higher, later stage of modernism: the city as the administrative center of the bureaucratic, technocratic state. . . . [I]n 1922, he proposed "A Contemporary City for Three Million People." This was a breathtaking, totally modern vision of spare, undecorated skyscrapers, evenly spaced in a park, that astonished the people of Paris and that still seems futuristic today.

. . . By announcing that his city would house 3 million people – about 100 times the population of Letchworth – he consciously flew directly in the face of the Garden City advocates while, at the same time, advocating many of their own ideals: simultaneously decongesting cities while maintaining their density. By proposing the use of skyscrapers, he incorporated a new element associated with the crass business culture of America as a solution for European urbanism. And, in 1925, he boldly announced a new version of the Contemporary City plan, the Plan Voisin, that was to be built on a site in the middle of Paris, previously cleared and leveled by bulldozers! The popular response was, of course, outrage, but Le Corbusier became instantly famous, a spokesman for a new, uncompromising modernism.

. . . [A]fter [World War II], . . . Corbusian principles were adopted by governments worldwide as a quick and easy response to the demands of reconstruction. . . . Today, the skyscraper in the park (as often as not reinterpreted as the skyscraper in the parking lot!) is one of the standard and ubiquitous realities of modern cities everywhere. . . .

Frank Lloyd Wright's alternative vision

While Le Corbusier promoted a vision of the city of tomorrow that embraced skyscraper development . . . Frank Lloyd Wright was the prophet of middle-class urban flight and automobile-based sprawl suburbia. . . . Wright advocated a naturalistic architectural style, as well as a vision of urbanism, that were totally at odds with Le Corbusier's hard-edge cubist conceptions.

Announced as early as 1932 in *The Disappearing City*, Wright's Broadacre City allocated a minimum of one acre per person, with no large urban concentrations whatsoever. . . . Broadacre City would be family-based, and Wright designed an extraordinary small house with an attached carport – the Usonian house – that subsequently became the model for millions of suburban houses in the decades following World War II. . . . [T]he Broadacre model, with its emphasis on the automobile and the telephone as annihilators of space and time, was prophetic of a new urban/suburban reality that would dominate the planning of the future.

Clearly, neither Frank Lloyd Wright nor Le Corbusier would be pleased by what world urban centers and middle-class suburbs actually became in the last half of the twentieth century. Residential lots were too small, the lives of the residents too conformist, to match Wright's Jeffersonian–Emersonian standards. And Corbusian reality never really achieved the purity and sublimity of the Corbusian dream. But the regionalism and decentralization proposed by the Garden City and New Town advocates now faced two rival approaches that would help to define the urbanism of the twentieth century and the traditions of modern urban planning.

PLANNING AND THE GREAT DEPRESSION

Cities and the crisis of capitalism

Throughout the 1920s and 1930s, the example of the Soviet Union was a powerful force in planning theory. There, few real innovations were accomplished in the area of urban planning, but planning that directed the entire society and economy, including the provision of great public works and new community development, was incorporated into a series of sweeping Five Year Plans. . . .

In all the capitalist democracies, the Great Depression of the 1930s called for a fundamental reevaluation of the relationship between government and the existing social order. Faced with near-total economic collapse – and properly alarmed by the rising tide of totalitarianism elsewhere – democratic governments in Europe and America sought new ways to stabilize themselves, to protect the lower strata of their populations from utter destitution, and to invest in massive new programs of social reform and infrastructure development. In the United States, Franklin Roosevelt's New Deal included a Public Works Administration that constructed thousands of post offices, courthouses, and hydroelectric dams throughout the nation in an impressive and uniform federal style. . . . This was the climate in which urban planning . . . made great strides.

Modern housing for the Depression poor

One of the most important figures in New Deal urban planning was Catherine Bauer (1905–64), [who] . . . profoundly affected U.S. housing and urban development policy throughout the 1930s and 1940s. After studying post-World War I housing in Europe, Bauer felt that the United States lagged far behind – four million "modern" postwar housing units built in Europe between the end of World War I and the beginning of the Great Depression compared to the paltry parallel U.S. record of no more than 10,000 comparable units completed during the same time. . . . Bauer was a true modernist with a faith in large-scale, rationalized housing using the most advanced building methods and materials – cement slabs, glass, and iron. . . . She saw housing units as intimately related to schools, shops, laundries, public open space for recreation, and gardens. . . . Bauer . . . played a leading role in formulating and securing passage of the critical U.S. housing Acts of 1937 and 1949 which created the U.S. Public Housing and Urban Renewal programs.

Patrick Abercrombie and the Barlow Report

In Britain, Patrick Abercrombie (1879–1957) . . . was appointed a professor of civic design at the University of Liverpool in 1914. In that position and as editor of *Town Planning Review* he established a reputation as Britain's leading academic planning theorist. Abercrombie was also a practitioner who developed many town and country planning schemes. . . .

[. . .]

Prime Minister Neville Chamberlain had long supported regional planning and Garden Cities. In 1937, he appointed a Royal Commission on the Distribution of the Industrial Population, popularly called the Barlow Commission, which undertook a monumental review of the location of industry and housing throughout Britain. While the Commission's concerns were to develop fundamental policy for industrial location that went well beyond immediate strategic concerns it was the danger of industrial concentration at the outbreak of the war and the perceived need for strong, centralized planning for postwar reconstruction that made national-level city and regional planning possible in Britain.

[. . .]

The great premises and programs of the Barlow Report were reiterated in Patrick Abercrombie's historic Plan for Greater London. The Abercrombie Plan called for the creation of New Towns outside of a decongested, greenbelted London. It became the centerpiece of the new Labour Party government's social policy. . . .

In Britain, and elsewhere in Europe, planners saw regionalism and New Towns policies, along with parallel increases in welfarism, that helped in the rebuilding process that was the inevitable work of postwar reconstruction. And in the United States, the 1949 housing Act, strongly influenced by Bauer, called for an expansion of public housing and instituted urban renewal. As if continuing the wartime total mobilization, massive new efforts at slum clearance and inner-city redevelopment were undertaken under this important legislation. Large-scale inner-city reconstruction projects borrowed heavily from Le Corbusier's ideas. And postwar prosperity also brought an extraordinary expansion of suburban tract-home communities, borrowing the energy and focus of wartime mobilization and applying them to domestic needs. Broadacre City became Levittown.

[. . .]

[T]he great accomplishments of early city planning must not be overlooked or undervalued. The great urban parks still enhance the lives of millions and constitute an incalculable asset for the residents of great cities. . . . Both the elegant civic centers created by the City Beautiful planners and the comfortable, sensitively designed garden suburbs built by the New Town developers of the 1920s remain models for emulation today. And the many dedicated architects, landscape designers, legal experts, social reformers, environmental activists, and others who contributed to the professionalization of modern urban planning deserve both the interest and respect of subsequent generations of urban specialists.

[. . .]

8

Reflections and Research on the U.S. Experience

From *The Profession of City Planning: Changes, Images and Challenges, 1950–2000* (2000)

Michael B. Teitz

Editor's introduction

While this selection celebrates the thirty years between 1965 and 1995 as a Golden Age of American urban and regional planning, the field has had many golden years as it evolved from its Progressive roots in efforts to improve slum housing, beautify cities, and deal with rampant growth. In the essay "Modernism and Early Urban Planning" in this *Reader*, Richard LeGates and Frederic Stout detail its development. This essay outlines the contemporary expression of these turn-of-the-century concerns, now labeled local economic development, urban design, and land use planning. It also argues that in the 1960s the heightened interest in environmental issues, the expansion of public participation in government decision making, and the emergence of professional planners to manage these changes have contributed to the flourishing of the field.

The institutional history of the American planning movement and the development of its professional practice reveal not only continuity in the field's core concerns but also a drive for self-improvement. As Eugenie L. Birch's article "Advancing the Art and Science of Planning, Planners and their Organizations" in the *Journal of the American Planning Association* (46: 1, 1980) relates, the U.S. planners' organizations originated in the early twentieth century when an amorphous group of public-spirited citizens became interested in cities, their rapid expansion, and the concomitant problems of slums, chaotic land uses, and environmental degradation. Although they directed their efforts at local affairs, they frequently assembled to share common concerns at national meetings of the National Conference of Charities, the American Civic Association, the American Institute of Architects, and the Municipal Arts Society of New York. They soon decided to convene their own forum, organizing in 1909 the first Conference on City Planning, in Washington, DC. The attendees came from all over the United States – New York settlement houses; the Hartford, Connecticut, Planning Commission; the Commercial Club of Chicago; and professional offices in Boston, Detroit, Kansas City, and San Francisco. This first meeting was so noteworthy that the U.S. Senate Committee on the District of Columbia recorded its sessions.

Buoyed by this success, the fledgling group formalized the organization as the National Conference on City Planning (NCCP). The NCCP grew to about 400 members in ten years, held annual meetings, hired an executive director, and published a quarterly magazine, *The City Plan*, as well as the proceedings of its annual meetings. Supported by the Russell Sage Foundation, which also provided substantial funding for the Regional Plan Association of New York and *The Pittsburgh Survey* (the first comprehensive, social science-based study of an American industrial city), the NCCP focused on general education and promoting states' adoption of enabling legislation for planning. Six years after its founding, it reported success in ten states where more than 100 planning commissions functioned.

By 1917, a subgroup of the NCCP formed, primarily comprising private sector-based landscape architects, civil engineers, lawyers, architects, land economists, and realtors. Named the American City Planning Institute, this affiliated unit convened for technical discussions on urban and regional planning. It was the field's first professional association and eventually evolved into the American Institute of Certified Planners (AICP).

While the founders saw their institute as an apolitical membership organization, they had no idea that it would grow to its current strength of 39,000 APA and 14,000 AICP members. Nor could they have envisioned the range of activities undertaken by today's planners. No doubt they would be gratified to see that their core concerns have prevailed over several generations and that their aim of integrating the skills of several professions to confront metropolitan issues has been realized.

In the spirit of the profession's founders, this selection challenges planners to continue to sharpen their thinking. It urges further improvement in the field, and indeed improvement is already occurring, especially in the land use area, now assisted by geographic information systems (GIS) technology. But, says the author, it is time to come to grips with some fundamental issues: the universal appeal of suburban life, the distinct differences between older and newer cities (those that cannot annex more land and those that can) and associated concerns revolving around appropriate levels of density in outlying areas, the means to mend rifts between advocates of center-city and neighborhood economic development, and the metropolitan imbalances within and between cities and suburbs.

The author, Michael Teitz, is Professor emeritus of City and Regional Planning, College of Environmental Design, University of California, Berkeley, and Senior Fellow, Public Policy Institute of California. (He previously was Research Director of the Institute.) His most recent publication (with Elisa Barbour) is *Blueprint Planning in California: Forging Consensus on Metropolitan Growth and Development* (San Francisco: Public Policy Institute, 2006).

For more on the American planning profession, see Mel Scott's *American City Planning since 1890* (Berkeley, CA: University of California Press, 1971), a history commissioned for the fiftieth anniversary of the founding of the American Institute of Planners; Jay Stein's *Classic Readings in Urban Planning* (Chicago: APA Planners Press, 2004), which has a number of readings discussing the emergence of the profession; and Donald A. Krueckeberg's (ed.) *The American Planner: Biographies and Recollections*, second edition (New Brunswick: Center for Urban Policy Research, 1994), a collection of profiles of important planners, providing a human dimension to the field. A classic monograph on the planning profession is Donald A. Schon's *The Reflective Practitioner: How Professionals Think in Action* (New York: Basic Books, 1983). Jon Teaford's *The Unheralded Triumph: City Government in America, 1870–1900* has a chapter on the rise of urban professions, offering an excellent contextual explanation for the emergence of urban and regional planning. Two articles in *Housing Policy Debate* present a lively interchange on the origins of today's planning environment: Robert Fishman's "The American Metropolis at Century's End: Past and Future Influences" (11: 1, 2000, pp. 199–213) and Robert A. Beauregard's "Federal Policy and Postwar Urban Decline: A Case of Government Complicity?" (12: 1, 2001, pp. 129–51).

Some future observer may well look back at the thirty years from 1965 to 1995 as a Golden Age of American planning. Despite the failure of some federal planning efforts, continuing decline in the central cities, and denunciations by critics of the quality of the suburban built environment, this period saw the greatest expansion of operational power on the part of urban planners since the profession emerged in the late nineteenth century. That this occurred is not simply the result of the growth of income and population, with its concomitant urbanization and infrastructure development after World War II, although those are certainly critical influences. For planning as an activity recognized and supported by public funds, we must also look to two important trends.

The first trend was the increase of environmental concerns and public awareness, which gave rise in the 1960s to key national and state legislation for environmental protection. These laws lent status to groups formerly excluded from legally challenging development, thereby changing the planning process profoundly. They also required extensive data gathering and analysis for the purpose of assessing the environmental impacts of proposed developments, leading to the formation of new planning-related organizations. In short, the environmental revolution empowered citizens and organized environmental groups, as well as created new demand for planners, both those working for public agencies and those working for nongovernmental organizations. By the late 1980s, the process of development was, in many respects, so transformed as to be unrecognizable from that of the 1960s.

A second, closely related trend was a shift in the relative power between local citizens, on one hand, and developers and local governments, on the other. Many factors have contributed to this change. Among them are the widely publicized negative effects of urban renewal, a federally supported program in the 1950s that generated huge opposition; the efforts at empowerment by the War on Poverty of President Lyndon Johnson in the 1960s; and the widespread growth in the local environmental awareness and opposition to growth and change at the local level. Although the level of power of the organized citizenry varied widely from place to place, their role in planning decisions had increased substantially by the 1980s, especially in their capacity to stall or prevent development. Paradoxically, this shift increased both the demand for and influence of local planners, who often found new allies in local planning disputes and who began to develop new roles in intermediation and consensus building in complex, conflictual situations. Planning and development became a complex process throughout the United States, and planning emerged as a key profession in managing this shift in power. . . .

Behind these movements in planning are profound economic and social changes in American society. Beginning in the 1970s, these shifts have given rise to major debates, both about the scale and the significance of the changes, as well as appropriate social responses to them. Arguably the most important of the changes has been the decline in the rate of growth of productivity that has occurred over the past twenty years.

[. . .]

Changes in the American economy and society have powerful implications for planning. Despite slowing productivity, income gains and a widening income distribution have permitted continuing suburban development. The impact of this development, both in older metropolitan peripheries and in newer regions of growth, has been to generate environmental pressure and demands for planning. At the same time, the larger changes have had critical effects on older metropolitan areas and their inner-city cores.

Metropolitan core cities are a traditional focus for planning, and their fate is closely tied to the field. Immigration, technological change, and competition have created a mixed picture for the central cities in the 1990s. Some have been prospering; for others, their long-term population decline appears to have slowed, and new sectors have emerged; for still others, conditions are worse than ever. . . . Clearly, when we speak of the central or inner cities of the United States, we must now distinguish carefully between those newer cities still able to annex land or able to attract population, and those either declining or in a condition of stability or slow growth.

[. . .]

. . . [T]he condition of the inner cities reflects their position within the larger economic environment. For those able to participate effectively in the sectors of growth over the past two decades, the picture is relatively positive. In this, cities resemble the suburbs of almost all areas in the United States. For those cities unable to participate, the result has been much more problematic. Many of them have found themselves attempting to cope simultaneously with loss of manufacturing employment, rapid immigration, and the increasing isolation of the poorest groups within the population. That dilemma has attracted a resurgence of research and policy interest about questions of poverty. These debates have influenced decision makers and provided the context of much that planners have thought and done in relation to inner-city problems. . . .

EVOLUTION AND DEBATE: LAND USE PLANNING AND LOCAL ECONOMIC DEVELOPMENT

The fortunes of planning have long been associated closely with those of development. This is evidenced by the continued importance of the traditional core of the field, local land use planning, and the new focus within the field on local economic development.

Land use planning

It is now a quarter of a century since the suburbs exceeded the aggregate population of central cities in U.S. metropolitan areas. With that development has come a "dominant vision" among ordinary people of the appropriate form of settlement comprising geographically unconstrained low-density development, detached single-family housing, widespread ownership of automobiles, low-rise workplaces with automobile access, small communities that are politically autonomous to a substantial degree, and an absence of visible poverty, brought about by the concentration of the poor in older, central-city areas. America is a suburban country, at least in terms of the environment inhabited by most of its people. The planning profession has never entirely come to terms with this fundamental change, despite the reality that most planners now deal with problems and issues in a suburban context. We may attribute this, in part, to the grievous problems of the older central cities, which deservedly continue to attract much attention from planners; but there are also serious differences among planners and urban analysts in their views of the automobile and its role in the future structuring of American cities.

[. . .]

Land use planning has . . . shown a new burst of creative energy in the past decades. This development has been stimulated by the larger contextual changes described above, but it has also come from the innovative work of practitioners and decision makers in American communities. . . .

In . . . land use planning . . . the general plan [has evolved] into . . . hybrid forms that fit the contemporary situations in communities. . . . Such plans are likely to be the product of much more interactive and participatory processes [than in the past], owing to the use of new technology and increased interest and participation by citizen groups. They are continually being created and redesigned at the local level, thanks to the ingenuity of professional planners and citizens.

The changes in process, which stem from the rising awareness of the environmental and fiscal impacts of growth, have been especially important at the state level, where there have been several efforts to go beyond individual community efforts to control development in ways that lead to mutually adverse outcomes. Two aspects of this trend are especially important – substance and process. Substantively, states as different as Florida, New Jersey, Oregon, and Vermont have sought to manage growth and development to a degree far exceeding anything seen in the [past]. Their success has been mixed, but the very fact of the effort marks an important evolution of planning in the United States, in which there is a substantial restructuring of governance and some evidence of an improvement in the quality of local plans.

The improvement in quality of land use planning has, in part, been facilitated by the rapid adoption of Geographic Information Systems (GIS) as a planning tool. Planners have responded enthusiastically to this technological advance – and for good reason. That GIS has the potential to improve operational effectiveness in relation to such things as data accuracy, availability, and access seems evident to planners in their day-to-day work. Perhaps the most intriguing aspect of GIS is its potential to change the character of analysis for decision making. From the outset of the use of computers in planning in the United States, researchers and practitioners have envisioned their capacity to provide information rapidly, to model and simulate the urban environment, and to permit the construction of real-time alternatives in the course of the planning process. The first round of this effort, in the 1960s, produced basic advances in understanding but did not live up to its promise. Nonetheless, in the succeeding two decades, a quiet development of models continued, especially in metropolitan transportation planning. The rapid improvement of computing speed and memory capacity in microcomputers, an astonishing decline in the cost of hardware, together with the creation of rapidly improving GIS software, have changed the situation to a remarkable degree. Not only have large-scale models enjoyed a modest revival, but the possibilities created by imaginative use of GIS have begun to permit new types of urban models that can incorporate constraints at a level of detail that begins to approximate that needed by planners and decision makers in practice.

. . . The increasing complexity of land use decisions and the growing number of players in the decision process have led to new forms of planning, particularly through negotiation and consensus building. Explicit processes for building consensus have emerged from experience in negotiation and through ideas from many sources. For example, from corporate strategic planning we see the importance of the identification and commitment of stakeholders, and from the field of dispute resolution, new forms of conflict management that seek to bypass the traditional political and judicial systems. Given the high stakes and strong concern among parties over land use decisions, this seems a natural tendency that is now beginning to make itself felt. The evidence for the value of consensus-building processes for land use planning certainly is not complete, but there are good arguments for its serious consideration in the field coming from what historically have been vastly different positions.

One source of support comes from planning theory, which also has been evolving. Theorists . . . are advocating the idea of communicative action. . . . [T]he key element . . . is its focus on practice as the source of knowledge. . . . [T]o understand a field of practice, it is first necessary to study it closely, listen to the language of its practitioners and those who are affected by it, and take seriously what they have learned from experience. Prescriptively, good practice carries this further through utilization of this knowledge in planning processes that draw on and respect this knowledge, principally in the search for consensus through structured but open and participatory dialogue.

Attractive as the ideological content of communicative action may be, especially to academics and students who do not have to deal with the realities of planning in a capitalist market economy, one may doubt that this is the only reason for its quiet advance. . . . [I]ts appeal is . . . rooted in attention to practice and to the methodology of building agreement on complex issues.

[. . .]

Debates over land use planning

The evolutionary trends in urban planning already discussed are accompanied by powerful, sometimes conflictual debates over some critical questions. No dynamic professional field is without such issues; they mark the boundary of what is known and believed about practice in relation to the changing world in which it is carried on. . . . [T]hese issues [are] first, metropolitan growth and transportation and, second, urban design. . . .

Most planners probably are close to a “balanced growth” perspective on metropolitan growth. Essentially, this view accepts the reality of low density, with the automobile as the dominant means of transportation; but it seeks alternatives. One reason is that the suburban metropolis is seen as inefficient in terms of transportation, as evidenced by complaints over suburban traffic congestion and commute times; the other is that the suburban metropolis is environmentally problematic, especially because of air pollution.

Higher-density centers consisting of employment and commercial clusters are now emerging within the low-density metropolis as employment decentralization follows population. Popularly identified as “edge cities,” these concentrations are found almost universally in American metropolitan areas. It is argued that they represent a new form of development that means higher density and a more urban character in suburban environments. In an automobile-dominated system, the growth of suburban concentrated employment has powerful implications for transportation effectiveness. Traffic congestion has appeared in places where it was never previously experienced, with strong political repercussions. . . . Among planners, spatial imbalance between jobs and housing is widely regarded as a critical contributor to traffic congestion in the suburbs and as a source of inaccessibility to employment for those segments of the population, especially the urban poor, who do not drive to work. . . . The possibility that traffic problems may be mitigated by land use change appeals to those opposed to low-density development for other reasons.

The second perception critical to the balanced-growth view is linked to suburban employment growth but is driven heavily by natural and social environmental concerns. It emphasizes the effects on environmental quality of metropolitan development that is very large in scale and auto-dominated. Both federal and state legislation have sought to improve air quality by a variety of means (for example, the creation of air quality management districts for critical air basins). After the obvious measures to control point-source pollution have been taken, it rapidly becomes evident that auto emissions are a major contributing factor. Among the ways to curb such pollution is the possibility that a different, denser, urban form might yield lower levels of auto travel, either through proximity of origins and destinations or through increased feasibility of alternative means of transportation. Planners tend to see such arguments in a positive light. . . .

These views are not held solely by urban planners; they have been widely reflected in research, in political debates, and in legislation, to the point of becoming the conventional wisdom. They are visible in much of the federal legislation affecting metropolitan development in recent years. Both the 1991 Intermodal Surface Transportation Efficiency Act (ISTEA) and the Clean Air Act Amendments of 1990 require that land use considerations be taken into account in planning transportation investments. Although there are always many reasons for the adoption of specific programs in the American political system, there can be little doubt that the decentralization of metropolitan employment, declining air quality, suburban traffic congestion, and energy consumption have all been important arguments in the passage of legislation that has directed huge public investments in new heavy and light rail systems.

Not everyone agrees with this version of the future of urban development. . . . A few critics . . . argue that the automobile-based metropolitan area is both efficient and desirable, and that public investment in fixed-rail systems is particularly unjustified. By and large, the planning profession and federal, state, and local legislatures took little notice, although such a view evidently has been the basis of the real politics of suburban development since the 1950s. . . .

[. . .]

Debates over urban design

The debate over the form of the emerging metropolitan areas and their transportation needs has been echoed in the realm of physical design. Dissatisfaction with suburban environments has long been expressed from many quarters, the most recent probably being the feminist critique of the single-family house and suburban neighborhood as isolating and inappropriate for women in this era. In the past decade, however, a new school of design for suburban development has emerged, harking back to much older roots in the idea of community and the virtues of density. This approach, known as "neotraditional" or "new urbanist" planning and design, has been created largely by architects, notably [the husband and wife team Andres] Duany and [Elizabeth] Plater-Zyberk and [Peter] Calthorpe. Its hallmark is the design of housing complexes intended to recover a mixture of small-town and urban values seen as having been lost in the low-density suburb. In its exemplar – Seaside, Florida, designed by Plater-Zyberk and Duany – the style involves a self-conscious choice of historic forms, both in housing and subdivision layout, that is reminiscent of the romantic designs sometimes seen in older suburbs. However, there is an underlying logic that uses the single-family house, grid street pattern, and relatively dense layout of the late nineteenth-century American small town to attempt to recapture perceived benefits of density. Increased social interaction and "neighborhood" quality asserted to come with higher density are the primary benefits intended. The ghosts of Ebenezer Howard and Clarence Perry are abroad, albeit in an inverted form with respect to density. . . .

Whether New Urbanism works on its own terms, and whether it can meet the test of the market, have yet to be determined fully. If Seaside and its few imitators were all that comprises neotraditionalism, there would be little point in viewing it as anything other than another design fad. But there is an interesting link to the question of planning for metropolitan development discussed in the previous section. That connection appears in the work of Calthorpe and others who have connected the idea of redesigning for density with changes in transportation structure. . . . An important means of achieving this is to link . . . developments with transit nodes and create substantial communities that are both suburban and transit-oriented.

Evidence from studies of the impacts of new transit systems in major metropolitan areas suggests that densification does indeed occur around stations, as would be expected in a market environment, but that its appearance depends upon local political attitudes to development and may not occur in close proximity. To what extent that development will permit or encourage pedestrian, or non-automobile, travel remains uncertain, but . . . recent work suggests that, with appropriate planning, transit villages may be part of the future as well as the past.

LOCAL ECONOMIC DEVELOPMENT

In the past decades, local economic development has quietly emerged as a second major sphere of interest and activity in urban policy.

[. . .]

The traditional approach to local economic development in the United States has relied on the idea that growth requires inward capital investment to develop local resources, utilize local labor, and create sales to nonlocal markets. . . . In essence the . . . conceptual structure argues that localities faced with economic difficulties [must engage] the active participation of local governments in partnership with the private sector. . . . [This] message [is] attractive to planners, reinforcing their tradition of local action. It also appealed to those states and localities, particularly the inner cities [with] . . . little prospect of attracting major [outside] investment. . . . Especially appealing [is] the notion of partnership between local governments and business in the process of local economic development. This idea had been given currency by the federal Urban Development Action Grant (UDAG) program put in place during the Carter administration in the late 1970s. Under the UDAG program, the federal government provided funds for local economic development projects that were to be leveraged by private investment in partnership with local governments. . . . [D]espite its termination, the UDAG program was important in moving urban economic development into a partnership mode and away from simple tax incentives. . . .

. . . As economic development became more widely recognized as a part of local government activity, so the question of its placement within the governmental structure arose, and its participants began to form quasiprofessional organizations. For planning, this has raised a number of issues, particularly over what the functional and organizational relationship of planning and economic development should be within government. The outcome is by no means clear. Functionally, the two are often in conflict, as citizens use the planning process to slow down or block development, or developers seek to override planning constraints. Organizationally, in some cities, economic development and planning are separate departments; in others, economic development resides in the mayor's office, reflecting the importance of high-level interaction in the progrowth coalition politics; in still others, planning and economic development have been subsumed into larger departments, often also including urban redevelopment agencies. What is evident is that planning and economic development in the 1990s are intimately related in tension as well as in mutual support.

[. . .]

For the older central cities, local economic development is especially important. Faced with serious economic problems, they have grasped at nostrums of all kinds, often with disappointing results. The challenge of economic development is complex, however. Elected officials are striving to overcome the dominant images of poverty and its associated ills, which leads them to seek development that will "turn the city around," a euphemism for bringing back the middle-class population. To do this, they have emphasized the development of advanced sectors and the renewal of central business districts. They must also respond to the needs and political demands of impoverished groups who clamor for equitable outcomes, especially through community development. Trying to square this circle is not easy. Both thrusts have generated highly conflictual responses in planning, epitomized by central business district and community development. In practice, cities have channeled resources into two different types of local development: the redevelopment of the central business district, and community development in inner-city neighborhoods. Within each approach, there is some debate over the efficacy of policy, although research is generally not conclusive. Perhaps the most interesting issue not explicitly spelled out in the literature concerns whether the two approaches are complementary or conflicting in their development goals.

[. . .]

CHALLENGES TO PLANNING

No matter how heated the debates within a professional field, the existence of the activity itself is rarely called into question.

[. . .]

Though small in comparison to other professions, city planning has a lively intellectual presence in the universities and a genuine set of tasks in practice. Its academics and practitioners maintain a tradition of questioning and discontent with the ability and willingness of the profession to meet the standards and levels of performance that they would like, but the field progresses nonetheless. Its character reflects major trends in American political, economic, and social life; but it also has its own blend of idealism and professionalism that sets it apart from other groups in the urban policy world. At this point, the field is challenged by serious issues that range from debates over the nature of the city itself to the ideological acceptability of public intervention and regulation of development. Among the most significant for the long term are the questions of the future of the central cities and the gulf between the races in American urban life.

Overall, planning has seen growing sophistication, in both practice and research, in the fields of land use planning and economic development. But the record of planning has been little better than that of any other profession in addressing inner-city issues. In some respects, notably the disasters of public housing and urban renewal in the 1950s and 1960s, it is worse, clearly contributing to the process of deterioration. It may be fair to say, however, that planners did learn from those errors. Planners also learned much from rethinking local economic development and from the community development efforts of the 1960s through the 1980s. Although they did not discover how to make community economic development effective, there can be no doubt that local communities, even of the poor, now have a stronger voice in the planning and development process. Ironically, though, the emergence of new voices also presents a major threat to planning, as evidenced by the increasingly antiregulatory stance of property owners.

At the beginning of the twenty-first century, modern professional planning in the United States is about 100 years old. It is a unique endeavor, both in its focus on the urban environment and in its idealism about improving the quality of life for people. As we might expect, the record of achievement is mixed. Nonetheless, the idealistic strain, rooted in the American Progressive tradition, remains evident within the field, although, as noted above, paying less attention to the inner cities and the poor. The next hundred years will bring new challenges to planning, both from the global economic transformation that is occurring and from the growing domination of social life by market forces. Planning as a profession has built the technical capacity to engage those changes in a way that supports its traditional values. Whether it can continue to be viable as a field and have a significant impact will depend on the next generation of planners.

9

Practitioners and the Art of Planning

Journal of Planning Education and Research (2001)

Eugénie L. Birch

Editor's introduction

Professions have several qualities that identify them. They possess expertise in a specialized field, a means of transmitting that expertise to new professionals, a code of ethics, and a concern for the public interest. All this is true of the urban planning profession. Over time, planners have identified and enlarged their expertise, refined their educational procedures (including establishing an accrediting body to approve university planning programs), adopted a code of ethics, and probed the nature of public interest applied to their field.

Three professional planning groups have helped the field develop in these areas. They are the American Planning Association, the American Institute of Certified Planners, and the Association of Collegiate Schools of Planning. The American Planning Association (APA) numbers more than 39,000 members drawn from the professional and civic communities. It holds annual meetings and has a broad list of publications ranging from its monthly *Planning Magazine* to subscriber-based technical reports such as the *Planning Advisory Service*. It sponsors forty-seven local chapters and twenty-one subject-oriented divisions, each pursuing specialized interests within the field. The APA's professional arm, the American Institute of Certified Planners (AICP), has 14,000 members admitted via examination and recognizes distinguished professionals through admission to its College of Fellows. The AICP oversees the professional code of ethics and certification maintenance programs to ensure the integrity and competence of its members. The Association of Collegiate Schools of Planning (ACSP), whose membership encompasses North American universities offering training in urban and regional planning, focuses on educational matters. Together the APA, AICP, and ACSP sponsor the Planning Accreditation Board (PAB), charged with administering the profession's accreditation program. The PAB has accredited more than seventy programs through a rigorous review that ensures minimum standards.

As planners acquire expertise, journals, books, technical newsletters, and other forms of media transmit that knowledge. The field's two journals of record – the APA's *Journal of the American Planning Association* (dating from 1915) and the ACSP's *Journal of Planning Education and Research* (dating from 1981) – play an important role in disseminating knowledge. The American Planning Association publishes books through its Planners Press, and other trade publishers are active, including Island Press, John Wiley and McGraw-Hill. Increasingly, the internet has become a vehicle for communicating new information. Educational programs including those that award undergraduate and postgraduate degrees provide a formal channel for passing on the knowledge and ethics of the field.

One of the greatest challenges of any profession is defining the "public interest," or common well-being. This is certainly the case of urban and regional planning, a field that deals with competing uses for land, a field that balances private property rights and public needs, a field that mediates among varied local (block, neighborhood, district) and broader (citywide, regional) interests.

Planners have long discussed how the public interest plays out in their field. They start with definitional questions. Is the public interest "what is good for everybody," or is it "the sum of individual preferences"? In the good-for-everybody scenario, government fiats or deliberative processes determine the course of planning. In the sum-of-preferences model, the market becomes the arbiter. Regardless of the definitional choices – which planning theorists have been reluctant to settle – the field has attempted to deal with the idea of public interest by developing the planning process, focusing on making decision making participatory and transparent, and giving voice to the multiple interests that exist in any planning arena. For a taste of how the definition of public interest has changed over time, see Martin Meyerson and Edward C. Banfield's *Politics, Planning and the Public Interest* (New York: Free Press, 1955) and Heather Campbell and Robert Marshall's "Utilitarianism's Bad Breath? A Reevaluation of the Public Interest Justification for Planning" in *Planning Theory* (1: 2, 2002, pp. 163–87).

In addition to serving the public interest, however defined, professionals in many fields also refer to the application of their expertise as plying the "art and science" of (name of field). In other words, professionals not only possess certain expertise, they exercise judgment, vision, or wisdom in applying it. Urban planners have routinely used the art-and-science phrase in describing their work. For example, they placed it in the mission statement and by-laws associated with their professional organizations, variously named the American Institute of City Planners (1917–38), American Institute of Planners (1938–78), and American Planning Association (1978 to the present). Amended two times since its codification in the 1930s, the mission statement has transposed "art" and "science." In the early version, science preceded art, and in the later versions art came first and remained there. In the most recent version of the APA mission statement the phrase has vanished entirely but surfaces in the AICP description with art preceding science.

This flipflopping reveals practitioners' interest in assuring the public that planning is indeed a profession based on scientifically derived expertise. This concern emerged in the initial organizing phase, as proponents of planning worked to distinguish it as a profession separate from its forbears – originally architecture, landscape architecture, civil engineering, real estate, and law; later, public administration and management and the social sciences; and still later, systems analysis, geospatial technologies, and communications, including consensus building, conflict negotiation, and dispute resolution. Each of these disciplines has contributed to the specialized knowledge of urban and regional planning. And blending and balancing the various elements has called upon the creative or "arts" side of the profession.

This essay traces the evolution of the field's expertise, especially that part regarded as art. It is one of three essays commissioned at the turn of the twentieth century by Bruce Stiftel, then president of the Association of Collegiate Schools of Planning – the other two were Lewis Hopkins's "Planning as a Science: Engaging Disagreement" and Linda Dalton's "Weaving the Fabric of Planning as Education." The triad, with a set of comments by leading planning educators, was published in the *Journal of Planning Education and Research* in 2001.

Eugénie L. Birch, the author, is the Lawrence C. Nussdorf Professor of Urban Research and Education in the Department of City and Regional Planning, School of Design, University of Pennsylvania, and founding codirector of the Penn Institute for Urban Research. She is the editor of this *Reader*. She has published widely on the planning profession – its history and its expertise – and on housing. In 2008, she completed the survey *Who Lives Downtown* for the Brookings Institution's Metropolitan Policy Program as part of a longitudinal study of forty-five U.S. cities. She is working with Gary Hack on the eighth edition of *The Practice of Planning*, known as the "Green Book," the handbook copublished by the International City Management and the American Planning Association.

Beyond the Green Books and other earlier writings like the Harvard Series in City Planning, publications advancing the expertise of urban and regional planning have proliferated in the past few decades. Notable works are: Larz Anderson's *Planning the Built Environment* (Chicago: APA Planners Press), the American Planning Association's *Planning and Urban Design Standards* (New York: Wiley, 2006), Alexander Garvin's *The American City: What Works, What Doesn't* (New York: McGraw-Hill, 2002), and, from Lloyd Rodwin and Bishwapriya Sanyal (eds), *The Profession of City Planning: Changes, Images, and Challenges, 1950–2000* (New Brunswick, NJ: Center for Urban Policy Research, 2000).

This article probes the meaning of the phrase "art of planning" as envisioned by its practitioners – those who work in the field and those who teach and research in academia. . . . Throughout the ninety-year life of the planning profession in the United States, these two groups have pushed the field forward to encompass current and evolving concerns, addressing them from their respective positions. . . . Over time, they have generated new knowledge through their practices. . . .

This article employs three meanings for "art," ones drawn from the term's dictionary definition. The first is "production or expression according to aesthetic principles" or, in shorthand . . . , *design*. In planning, this ranges from the physical planning or urban design involved in the arrangement of land and buildings to the creation of visions for ideal communities. The second is "the principles or methods governing any *craft* or branch of learning." This covers planning techniques and includes the understanding of legal, quantitative, social science, geographic, or other substantive concepts pertaining to the interests of the field. The third is "exceptional skill in conducting any human activity" or *presentation*. This is an allusion to the planner's personal attributes or skills and refers to everything from the ability to oversee the planning process, explain and represent the planning product (plans, implementation programs), and develop the judgment, discretion, practical reasoning, calculation, and prudence in carrying out planning. Over time, planners have given differing interpretations to these facets.

In exploring the changing meanings of the art of planning, the article . . . looks at art as it has appeared in planning literature in the past sixty years. . . . [It] will examine quasi-official offerings: first, focusing on the early volumes of the Harvard City Planning Series (eighteen books published from 1933 to 1973 and edited by Henry Vincent Hubbard, the "father" of planning education); and second, examining the so-called "Green Books" (handbooks published about every ten years from 1941 to the present by the International City/County Management Association [and later jointly with the American Planning Association], whose successive versions encompassed the growing knowledge and techniques of the field. . . .

1930S TO 1960S: SCIENCE AND ART OF PLANNING

[. . .]

[T]he field had been evolving from 1917, when twenty-four men founded the . . . American Institute of City Planning. Originally dominated by private consultants, largely dedicated to producing municipal master plans and their implementing documents, by 1938 the profession numbered about one hundred. That year the American Institute of City Planning members deleted the word "city" from the organization's title, making it American Institute of Planners (AIP). By then, it had adherents who were likely to be public servants involved not only in local affairs but also in national land-based programs relating to housing, resource management, and settlement patterns. The massive New Deal efforts designed to address widespread Depression-caused unemployment through infrastructure construction had opened up many new avenues for planners. For example, they participated in determining dwelling-unit and site-plan standards for the Federal Housing Administration, land arrangements for the Greenbelt Towns, the nature and type of databases for the National Resources Board, and regional and town site development specifications for the Tennessee Valley Authority.

In this work, they drew heavily from design, engineering, and law and also incorporated knowledge from other areas, especially the social sciences, for political science, sociology and economics flourished in the 1930s. For example, they drew from the Chicago School of Sociology, which was at its height in explaining the fundamentals of urbanism, and from Keynesian economics, whose "pump priming" principles guided New Deal policy.

While under these conditions, the design and craft facets of the field expanded dramatically, the presentation approaches basically remained unchanged from the earliest days of expert-driven diagnosis and prescription accompanied by the graphic and textual material of planning reports. As planners celebrated their mastery of the technical aspects of the field they viewed themselves as apolitical and neutral. . . . For them, the planning processes took place in their offices or on site with little or no participation of the people for whom they were planning. They presented their output, paper master plans, site plans, and regulatory programs ready for implementation to their respective audiences as finished products.

[. . .]

To codify their expertise, planners produced textbooks and monographs capturing the art (design, craft, and presentation) of planning from their own experiences. . . . [P]ublished under the aegis of the Harvard City Planning Series, their art was land-based and aimed to advance new ideas about site planning, traffic management, and zoning. For example, in *Urban Land Uses* (1932), Harland Bartholomew, principal of one of the nation's largest consulting firms, collected data on twenty-two cities to suggest how to distribute uses within a zoning ordinance. Ever since *Euclid v. Ambler* upheld municipal jurisdiction in zoning in 1926, American cities adopted this powerful instrument with great enthusiasm but limited information. Until the publication of Bartholomew's book, there was little empirical knowledge about the relationship among land uses, much less how to allocate them for a whole city. And in *Design of Residential Areas* (1934), Thomas Adams, executive director of the Regional Plan Association of New York, demonstrated the latest in housing design and site planning, espousing the benefits of large-scale settlements planned to incorporate not only the dwelling units but also the arrangement of open space and streets in the so-called "neighborhood unit." His vision incorporated a comprehensive and unified approach.

While planners did not explicitly write about the presentation techniques needed to succeed in their field, they did assemble lists of exemplary planning products. When Henry-Vincent Hubbard and his wife, Theodora Kimball Hubbard, a Harvard librarian, assessed professional advances in *Our Cities, Today and Tomorrow: A Study of Planning and Zoning Progress in the United States* (1929), they demonstrated how planners defined their evolving product and, by implication, their presentation. They started with City Beautiful plans, such as Daniel Burnham's *Plan for Chicago* (1909), showing their systems for transportation and open space and the design of civic centers, and moved to City Efficient plans, such as the *Official Plan of the City of Cincinnati* (1925), highlighting their more fine-grained analysis of the traditional areas and addition of land use analysis, zoning recommendations, and capital budgeting proposals. They argued that, over time, the planners' skills encompassed not merely the ability to offer graphic and textual views but also included the fusion of assembled data and information into a unified approach that allowed the many parts to function efficiently and in a complementary fashion.

In the early 1940s, the Institute for Training in Municipal Administration, an ICMA subsidiary, asked Cincinnati planner Ladislas Segoe to develop instructional materials for the field. The result was *Local Planning Administration* (later known as the Green Book, which denoted the color of its cover). As with the writers for the Harvard City Planning Series, the author, who had a renowned planning practice, wrote from experience. Later cited as a "godsend" and "undoubtedly the most influential planning book in the United States during the first half of the twentieth century," this state-of-the-art compilation was really a looseleaf notebook containing almost 700 single-sided mimeographed pages, maps, and tables. It described the role of planning in local government, the studies required for a city plan, the contents of such a plan, implementation devices, and planning agency administration. Above all, it perpetuated the idea of the planner as a technical expert who diagnosed his or her clients' ills and prescribed their solutions.

Fueling the development of the art of planning from the 1940s to the 1960s were academic leaders from the small number of universities that offered degrees in city planning. Among them were Frederick J. Adams, head of MIT's Department of City and Regional Planning; G. Holmes Perkins, chairman of Harvard's Department of Regional Planning and later dean of Penn's Graduate School of Fine Arts; and Walter Blucher, executive director of the American Society of Planning Officials (ASPO) and first director of the University of Chicago's Planning Institute. Working on AIP's successive education committees, they fleshed out art in curriculum development in their own institutions.

[M]ost programs lodged with sister disciplines, architecture and landscape architecture, in schools of design firmly dominated by practitioner outlooks. In this environment, design, craft, and presentation focused narrowly on the physical, land-based and applied issues specified in the Green Book. Although students had classes tempered by the social sciences, they spent most of their time in studios. Their instructors engaged extensively in field practice in addition to teaching.

[. . .]

Within a decade of its first issuance, the Green Book was obsolete. So ICMA reissued hardbound texts . . . in 1948 and 1959 to cover the old and new material. In their thirteen chapters, 337 and 467 pages, respectively, the new Green Books covered the traditional topics of the earlier version and included a new concern: urban renewal. Howard Menhinick, director of the Department of Regional Studies, Tennessee Valley Authority, and a faculty member of Harvard's School of Planning from 1929 to 1937, singlehandedly put together the 1948 edition. But by 1958, the next Green Book editor, Mary McLean, director of research for ASPO, called in twelve authors, about three-quarters drawn from practice, to help with the update.

On the whole, these volumes expanded the design and craft aspects of the field but reiterated the presentation formats of the past. This is best illustrated in the discussions of slums and blighted areas that emerged in the late 1940s and continued into the 1950s. While both editions acknowledged the complexities of the surrounding social conditions, they asserted that the planner had expertise only in the physical environment and could lend this skill to multidisciplinary teams attacking obsolete land uses. In the 1948 edition, Menhinick appended four pages devoted to urban redevelopment to a much longer chapter on district planning. This small section focused primarily on land assembly and administrative structure and speculated about potential issues revolving around determining property values, land tenure arrangements, reuse of condemned land, population displacement, and relocation.

By 1959, urban renewal merited an entire chapter and a dominant place in the book. With thirty-nine pages on the subject, author Marion Massen, a staff member of the National Association of Housing and Renewal Officials (NAHRO), defined it in terms of design and craft, focusing on physical layouts, legislative requirements, model administrative structures, examples of blight analysis studies, and before/after photographs. She discussed redesign and density patterns, tie-ins with highway planning, and the need for a metropolitan approach to redevelopment.

THE ART OF PLANNING, SIXTIES-STYLE

Just as New Deal activities had prompted the examination of the profession in 1938, the implementation of postwar programs would cause heavy soul searching in the 1960s. . . . The Green Book barely kept up to date in its two revisions. The implementation of the Housing and Slum Clearance Act (1949) and its major amendments (1954, 1961, 1968), the Federal-Aid Highway Act (1956), the Economic Opportunity Act (1965), and the Demonstration Cities Act (1966) began to transform American settlement patterns and approaches to land development. For example, combined with existing market tendencies toward decentralization, the highway legislation contributed to the dominance of the suburbs. (By 1970, more people would live in these low-density settlements than in cities or rural areas.) The execution of slum clearance and urban renewal programs turned out to be more complicated than the mere updating of land uses originally envisioned by policy makers. Regional demographic shifts, industrial restructuring, and "edge city" growth, all having early twentieth-century roots, also began to blossom in the 1960s.

Finally, heightened consciousness of racial inequality, peaking with publication of the *Report of the National Advisory Commission on Civil Disorders* (1968), bitter divisions revolving around the U.S. involvement in the war in Vietnam, and discovery of the negative environmental impacts of large public works, especially in transportation and urban renewal projects, fueled the planners' professional reassessments that took place in a turbulent nation struggling with civil disorder and a massive loss of confidence in the federal government. Moreover, with their focus on city and suburb and their claimed expertise in urban structure and land use, planners were often at the heart of the debates on domestic issues. They had contributed not only the intellectual capital for many of the era's policies and practical techniques for their implementation but also had generated extended critiques of these approaches as they evaluated the programs.

[. . .]

In this period, education in planning was also changing. It was no longer the sole province of a few schools. By the end of the 1960s, more than seventy universities offered advanced degrees in planning, usually a Master's of City Planning.

[. . .]

By the late 1960s, the profession called for a new Green Book. Its editors, William I. Goodman and Eric C. Freund, educators at the University of Illinois who both had extensive infield practice experience, entirely recast *Local Planning Administration* . . . [and changed] the title to the loftier *Principles and Practice of Urban Planning.* In assembling the volume, they amplified the definitions of the art of planning, especially in the craft and presentation areas. At the same time, they expressed a new level of tentativeness about the profession, probably responding to the criticisms it began to weather in this period:

> The urban environment obviously means many things to different people. It is organic, metropolitan, overcrowded, value-laden, pluralistic, and diverse. . . Planning and urban management, relatively young professions, have an exciting future at a time when change is a way of life.

Despite the caution, this 620 page tome was a third larger than its predecessor and had twenty chapters contributed by twenty-four authors, 70 percent drawn from field practitioners. While it elaborated the traditional topics of the previous edition more fully – urban renewal still had its own chapter, now situated at the end of the book, pushed out by other interests – it had new entries on systems analysis and planning and the public. Where earlier versions had a single chapter on planning methods, this one had nine. The editors gave increased attention to methods as applied to the traditional functional areas, including city design, land use, and transportation, and added others such as social welfare planning and computer-based planning information systems. In two new chapters, one recognizing the contemporary advocacy planning and the other identifying aspects of today's communicative planning, the editors also showed a different sensibility to presentation, inching toward recognizing the impossibility of the planner's being a neutral technician.

[. . .]

THE ART AND SCIENCE OF PLANNING IN THE SEVENTIES

Interest in the field of urbanism ran high throughout the 1970s, especially after President Lyndon Johnson signed legislation creating the cabinet-level Department of Housing and Urban Development in 1965. The federal presence in cities, redefined by President Richard M. Nixon in his 1974 creation of block grant funding, remained important. For example, in the mid 1970s intergovernmental transfers, primarily federal aid, generated 30 percent to 65 percent of the budget revenues in the nation's largest cities. And of particular note was the emergence of environmental planning, fostered by the passage of the National Environmental Policy Act (NEPA) in 1969 and subsequent legislation regarding water and air pollution. In other areas, revisions to the Federal Highway Act recognized mass transit, and passage of favorable local, state, and federal taxes and regulations boosted historic preservation and affordable housing production. Practitioners, both in the field and in academia, led many initiatives in these fields.

There was an outpouring of literature stretching from planning theory to suburban sprawl. The authors were not always planners, but the topics certainly informed practitioners' thinking about the art of planning.

[. . .]

Concurrently, university programs proliferated. . . . By the end of the 1980s, about eighty schools had degree offerings. The demand for professors to fill teaching positions joined with the rigor of university hiring policies led many to appoint Ph.D.-bearing scholars who had little or no academic training in planning or experience in field practice. Drawn heavily from the social sciences, these scholars applied their disciplines to planning. Trained in scientific methods, they tended to be skeptical of professions and turned their interpretations of the art of planning to developing planning theories and evaluative studies of current and past planning activities. In general, they did not contribute to technically based fieldwork. In fact, the academy began to reject applied work as a legitimate scholarly activity.

[. . .]

In this environment, the ICMA and the APA issued the fifth Green Book, with its editors' actually needing a two-volume set to cover the material. The editors, Frank So, APA deputy director, and Judith Getzels, APA director of research, removed the word "principles" from the title, calling it *The Practice of Local Government Planning* (So *et al.* 1979). (Seven years later, So, Hand, and McDowell produced *The Practice of State and Regional Planning*, 1986.) Shaped and shaken by a decade of knowledge creation and criticism, this 676 page volume had twenty-one chapters contributed by thirty authors, three-quarters of whom were field practitioners. In this version, however, the editors used two chapters, "Values of the Planner" and "Historical Development of American City Planning," to advance the idea that the planner, contrary to decades of prior teaching, was not a neutral technician but a professional who was influenced by personal values and the context of the times. . . . [T]hey had chapters on the enlarged domain of planning, pointing to some new directions in "Planning for the Arts" and "Planning Educational Services" and forging the area of economic development in a three-part, thirty-seven-page chapter. Remarkably, they did not offer a chapter on environmental planning despite the passage of the NEPA ten years earlier, but folded reference to the subject in a half-page statement at the end of "Utilities Services." Finally, they reflected current thinking on urban renewal/redevelopment by subsuming the topic in three chapters, "Maintenance and Renewal of Central Cities," "Planning for Diverse Human Needs," and "Planning for Urban Housing."

THE ART OF PLANNING IN THE EIGHTIES AND NINETIES

. . . [I]n the 1980s and 1980s . . . the literature continued to grow.

The degree programs, now numbering about 100, continued to focus on master's and doctoral education.

[. . .]

Publication of two editions of the Green Book bracketed the next two decades. Frank So and Judith Getzels remained as editors for the sixth edition (1988), a reduced and reshaped version of the 1979 volume. Cutting it by about 20 percent, they organized the material into sixteen chapters (down from twenty-one) written by nineteen authors. This time, they began to rely more heavily on academics for content. (In former versions, field practitioners had comprised at least 75 percent of the contributors; here, they comprised just more than 60 percent, with academic practitioners constituting the remainder.) And they focused the material on a narrower definition of the art of planning. For the core of the book, they compressed methodology and policy discussions into nine chapters tied to the functional specializations, including a new one on environmental planning, and others on urban design, transportation, and economic development. They put social policy interests in two sharply defined chapters: "Social Aspects of Physical Planning" and "Planning for Housing." Finally, they reshaped the "Values of the Planner" chapter, adding short sections on ethics and negotiation and conflict resolution. In fall 2000, ICMA and APA issued the seventh Green Book, now having three editors – two academics, Charles J. Hoch, University of Illinois, Chicago Circle, and Linda C. Dalton, California Polytechnic University, San Luis Obispo; and one field practitioner, Frank S. So, executive director, APA. Having 497 pages (down from its 554 page predecessor) with seventeen chapters, its thirty-two authors were predominantly academic practitioners. This slant affected the content as the editors added contemporary findings to time-honored topics and opened up new areas hitherto uncovered. Regrettably, they eliminated the planning history chapter. But they did add a much-needed description of the state of the field to date. They included a sophisticated planning theory chapter and expanded the planning information section. They solidified the position of environmental planning with three chapters on the subject and focused on urban design; land use; transportation; and environmental, housing, and community/economic development as the core functional areas. They repeated the coverage of the traditional implementation strategies of land use regulation and budgeting and finance. They gave new emphasis to presentation with two chapters, "Building Consensus" and "Communities, Organizations, Politics and Ethics," that delved more deeply than ever before into the new ways that planners participated in the planning process.

THE ART OF PLANNING IN THE MILLENNIUM

As this account has related, the art of planning, as seen in the . . . successive editions of the Green Book, is composed of three elements: design, craft, and presentation. While the markers (Green Books) used in this discussion are reactive, consensus-driven documents that compress, codify, or summarize what their authors, practitioners of planning, consider important, they derive from the continuing flow of scholarly literature and field experience. Over time, planners have amplified or modified the definition of the three facets of the art of planning, adding new areas of expertise and redefining their role in exercising this knowledge. A singular result of the changes has been the production of an entirely different kind of planner than in the past. What planners do and how they do it has moved from "neutral" technicians who prescribe general plans or calculate specifications for projects for municipalities to experts, cognizant of their own and others' value systems, who blend design, craft, and presentation in numerous roles, often based on considerations of land and place, but enriched with knowledge and skills emanating from a variety of disciplines. For example, planners now practice in areas that include public development with its blending of design, finance, and negotiation; community and economic development with its expertise in negotiation, business, and housing and community organization; environmental planning and growth management with its unity of land use, infrastructure investment, and regulation; transportation planning with its understanding of land use, finance, and travel behavior. Today, planners are strategists, regulators, program evaluators, and project managers.

[. . .]

Furthermore, their serious, well intentioned self-examination has yielded a profession that self-corrects often, adapting to change gracefully. Sometimes this quality has created a level of self-consciousness that has been self-defeating. But on the whole, its practitioners – both in the field and in the academy – share a set of common values as reflected in the . . . Green Book contents.

[. . .]

These are a few ideas, designed to stimulate discussion and action. As practitioners of this generation develop the art – design, craft, and presentation – of planning for the next century, they should make such progress that when a new intellectual history of planning appears at the brink of the next millennium, its author will refute Sir Peter Hall's assertion (1988, 1989) that the field's most significant ideas emerged in the last half of the nineteenth and the first decades of the twentieth centuries. Hopefully, when other observers examine the profession's mission statements and Green Books (or whatever markers they choose), they will find proof that the late twentieth and early twenty-first centuries made up a most productive era for advancing the art of planning.

10

The Shapes of Suburbia

From *Building Suburbia: Green Fields and Urban Growth, 1820–2000* (2003)

Dolores Hayden

Editor's introduction

Another way to look at the suburb is through its long history, one dating to the Romans, surfacing in the Middle Ages, and blossoming in the industrial age with the emergence of the modern Anglo-American suburbs in the late eighteenth-century in England and the early nineteenth century in the United States. As industrialization created economically prosperous classes and supported transportation advances that enabled some to flee to suburban, single-family homes, designers and home economists articulated what they envisioned as the ideal environment. For example, landscape architect Andrew Jackson Dowling published *Cottage Residences* (New York: Putnam, 1842), a pattern book of model houses, and educator Catherine Beecher composed *A Treatise on Domestic Economy for the Use of Young Ladies at Home and at School* (Boston, MA: Thomas Webb, 1842) delineating gender roles for members of suburban households. Together, they laid the foundation for the American Dream (although that term was not coined until the 1930s), which stood for a house, a wife or husband, two children, and financial security – the rewards of hard work aspired to by middle-class families and a vast number of immigrants who came to the United States in search of the good life. In the twentieth century, mass production of housing, the invention of the car, and the introduction of cheap mortgages expanded suburban possibilities to an ever-widening group.

From the mid nineteenth century onward, designers and, later, urban and regional planners crafted suburban plans reflective of the needs of their times. Their designs had several characteristics in common: curvilinear streets in parklike settings, communal open space, a predominance of single-family housing (sometimes clustered), and, in the early suburbs, a town center and a rail or rapid-transit connection to a nearby center city. This selection presents seven notable model plans: Riverside, Illinois (1868), by Central Park designer Frederick Law Olmsted; Shaker Heights, Ohio (1912), underwritten by the Van Sweringen brothers, the Midwestern railway barons; Radburn, New Jersey (1929), designed as a "Town for the Motor Age" by a New York-based team headed by Clarence Stein and Henry Wright; Reston, Virginia (1963), and Columbia, Maryland (1964), founded by real-estate developers Robert E. Simon and James Rouse, respectively; Levittown (1951), undertaken by the Levitt brothers, who learned the prefab construction trade by building wartime housing; Irvine, California (1971), by Los Angeles architect William Periera; and Celebration, Florida (1994), a "New Urbanist" conception of New York firms Robert A. M. Stern and Cooper Robertson.

The majority of American suburbs did not hew to the models listed above. Financiers, local government leaders, federal mortgage insurance administrators, and developers had their reasons for pursuing the fragmented land settlement patterns that have been the norm. (See Plate 1.12.) Primary among the reasons for rejecting large-scale planned communities were their high upfront costs, especially the infrastructure expenses for roads, town centers, and communal open space. The typical developer simply did not have the resources to make these investments, nor did the local governments that had jurisdiction over development within their boundaries. Municipal zoning codes – laws regulating the use, and intensity of use, of land – militated against plans that called for mixing homes and stores or clustering housing. The features of the model plans that did survive were the curvilinear streets and single-family dwellings.

With its construction of 47,000 miles of interstate highways and support of Federal Housing Administration mortgage insurance and other financing activities, the federal government played an enormous role in spurring suburban development.

Dolores Hayden, Professor of Architecture, Urbanism, and American Studies at Yale University, is the author of this piece. Her writings include *The Power of Place: Urban Landscapes as Public History* (Cambridge, MA: MIT Press, 1995), *The Grand Domestic Revolution: A History of Feminist Designs for American Homes, Neighborhoods, and Cities* (Cambridge, MA: MIT Press, 1981), *Redesigning the American Dream: Gender, Housing, and Family Life* (New York: Norton, 1985, revised and expanded 2002), and *A Field Guide to Sprawl* (New York: Norton, 2004).

The suburbs are enormously interesting to a number of writers, including historians, social scientists, urban and regional planners, and journalists. Kenneth Jackson's *Crabgrass Frontier: The Suburbanization of the United States* (New York: Oxford University Press, 1985) provides a general survey and has a brilliant section describing the Levitt brothers' construction methods. Robert Fishman's *Bourgeois Utopias: The Rise and Fall of Suburbia* (New York: Basic Books, 1987) tracks British and American suburban development. Sam Bass Warner's *Streetcar Suburbs: The Process of Growth in Boston, 1870–1900* (Cambridge, MA: Harvard University Press, 1962), Marc Weiss's *The Rise of the Community Builders: The American Real Estate Industry and Urban Land Planning* (New York: Columbia University Press, 1987), and Patricia Burgess, *Planning for the Private Interest: Land Use Controls and Residential Patterns in Columbus, Ohio, 1900–1970* (Columbus, OH: Ohio State University Press, 1994) focus on the local mechanics of suburban development. Recent additions to the literature on suburban history are Jon C. Teaford, *The American Suburb: The Basics* (New York: Routledge, 2008), and Kevin M. Kruse and Thomas J. Sugrue (eds), *The New Suburban History* (Chicago: University of Chicago Press, 2006). Two classics document suburban life in the 1950s and 1960s: William H. Whyte's *The Organization Man* (New York: Simon & Schuster, 1956) and Herbert Gans's *The Levittowners: Ways of Life and Politics in a New Suburban Community* (New York: Vintage Press, 1967). An account of a more recent suburban development, Celebration, Florida, is Douglas Frantz and Catherine Collins's *Celebration, U.S.A.: Living in Disney's Brave New Town* (New York: Holt, 1999).

Flying across the United States, airline passengers look down on dazzling, varied topography, yet from Connecticut to California, monotonous tracts of single-family houses stretch for miles outside the downtowns of major cities. Subdivisions interrupt farms and forests. They crowd up against the granite coast of Maine and push into the foothills of the Rocky Mountains. Next to residential areas lie highways, shopping malls, and office parks. They overwhelm small town centers. More Americans reside in suburban landscapes than in inner cities and rural areas combined, yet few can decode the shapes of these landscapes or define where they begin and end.

Demographers still describe suburbs as "the non-central city parts of metropolitan areas," a negative definition, but suburbia has become the dominant American cultural landscape, the place where most households live and vote. Describing suburbia as a residential landscape would be wrong, however, because suburbs also contain millions of square feet of commercial and industrial space, and their economic growth outstrips older downtowns. Most confusing of all, suburbia is the site of promises, dreams, and fantasies. It is a landscape of the imagination where Americans situate ambitions for upward mobility and economic security, ideals about freedom and private property, and longings for social harmony, and spiritual uplift.

For almost 200 years, Americans of all classes have idealized life in single-family houses with generous yards, while deploring the sprawling metropolitan regions that result from unregulated residential and commercial growth. With no national land use policy in the United States, single-family housing has often driven suburban planning by default. Between 1994 and 2002, real estate completed about 1.5 million new units of housing every year, most of them suburban single-family houses. The production of millions of houses involving massive mortgage subsidies by the federal government, huge expense to individual families, and extraordinary profits for private real estate developers has largely configured Americans' material wealth and indebtedness, as well as shaped American landscapes. The metropolitan building process holds the key to many aspects of American culture, yet few know its social and spatial history.

. . . [An] account of suburbanization since 1820 [explores] how entrepreneurs and residents have transformed fields, meadows, and woods into habitable space. The speed and spatial scale of land development have increased with each decade. In the earliest years of mercantile capitalism, a few suburban entrepreneurs launched isolated experiments in subdividing property and building new communities with the help of family and friends. Some real estate developers and boosters began to work together, forming political alliances called "growth machines." Between 1870 and 1920, at the height of industrial capitalism, developers extended their reach and promoted urban peripheries systematically, often working in partnership with transit owners, utilities, and local government. After the rise of a powerful real estate and construction lobby in 1925, the federal government took a major role largely through tax, banking, and insurance systems in subsidizing private development of residential and commercial property on a national basis. By the mid-1950s, federal tax supports for commercial developers and direct federal support for highways provided incentives for unchecked growth on a scale that earlier entrepreneurs could never have imagined. By the 1980s, state and local governments also frequently supported private commercial development with direct subsidies.

Within the process of building, the forms have changed every few decades, and the history of suburban construction can be understood as seven vernacular patterns. Borderlands began around 1820; picturesque enclaves around 1850; streetcar buildouts around 1870; mail order and self-built suburbs around 1900; mass-produced, urban-sitcom suburbs around 1940; edge nodes around 1960; and rural fringes around 1980. All of these patterns survive in the metropolitan areas of 2003. Many continue to be constructed.

Each pattern is defined by distinctive development practices, building technologies, marketing strategies, architectural preferences and environmental attitudes. Despite some mid twentieth-century claims that suburbia is a classless place, in each era of suburban life, economic class has affected residents' employment options, commuting chokes, lot sizes, and house sizes, as well as favored shapes for houses, porches, and yards. There are working-class, middle-class, and upper-class configurations intertwined with the seven suburban patterns.

Most previous accounts of suburban history have been organized around improvements in transportation technology, and explicitly or implicitly the authors suggest that transportation technology made residential growth inevitable. Categorizing places by commuters' choices – railroad suburb, streetcar suburb, automobile suburb – also leads to a focus on middle-class and upper-class male breadwinners and their housing. In contrast, this book highlights the complex relationships between real estate entrepreneurs and a wide range of suburban residents and workers. It explores the interplay of natural and built environments, considers women's and children's lives as well as men's, discusses working-class houses and yards as well as affluent ones, and explains why suburbia has been of great interest to political lobbyists.

Many different kinds of visual source materials reveal the precise shapes of suburbs, including maps, plans for towns, designs for houses, and photographs of households. The built places themselves provide material evidence . . . documented in both architectural and aerial photography. Low-level, oblique-angle aerial photography is especially useful for capturing the scale of recent developments in relation to older patterns.

THE TRIPLE DREAM AND THE GROWTH MACHINES

The "American dream" is embedded in these seven evolving patterns of suburban development.

Unlike every other affluent civilization, Americans have idealized the house and yard rather than the model neighborhood or ideal town. From the beginning the dream conflated piety and gender-stereotyped "family values." The ideology of female domesticity, developed in the United States during the same era when suburban borderlands were first attracting settlers, elevated the religious significance of woman's work, defined as bearing and rearing children in the strong moral atmosphere of a Protestant home set in a natural landscape. The single-family house was invested with church-like symbols as a sacred space where women's work would win a reward in heaven. Catholic and Jewish immigrants also tied domesticity to religion.

The ideology of female domesticity, popular since the 1845, was wedded to a cult of male home ownership, extended to include working-class men around the 1875. Over the years, developers embellished the religious imagery. In 1921 an editorial writer for the *National Real Estate Journal* told readers that the Garden of Eden was the first subdivision. While Eden also took the fancy of the editors of *American Architect and Building News*, who claimed that Adam and Eve built their home in Short Hills, New Jersey, a perfect town, many more developers have sited their new houses in heaven. An angel with a sword of justice delivered developer Samuel Gross's "home at $10 a month" to a Chicago workman. A *New Yorker* cover showed a new house floating on pink clouds, above a husband, wife, and child ascending into the sky in 1946, holding their blueprints. (The artist, Constantin Alajalov, included one sharp detail: outside the back door of this upper middle-class house, an African-American cook chats with a Fuller Brush salesman.) More recently, heavenly notes were sung by architects Andres Duany and Elizabeth Plater-Zyberk when they announced "The Second Coming of the American Small Town" at Seaside, Florida. A memoir by D. J. Waldie evoked Lakewood, California, as *Holy Land*.

Occasionally, developers have relocated the sales pitch for heaven in the secular landscape of happiness. William F. Chatlos built 1,000 "Happiness Homes" in Williston Park, Nassau County, Long Island, in the 1920s. His three-bedroom Tudor and Dutch Colonial houses occupied an alphabetical grid of streets named for prominent colleges and universities, including Amherst, Brown, and Cornell.

In 1945, advertising copywriters for General Electric promoted purchasing a home as "an adventure in happiness." Listing many electrical appliances, GE told veterans, "It's a promise!" Most Americans want to believe in a "Happiness Home." In the 1990s many flocked to buy houses in "Celebration," a real estate development by the Walt Disney company, previously known for building theme parks advertised as "the happiest place on earth."

The dream of the single-family suburban house implies isolation, lacking physical and social context. The double dream expressed by women is house plus community and neighborhood sociability. Others have proposed a different double dream, a house set in unspoiled nature. The result is a triple dream, house plus land plus community, the kind of neighborhood space represented in Phyllis McGinley's charming quatrain about Larchmont, New York: "I know a village full of bees / and gardens lit by canna torches, / where all the streets are named for trees / and people visit on their porches." This triple dream encompasses both the private and public pleasures of peaceful, small-scale residential neighborhoods.

Developers and builders of suburbs have expressed more specific, monetary aims. For them, suburban real estate has been important for its price and profitability. Conflict has characterized every era of development as green fields have attracted residents to the peaceful outskirts of cities but also drawn promoters. In addition to those directly involved in the building process, the boosters of growth have included lawyers, owners of suburban transportation companies (including ferries, railroads, and streetcars), owners of utilities, and owners of local newspapers, supermarkets, and big-box stores. They have handled house purchases, punched commutation tickets, generated electricity, increased total circulation, marketed cornflakes, and sold screwdrivers. Once a suburban area is established, growth promoters usually seek greater and greater levels of density. The residents' hope of unspoiled nature fails because open land vanishes with increased development. Their hope of community is betrayed when tracts of houses, hyped as ideal "communities," lack social and economic centers, parks, schools and necessary infrastructure. Taxpayers must scramble to supply what residential developers have failed to provide. Commercial developers promote their own proposals, often disrupting the scale of existing neighborhoods even more. Contestation between residents who wish to enjoy suburbia and developers who seek to profit from it lies at the heart of suburban history.

11

Ramapo Plus Thirty: The Changing Role of the Plan in Land Use Regulation

From *Urban Lawyer* (2003)

Edward J. Sullivan and Matthew J. Michel

Editor's introduction

Since urban and regional planners focus on land development, they are attentive to the legal issues of land use. The U.S. legislative and judicial systems have framed land use and property disposition practices especially as they apply to private ownership. State legislatures and courts oversee local government use of the police power (that is, making regulations to protect people's heath, safety, and general welfare), and they distinguish between the police power and takings (depriving a landowner of the use of his or her property). In the former case, the property owner must conform to the law; in the latter, the government must compensate the property owner.

In urban and regional planning, local governments employ the police power for three types of activities: developing comprehensive plans; regulating land use through zoning; and managing the amount, timing, and direction of growth. Usually municipalities exercise these powers, but in some cases regional and state governments do. According to Rolf Pendall, Robert Puentes, and Jonathan Martin in their survey *From Traditional to Reformed: A Review of the Land Use Regulations in the Nation's Fifty largest Metropolitan Areas* (Washington, DC: Brookings Institution Metropolitan Policy Program, 2006), 85 percent of the localities studied had comprehensive plans and 91 percent had zoning ordinances, but far fewer had growth management tools – ranging from 2 percent to 38 percent for different types of controls.

The rapid acceptance of zoning and comprehensive planning is due to a number of phenomena. Zoning dates to 1916, when the first comprehensive ordinance was passed, in New York City. Comprehensive planning was not widely adopted until the 1970s. As explained in this selection, one reason zoning spread so early and speedily was the clarity and rationality of the model law developed by the U.S. Department of Commerce, the Standard State Zoning Enabling Act (SZEA, 1926), contrasted with the more ambiguous Standard City Planning Enabling Act (SCPEA, 1928). The authors of the SZEA and the SCPEA included prominent planner Frederick Law Olmsted, Jr, the heir to his father's design firm; lawyer Edward M. Bassett, the author of the New York Building Zone Resolution, 1916; and lawyer Alfred Bettman of Cincinnati, whose *amicus curiae* brief to the Supreme Court would support the positive decision described below.

Another reason for zoning's quick acceptance is the language of the U.S. Supreme Court decision *Village of Euclid, Ohio v. Ambler Realty Company* 272 US 365 (1926) upholding zoning. In this 1922 case, Euclid, Ohio, a largely undeveloped fourteen-square-mile suburb of Cleveland with a population of 5,000, had adopted a zoning ordinance that mapped the plaintiff's sixty-eight-acre vacant site as residential. Desiring to sell the land for industrial use, the plaintiff complained that the zoning classification deprived him his Fourteenth Amendment rights of due process and equal protection under the law. The Supreme Court, in rejecting his claim, labeled the law "a *comprehensive* zoning *plan*" (emphasis added), outlined how it had taken into account all of the village property in its six use categories, and asserted that "its governing authorities, presumably representing a majority of its inhabitants and voicing their will, have determined, not that industrial development shall cease at its boundaries, but that the course of such development shall proceed within definitely fixed lines." The court's language implied that the zoning ordinance was, in effect, a plan, arrived at through democratic processes, a belief that courts would promulgate for the next three or four decades. Under these conditions, many municipalities saw little need to adopt an official plan.

Nonetheless, urban and regional planners worked to convince municipalities and regions to undertake comprehensive plans, succeeding in a number of places. Leading examples are the *Plan of Cincinnati* (1925) and the *Regional Plan for New York and its Environs* (1929), both produced by private, good-government planning associations. The Cincinnati plan received official recognition when the city's planning commission adopted it. The *Regional Plan* remained an unofficial (but influential) document. In the main, urban and regional plans of this era were voluntarily adopted and advisory with little legal force.

However, after World War II, federal and state governments and the courts began to vest plans with more importance. The Housing Act of 1954 and the Federal-Aid Highway Act of 1962 mandated local planning as a condition of funding. In the 1970s some states – notably California, Florida, and Oregon – required localities to produce plans. Finally, in two influential cases, the federal courts upheld zoning ordinances that contained local growth management devices when they accorded with a comprehensive plan.

In *Construction Industry Association of Sonoma County v. City of Petaluma* (522 F. 2d 897 (9th Cir), cert. denied 424 US 934 [1976]) the court allowed Petaluma to cap housing permits according to its Residential Development Control System because the program conformed to the city's Environmental Design Plan of 1972. In *Golden v. Planning Board of the Town of Ramapo* (30 N.Y. 2d 359, 265 N.E. 2d 291, appeal dismissed, 409 U.S. 1,003 [1972]), the court upheld provisions in the city's zoning ordinance that required a special permit for housing development linked to having adequate public facilities. The existence of a capital improvement plan incorporated into a comprehensive plan convinced the court that the city would issue permits in a rational, judicious manner.

This second decision was a huge victory for the fast-growing suburbs seeking to control expansion through planning. Ramapo was overwhelmed by its rapid population surge – 70 percent in the 1950s and another 119 percent in the following decade (from 21,000 to 77,000 over the twenty-year period). In 1966, it crafted a comprehensive plan outlining future development, and, three years later, adopted the zoning ordinance that became the subject of the court case. In the following decades the regulation successfully slowed growth to 16 percent in the 1970s and 5 percent in the 1980s. But, more important, this case, along with *Petaluma*, gave standing to the comprehensive plan, encouraging a number of other places to require adoption. Today, it is the legal foundation of contemporary growth management strategies.

In addition to tracing the evolution of zoning and planning, this selection discusses the impact of *Ramapo* on state land use law. Its principal author, Edward J. Sullivan, is a partner in the Oregon firm Garvey Schubert Barer, specializing in real estate, land use, and affordable housing and community development law. His most recent publications are "Year Zero: The Aftermath of Measure 37," 36 *Environmental Law* 131 (2006), and "Answered Prayers: The Dilemma of Binding Plans" in Daniel Mandelker (ed.), *Planning Reform in the New Century* (Chicago: APA Planners Press, 2005). Matthew J. Michel is an attorney with Beery Elsner & Hammond, Portland, Oregon, practicing local government law.

For additional reading see "Whatever Happened to Ramapo?" *Planning Magazine*, December 2003, and William Fulton *et al.*, *Growth Management Ballot Measures in California* (Ventura, CA: Solimar Research Group, 2002), both of which discuss the impact of the two cases. Author of the Ramapo regulations Robert H. Freilich argues forcefully for additional strengthening of comprehensive plans in "Smart Growth Planning in the Twenty-first Century" in Daniel Mandelker (ed.), *Planning Reform in the New Century*, cited above. Basic textbooks on land use law and growth management offer more information about the importance of comprehensive plans for contemporary urban and regional planning. Examples are: Julian Conrad Juergensmeyer and Thomas E. Roberts, *Hornbook on Land Use Planning and Development Regulation Law* (St Paul, MN: West Publishing, 2003), and Arthur C. Nelson and James B. Duncan, *Growth Management Principles and Practices* (Chicago: APA Planners Press, 1995).

I INTRODUCTION

The role of the comprehensive ("master" or "city") plan in the land use regulatory process varies by state. Despite finding their common, historical roots grounded in model language, the circumstances and experiences among and within the states vary in the significance each gives to the comprehensive plan within the land use system.

The confusion over the role of and the need for a comprehensive plan arose from Section 3 of the *Standard Zoning Enabling Act* (SZEA), promulgated by the U.S. Department of Commerce in 1926. The SZEA is model legislation for states to use in enabling local government to undertake comprehensive land use regulation. By the end of 1927, over half of the states adopted the SZEA in one form or another. Section 3 of the SZEA requires that zoning be "in accordance with a comprehensive plan." Interestingly, the term "comprehensive plan" is not defined, and that lack of definition has caused continual confusion for local planners and the courts. . . .

Adoption by over forty states of a companion piece of model legislation to the SZEA, the *Standard City Planning Enabling Act* of 1928 (SPEA), planted further seeds of confusion. The SPEA Model Code affords local governments the option to preserve future rights of way with a "city" or "master" plan, but does not use the term "comprehensive plan" found in Section 3 of the SZEA. This raises the question of whether the Acts are functional equivalents or distinct land use documents. Further, the SPEA bypasses state-level planning and delegates land use regulation authority to local governments. This calls into question whether a plan constrains or precedes zoning regulations.

Due to this lack of clear instruction, America's state capitals and city halls have questioned the need for a plan at all. Some accepted the notion of adopting a long-term planning document as support for a zoning scheme with varying degrees of enthusiasm. Still others embraced a master plan wholeheartedly and in blind faith. Because it was easier and permissible to adopt precise land use regulations rather than to adopt a future land use planning vision, most local governments undertook zoning regulation schemes unconstrained by any long-term planning. Planning was innocuous and, to the extent that long-term planning was provided, it often gathered dust on municipal office shelves or languished in broad, nonbinding general aspirations.

In addition, state courts struggled with SZEA/SPEA language when applying planning laws incongruous with zoning ordinances. Most local governments either did not adopt a plan before undertaking zoning regulations or zoned (or more likely rezoned) land inconsistent with the adopted plan. State court judges were then faced with invalidation of zoning or finding another way to justify local action. Often this resulted in upholding zoning through conflation of the "comprehensive plan" language with the existing zoning map. Too often, this view resulted in judges accepting any argument of consistency with a land use plan whether or not a land use plan existed. Often the plan described the mere apparition of consistency with a land use plan, revealing itself to the judges examining the challenged regulation if to no one else.

This approach was rejected in the case of *Golden v. Planning Board of the Town of Ramapo*, a watershed case in which the New York Court of Appeals upheld a local government's reliance on the comprehensive plan as law rather than merely aspirational goals as a basis for planning. Before *Ramapo*, and beginning in the 1920s, local-level planning occurred in only a limited number of circumstances. The long-term planning was used in urban renewal areas for federal grant consideration and, although rarely, by local governments choosing to draft a comprehensive plan. One of the boldest local government planning experiences of the late 1960s took place in Ramapo, New York, and the adoption of its comprehensive plan. The role of this comprehensive plan based on the sufficiency of infrastructure was challenged . . . as an unconstitutional taking. Ultimately, the New York Court of Appeals upheld the scheme and the U.S. Supreme Court denied *certiorari* on the takings issue. Ramapo's sequential growth scheme was a novel answer to the confusion over how zoning regulations could be made in accord with a comprehensive plan.

. . . Since *Ramapo*, many state legislatures have revisited the SZEA and SPEA models to make local governments agents for implementing a series of state land use policies or goals that include, among other things, the consideration of the existing public infrastructure. Some states have departed completely from the original model codes, enacting a land use code unique to the state's land use needs but still retaining the long-term planning role of the comprehensive plan. Yet some states continue to use the original SZEA/SPEA model codes and view comprehensive plan policies as ideals rather than legal requirements. This evolving realignment of land use authority replaces the confusion of the comprehensive plan's role in land use planning with state-mandated policies regarding conservation of natural resources, prevention of sprawl and other policies of the day.

To expand on *Ramapo*'s legacy of activating comprehensive planning in the states, this article updates a previous article on the role of the comprehensive plan in land use regulation written in 1975 that described the state court trends in land use decisions applying to a comprehensive plan. In the 1975 article the authors suggested that a state court applies one of three general approaches to interpret the relationship of a plan to land use regulation: Unitary, Planning Factor or Planning Mandate approaches.

By 1975, most state courts adhered to the Unitary approach to applying the SZEA "[zoning] in accordance with a comprehensive plan" language that the plan was either the zoning map or some coherent growth principle existing either within or outside the land use regulations themselves. Despite criticism by some academics, this approach prevailed when a developer or homeowner contested a land use regulation that had no separate land use plan. The courts reviewed a land use regulation for arbitrariness relative to the existing zoning regulations to the exclusion of any planning land use policies. Courts employing a Planning Factor approach gave some legal authority to a plan, but not dispositive authority.

The 1975 article advocated the Planning Mandate approach, i.e., that the plan be given dispositive weight as a "quasi-constitutional" document; that it be required as a precondition to and must be consistent with all subsequent land use regulation. Further, the 1975 article predicted that land use policy, quasijudicial evaluation of plan conformity to land use regulations, and issues of plan interpretation.

. . . Before *Ramapo*, New York law was based on the SZEA/SPEA model: expressly allowing population density regulations to limit development but not planning based on the "'sequen[ce]' or 'timing'" of accompanying public infrastructure. Up to this point, the New York courts had been applying a unitary approach to evaluate a land use action, placing little weight on any planning standards found in local regulations as a basis for local action.

However, the New York Court of Appeals decision in *Ramapo* elevated land use planning above mere regulations to a separate and independent factor by which to measure and evaluate land use regulations. The flexible time restrictions for development within Ramapo's eighteen-year planning period were key in the court's decision to accept Ramapo's plan. Developers could either provide the infrastructure or wait for the town to build infrastructure; thus, the restrictions did not foreclose development. . . . Sustainability based on an evaluation of and sufficiency of the public infrastructure was viewed as implicit as part of directing growth for the benefit of local residents. Rejecting the Unitary approach, the court in *Ramapo* adopted the Planning Factor approach to analyze the plaintiff's taking claim. Ramapo's plan for sequential development added a new temporal element to federal takings law by articulating what was a reasonable use over a reasonable period of time. The comprehensive plan provided a legal basis for reasonably delayed development. . . . [The] deliberate implementation strategy connecting infrastructure planning to land use regulation took advantage of the existing comprehensive plan requirement, and thereby elevated the plan to a legal requirement despite the lack of guidance found in the vague and incongruous SZEA/SPEA statutory framework.

This emphasis on quality planning based on available public infrastructure was an exciting new development in planning law that revitalized the field and suggested that a planning vision could be enforced by land use regulations. . . .

The comprehensive plan's variety of roles in land use regulation is the result of land use legislation and a diaspora in statutory interpretation of the SZEA/SPEA models. The progress of the various states' legislatures in amending their comprehensive plan requirements and courts in recognizing the hierarchical relationship of planning to land use regulation through Unitary, Planning Factor and Planning Mandate [is highlighted by] . . . the *Ramapo* decision [that] sets the stage as a landmark transition in the comprehensive plan as law from a Unitary to a Planning Factor approach taken by the various state courts when interpreting the SZEA/SPEA language. . . . [T]he Unitary approach is slowly being replaced with Planning Factor and Planning Mandate approaches. This trend gives a greater role to the plan as a factor when evaluating the validity of a land use action, and in some cases the plan can be a dispositive factor. One of the principal reasons for providing the comprehensive plan with such respect in recent years is the position taken by the American Planning Association in its Smart Growth project.

[. . .]

II THE CART BEFORE THE HORSE: A UNITARY APPROACH

[. . .]

The historical accident of zoning occurring before planning in land use law put the cart before the horse; planning should drive zoning to create coherent land development. A unitary zoning and planning doctrine makes no improvement on land use planning to withstand the political winds of asserted property rights and local politics. Further, treating the existing zoning regulations as a comprehensive plan *in toto* makes for a single, and elliptic, standard of proof for a municipality's compliance with its comprehensive plan, to wit: a proposed zoning regulation is valid if it complies with the existing zoning scheme. This deferential standard puts zoning ahead of planning because local land use decisions need only comply with existing zoning regulations, virtually eliminating the need or rationale for any long-term planning requirement.

[. . .]

With the several states beginning from the SZEA/SPEA models, the relationship between zoning and land use planning could be easily tolled among the states by observing whether a state's code created a separate plan, and also noting the legal significance the plan garnered in challenges to zoning decisions. Overall, the number of states under the Unitary approach has declined, and the Planning Factor approach appears to have absorbed those states moving away from the Unitary approach. While the courts may provide some explanation for this migration in legal significance of the comprehensive plan, there is an undercurrent of legislative change to the original SPEA adopted by the states. These amendments to the original SPEA model provide interesting variations of how the comprehensive plan has a varying definition and authority based on a distinction between zoning and planning.

[. . .]

The most active category of states is the Planning Factor category. The emergence of the Planning Factor approach appears to be driven equally by judicial interpretation and legislative change. At one end of the Planning Factor spectrum are states maintaining a structural similarity to the original SPEA by defining the comprehensive plan within the existing zoning scheme. At another end are states adopting "smart growth" initiatives. These initiatives are significant variations on the SPEA that explicitly require communities to develop a separate comprehensive plan that is then given legal significance in reviewing zoning decisions.

[. . .]

The *Ramapo* court applied a Planning Factor approach by looking to the master plan for legal justification of land use restriction. New York's land use statutes did not change; the master plan occupied an existing legal standard lost in the confusion of how a proper comprehensive plan could operate. Prior attempts at planning used comprehensive plans as a tool to forestall growth. The Ramapo plan resolved the confusion by showing that a plan manages growth according to reasonable infrastructure development. The Ramapo plan gave new life to the statutory "well considered," comprehensive plan requirement by articulating a reasonable basis for land restrictions. The court recognized the significant energy put into the Ramapo plan that included "a four-volume study of the existing land uses, public facilities, transportation, industry and commerce, housing needs and projected population trends." The master plan's capital planning and improvement program gave tangible and reasonable bases for delaying development.

The role of the master plan as law in New York expanded after the *Ramapo* decision. Ramapo's plan was a document separate from the zoning ordinance;

a legal standard separate from the rationality of the existing zoning ordinance. It showed how planning could be a legal factor in land use law by presenting a reasonable basis for comprehensive regulation . . . The *Ramapo* court applied the plan as a factor that justified controlled growth in the legal analysis of the sequential development scheme. Thus, the shift from a Unitary approach to a Planning Factor approach increased the legal significance of the plan as law.

III CATEGORIZING THE ROLE OF THE COMPREHENSIVE PLAN THROUGH STATE LEGISLATION AND CASE LAW

As Ramapo demonstrated, a separate, comprehensive planning document can be a legal standard for land use actions. New York mandates a plan but it need not be separate from an existing comprehensive zoning ordinance. In the other states, plans may be advisory or mandatory according to state law, but the form that the plan may take remains unsettled among the state experiences. Generally, a master plan separate from a comprehensive zoning scheme provides a legal standard for land use actions which leads the state court to apply a Planning Factor or Planning Mandate approach to land use actions. The most interesting shift in planning law since *Ramapo* has been away from the Unitary category of jurisdictions toward the Planning Factor and Planning Mandate categories.

This shift has been largely guided by state land use planning enabling laws that allow or even compel local governments to engage in the comprehensive planning process. The court may then give that comprehensive plan a role in deciding the reasonableness of the regulation. The comprehensive plan as a separate advisory document has emerged as a modern trend in land use planning. Such a plan introduces a reasonableness and compatibility standard beyond arbitrary and capricious judicial deference.

The American Planning Association's *Growing Smart Legislative Guidebook* provides a nationwide snapshot of each state's legislative amendments to the original enabling language in the SPEA. The *Guidebook* divides states' efforts at land use planning reform into four categories of progressively significant departures from the original SPEA language. A majority of states still retain the original SPEA plan language with slight or moderate variations. In contrast, eleven states have completely abandoned the SPEA language; for example, Oregon created its own unique land use system with enabling language that make planning an integral part of land use regulation.

Table [11.1] charts the states on the legal significance of the comprehensive plan under Unitary,

Table 11.1 Legal and statutory significance of comprehensive plans

JUDICIAL INTERPRETATION					
	Planning mandate			CA, WI	DE, FL, OR, WA
	Planning factor	AK, CO, IA, IL, KS, MT, NC, NE, SD, VA, WY	NV, NY, MN	ID, KY, MA, PA, SC	GA, HI, ME, MD, NJ, RI, VT
	Unitary	AL, IN, LA, MO, ND, NM, OH, OK, TX, WV	AR, CT, MS, MI, UT	AZ, NH	TN
		SPEA	**Moderate, significant changes**	**Many, significant changes**	**Total revision**
	ENABLING LEGISLATION				

Planning Factor, or Planning Mandate approaches against the states statutes implementing planning law as it deviates from the original SPEA. The table provides an easy summary of the comprehensive plan as law. Table 11.1 conveys national trends in planning towards the master plan as law in making land use decisions. First, reading across the table along the "Unitary" row shows that the Unitary approach decreases as states amend and revise their planning laws. Conversely, amending the state planning laws trends toward a more significant role of the plan as law. Further, reading along the horizontal, "Enabling Legislation," axis supports this conclusion. When the SPEA changes in states, moving left to right on Table 11.1, the case law trends reflect an increase in the legal significance of the plan as courts begin applying the Planning Factor and Planning Mandate approaches.

Second, minor changes to the SPEA model language do not necessarily result in a plan receiving greater legal significance upon judicial review, i.e. a state court applying a Unitary approach is equally likely to adopt a Planning Factor approach as not when the state legislature only tinkers with planning law. Conversely, more extensive changes to the language of the SPEA towards courts placing increased legal significance on a master plan. These two trends may be seen by vertically bisecting Table 11.1 between the "Moderate, significant changes" and "Many significant changes" columns. States on the left half have little if any SPEA changes and are generally split between Unitary and Planning Factor approaches. States on the right half have made greater SPEA changes and state courts generally apply a judicial interpretation recognizing the plan as a factor or as law. Therefore, a planning statute may change without a corresponding change in judicial interpretation; the degree of statutory change is a significant indicator of a change in judicial interpretation.

[. . .]

Several states have significantly augmented their planning statute with planning elements and standards while retaining the general statutory framework of the original SPEA. Common statutory requirements added to the original SPEA language include land use, housing, agricultural and open space preservation, transportation, community facilities, implementation guidance, and general policy statements.

Legislative changes to the original SPEA language generally come under the planning rubric of Smart Growth initiatives developed in response to suburban sprawl, diminishing farmland, and affordable housing. Some states have adopted parts of the Smart Growth comprehensive and regional approach while retaining the traditionally local role in land use planning. This empowers municipalities to integrate community planning ahead of zoning to correct the historical error of zoning preceding planning. These many significant changes to state planning laws create the potential for concomitant changes to past judicial interpretation of what is "in accordance" with a comprehensive plan.

[. . .]

IV *RAMAPO'S* LEGACY: THE COMPREHENSIVE PLAN IS LAW IN PLANNING MANDATE STATES

Although the Ramapo plan was conceived in the context of the New York court's unitary approach to planning, the *Ramapo* decision transformed New York into a Planning Factor state, i.e. one in which a plan is a factor in evaluating the validity of zoning regulations. It is doubtful that the scheme would have survived judicial scrutiny had there been no plan or capital improvements program by which the town could have justified the stringent regulations it imposed on land for an eighteen-year period.

[. . .]

It was the constellation of the town's land use and capital improvements plans that enabled it to convince the New York Court of Appeals that it would "provide a balanced cohesive community dedicated to the efficient utilization of land."

Ramapo has led the way in many areas of planning, particularly in dealing with the phasing and timing of growth, as well with the issue of the existence and concurrency of public infrastructure. Some states, like Florida and Washington, require a showing of concurrency at the time of development or within a fixed period thereafter. Others, like Oregon, require planning for future uses and the public facilities and services to support those uses. In each case, the relationship between planning and public services and facilities is a known quantity. All of these efforts may be traced back to *Ramapo*.

[. . .]

V CONCLUSION

Ramapo made the connection between comprehensive planning and the requirement for adequate public facilities and services. The town's land use plan was tied to the provision of roads, drainage facilities, water and sewer lines, and public facilities such as schools and police and fire stations. The vision, for an integrated and serviced community, was guaranteed by the requirements that public services and facilities are either in place or committed to installation by a date certain.

A Unitary/Planning Factor cohort and a Planning Factor groups, distinguishable by statute and judicial decision, allows national planning law trends to emerge. Table 11.1 demonstrates that a positive correlation exists between changes to the original SPEA language and the significance of the plan as law.

As states begin to tinker with their planning law they invariably put greater significance on planning. Legislative activity in planning and development is a key factor in understanding how planning law has developed. State court decisions interpreting comprehensive plans generally defer to local expertise and follow closely the statutory legal environment for land use decision making. Therefore, state planning codes are the change agents for the future of planning law.

12

Anglo-American Town Planning Theory since 1945: Three Significant Developments but no Paradigm Shifts

From *Planning Perspectives* (1999)

Nigel Taylor

Editor's introduction

Underlying any discipline involving scientific inquiry is a set of theories that predict behavior and provide a framework for the field's knowledge and, ultimately, practice. Its practitioners develop these theories by articulating hypotheses (expressing causal relationships among different factors or variables) and proving or disproving them through experimentation and empirical observation. If proved correct, the hypothesis becomes theory; if not, it is rejected and the researcher develops a new one, repeating the cycle until the relationship proves true and other researchers can replicate the results. When research reveals evidence that disproves an existing theory, it stimulates new ways of thinking, or a new paradigm, about a given problem. This selection reviews the postwar planning theory paradigm.

There are differences of opinion regarding basic assumptions about planning theory. Some theorists maintain that urban and regional planning address "wicked problems," or situations that defy traditional theorizing built on expertise; others believe planning issues are vexing but subject to such theorizing.

One source of this dichotomy is the expansive vision of planning that emerged in the Depression. At that time, national governments around the world attacked economic failure by launching large-scale public works in order to "prime the pump," or stimulate the economy, a Keynesian strategy. In the United States, New Deal activities gave credence to the effectiveness of planning – the choice of public works projects, for example, called for rational problem analysis and solution. One seminal theorist, Rexford Tugwell, a member of Franklin Delano Roosevelt's Brain Trust, convinced F.D.R. to create a Resettlement Administration to undertake a project known as the Greenbelt Towns to relocate distressed families in self-sufficient suburban towns combining the best features of city and country life.

Through this and other New Deal work, Tugwell became convinced that modern government needed a fourth power, planning. He tested this idea at the local level, serving as the first chair of the New York City Planning Commission, and at a higher level, as governor of Puerto Rico, where he started planning efforts for the whole island, a type of national planning based on social science analysis. He championed the concept in a planning program he founded and headed between 1946 and 1957 at the University of Chicago. This small but extremely influential program nurtured such scholars as Martin Meyerson and Harvey Perloff, who subsequently became leaders in planning education through the 1960s and 1970s. They tested this broader view of planning, focusing their research on its use in improving public administration, democratic decision making, and embracing social and economic problems as well as the traditional land-based concerns.

This work provided a foundation for an ongoing dialogue related to planning theorists who viewed planning problems as defying solutions that could be reached using rationality and professional expertise. Instead, they turned to the political process as a theoretical foundation. Horst W. J. Rittel and Melvin M. Webber, who wrote the essay "Dilemmas in a General Theory of Planning" in *Policy Sciences*, 4 (1975), argued that planning issues have few boundaries, are unique, have no correct or incorrect answer (and all answers are value-ridden anyway). They built their case on three published works: Nobel Prize-winning economist Friedrich Hayek's *The Road to Serfdom* (Chicago: University of Chicago Press, 1944), a study of centralized economic planning in collectivized societies that asserted that the economic planners could never have enough information to make proper decisions and argued that the democratic process was the best broker of planning problems; Edward Banfield and Martin Meyerson's *Politics, Planning and the Public Interest: The Case of Public Housing in Chicago* (Glencoe, IL: Free Press, 1955), which explored conflict resolution and the definition of the so-called public interest; and Alan Altshuler's *The City Planning Process: A Political Analysis* (Ithaca, NY: Cornell University Press, 1965), a study of transportation and land use planning in Minneapolis and St Paul, Minnesota, that rejected the generalized goals of comprehensive planning as politically unviable and supported instead an approach focusing on specific, definable goals more easily understood and implemented. These works launched several generations of planning theorists who put forth ideas related to the process of planning in democratic societies. Today's expression of this school of thought is "communicative" planning. Two key books on this topic are John Forester's *Planning in the Face of Power* (Berkeley, CA: University of California Press, 1989) and Patsy Healey's *Collaborative Planning: Shaping Places in Fragmented Societies* (Vancouver, BC: University of British Columbia Press, 1997).

While this group of scholars of planning theory forged ahead, others took a look at explanations of urban growth and development, especially in the fields of transportation, housing and community development, and urban design. For example, researchers Robert Mitchell and Chester Rapkin, who developed a theory of land use and transportation now known as the "transportation–land use connection," published *Urban Traffic: A Function of Land Use* (New York: Columbia University, 1954). This book showed that by analyzing land use and its users, and by measuring quantifiable factors like age, gender, income, and family status, a planner could predict trips and assess future travel demand, the basis of transportation planning. In the ensuing years, various researchers utilized this theory to create computer models, now used by Metropolitan Planning Organizations (MPOs) to develop plans. One of the most significant refinements to the theory is Robert Cervero's observations about the imbalances between employment and residences and their causal relationship to suburban traffic congestion. These findings, reported in a prize-winning article, "Jobs–Housing Balancings and Regional Mobility" (*Journal of the American Planning Association*, 55: 2, 1989), are now ingrained in Smart Growth plans that seek to minimize travel by locating housing and jobs in close proximity.

This selection addresses the emergence of the two approaches to postwar planning theory, one dealing with "how" society makes planning decisions and the other attempting to understand "what" causes urban growth and change. It argues that they are interrelated and demonstrates their connections. Understanding the validity of both approaches should be part of the education of today's urban and regional planner. For example, in this *Reader* Paul Davidoff's "Advocacy and Pluralism in Planning" exemplifies the "how" approach, and Kevin Lynch's "Dimensions of Performance" falls into the "what" category.

Nigel Taylor, author of this selection, is Principal Lecturer, Planning and Architecture, Faculty of the Built Environment, University of the West of England. His *Urban Planning Theory since 1945* (London and Thousand Oaks, CA: Sage Publications, 2004) expands on this selection, connecting the procedural and substantive areas of planning theory.

Most writings on the planning theory that explores the process of decision making in planning are essays and journal articles that have been collected in books. One of the first collections is Andreas Faludi's *A Reader in Planning Theory* (London: Butterworth Heinemann, 1973). More recent collections are Scott Campbell and Susan S. Fainstein, *Readings in Planning Theory* (London: Blackwell, second edition 2003), and Seymour Mandelbaum, Luigi Mazza, and Robert W. Burchell, *Explorations in Planning Theory* (New Brunswick, NJ:

Center for Urban Policy Research, 1996). For theories dealing with planning specific types of development, see work related to the specialty: in transportation, see Hank Dittmar and Gloria Ohland, *The New Transit Town: Best Practices in Transit-oriented Development* (Washington, DC: Island Press, 2004); in regional planning, see Peter Calthorpe and William Fulton, *The Regional City* (Washington, DC: Island Press, 2001); and for an overview of current thinking on a range of planning problems, see Jonathan Barnett (ed.), *Planning for a New Century* (Washington, DC: Island Press, 2001).

INTRODUCTION

Over the fifty-year period since the end of the Second World War there have been a number of important shifts in town planning theory . . . as it has developed in Britain and North America . . . clarifying what kind of an activity town planning is (and hence what skills are appropriate to its practice). . . .

In studies of the history of ideas, it has become fashionable to describe significant shifts in thought as "paradigm" shifts, and some planning theorists have applied this concept to changes in town planning thought since 1945. Therefore, in addition to offering an account of the main shifts in (Anglo-American) town planning theory since 1945, I shall also assess whether it is appropriate to describe these changes as paradigm shifts. . . .

THE IDEA OF PARADIGMS AND PARADIGM SHIFTS

The use of the term "paradigm" to describe major shifts in thought was first coined by Thomas Kuhn in his account of the history of scientific thought. Before Kuhn, it was widely assumed that scientific knowledge had grown steadily through history as more and more empirical evidence of phenomena had been accumulated. Kuhn's examination of the history of science led him to conclude that this gradualist, evolutionary view of the advance of scientific knowledge was misleading. For, according to Kuhn, if we examine any branch of science, we find that there are certain fundamental theories, conceptions or presuppositions which hold steady for very long periods – often for hundreds of years. These settled views of the world become so fundamental to people's whole conceptual scheme of reality that it is extremely difficult (and in some cases impossible) for most people to think of reality as being different; that, indeed, is why such views are fundamental. Because these fundamental theories constitute people's view of the world (or a significant part of it), they are, literally, "world views", and it is these enduring world views which Kuhn describes as "paradigms". [An example of a paradigm] in the history of science would be the pre-Copernican view that the Earth was flat and at the centre of the Universe. . . .

Kuhn's account of the history of science allows that, through the period in which any given paradigm prevails, advances in scientific theory still occur as a result of empirical research which uncovers fresh evidence about phenomena. . . . During the long periods of history in which a given paradigm prevails, scientists are often aware of empirical evidence which does not "fit" the prevailing paradigm, which the prevailing paradigm seems unable satisfactorily to explain. However, according to Kuhn, most scientists do not allow this "contrary" evidence to unseat their adherence to the paradigm on which they rely to explain phenomena in the world. Rather, they tend to "turn a blind eye" to these puzzling phenomena, often in the belief that one day someone will succeed in explaining the seemingly "anomalous" evidence within the framework of the given paradigm. The great scientists in history, however, have typically been curious about anomalous evidence which a prevailing paradigm is unable to explain and, as a result, have created a radically new account of the world which succeeds in explaining the hitherto puzzling evidence as well as the evidence previously explained by the "old" paradigm. This new fundamental theory amounts to a whole new conceptual scheme, worldview, or paradigm. In Kuhn's terms, then, a paradigm shift is a revolutionary shift in thought, because a whole way of perceiving some aspect of the world is overturned and replaced by a new theoretical perspective. [An example is] the shift from viewing the Earth as flat and at the centre of the Universe to seeing it as round and orbiting the Sun. . . .

It should be clear from this account that, for Kuhn, paradigm shifts are fundamental theoretical changes. It is this, which explains why paradigm shifts typically occur infrequently in the history of science. Any given paradigm, once established, shapes the whole way a scientific community (and beyond that, the general public) views some aspect of the world, and tends to endure for centuries, not just decades. . . .

[. . .]

TWO SIGNIFICANT SHIFTS IN THE WAY TOWN PLANNING HAS BEEN CONCEIVED: FROM THE PLANNER AS A CREATIVE DESIGNER TO THE PLANNER AS A SCIENTIFIC ANALYST AND RATIONAL DECISION MAKER

For almost twenty years following the Second World War, town planning theory and practice was dominated by a conception, which saw town planning essentially as an exercise in physical design. In fact, this view of town planning stretched back into history, arguably as far as the European Renaissance, arguably even further back than that. Its long historical lineage is shown by the fact that, for as far back as we can see, what came to be seen and described as town planning was assumed to be most appropriately carried out by architects. Indeed, such was the intimate connection between architecture and town planning that the two were not distinguished throughout most of human history. Thus what we call town planning was seen as architecture, its only distinctiveness being that it was architecture on the larger scale of a whole town, or at least part of a town, as distinct from individual buildings.

This conception of town planning as "architecture writ large" persisted down to the 1960s, as was shown by the fact that most planners in the post-war years were architects by training, or "architect-planners". Indeed, because of this, the established professional body for architects in Britain, the Royal Institute of British Architects (RIBA), resisted the establishment of a separate professional body for town planning, arguing that town planning as a practice was already "covered" by themselves. The close link in the post-war years between design and town planning, and hence between architecture and town planning, also explains why at this time aesthetic considerations were regarded as central to town planning. Like architecture, town planning was viewed as an "art", albeit (again like architecture) an "applied" or "practical" art in which utilitarian or "functional" requirements had to be accommodated. . . .

Against this background, the bursting on to the scene in the 1960s of the systems and rational process views of planning represented a rupture with a centuries-old tradition, and so might well be viewed as a Kuhnian paradigm shift. . . .

It is worth noting, in passing, that the systems and rational process views of planning are conceptually distinct, and so really two theories of planning, not one. Thus the systems view was premised on a view about the object that town planning deals with – towns, or regions, or the environment in general – were viewed as "systems". By contrast, the rational process view of planning was concerned with the method or process of planning itself, and in particular it advanced an "ideal type" conception of planning as a procedure for making instrumentally rational decisions. But setting aside this important distinction, both the systems and rational process views of planning, taken together, represented a radical departure from the then prevailing design-based view of town planning. This shift in planning thought can be summarized under four points.

- First, an essentially physical or morphological view of towns was to be replaced with a view of towns as systems of interrelated activities in a constant state of flux.
- Second, whereas town planners had tended to view and judge towns predominantly in physical and aesthetic terms, they were now to examine the town in terms of its social life and economic activities. . . .
- Third, because the town was now seen as a "live" functioning thing, this implied a "process", rather than an "end-state" or "blueprint" approach to town planning and plan making.
- Fourth, all these conceptual changes implied, in turn, a change in the kinds of skills, or techniques,

which were appropriate to town planning. For if town planners were trying to control and plan complex, dynamic systems, then what seemed to be required were rigorously analytical, "scientific" methods of analysis.

Overall, the shift in planning thought brought about by the systems and rational process views of planning can be summed up (albeit rather crudely) by saying that, whereas the design-based tradition saw town planning primarily as an art, the systems and rational process theorists suggested that town planning was a science. For, on the one hand, the analysis of environmental systems (regions, cities, etc.) involved systematic empirical – and hence "scientific" – investigation and analysis of interrelationships between activities at different locations. And, on the other hand, the conception of planning as a process of rational decision making was also commonly equated with being "scientific". . . .

However, significant though this shift in town planning thought undoubtedly was, it remains open to question whether it should be likened to a Kuhnian paradigm shift. Although the shift in planning thought described above did have the effect of marginalizing considerations of design and aesthetics in planning theory for about twenty years, in British planning practice (especially in the development control sections of planning authorities) planners still continued to evaluate development proposals partly in terms of their design quality and aesthetic impact. . . . In planning practice, therefore, the design-based conception of town planning was not completely superseded by the change in theoretical perspective described above.

Admittedly, these observations are about planning practice, not planning theory. But the continuation of the physical design view of planning in planning practice was also theoretically significant. For it drew attention to the fact that, at the level of "local" planning at least, many planners continued to believe that the physical form and aesthetic appearance of new development were important concerns of town planning. And although there were lessons for small area "local" planning in systems and rational process thinking (e.g. in giving greater consideration to the social and economic effects of development proposals; in approaching local planning as a rational process, etc.), these lessons could be accommodated within an essentially traditional design-based conception of planning. At the level of local planning, therefore, the shift in planning thought described above did not replace the physical design view of planning; it was not a paradigmatic shift in that sense. Furthermore, the continuing relevance of the physical design-based conception of town planning to town planning theory has been shown by the revival of theoretical interest in questions of urban design in the 1980s and 1990s.

It was therefore primarily at the broader, more strategic level of planning that the design-based view of planning was supplanted by the changing conception of planning ushered in by the systems and rational process views of planning. In fact, the main shift in planning thought brought about by the systems and rational process views of planning was in clarifying a distinction between strategic and longer-term planning on the one hand, and "local" and more short-term planning on the other. And it was at the former, strategic level of planning that the altered conception of town planning brought about by systems and rational process thinking was most relevant. In retrospect, then, the shift in town planning thought in the 1960s described above was not a wholescale revolution which completely ousted the incumbent design-based view of town planning; rather, the systems and rational process views of planning "added" to the design-based view. The shift in planning thought described above was not therefore a revolution in thought comparable to the paradigm shifts in scientific theory described by Thomas Kuhn.

FROM THE PLANNER AS TECHNICAL EXPERT TO THE PLANNER AS A MANAGER AND "COMMUNICATOR"

Although there were marked differences between the traditional conception of town planning as an exercise in physical planning and design, and the conception of town planning as a rational process of decision making directed at the analysis and control of urban systems, there was one thing that both these views had in common. Both presumed that the town planner was someone who possessed, or should possess, some specialist knowledge and skill – some substantive expertise – which the layperson did not possess. It was this, which qualified the planner to plan. And since a central condition of professionalism is the possession of some specialist knowledge or skill, it was this, too, which justified any claim town planners might make to constitute a distinct "profession".

Clearly, views about the content of the specialist skill appropriate to town planning varied according to which of the foregoing conceptions of planning were adopted. Under the traditional design-based view of town planning, the relevant skills were seen to be primarily those of aesthetic appreciation and urban design. Under the systems and rational process views, the skills were those of scientific analysis and rational decision making. But still, under both conceptions, the town planner was conceived as someone with a specialist knowledge, understanding, and/or skill. However, this whole idea of the planner as someone with some substantive expertise came in its turn to be challenged by an alternative view of town planning.

This challenge emerged again in the 1960s, when it came to be openly acknowledged that town-planning judgments were at root judgments of value (as distinct from purely scientific judgments) about the kinds of environments it is desirable to create or conserve. Once this view was taken, it naturally raised the question of whether town planners had any greater "specialist" ability to make these judgments than the ordinary person in the street. Indeed, people's experience of much of the planning of the 1960s – such as comprehensive housing redevelopment or urban road planning – seemed to indicate not. The emergence of the view that town planning was a value-laden, political process therefore raised not so much the question of what the town planner's area of specialist expertise should be, but, more fundamentally, the question of whether town planning involved any such expertise at all.

From this radical questioning of the town planner's role, there developed a curious bifurcation in planning theory, which has persisted to this day. On the one hand, some planning theorists have continued to believe that the practice of town planning requires some specialist substantive knowledge or skills – be it about urban design, systems analysis, urban regeneration, sustainable development, or whatever. On the other hand, there has developed a tradition of planning thought, which openly acknowledges that town-planning judgments are value-laden and political. As noted above, one conclusion, which might have been drawn from this, would be to reject entirely the idea that town planning involves, or requires, some specialist expertise, and indeed, some "radical" planning theorists have flirted with this view. However, most planning theorists who have openly acknowledged the value-laden and political nature of planning have developed an alternative line of thought. This rejects the idea that the town planner is someone who is specially qualified to make better planning decisions or recommendations – because what is "better" is a matter of value, and (so the argument goes) planners have no superior expertise in making value judgments over environmental options. However, the view is still taken that the town planner possesses (or should possess) some specialist skill, namely, a skill in managing the process of arriving at planning decisions and facilitating action to realize publicly agreed goals.

Through the 1970s and 1980s, therefore, a tradition of planning theory emerged which viewed the town planner's role (and hence his or her "professional" expertise) as one of identifying and mediating between different interest groups involved in, or affected by, land development. In this way, the town planner was seen as someone who acts as a kind of cypher for other people's assessments of planning issues, rather than someone who is specially qualified to assess these issues him – or her – self. The town planner was viewed as not so much a technical expert (i.e. as someone who possesses some superior skill to plan towns), but more as a "facilitator" of other people's views about how a town, or part of a town, should be planned. To conceive of the town planner like this as a kind of manager of the process of making planning decisions could easily conjure up an image of the planner as a grey-suited chairperson of meetings. But tacked on to this view went a more particular ideological commitment which made it more appealing and inspiring to idealists in the profession, namely a commitment to ensure that the process of planning was open and democratic, especially to disadvantaged or marginalized groups who tended to be ignored or overridden in decisions about land development.

An early version of this theory of town planning, and of the planner's role, was Paul Davidoff's "advocacy" view of planning in the 1960s. The most recent version is the communicative planning theory inspired, particularly, by Habermas's theory of communicative action. In this, the skills of inter-personal communication and negotiation are seen as central to a non-coercive, "facilitator" model of town planning. Indeed, in relation to involving the public in planning, it has even been suggested that the kinds of inter-personal skills needed by the communicative town planner are those of the listener and the counsellor.

[. . .]

This view of the knowledge and skills relevant to town planning is a far cry from the view that the specialist skill of the town planner resides in being either an urban designer or a systems analyst.

In relation to this view of the planner as a manager, communicator and "facilitator" of planning decisions, it is also relevant to note here the emergence of a concern with implementation during the 1970s and 1980s. . . . [This] presented a critique of the rational process model of planning because it tended to emphasize the task of making decisions over that of taking action. This . . . drew attention to the much overlooked fact that, frequently, the most carefully thought-through public decisions and policies did not actually result in the necessary action to realize their intentions. This was because, in attending primarily to the business of making decisions about appropriate policies and plans, insufficient attention was given to the problems of how these policies and plans might get implemented. A concern with implementation . . . thus became a central preoccupation of some planning theorists in the 1970s and 1980s.

Although this concern with implementation had rather different roots from the view of planning as a political activity described above, it issued in much the same conception of planning and the skills appropriate for it. For the general conclusion of the implementation theory of the 1970s and 1980s was that, to be an effective implementer of public policies and plans, planners needed to become effective at networking, communicating and negotiating with other agents involved in the development process. In short, theoretical reflection on the problem of implementation led also to a view of the planner as one who should be a capable manager, "networker" and communicator.

Nowadays, many planners do describe themselves as managers and facilitators of the process of planning rather than as people with a special expertise in planning towns. So the shift in town planning thought described here has undoubtedly been significant. Hence . . . the description of "communicative action and interactive practice" as the new "paradigm" of planning theory. However, once again a word of caution is in order before it is too readily assumed that this change in planning thought represents a Kuhnian paradigm shift. For it is possible to imagine some kind of rapprochement between the two views described here. Thus one could adopt a view of the town planner as one whose role is primarily that of a communicator and negotiator, but where, in communicating and negotiating with others, the planner also brings to bear some specialist knowledge which, for example, would enable him or her to point out the likely consequences of development proposals on the form and functioning of a town. Such a model of the town planner would be akin to that of, say, civil servants who are experts in economic matters, and who impart their specialist economic understanding to those they advise who make decisions about economic policy. To be effective as an adviser, such a town planner would have to be skilled in communicating and negotiating with others, but he or she would also have to possess some specialist knowledge to bring to the communicating table to assist others in arriving at planning decisions. The alternative "substantive" and "procedural" conceptions of town planning described in this section are therefore not as fundamentally at odds, or "incommensurable", as the paradigm shifts in the history of science described by Thomas Kuhn.

MODERNIST AND POSTMODERNIST PLANNING THEORY: A SHIFT IN NORMATIVE PLANNING THOUGHT

According to some commentators, since about the late 1960s there has been a significant shift in Western thought and culture from "modernism" to "postmodernism", and some view this as so fundamental as to constitute a shift in world view, or a paradigm shift. This alleged paradigm shift has a special bearing on town planning, because modern architecture and town planning have jointly provided one of the main "sites" where the shift from modernism to postmodernism is supposed to have taken place. . . .

At one level, postmodernism can be viewed as a movement opposed to the styles of art and design associated with the modern movement. Thus in architecture, postmodernists rebelled against the aesthetic minimalism and anonymity of the plain geometrical buildings (and comprehensive planning schemes) of the modern movement, and against the modernist dogma of functionalism, which had legitimized this stripped-down architecture. Postmodern architects therefore sought to "bring back style" to contemporary buildings to enrich their aesthetic content and give them "meaning". . . .

In relation to town planning, Jane Jacobs . . . expressed a similar preference for complexity in the city, in opposition to the simplified order which modern town planning theorists like Ebenezer Howard and Le Corbusier had advocated. Jacobs berated the simplemindedness of single use zoning and "comprehensive" redevelopment in urban areas, both of which showed little understanding of, or regard for, the delicate social and economic fabric and vitality of existing city areas. Instead, she advocated mixing land uses, and leaving many so-called slum areas alone to "unslum" themselves. . . .

These architectural and planning ideas certainly represented a departure from the prevailing "modernist" orthodoxy. But, according to some accounts, the shift from modernism to postmodernism goes deeper than just a preference for greater "complexity" in architecture and town planning. For, it is said, underpinning and driving the modern movement in architecture and town planning (and other developments in modern western culture) has been a more fundamental intellectual orientation, which has its roots in the European Enlightenment of the eighteenth century. This Enlightenment "world view" has been characterized as an optimistic belief in human progress based on analytical reason and scientific understanding. Quite apart from its "machine aesthetic", modern town planning thought and practice has been seen as expressing this more general Enlightenment world view or paradigm, and correspondingly, postmodern planning theory is seen by some as representing a break with this intellectual tradition.

[. . .]

. . . [M]odernist planning has relied largely on "a mastery of theory and methods in the social sciences", so that planning knowledge and expertise is "grounded in positive science, with its propensity for quantitative modeling and analysis". Second, . . . this model of planning is also equated with a "top-down" . . . normative model of planned action, in which . . . the state is vested with the authority to undertake town planning. Associated with this, modernist planning has sought to be "comprehensive", and this is displayed, for example, in the idea that planning should aim to realize what is in the overall public interest.

[. . .]

Broadly, postmodernists argue that the world and our experience of it are far more complex and subtle than has typically been realized in the Modern Age. In relation to cities and the environment, postmodernists claim that people's experience of places, and from this the qualities of places, are much more diverse and "open" than was implicit in many modernist schemes for improving the city – and especially in the bombastic simplicities of modernist architectural schemes for the ideal future city, such as Le Corbusier's radiant city. In place of the modernist architect's and planner's emphasis on simplicity, order, uniformity and tidiness, postmodernists typically celebrate complexity, diversity, difference and pluralism. This echoes Jane Jacobs's . . . celebration of urban complexity and diversity more than thirty years ago. Postmodernists therefore argue that there can be no one type of environment, which is ideal for everyone, no singular conception of environmental quality. Thus whilst some may continue to hold as a planning ideal Howard's vision of Garden Cities, others will prefer the buzz and excitement of . . . the "teeming metropolis" [and the] . . . view that the ideal postmodern city is a multicultural city.

[. . .]

But do these postmodern values amount to a paradigmatic break with the values of so-called modernist planning? Two points can be made to dispute this. First, in relation to the general value stance of diversity and pluralism espoused by postmodernists, this can be taken to an extreme where any "difference" is accepted or permitted; in other words, a position of complete moral and political relativism. And such an extreme ethical relativism is open to criticisms similar to those raised earlier against postmodern epistemological relativism. To be sure, we may endorse a more complex, variegated and "multicultural" conception of environmental quality, but this need not exclude a commitment to some overarching – even universal – normative principles of planning. For example, shouldn't town planning, wherever it is practised, do what it can to help bring about economically and environmentally sustainable development, development which is not socially divisive, and development which is experienced as an aesthetic delight? . . .

[. . .]

The idea, then, that either a kind of postmodern liberalism or some version of community planning constitutes a paradigmatic break with social democratic politics is, to say the least, premature. . . .

CONCLUSION: CHANGE AND CONTINUITY IN TOWN PLANNING THOUGHT

Unquestionably, there have been significant changes in town planning thought since the end of the Second World War, and (for all the talk of the gulf between planning theory and practice) these changes have affected planning practice. Thus town planners now operate with a very different conception of town planning, and bring to it quite different skills, from the architect-planners of fifty years ago. The idea, for example, of town planning as an exercise in "managing the process of arriving at planning decisions" and "negotiating agreements for action", and hence of the town planner as someone who is primarily a communicator and a "facilitator", would not really have occurred to early British post-war planners. . . . Contrariwise, the idea that the prime task of town planners is to undertake surveys of the physical and aesthetic characteristics of towns, and then sit in front of drawing boards drawing up master plans, would strike most present-day planners as hopelessly limited, naive and outmoded. Similarly, the "modernist" ideas that better cities can be created by drastic surgery and "comprehensive" development has long been discredited.

These are significant changes in people's conception of the activity of town and country planning, and the values, which should inform it. And there is no harm in describing these changes as "paradigm shifts", so long as we appreciate that we are employing this term in a fairly loose or weak sense. However, in anything like the strong sense of the term as used by Thomas Kuhn to describe revolutionary changes in thought in the history of science, it is doubtful whether any of the changes in town planning thought since 1945 are appropriately described as paradigm shifts. For, in the strong sense of the term, a paradigm shift is marked by a fundamental change in world view – a kind of *Gestalt* switch in people's whole conceptual scheme. And whilst there have been significant discontinuities in planning thought since 1945, there have also been continuities across these changes.

For example, whilst, on the face of it, the shift from the physical design view of town planning to the systems and rational process views of planning might look like a change of world view (and hence a Kuhnian paradigm shift), the systems and rational process views of planning did not completely supersede the urban design conception of town planning. Thus, at least at the "local" level of small area or "district" planning and the control of development, good urban design (and design control) is still regarded as central to good town planning; indeed, a concern with urban design within town planning has experienced something of a revival since the mid 1980s. Arguably, a more likely candidate for a Kuhnian paradigm shift has been the shift from a view of the town planner as an expert to the planner as a manager and facilitator – the shift, in other words, away from a view of the planner as the supplier of answers (in the form of "master" plans) to that of the planner as someone skilled at eliciting other people's answers to urban problems and somehow "mediating" between these. But even this view of town planning as a species of "communicative action" is compatible with the view that town planners should possess at least some area of expertise, for example about the likely effects of proposed changes to urban environments. Similarly, the postmodern emphasis on complexity, difference and relative values is not completely incompatible with a commitment to some overarching, universal principles of environmental quality, still less with a reliance on (sophisticated) reason and scientific understanding.

Looking back over the changes in town planning theory since the end of the Second World War, some planning theorists have suggested that planning theory has fragmented into a plurality of diverse and even incompatible theoretical positions or "paradigms". However, this article has expressed scepticism with this idea. Although there have been significant shifts in planning thought since 1945, there have also been significant continuities. Indeed, the shifts in town planning thought over this period can be regarded as developmental rather than as ruptures between incompatible paradigms of planning. In other words, the changes to town planning theory described here can equally well be viewed as "filling out", and thereby enriching, the rather primitive conception of planning, which prevailed in the immediate post-war years. On this account, the story of town planning theory since 1945 has been one of developing sophistication as we have learnt more about the greater complexity of urban environments and the diverse values of different communities.

13

New Directions in Planning Theory

From *Urban Affairs Review* (2000)

Susan Fainstein

Editor's introduction

While theories of urban and regional planning are products of their times, three common themes unite them. They are: a focus on people and social reform, attention to space and place, and views on political and economic processes. All of the theories endorse, in one form or another, a broad, systematic or comprehensive approach. Any one of these three themes might dominate a period's planning practices, but they are all present to a greater or lesser extent, sometimes competing for planners' attention, sometimes advocated by a subgroup of planners. This selection outlines three schools of planning theory now flourishing.

A look at planning history places theory in context and shows the persistence of the three themes. The 100 years since the profession's beginnings can be divided into five periods, each roughly two decades. They are: "The Roots of Modern Planning," 1890s to 1920; "Planning Becomes a Profession," 1920s to 1940s; "Planning Explodes," 1950s to 1970s; "Planning Revamped," 1970s to 1990s; and "Planning Revisits its Roots," 1990s to present.

From the 1890s to the 1920s the varied elements of the planning movement began to emerge to address urban problems, diagnosed as population congestion, especially in the overcrowded, disease-ridden slums; chaotic development of downtowns and their surrounds; and ineffectual or corrupt municipal government. Planners theorized that boldly creating healthful, beautiful, and ordered cities – either by repairing existing conditions or building new cities – would solve these problems. Some had plans to rid cities of slums through clearance or the enforcement of housing codes, believing that eliminating slums would deal with health, crime, and other maladies. (This was the people-and-social reform stream.) Others had plans to make cities beautiful by adding broad avenues, monumental civic centers, and waterfront promenades – in the belief that their big plans would inspire a sense of pride and good citizenship among the inhabitants. (This was another way of thinking about the role of beauty and, indirectly, the role of good government in managing urban places.) A third group devised rules and documents, zoning and master plans, hypothesizing that straightforward political processes that ordered land not only promoted good health but also protected real estate values and prevented ugly or dysfunctional land use, thus enabling a city to promote the general welfare of its residents. (Here the political and economic theories are in play.)

In the two decades between the World Wars, urban and regional planners refined these theories as they developed professional practices. Seeking to define their expertise, they still emphasized people/social reform, space/place, and political/economic concepts, but interpreted them differently. Health was paramount in the proposals for decentralizing people to self-contained places like Radburn, or in regional plans that saw open space as the "lungs" of metropolitan areas. By devising different spatial arrangements, planners attempted to address recently articulated negative social effects of urban living (the lack of community and personal identity). The "neighborhood unit," for instance, postulated that grouping houses around a school or community center would enhance social interchange. Using population projections to predict and locate future growth, planners adopted standards for the quantity of land needed for different activities. To create the master plans of the 1920s, they invented a coherent, multistep process, later labeled the "rational decision making" model. In its bare form, the rational model incorporated a five-step cycle (establish goals and objectives, collect and analyze data to identify problems, outline alternatives to solve the problems, select and implement the best alternative, and evaluate the outcome, assuring that it meets the goal and objectives, revising approach as changes occur). It guided, and still guides, professional practice.

In the 1950s to 1970s the urban–comprehensive–people–space–politics theories of the earlier period were expanded and refined. For example, with the rise of mandated comprehensive planning – resulting from the Housing and Slum Clearance Act of 1949 and the Federal-Aid Highway Act of 1962 – and the subsequent outpouring of projects and later research on their strengths and weaknesses, groups of planners delineated the three themes more clearly than ever, challenging older theories, and laying the foundations for the theories discussed in this selection. The long-held belief that social improvement would occur by dealing with physical conditions – through slum clearance or separation of land uses – came into question. This led to revisions in the basic tools of zoning (adding inclusionary elements or allowing mixed uses), and in the contents and scope of comprehensive plans to redefine space and place. The rise of advocacy planning, expanded citizen participation, and a suggested redefinition of the role of the planner – from technical expert to a manager of consensus – resulted in a proliferation of theoretical writings, still focused on the three themes.

The fourth era of planning history, from the 1970s to 1990s, was a tumultuous time for a cadre of scholars who now called themselves planning theoreticians. They consciously set about developing a body of knowledge for their field, mainly focusing on people/social reform and political/economic processes. The most extreme of these theorists rejected space and place theories as unserviceable, turning to systematic critiques of social injustices wrought by market economies. Their work drew sharp criticism, exemplified by the flare-up caused by Sir Peter Hall in his provocative essay "The Turbulent Eighth Decade: Challenges to American City Planning" in the *Journal of the American Planning Association* (55: 3, 1989). Reflecting on the evolution of U.S. planning, Hall charged: some "radical critics . . . flatly denied that, given the system, any positive theory of planning action was necessary or desirable. Others retreated to bold but largely vacuous rhetoric by the end of the 1970s planning theory had become divorced from planning practice. And the top products of the nation's top planning schools were becoming increasingly uninterested in the day-to-day activity of planning." While this commotion attracted much attention among some groups in the planning academy, other theoreticians bore down on space/place components, bringing new research to bear on economic development, urban design, environmental planning, and transportation, adding new conceptual approaches to the field. Principles of community-based development, transit-oriented design, conservation easements, and transfer of development rights were among the era's breakthroughs based on practice.

Planning theoreticians of all types began to rethink their approaches in the era of planning history that started in the late 1980s and early 1990s and continues to this day. Issues of space and place reentered all types of planning theory discussions, but with new twists, as outlined in Susan S. Fainstein's selection below.

Fainstein, professor in the Department of Urban Planning and Design, Graduate School of Design, Harvard University, has written extensively on planning theory from a political economy perspective. Additionally she has pursued interests in real estate development and diversity in planning in her books: *The City Builders, Restructuring the City: The Political Economy of Urban Redevelopment* (London: Blackwell, 1994), with the revised and expanded version published as *The City Builders: Property Development in New York and London, 1980–2000* (Lawrence, KS: University Press of Kansas, 2001). She also coedited *Gender and Planning: A Reader* (New Brunswick, Rutgers University Press, 2005) with Lisa J. Servon.

To further explore planning theory from the Fainstein perspective, see *Readings in Planning Theory* (London: Blackwell, 2003), ed. with Scott B. Campbell, *Readings in Urban Theory* (London: Blackwell, 2002), ed. with Scott B. Campbell, and *Divided Cities: New York and London in the Contemporary World* (London: Blackwell, 1992) with Ian Gordon and Michael Harloe.

In this article, I discuss and critique contemporary planning theory in terms of its usefulness in addressing what I believe to be its defining question: What is the possibility of consciously achieving widespread improvement in the quality of human life within the context of a global capitalist political economy? I examine the three approaches . . . (1) the communicative model, (2) the New Urbanism, and (3) the just city. In my conclusion, I defend the continued use of the just-city model and a modified form of the political economy mode of analysis that underlies it.

The first type, sometimes called the collaborative model, emphasizes the planner's role in mediating among "stakeholders" within the planning situation; the second, frequently labeled neotraditionalism, paints a physical picture of a desirable city to be obtained through planning; and the third, which derives from the political economy tradition, although also outcome-oriented, is more abstract than the New Urbanism, presenting a model of spatial relations based on equity. This typology of planning theories is not exhaustive – there remain defenders of the traditionally dominant paradigm of the rational model, as well as incrementalists who base their prescriptions on neoclassical economics, and Corbusian modernists, who still promote formalist physical solutions to urban decay. Nor are the types wholly mutually exclusive – each contains some elements of the others, and some theorists cannot be fit easily into one of the types. Nevertheless, each type can claim highly committed proponents, and each points to a distinctive path for both planning thought and planning practice.

Differences among the types reflect the enduring tension within planning thought between a focus on the planning process and an emphasis on desirable outcomes. In the recent past, neither tendency has fully dominated because theoretical orientations toward process and outcome have respectively affected different aspects of practice. Thus the concept of the rational model represented an approach based wholly on process with little regard either to political conflict or to the specific character of the terrain on which it was working . . . This model has provided the metatheory for planning activity in the decades since the 1960s, incorporating the faith in scientific method that swept through the social sciences during the Cold War period. Within planning practice, it has primarily been used for forecasting impacts and for program evaluation. At the same time, however, as the rational model held sway among theorists, planning practitioners engaged in the development of zoning and environmental regulations, upholding an atheoretical, physical outcome-oriented vision. . . . Outcome-oriented physical planning has left its mark on metropolitan areas in the form of urban renewal, low-density development, and spatial and functional segregation.

Although the rational model and the physical master plan were the dominant late twentieth-century modes of planning practice throughout the world, they did not escape a powerful critique. Their opponents, who decried the distributional consequences of these approaches . . . persistently inquired into who benefited from planning effort. . . .

The recent theoretical moves involved in the typology sketched earlier represent a reaction both to previously dominant modes of thought and also to events "on the ground." Thus the communicative model responds to the imposition of top-down planning by experts . . . , the New Urbanism is a backlash to market-driven development that destroys the spatial basis for community, and the just-city formulation reacts to the social and spatial inequality engendered by capitalism. . . .

THE COMMUNICATIVE MODEL

[. . .]

Within communicative theory, the planner's primary function is to listen to people's stories and assist in forging a consensus among differing viewpoints. Rather than providing technocratic leadership, the planner is an experiential learner, at most providing information to participants but primarily being sensitive to points of convergence. Leadership consists not in bringing stakeholders around to a particular planning content but in getting people to agree and in ensuring that, whatever the position of participants within the social-economic hierarchy, no group's interest will dominate.

[. . .]

THEORETICAL AND PRACTICAL DEFICIENCIES

[. . .]

The communicative theorists make the role of the planner the central element of discussion. Both the context in which planners work and the outcome of planning fade from view. . . . [I]n communicative planning theory the spotlight is on the planner. Instead of asking what is to be done about cities and regions, communicative planners typically ask what planners should be doing, and the answer is that they should be good (i.e., tell the truth, not be pushy about their own judgments). Like the technocrats whom they criticize, they appear to believe that planners have a special claim on disinterested morality. . . .

The present trend among communicative planning theorists is to avoid broad examinations of the relationship between planning, politics, and urban development. Much recent work in planning theory has been devoted to examining the meanings of planners' conversations with developers and city officials, deconstructing planning documents, and listening to planners' stories. . . .

[. . .]

. . . Yet even if we accept the premise that the purpose of planning theory is simply to tell planners what they ought to be doing, such knowledge depends on an accurate appraisal of the situation in which planners find themselves. Explanatory theory allows the observer to identify the general characteristics of a situation, and these characteristics cannot be inferred simply through the examination of discourse. This is not to deny the usefulness of experiential learning or of case analysis in contributing to understanding. But it does mean transcending individual experience, placing cases in a broad context, making comparisons, and not limiting analysis to exegesis.

In addition to questions of method, communicative theory runs into the fundamental issues of pluralist theory. Communicative theorists avoid dealing with the classic topic of what to do when open processes produce unjust results. . . .

[British planner Patsy] Healey used the term collaborative planning to describe the process by which participants arrive at an agreement on action that expresses their mutual interests. She argued against a structuralist or political economy approach by contending that people do not have fixed interests. In other words, a particular structural position (e.g., capitalist) does not automatically produce a particular policy position (e.g., deregulation). Discussion can lead capitalists to understand how they could benefit financially from environmental regulation when they might reflexively have opposed any attempt to restrict their freedom to pollute. . . .

[. . .]

. . . [T]he . . . New Jersey State Plan . . . demonstrate[s] the efficacy of the communicative model. Here stakeholders from throughout the state participated in a series of meetings that produced a document targeting some areas for growth or redevelopment and others for conservation. Implementation depended on "cross-acceptance," whereby localities, rather than being forced to conform to the statewide plan, would agree to conduct their planning in accordance with it in return for certain benefits.

Yet . . . issues affected the implementation of the New Jersey State Plan. To start with, to win approval of the various participants in the planning process, the plan contained only weak requirements for the construction of affordable housing, suburban integration, and compact development, even though lack of housing for low-income residents, suburban exclusion of the poor and minorities, and lack of open space were identified as the principal problems that planning was supposed to overcome. Then, despite the moderate nature of the plan and the cross-acceptance process, its implementation has been halfhearted at best and often strongly resisted by local planning boards. The principal result of consensual planning in New Jersey has been the continuance of a system whereby the market allocates land uses.

[This example points] to one problem of communicative planning in practice – the gap between rhetoric and action. . . . A second practical problem of communicative planning is the lengthy time required for such participatory processes, leading to burnout among citizen participants and disillusion as nothing ever seems to get accomplished. . . . A third issue arises from the difficulties involved in framing alternatives when planners desist from agenda setting. Thus, for example, in Minneapolis, Minnesota, the city established a neighborhood planning process whereby residents formulated five-year plans for their neighborhoods and were allocated fairly substantial sums of money to spend. Planners assigned to facilitate the process were committed to a nondirective role and therefore only proposed actions when asked. The result was that some neighborhoods reached creative solutions, especially when participants were middle-class professionals, but others floundered in attempting to rank priorities and to come up with specific projects, sometimes taking as many as three years to determine a vague and hard-to-implement plan.

Finally, there is a potential conflict between the aims of communicative planning and the outcomes of participatory planning processes if planning is conducted within narrow spatial boundaries. The familiar specter of NIMBYism (not in my backyard) raises its head whenever participation is restricted to a socially homogeneous area. Communicative theorists are committed to equity and diversity, but there is little likelihood that such will be the outcome of stakeholder participation within relatively small municipalities. Organizing planning across a metropolitan area to encompass diversity of class, race, and ethnicity requires extending the process through multiple political jurisdictions to escape the homogeneity imposed by spatial segregation. The obstacles to involving citizens in metropolitan-wide planning, however, are enormous, and doing so means sacrificing the local familiarity that is the rationale for participatory neighborhood planning.

[. . .]

. . . [Finally,] city building for the benefit of nonelite groups requires empowering those who are excluded not just from discussions but from structural positions that allow them genuine influence. Ability to participate is one resource in the struggle for power, but it must be bolstered by other resources, including money, access to expertise, effective organization, and media coverage. Communicative theorists probably would not deny the importance of these resources, but neither do their analyses dwell on them. This omission constitutes the fundamental weakness of the theory.

THE NEW URBANISM

New Urbanism refers to a design-oriented approach to planned urban development. Developed primarily by architects and journalists, it is perhaps more ideology than theory, and its message is carried not just by academics but by planning practitioners and a popular movement. New Urbanists have received considerable attention in the United States and, to a lesser extent, in Great Britain. Their orientation resembles that of the early planning theorists – Ebenezer Howard, Frederic Law Olmsted, Patrick Geddes – in their aim of using spatial relations to create a close-knit social community that allows diverse elements to interact. The New Urbanists call for an urban design that includes a variety of building types, mixed uses, intermingling of housing for different income groups, and a strong privileging of the "public realm." The basic unit of planning is the neighborhood, which is limited in physical size, has a well defined edge, and has a focused center. . . .

The New Urbanism stresses the substance of plans rather than the method of achieving them. In practice, it has stimulated the creation of a number of New Towns and neighborhoods, of which Seaside, in Florida, is the best known. Fundamental to its development has been a critique of American suburbia. . . .

In this analysis, suburbia is responsible for far more than traffic congestion on the freeway and aesthetically unappealing strip-mall development. It is also the producer of crime and anomie.

In its easy elision of physical form with social conditions, the New Urbanism displays little theoretical rigor. Unlike other trends in planning, however, it is noteworthy for the popular response it has achieved. Although its appeal results partly from widespread dissatisfaction with suburban development and nostalgia for traditional forms, it also stems from the strong advocacy of its supporters, who have joined together in the Congress for the New Urbanism (CNU). The New Urbanists do not fear playing the role disdained by the communicative theorists – that of persuasive salespersons for a particular point of view and deployers of strategies aimed at coopting people. . . .

[Their leader, Andres] Duany did make a gesture toward participatory planning in his endorsement of citizen involvement in the charette, the lengthy design workshop that furnished the details of his developments. But one suspects that the purpose is as much cooptive as informative. . . . Duany and his confederates in the CNU did not fear distorted speech, nor did they shrink from using democratic procedures in responding to the public's stylistic preferences as a screen to achieve their desired sociospatial arrangements.

CRITIQUE

The New Urbanism is vulnerable to the accusation that its proponents oversell their product, promoting an unrealistic environmental determinism that has threaded its way throughout the history of physical planning. [Nonetheless, some have] praised certain aspects of the New Urbanism – its emphasis on public space, its consideration of the relationship between work and living, and its stance toward environmental quality. . . .

As a consequence of its spatial determinism, the New Urbanism runs into certain dangers. One frequently made criticism is that it merely calls for a different form of suburbia rather than overcoming metropolitan social segregation. Duany responded to this accusation by arguing that, because most Americans are going to live in suburbs, planners need to build better suburbs. Moreover, he contended that it is not his philosophy but, rather, political opposition and obsolete zoning ordinances that prevent him from working in inner cities. And indeed, the effort to overcome the environmentally destructive, wasteful form of American suburban development constitutes the most important contribution of the New Urbanism to the common weal.

The movement is less convincing in its approach to social injustice. . . . [T]he New Urbanism can commit the same errors as modernism – of assuming that changing people's physical environment will somehow take care of the social inequalities that warped their lives. To be sure, with its emphasis on community, it is unlikely to commit the principal sin of modernist redevelopment programs – destroying communities to put people in the orderly environments that were thought to enhance living conditions. The real problem replicates the one that defeated Ebenezer Howard's radical principles in the construction of Garden Cities. To achieve investor backing for his schemes, Howard was forced to trade away his aims of a socialist commonwealth and a city that accommodated all levels of society. The new urbanists must also rely on private developers to build and finance their visions; consequently, they are producing only slightly less exclusive suburbs than the ones they dislike. Although their creations will contain greater physical diversity than their predecessors', their social composition will not differ markedly.

. . . [T]he New Urbanist emphasis on community disregards "the darker side" of communitarianism. . . . [T]he enforced conformity of community blocks the creativity arising from diversity and conflict. . . . Two problems come to the fore here. The classic and more important dilemma results from the two-edged quality of community, which in providing emotional sustenance to its members, necessarily excludes others. A second problem arises within theories of planning and urban design that urge the creation of exciting locales: Is planned diversity an oxymoron? . . .

At the same time, relying on the market for an alternative to planning will not overcome the problem of homogeneity. The failure of the market to provide diversity in most places means that, if planners do not attempt to foster it, the outcome will be increasingly segregated neighborhoods and municipalities. . . .

Only a publicly funded effort to combine social groups through mixing differently priced housing with substantial subsidies for the low-income component can produce such a result. The New Urbanists seek to create housing integration but, in their reliance on private developers, are unable to do so on a sufficient scale or across a broad enough range of housing prices to have a significant effect. However, a serious effort to attract public subsidy for the low-income component of their communities would involve the New Urbanists in a political battle for which their architectural training and aesthetic orientation offer few resources. The appeal of Victorian gingerbread and Cape Cod shingle would not override the fear of racial and social integration.

For planning theory, the most interesting aspect of the New Urbanism is that its assurance of a better quality of life has inspired a social movement. Its utopianism contrasts with communicative planning, which offers only a better process. Thus there is a model of planning practice that is based not on the picture of the sensitive planner who listens and engages in ideal speech but on the messianic promise of the advocate who believes in a cause and eschews neutrality. As in all such cases, the benefits are exaggerated. But there is an attraction to the doctrine, both because of its hopefulness and because the places it seeks to create do appeal to anyone tired of suburban monotony and bland modernism.

THE JUST CITY

[. . .]

Just-city theorists fall into two categories: radical democrats and political economists. The former differ from communicative planning theorists in that they have a more radical concept of participation that goes beyond the involvement of stakeholders to governance by civil society, and they accept a conflictual view of society. They believe that progressive social change results only from the exercise of power by those who previously had been excluded from power. Participation is the vehicle through which that power asserts itself. The political economy group, upon whom I shall focus in this section and among whom I include myself, takes an explicitly normative position concerning the distribution of social benefits. . . .

The audience for this endeavor has remained vaguely defined. By inference, however, one can deduce that the principal target group is the leadership of urban social movements. Because political economic analysis mostly condemns policy makers for being the captive of business interests, it is addressed primarily to insurgent groups, to officials in progressive cities, and to "guerrillas in the bureaucracy". Whereas the communicative planning theorists primarily speak to planners employed by government, calling on them to mediate among diverse interests, just-city theorists do not assume the neutrality or benevolence of government. For them, the purpose of their vision is to mobilize a public rather than to prescribe a methodology to those in office.

A theory of the just city values participation in decision making by relatively powerless groups and equity of outcomes. The key questions asked of any policy by political economists have been "Who dominates?" and "Who benefits?" The "who" has typically been defined by economic interest, but economic reductionism is not necessary to this mode of analysis; evaluation of outcomes can also be conducted with regard to groups defined by gender, race, and sexual orientation. Nor does the emphasis on material equality need to boil down to an expectation that redistribution should proceed to a point at which there is no reward to achievement.

The characteristic weakness of socialist analysis has been its dismissal of economic growth as simply capital accumulation that benefits only capitalists. Socialist doctrine fails to mobilize a following if it only ensures greater equality without also offering improved circumstances for most people. The market model and neoliberalism have proved popular because they promise increases in affluence for all even if within the context of growing inequality. . . .

A persuasive vision of the just city needs to incorporate an entrepreneurial state that not only provides welfare but also generates increased wealth; moreover, it needs to project a future embodying a middle-class society rather than only empowering the poor and disfranchised. . . . [A] great deal more attention needs to be paid to identifying a formula for growth with equity. And such an approach has to take into account the perseverance of a capitalist world economy and the evident success, at least for the moment, of a liberalized U.S. economy.

Participation in public decision making is part of the ideal of the just city, both because it is a worthy goal in itself and because benevolent authoritarianism is unlikely. At the same time, democracy presents a set of thorny problems that have never been theoretically resolved. . . . Democratic pluralism, with its emphasis on group process and compromise, offers little likelihood of escape from dominance by those groups with greatest access to organizational and financial resources. Democratic rule can deprive minorities of their livelihood, freedom, or self-expression. Classic democratic theory deals with this problem through imbuing minorities with rights that cannot be transgressed by majorities. But what of the minority that seeks to exercise its rights to seize power and take away the rights of others in the name of religious authority or racial superiority? Democratic principles can easily accommodate ineffective or harmless minorities; they flounder when confronted with right-wing militias, religious dogmatists, and racial purists. Thus the appropriate criterion for evaluating a group's claims should not be procedural rules alone; evaluation must comprise an analysis of whether realization of the group's goals is possible and, if so, whether such realization leaves intact the principle of social justice. Democracy is desirable, but not always.

Within a formulation of the just city, democracy is not simply a procedural norm but rather has a substantive content (see Pitkin 1967). Given the existing system of social domination, it cannot be assumed that participation by stakeholders would be transformative in a way that would improve most people's situation. Consequently, deliberations within civil society are not ipso facto morally superior to decisions taken by the state. . . .

[. . .]

Applying the just-city perspective, one must judge results, and furthermore, one must not forget that the results attainable through public policy are seriously constrained by the economy. Thus, even when the principal concern is not economic outcomes but ending discrimination or improving the quality of the environment, economic interests limit possible courses of action. To go back to the example of the New Jersey State Plan mentioned earlier, its primary purpose was environmental protection, not social integration or redistribution of land and property. Nevertheless, its content was affected by the state's dependence on private investors for new development and its implementation restricted by fears of landowners that their property values would be adversely affected by growth regulation. Thus economic interests impinge on planning even when the economy is not its foremost object.

[. . .]

RESURRECTING OPTIMISM

The three types of planning theory described in this article all embrace a social reformist outlook. They represent a move from the purely critical perspective that characterized much theory in the 1970s and 1980s to one that once again offers a promise of a better life. Whereas reaction to technocracy and positivism shaped planning theory of that period, more recent planning thought has responded to the challenge of postmodernism. . . .

[. . .]

The progressives of the previous period spent much of their energy condemning traditional planning for authoritarianism, sexism, the stifling of diversity, and class bias. More recent theorizing has advanced from mere critique to focusing instead on offering a more appealing prospect of the future. For communicative planning, this means practices that allow people to shape the places in which they live; for New Urbanists, it involves an urban form that stimulates neighborliness, community involvement, subjective feelings of integration with one's environment, and aesthetic satisfaction. For just-city theorists, it concerns the development of an urban vision that also involves material well-being but that relies on a more pluralistic, cooperative, and decentralized form of welfare provision than the state-centered model of the bureaucratic welfare state.

At the millennium's end, then, planning theorists have returned to many of the past century's preoccupations. Like their nineteenth-century predecessors, they are seeking to interpose the planning process between urban development and the market to produce a more democratic and just society. The communicative theorists have reasserted the moral preoccupations that underlay nineteenth-century radicalism, the New Urbanists have promoted a return to concern with physical form, and just-city theorists have resurrected the spirit of Utopia that inspired Ebenezer Howard and his fellow radicals. Although strategic and substantive issues separate the three schools of thought described here, they share an optimism that had been largely lacking in previous decades. Sustaining this optimism depends on translating it into practice.

Planning today and yesterday

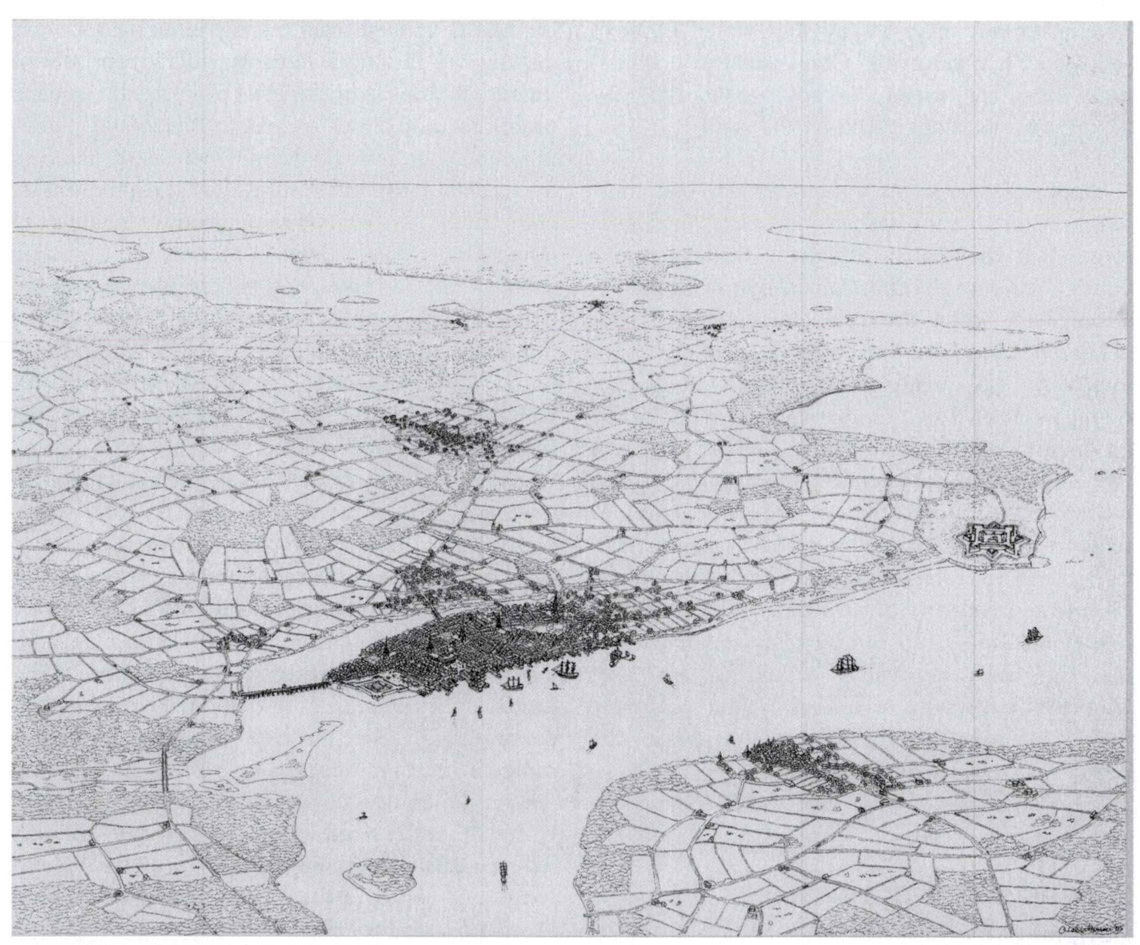

Growth of a typical Northeastern city: 1760. (Drawing by Andrew Whittemore)

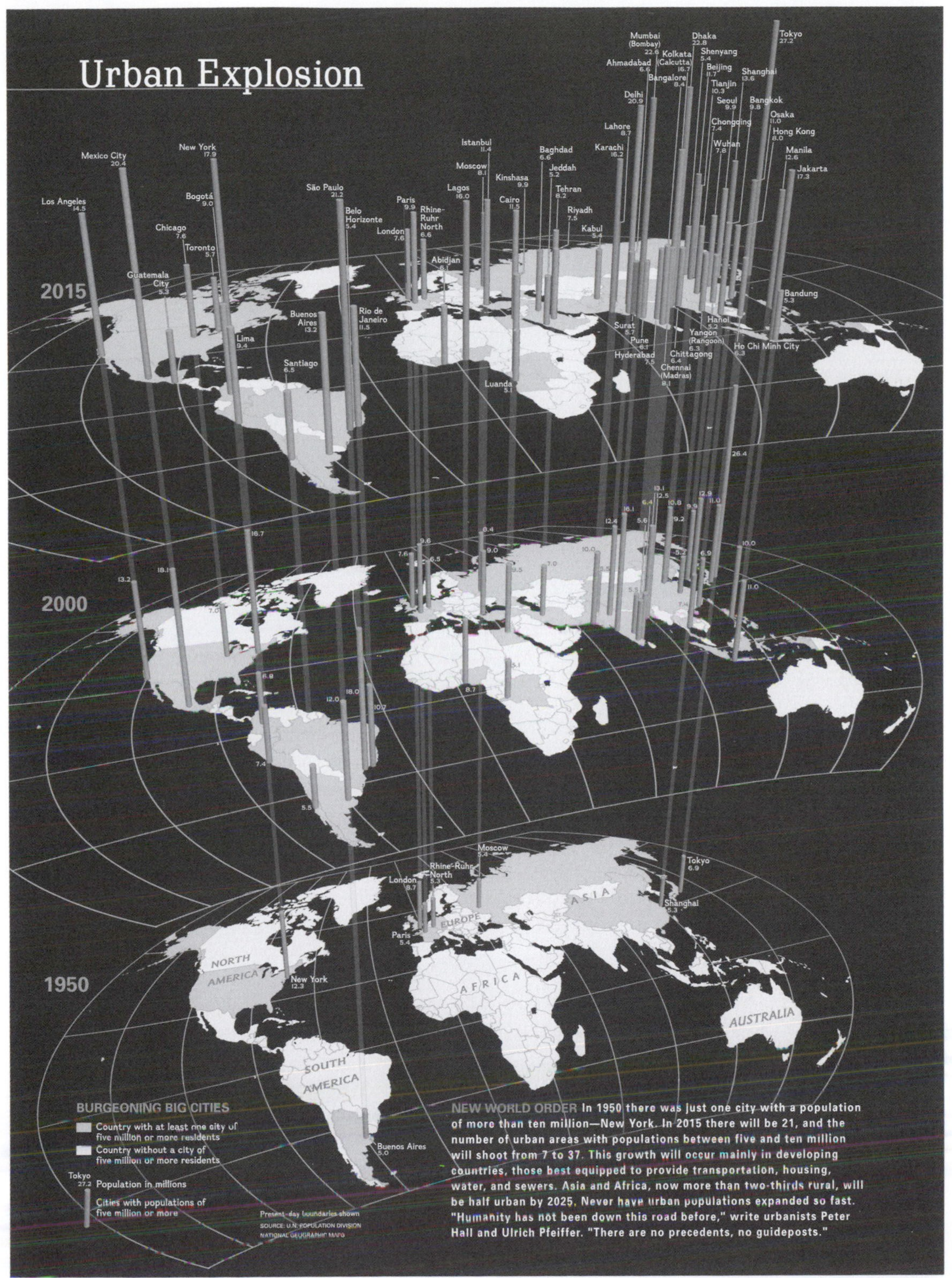

Plate 1 *World Urbanization*. In the century of the city, this image illustrates the location of today's urban inhabitants. By 2050, 5 billion or 60 percent of the world's 8.2 billion population will be urban. The cities in the global South will experience the greatest growth, with Asia and Africa anticipating the highest rates of increase. (Courtesy of the National Geographic Society)

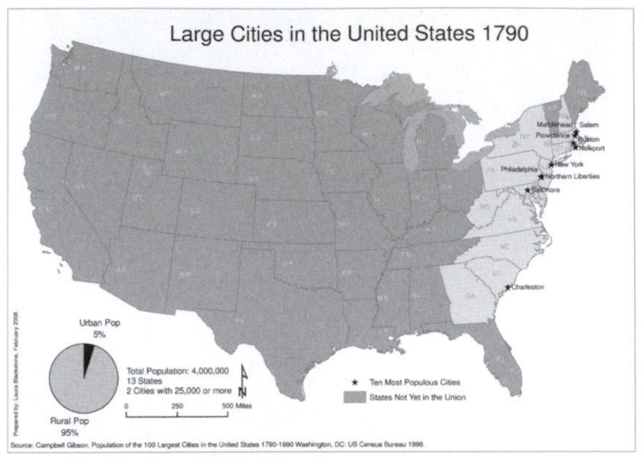

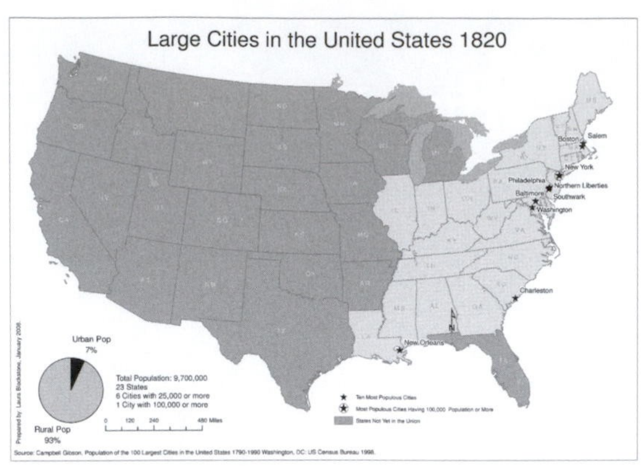

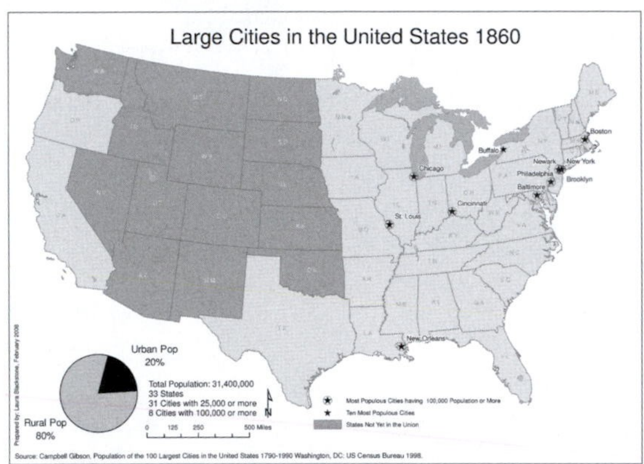

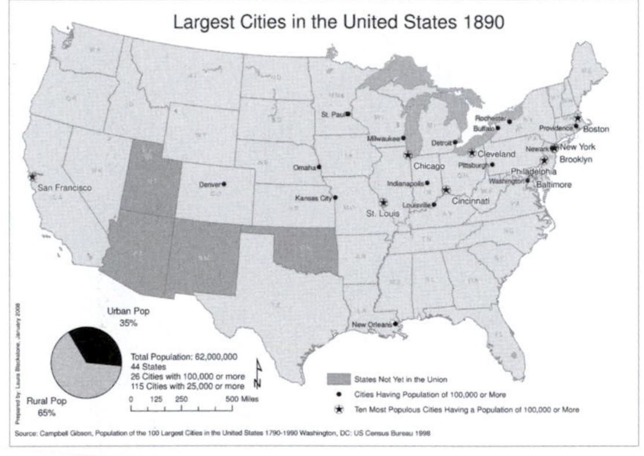

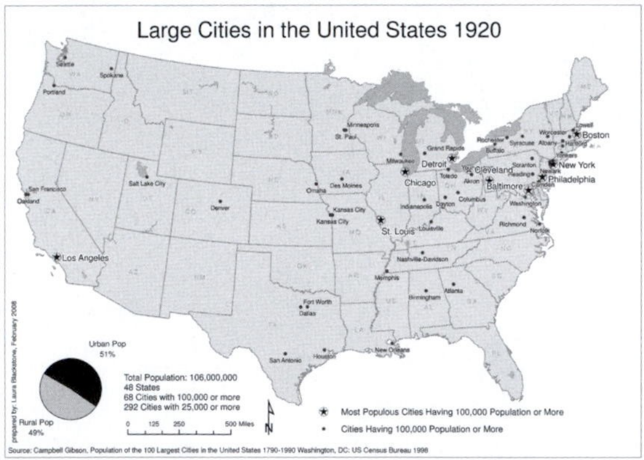

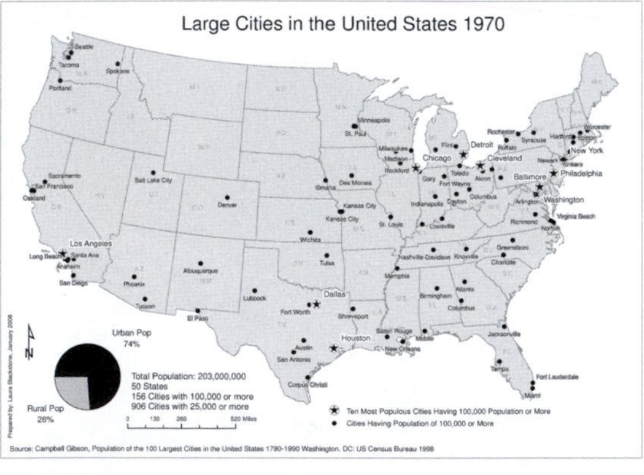

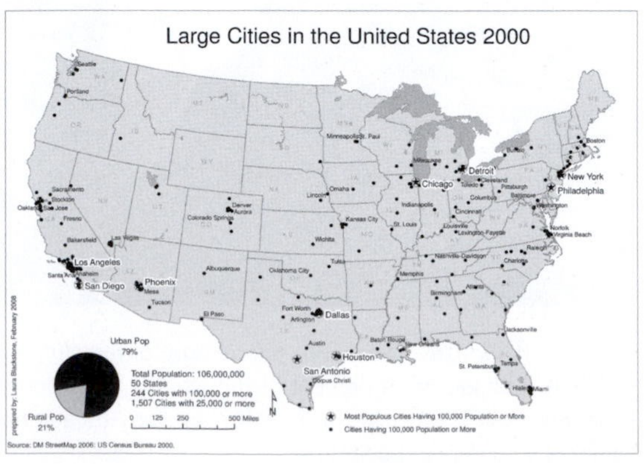

Plate 2 *U.S. Large Cities, 1970–2000.* Over time, the U.S. population has become increasingly urban, moving from 5 percent in 1790 to 51 percent in 1920 to 79 percent in 2000. Concurrently, large cities having a population of 100,000 or more grew from one (New York) in 1820 to 244 in 2000. (Courtesy Eugénie L. Birch)

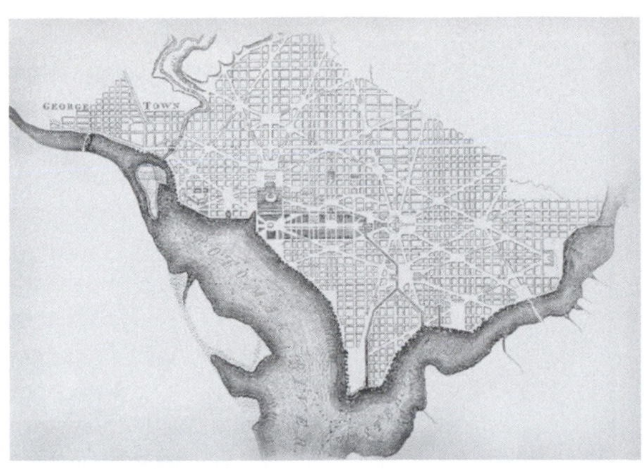

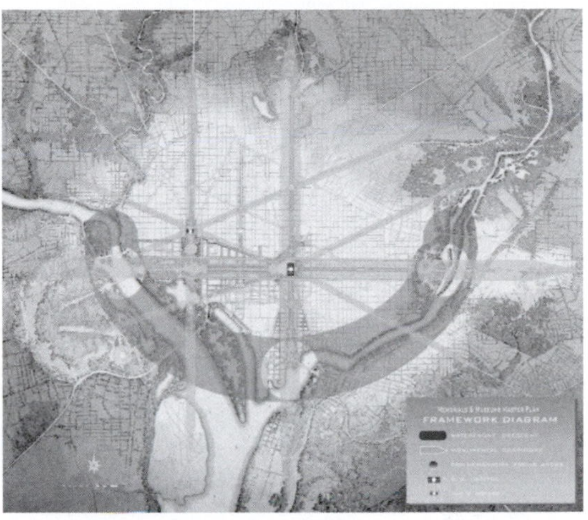

Plate 3 *Washington DC 1791 and 1997.* The L'Enfant plan for Washington 1791 (*left*) laid out a pattern for the nation's capital that modern planners are using as a foundation for future growth and development. The image of the city's Monumental Core Framework from *Extending the Legacy: Planning America's Capital for the Twenty-first Century* (1997) (*right*) illustrates how the National Capital Planning Commission, the agency charged with overseeing planning matters, envisions building from the past. (Courtesy National Capital Planning Commission)

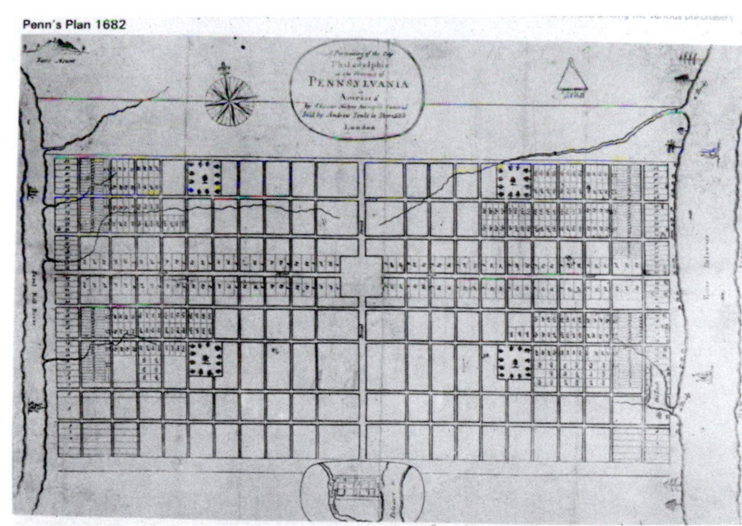

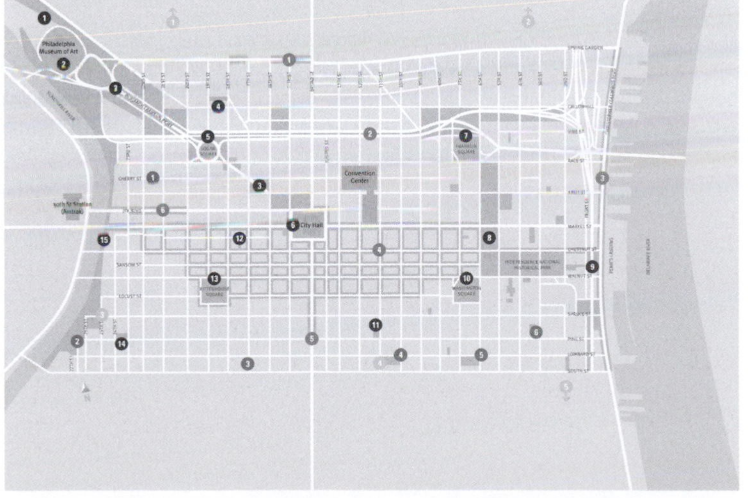

Plate 4 *Philadelphia Greenways.* Greening Philadelphia has a long history dating from the city's earliest plan, William Penn's "Greene Country Towne" (1683) encompassed 1,200 acres and incorporated a central square and four parks within its gridded streets (*upper*). Today, the Philadelphia Center City District (CCD) serves as a guardian of green, protecting the original plan and such later additions as the City Beautiful-era Benjamin Franklin Parkway (1917) and Independence Hall Mall and Society Hill neighborhood, the solid coloured areas on the map (lower). CCD also campaigns for greening downtown, as indicated by the dark lines along the streets in the lower map. (Courtesy Library of Congress and Philadelphia Center City District)

Plate 5 *New York City: Carver Houses, before and after*. Late nineteenth-century housing reformers aimed to relieve slum conditions exemplified by the block (*left*) on the notorious Lower East Side. They promoted building new neighborhoods with airy, well-lit apartments, fewer streets and ample open space landscaped with playgrounds, public art, and greenery, as exemplified by the Carver Houses (*right*) of 1958, a thirteen-building, 1,246 unit complex in Harlem. (Courtesy LaGuardia and Wagner Archives, La Guardia Community College/City University of New York)

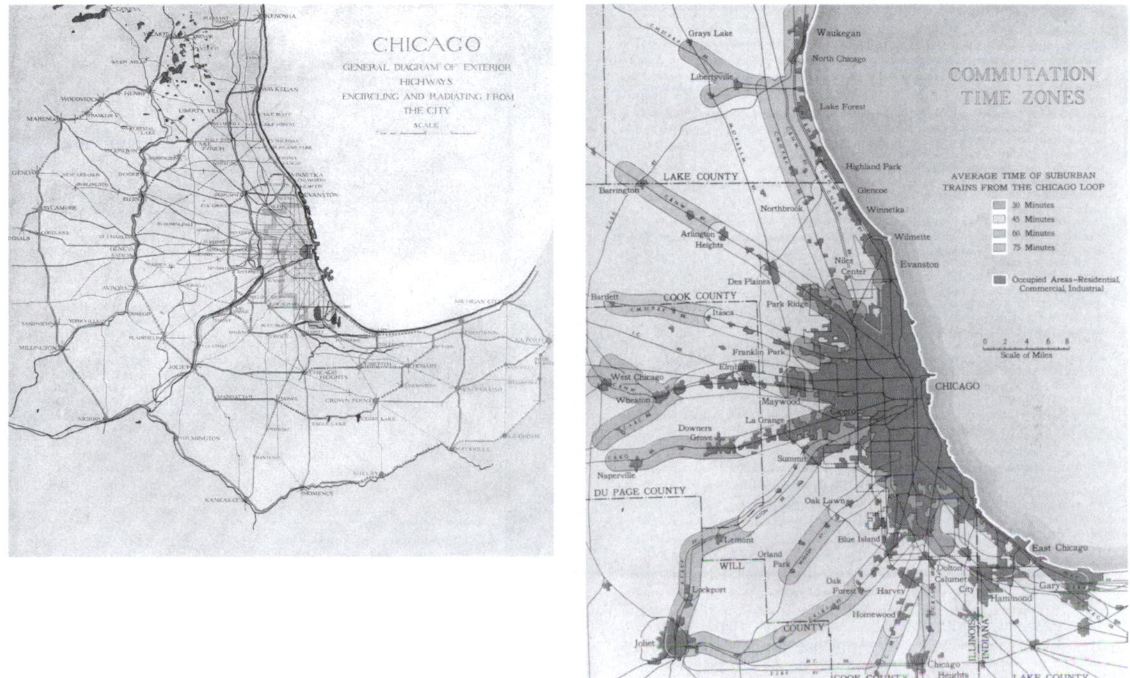

Plate 6 *Chicago: 1909 Plan for Chicago and 1999 Chicago 2020 Metropolis Plan.* Planning for the Chicago region began at the turn of the twentieth century when the city's Commercial Club commissioned local architect Daniel Burnham to produce the *Plan of Chicago* (1909) (*left*). Burnham proposed strengthening the city center, improving regional transportation and open space. One hundred years later, the Commercial Club sponsored an updated plan, *Metropolis 2020* (*right*) that reinforced the importance of metropolitan planning along the lines originally set by Burnham. (Courtesy Chicago Metropolis 2020)

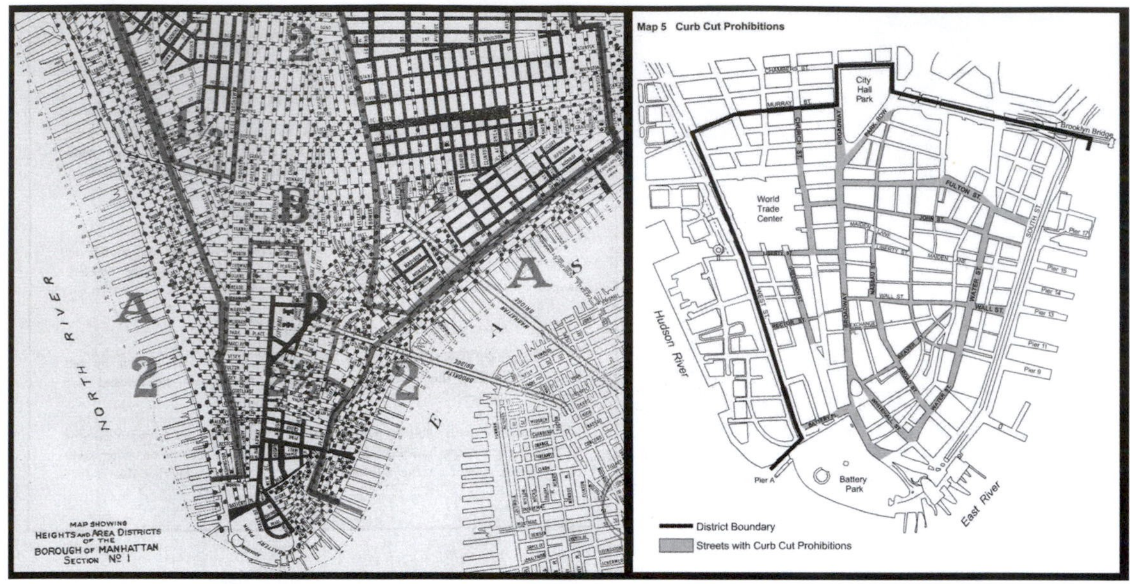

Plate 7 *Lower Manhattan Zoned: Two Eras.* New York City passed the nation's first comprehensive zoning ordinance in 1916 and has remained a pacesetter in using zoning to achieve planning goals. In 1916 (*left*) zoning established uses and density while protecting light and air in the heavily built-up Lower Manhattan. In 1998, it focused again on Lower Manhattan by accommodating several urban design concerns, including prohibiting curb cuts, as indicated by the heavy grey lines, in order to protect the area's pedestrian-friendly environment (*right*). The City amended the Lower Manhattan zoning in 2001 and 2007 in an effort to shape the area to accommodate a mixed-use downtown. (Courtesy New York City Department of City Planning)

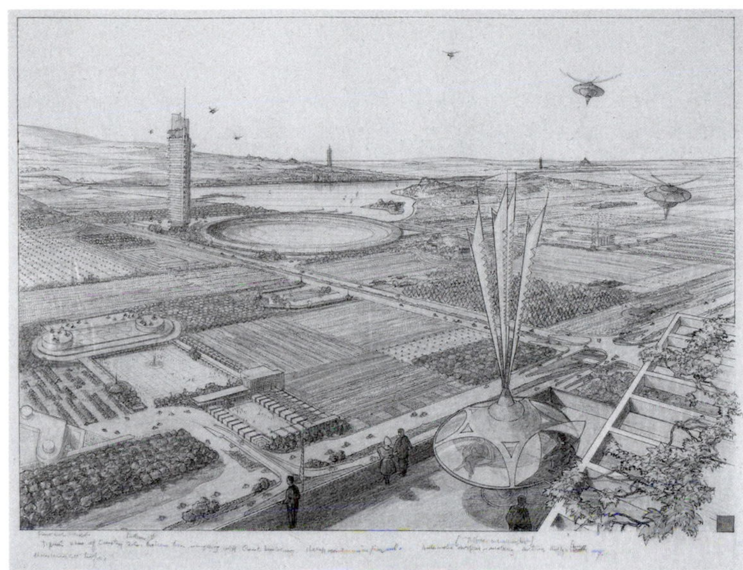

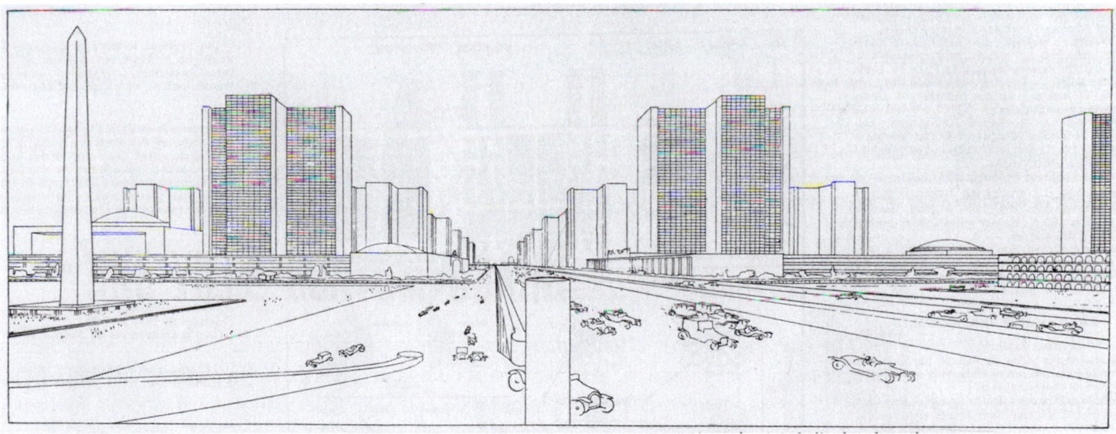

Plate 8 *Broadacre City and the Ville Contemporaine.* Frank Lloyd Wright's vision, Broadacre City (1932) (*upper*), dispersed single-family housing in verdant countryside laced with a modern street grid, civic amenities and active farms. (Courtesy Art Resource.) In contrast, Le Corbusier's *Ville contemporaine* (1922) and *Ville radieuse* (1935) (*lower*) called for high densities, with living and working taking place in skyscrapers, connected by high-speed highways. (Courtesy Fondation Le Corbusier/Artists' Rights Society)

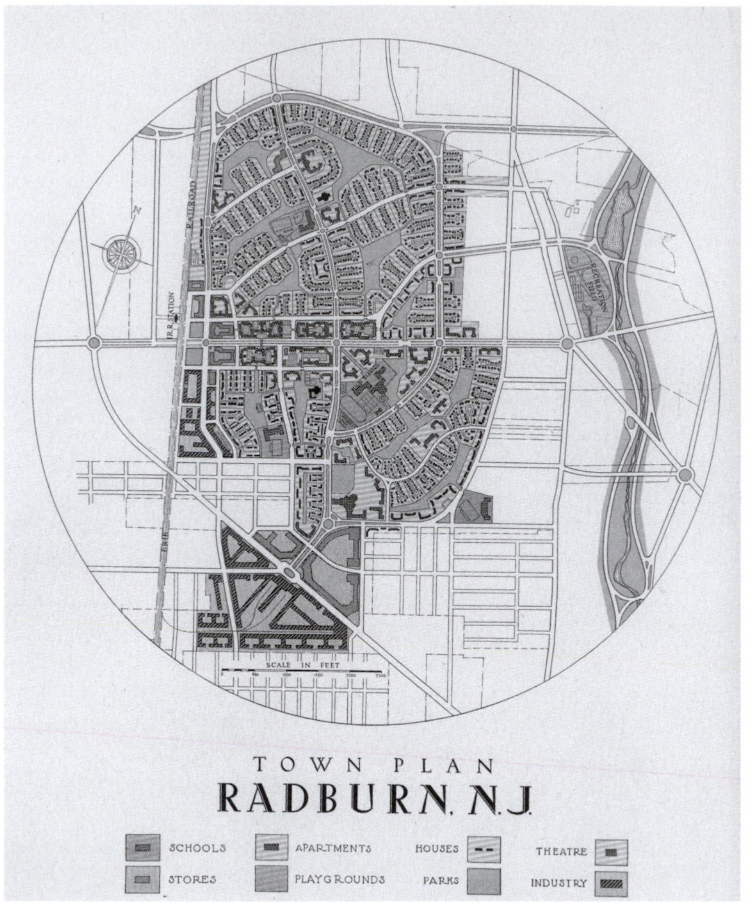

Plate 9 *The Radburn Plan*. The plan for Radburn, NJ, a garden city, incorporated important design innovations: the neighborhood unit, a residential district within walking distance of a school, and a unique street system, later called the "Radburn idea," that promoted the separation of different types of traffic and safe pedestrian ways. Radburn's sponsor, the City Housing Corporation, built only a small section of the planned community before the Depression forced it into bankruptcy. Extremely influential, its residential and transportation concepts were replicated in many later community plans. (Courtesy the Regional Plan Association, New York)

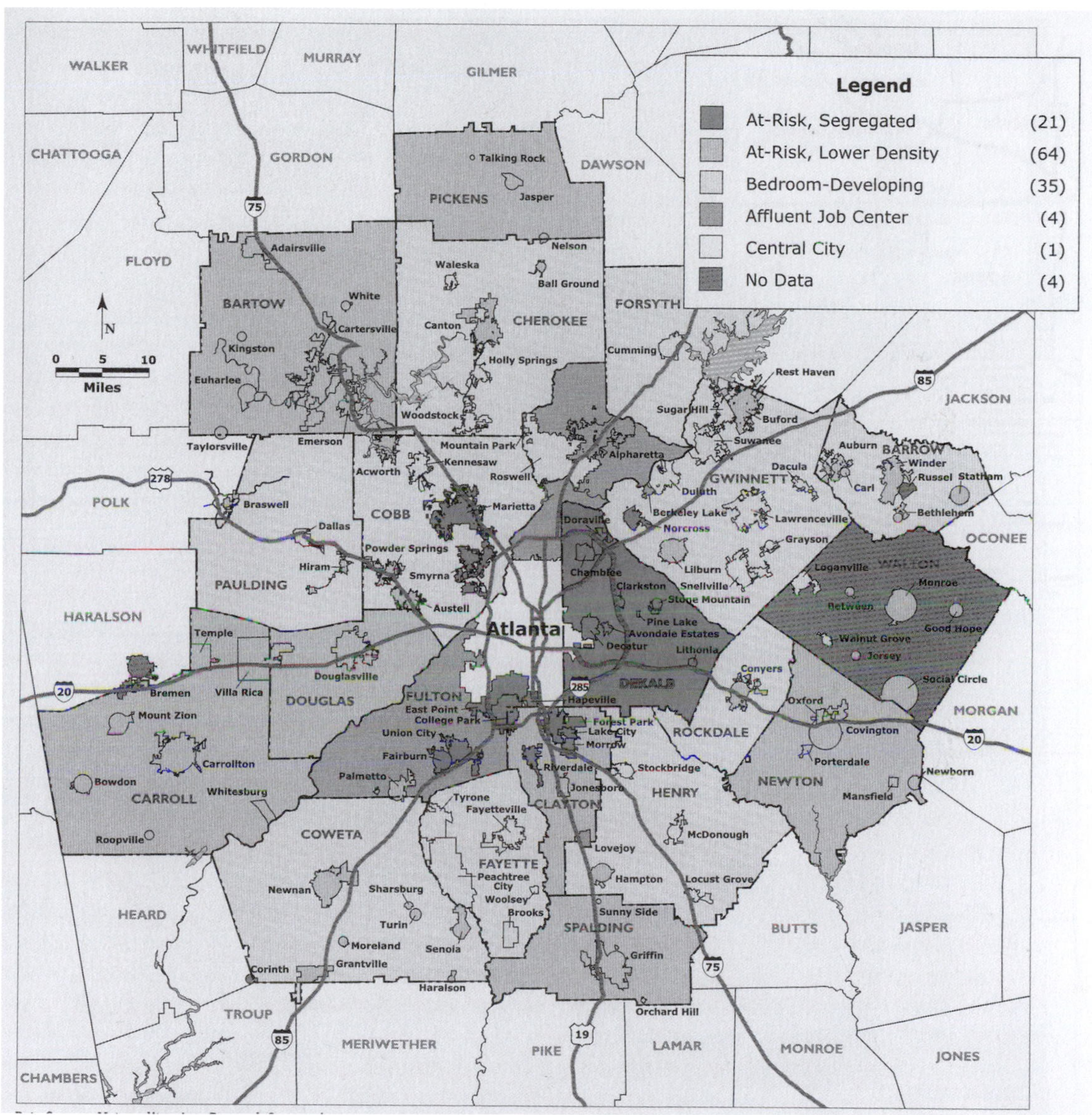

Plate 10 American Metropolitics *View of Atlanta*. The Atlanta metropolitan area, portrayed according to descriptive categories defined by Myron Orfield in *American Metropolitics*, is composed of towns, cities and counties whose populations vary according to income, educational level, age and race. These variations create differing needs for municipal services. These communities also have differing capacities to pay for the needed services, depending on the composition of their tax bases. (Courtesy the Brookings Institution)

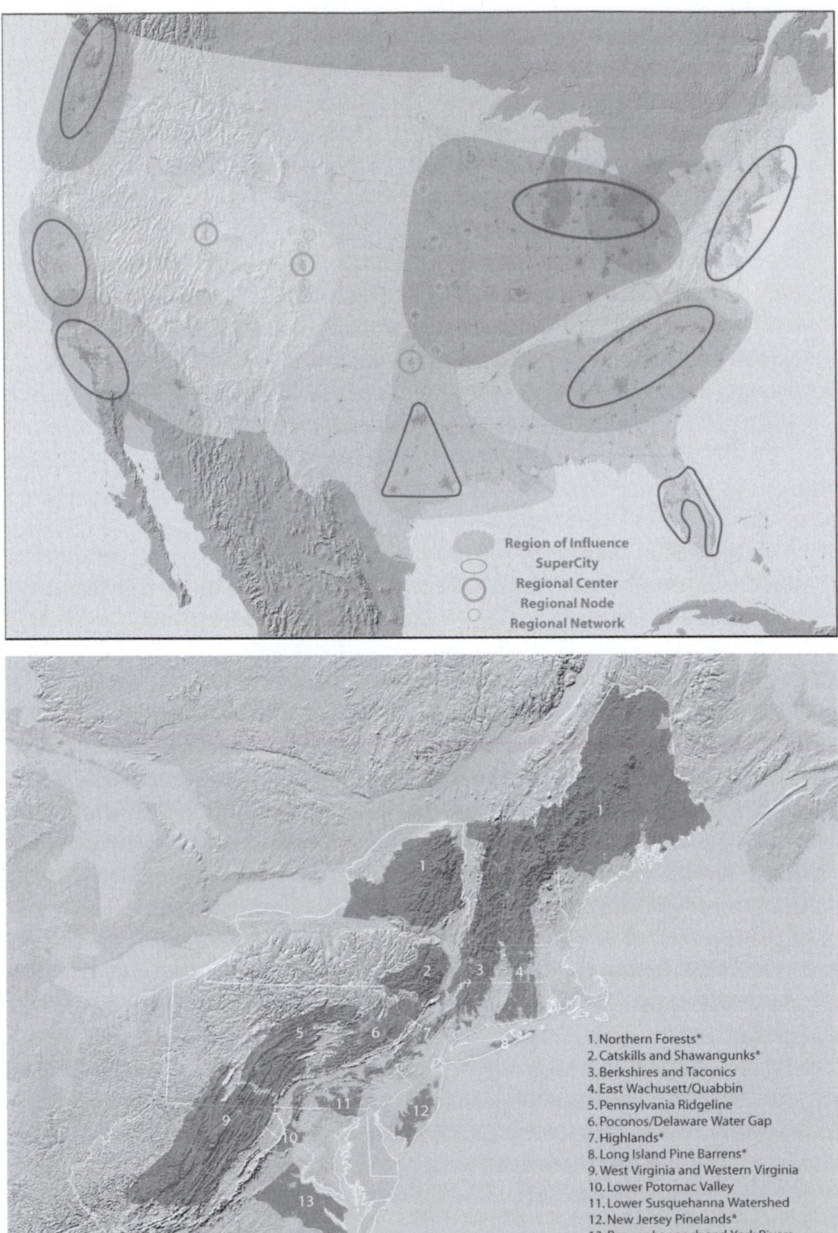

Plate 11 *U.S. Megaregions and the Northeast Megaregion Ecostructure Map.* Analysts with America 2050 anticipate that U.S. population, which will reach 400 million well before mid century, will concentrate in the ten megaregions illustrated on the map (*upper*). (Courtesy Regional Plan Association.) To remain economically competitive and sustainable, these agglomerations have to invest systematically in infrastructure and open space. The map of the Northeast megaregion (*lower*) outlines the environmentally sensitive areas needing attention in the near future. (Courtesy University of Pennsylvania)

Plate 12 *Sprawl, c. 2006.* Most observers define sprawl as scattered, low-density development, as shown in this small residential neighborhood. Commercial uses also contribute to sprawl as they spread throughout the landscape. For many city planners the word "sprawl" has negative connotations, because such development forms consume land randomly, require new roads – often disconnected, as seen above – and other municipal services that could be provided more efficiently if the development were continuous and compact. (Courtesy Emily Yuhas)

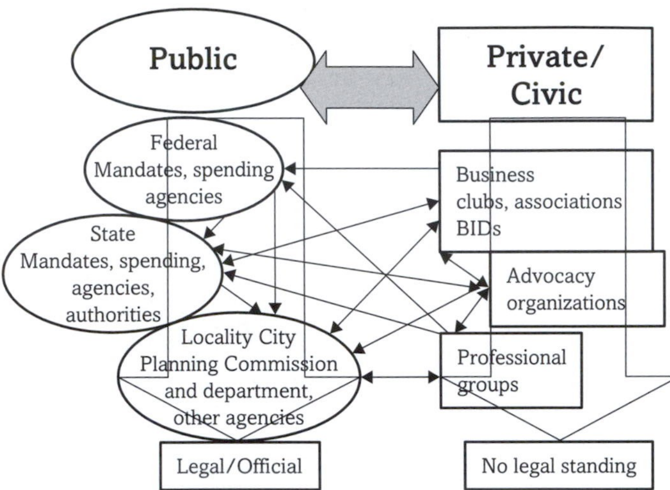

Plate 13 *Planning Agents.* In the United States many agents or participants are involved in urban and regional planning. They have complex interrelationships, as indicated by the arrows flowing around and among them on the diagram. They have varying sources of power; some have a legal mandate to plan, others have no legal standing but gain power through exercising their right to vote or lobby or express authoritative knowledge due to professional training. (Courtesy Eugénie L. Birch and Routledge/Taylor & Francis Group)

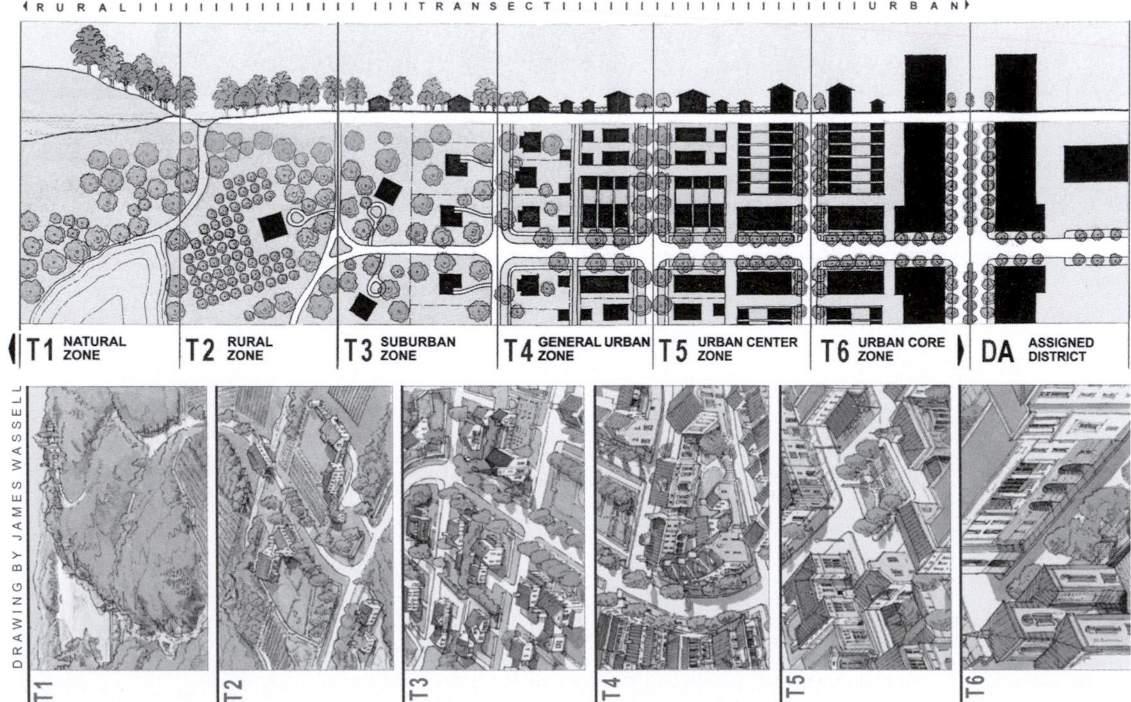

Plate 14 *The Transect.* The New Urbanists, a group identified with architects Andres Duany and Elizabeth Plater-Zyberk, conceive land and its use as a continuum or "transect" that starts with natural areas having the least development and ends with the urban core, having the most development. They distinguish six steps along the way by attributing different amounts of open space, streets, and buildings. This concept underlies their proposed zoning code revisions, labeled "form-based codes," that regulate form and allow mixed uses. (Courtesy Duany Plater-Zyberk & Co.)

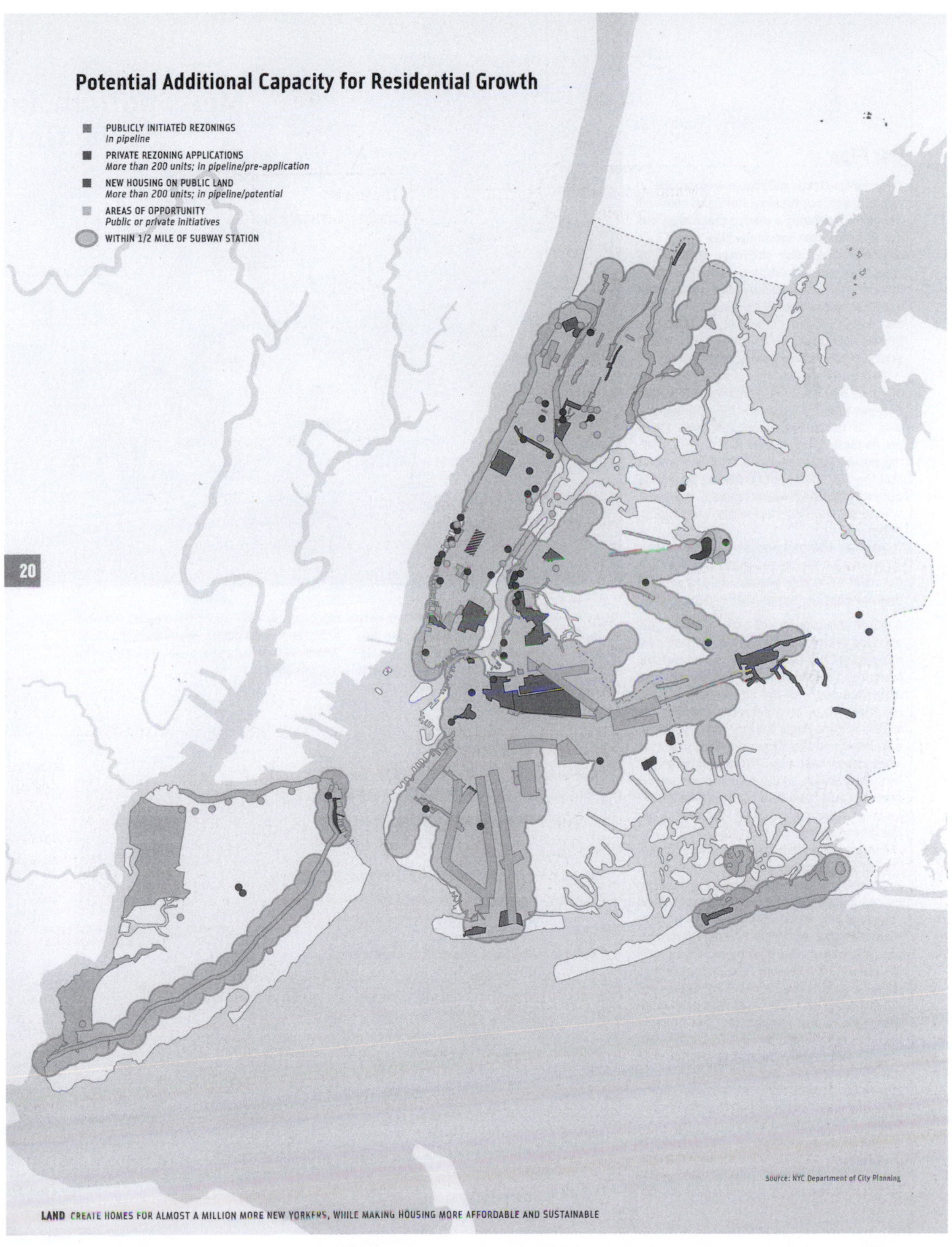

Plate 15 *PlaNYC 2030.* On Earth Day 2007, Mayor Michael Bloomberg unveiled *PlaNYC 2030*, an effort to address sustainable growth issues with advanced planning techniques. The plan shows how the city can absorb an anticipated increase of one million people by building more densely in areas close to its subway system, as illustrated in this image. It also calls for limiting vehicular traffic through congestion pricing, planting one million trees and assuring the presence of parks within walking distance of all residents. (Courtesy City of New York)

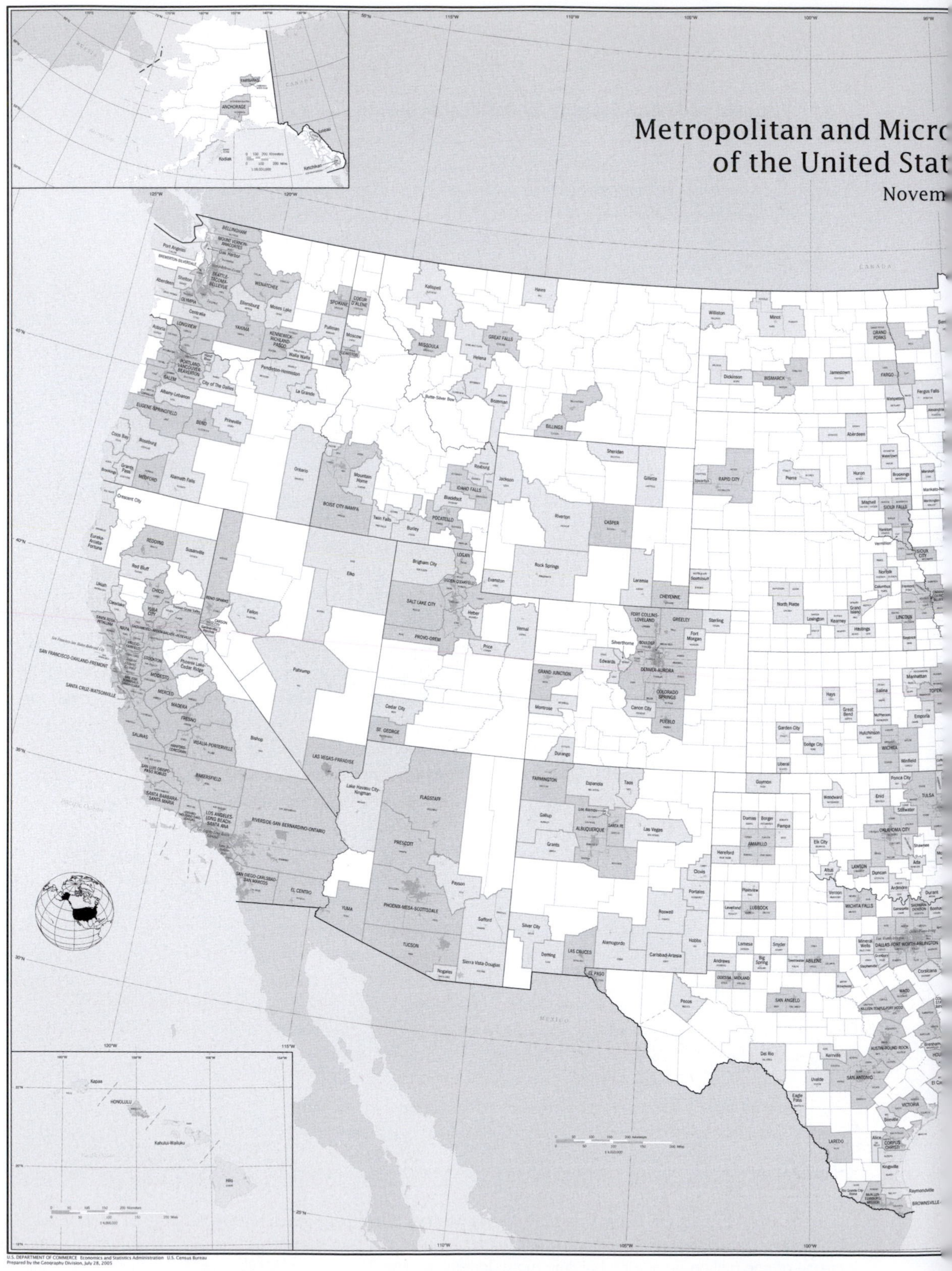

Plate 16 *Metropolitan and Micropolitan Areas, 2004.* The U.S. Bureau of the Census collects data on several kind of places, including cities and metropolitan areas. Since the 1950s, it has designated Metropolitan Statistical Areas, or metros (cities of 50,000 and their surroundings), to capture rising suburbanization. It has since added a new category,

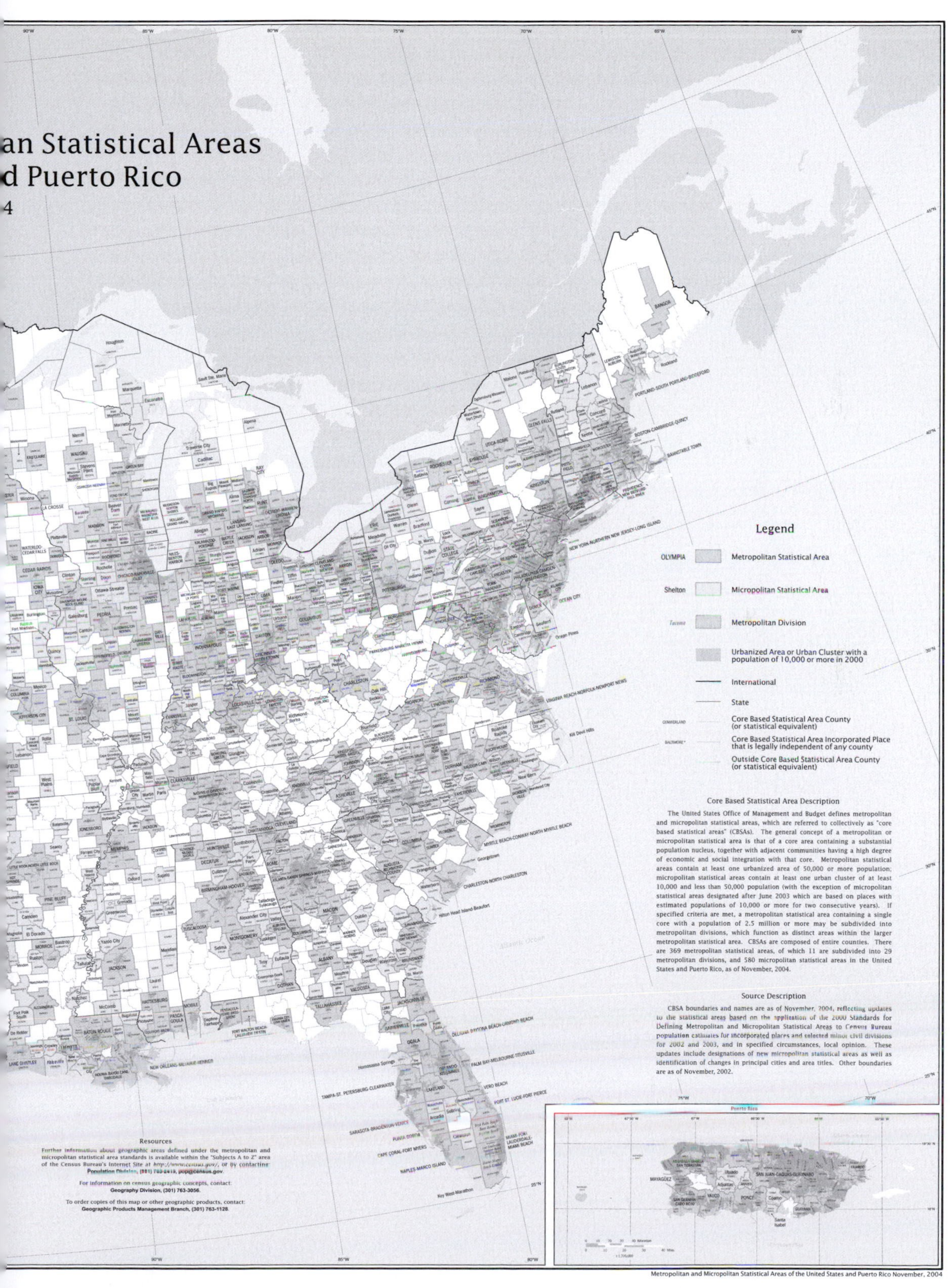

Micropolitan Statistical Areas, or micros (cities of 10,000 to 49,000 and their surroundings), to indicate current development patterns. This map shows the location of the nation's metros and micros. (Courtesy U.S. Bureau of the Census)

PART THREE

Classics in Urban and Regional Planning

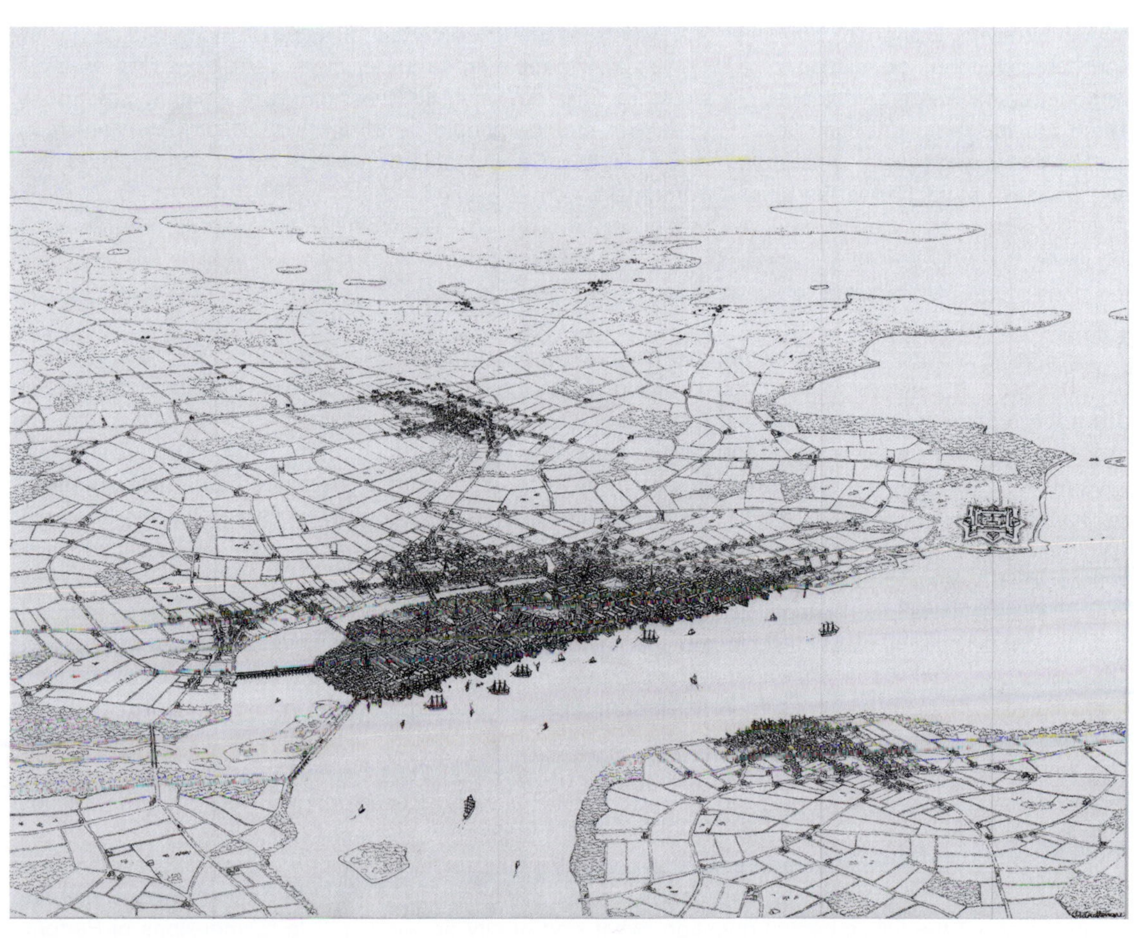

Growth of a typical Northeastern city: 1820. (Drawing by Andrew Whittemore)

INTRODUCTION TO PART THREE

Back in the early twentieth century, when American cities were exploding as a result of industrialization and immigration, a new class of literature emerged that reflected on these changes. Many of these works have become classics. In *Sister Carrie*, novelist Theodore Dreiser vividly traced the extraordinary life of a young woman who came to Chicago from the countryside. Journalist Upton Sinclair also used Chicago as the backdrop for *The Jungle*, a dramatic tale of an immigrant family's decline in their new home. And in publishing an article entitled "Urbanism as a Way of Life" (*American Journal of Sociology*, 44, 1938, pp. 1–24), University of Chicago professor Louis Wirth opened up the field of urban sociology.

Wirth tackled the question of physical determinism in his essay – specifically, how urban living affects human behavior. Using Chicago as his laboratory, he identified certain qualities of a city, like its density and heterogeneous population, and observed how people acted under these conditions. For example, when a person walks down a busy city street, he does not say hello to passers by – it simply is not done. There are too many encounters, and greeting a stranger would be an invasion of privacy. Wirth concluded that city dwellers, isolated from the familiar social contacts of a village or rural place, felt lonely and socially unattached in the urban environment.

In presenting this idea of urban anonymity, Wirth dropped a bombshell – no one had thought about city living that way before. His essay laid the foundation for urban sociology, a discipline still examining the phenomena Wirth called to our attention. And it became a classic, read by generations of students interested in urbanism.

So what is a classic? It is a memorable piece of writing that has changed people's views on essential questions. It conveys wisdom and knowledge. And, most important, it retains its hold on readers down through the years. A person might peruse a classic in her youth and find it compelling. When she revisits it later, it provides more or different insights. Why? The returned reader, now more experienced, sees new dimensions in the work. Or times may have changed so that the piece has a different meaning for her. The Italian novelist Italo Calvino captured the essential quality of classics when he wrote: They "exert a peculiar influence, they refuse to be eradicated from the mind . . . they conceal themselves in the folds of memory, camouflaging themselves as the collective or individual unconscious." (*Why Read the Classics?* New York: Vintage, 2001.)

In a scholarly arena, classics serve many functions. They provide a foundation for a branch of inquiry. They raise basic questions. They offer important research models. In the field of urban and regional planning, classics range from books to *amicus curia* briefs. Always read by planning students, they have become part of the field's collective knowledge, providing common vocabulary and reference points. The seven selections offered here are not the only classics in the field, but they are important ones, raising broad questions about planning: its *raison d'être*, underlying assumptions, motivations, and practitioners' roles.

The first three readings are addressed to a wide-ranging audience, from the professional planner to the citizen. Their authors want to help people across the spectrum develop a critical eye and cultivate judgment about the field's central question: what kind of city do we want? In "Dimensions of Performance," urban designer Kevin Lynch suggests that planners should aspire to an ideal – "the good city"

– and then offers ideas about what such an ideal might be, insisting planners develop performance standards. Having made these suggestions, Lynch raises a number of questions for planners to consider. Is "the good city" the same for everyone? How do people know if they have reached the ideal living environment? And how can planners help them evaluate these places?

When journalist Jane Jacobs wrote "Downtown Is for People," the exasperation that she would later express in her bestseller, *The Death and Life of Great American Cities* (New York: Random House, 1961), is quite evident. *Can't you get out of the clouds and look around?* she chides urban planners. *Can't you listen to the locals before you begin to plan? Can't you see what works in our cities?* The reader can almost see her pounding her head and saying, "It's the street, stupid!"

Don't be fooled into thinking that the next selection, "Home Remedies for Urban Cancer" by the public intellectual Lewis Mumford, is merely a book review. It is not. It is a thoughtful discourse about urbanism in which Mumford insists that cities can be too big. Planners need to consider this challenge when deciding whether they should limit city size, and how to manage urban density and open space and establish proper relationships among cities and their regions. Mumford reminds planners that they should know how to separate good urban design from bad, and [how to] recognize and stimulate progress and innovation in planning cities.

The last selections focus on making urban planning more effective within American society. They approach that puzzle from four directions: legal, political, environmental, and fiscal. In "The Master Plan: An Impermanent Constitution," lawyer Charles M. Haar explores the relationship between the legal system and urban planning, asking how we can use our laws to strengthen planning. By zeroing in on master plans he opens up a number of ideas, still being debated today, about the functions of such plans, and their use harnessing local government development tools – public works, zoning, subdivision controls, and street mapping – to mediate public and private interests. He suggests that planners have to decide whether they can realistically promote master plans that might tie the hands of future elected officials.

"Advocacy and Pluralism in Planning" by Paul Davidoff, also a lawyer, sheds light on equity issues. Davidoff urges planners to examine their own values because those values affect their plans and the communities they are planning. And he prods them to represent poor, the uneducated, and the disenfranchised in our society.

If Davidoff advocates for the voiceless, landscape architect/planner Ian McHarg speaks on behalf of the environment. In "The Metropolitan Region," he asks planners to think about a simple but perceptive question: does all land have equal value – aesthetic, monetary, suitability for development? Of course it doesn't, he says, then offers a means to determine what land is suitable for urban development, outlining how knowledge of historical geology, ecology, and hydrology helps make these determinations. He bids planners to think beyond political boundaries when they plan to consider the whole ecosystem in which a city lies.

Finally, in "Paying for the Plan," Robert Yaro and Tony Hiss argue that planners need to establish the real costs of their suggested initiatives and garner citizen support for called-for public investments. According to this planner/writer team, planners have yet to develop the means of quantifying the value of key planning projects like open space conservation and new mass transit lines, and they need to explore ways to pay for public goods, such as user fees. This last selection was published relatively recently, but its arguments are so strong that it will become, or perhaps already is, a classic in the field.

14

Dimensions of Performance

From *A Theory of Good City Form* (1984)

Kevin Lynch

Editor's introduction

Writings on urban form fall into two categories: descriptive and prescriptive. While this selection is prescriptive because its author, Kevin Lynch, outlines qualities of a desirable urban place, he derives these general rules from his testing descriptive work against his experience in practice. The book from which it comes – *A Theory of Good City Form* – is a landmark in urban and regional planning scholarship for its sensibility and clarity. It highlights the field's central intellectual concern: articulating the physical shape and arrangement of functions in cities and regions, a mission that distinguishes urban and regional planning from other disciplines.

Descriptive scholarship on urban form has many sources: social science, science, humanities, and design. Geographers and land economists, for example, study the layout of cities and regions on the landscape. In general, they classify patterns of and within cities, analyzing land use and transportation relationships. They characterize city groupings as linear (e.g. organized along transportation lines like a highway), clustered (e.g. bunched around a resource like a lake), and hierarchical or central place (networked cities surrounding a dominant city, such as Newark, New Jersey; White Plains, New York; and Stamford, Connecticut – all arrayed around New York City).

In capturing the relationships among different land uses within cities over time, these social scientists have advanced theories describing the internal organization of urban places, formulated from the 1920s to the present. With names often reflecting the city shape, the theories are:

- *Concentric ring*, 1925. A city is circular, relatively compact, has a downtown and a series of zones emanating outward in bands with each one dedicated to a particular land use.
- *Sector*, 1939. A city is star-shaped and has conjoined sectors growing along transportation axes from the central business district outward, with the most accessible land being the most desirable, usually dedicated to high-income residential use.
- *Multiple nuclei*, 1945. A city is a jigsaw puzzle with a downtown and several subcenters and associated uses that form specialized districts.
- *Dispersed*, 1963. A city is widespread and more land-consumptive than in earlier eras – not wholly centered on its downtown – reflecting land use patterns resulting from automobile use.
- *Edge city*, 1991. An urban place having downtown qualities due to the migration of employment (offices of a minimum of 5 million sq. ft and retail of a minimum of 800,000 sq. ft) from historic centers to suburban locations, often at highway intersections.
- *Peripheral city*, 1997. A metropolitan area with shopping and workplaces occupying a wide band along a beltway circling a downtown and surrounded by low-density suburbs at its outer edges.
- *Edgeless city*, 2001. A metro area with highly dispersed office clusters, no clear boundaries, and less than 5 million sq. ft of office space, with employment and retail scattered throughout a metropolitan region.

While geographers and land economists came up with the foregoing formulations to present urban patterns, other researchers – including urban designers, architectural historians, and historians – studied the morphology, or shapes, of cities as defined by lots, blocks, streets, and buildings. They, too, classified what they observed. For example, they cataloged cities as being orthogonal (having a checkerboard or regular block pattern) or organic (having an irregular block pattern usually determined by topography or natural features). They looked at the scale and grain of blocks, assessing their range, from coarse (large) to fine (small). They delineated public and private space (streets, plazas, parks versus buildings, gated enclosures). They documented historical development, looking at density and the conversion of buildings from one use to another over time. They also devised formulas for "reading" the city. One of the most famous is the one created by Kevin Lynch in his seminal work *The Image of the City* (Cambridge, MA: MIT Press, 1960), identifying five characteristics (landmarks, nodes, districts, edges, and paths) that aid people in navigating in a city.

Finally, sociologists and other social scientists have explored how cities – characterized by large size, density, and heterogeneity, and, later, by dispersed suburbs – affect human behavior. They, too, arrived at many conclusions, the most enduring of which posits that unique and identifiable physical arrangements in urban and metropolitan places contribute to an individual's sense of social connection and community links. This literature includes Louis Wirth's pioneering essay "Urbanism as a Way of Life" (*American Journal of Sociology*, 44: 1, 1938) and Robert D. Putnam's *Bowling Alone: The Collapse and Revival of American Community* (New York: Simon & Schuster, 2000).

Urban and regional planners have drawn from all theories to make recommendations about future urban form and patterns and create comprehensive plans for cities and regions. This selection is a leading example of a planner's abstracting guidelines for development from rich knowledge base provided by earlier scholars. While reading this piece, examine your own beliefs about what makes good city form.

Kevin Lynch (1918–84), trained as an architect and city planner, taught at Massachusetts Institute of Technology for thirty years, blending an academic career with practice. His firm, Carr Lynch Associates, worked on urban planning and design projects with the Boston Redevelopment Authority, Puerto Rico Industrial Development Company, Tufts-New England Medical Center, and cities throughout the United States, including Detroit, Minneapolis, San Francisco, Los Angeles, and San Diego. Of his seven books, *Site Planning* (Cambridge, MA: MIT Press, 1984) with Gary Hack has had three editions, while *A Theory of Good City Form* (Cambridge, MA: MIT Press, 1981) is the culmination of his life of observation, teaching, and practice. In 1995, two former students who are both professors of city planning – Tridib Banerjee, University of Southern California, and Michael Southworth, University of California, Berkeley – collected and interpreted his works in *City Sense and City Design* (Cambridge, MA: MIT Press, 1995).

For more on city form from the various disciplines, see Dave H. Kaplan *et al.*, *Urban Geography* (New York: Wiley, 2003); Jan Lin, *The Urban Sociology Reader* (London and New York: Routledge, 2005); Mark Gottdiener and Ray Hutchison, *The New Urban Sociology* (Boulder, CO: Westview Press, 2006); and Spiro Kostof, *The City Shaped: Urban Patterns and Meanings Through History* (New York: Bullfinch Press, 1991) and *The City Assembled: The Elements of Urban Form Through History* (London: Thames & Hudson, 1992).

DIMENSIONS OF PERFORMANCE

Performance characteristics will be more general, and the easier to use, to the degree that performance can be measured solely by reference to the spatial form of the city. But we know that the quality of a place is due to the joint effect of the place and the society which occupies it. I can imagine three tactics for avoiding the necessity of taking the entire universe into account in this attempt to measure city performance. First, we can elaborate those linkages between form and purpose which exist because of certain species-wide or human settlement-wide regularities: the climatic tolerances of human beings, for example, or the importance of the small social group, or the very general function of any city as a network of access. Second, we can add to the description of the spatial form of a place those particular social institutions and mental attitudes which are directly linked to that form and repeatedly critical to its quality. . . . Both of these tactics will be employed below.

Third and last, however, we must realize that it would be foolish to set performance *standards* for cities, if we mean to generalize. To assert that the ideal density is twelve families to the acre, or the ideal daytime temperature is 68°F, or that all good cities are organized into residential neighborhoods of 3,000 persons each, are statements too easily discredited. Situations and values differ. What we might hope to generalize about are performance dimensions, that is, certain identifiable characteristics of the performance of cities which are due primarily to their spatial qualities and which are measurable scales, along which different groups will prefer to achieve different positions. It should then be possible to analyze any city form or proposal, and to indicate its location on the dimension, whether by a number or just by "more or less." To be general, the dimensions should be important qualities for most, if not all, persons and cultures. Ideally, the dimensions should also include all the qualities which any people value in a physical place. (Of course, this last is an unbearably severe criterion.)

For example, we might consider *durability* as a performance dimension. (**But we won't. This is a red herring.*) Durability is the degree to which the physical elements of a city resist wear and decay and retain their ability to function over long periods. In choosing this dimension, we assume that everyone has important preferences about the durability of his city, although some want it evanescent and others would like it to last for ever. Furthermore, we know how to measure the general durability of a settlement, or at least how to measure a few significant aspects of durability. A tent camp can be compared to a troglodyte settlement, and, given the values of a particular set of inhabitants, we can tell you which one of them is better, or people can make that evaluation for themselves. They can also decide how much durability they are willing to give up in return for other identifiable values. Perhaps we can show that very low or very high durabilities are bad for everyone, and so we identify an optimum range. Although the linkage of durability to basic human aims is only a chain of assumptions, we believe that the assumptions are reasonable. Correlations of durability with preference exist, and people are content to use this idea as a workable intermediate goal. Meanwhile, its connection to city form – to such concrete physical characteristics as building material, density, and roof construction – can be explicitly demonstrated.

To be a useful guide to policy, a set of performance dimensions should have the following characteristics:

1 They should be characteristics which refer primarily to the spatial form of the city, as broadly defined above, given certain very general statements about the nature of human beings and their cultures. To the extent that the value set on those characteristics varies with variations in culture, that over time be valued dependence should be explicit. The dimension itself and its method of analysis should remain unchanged.
2 The characteristics should be as general as possible, while retaining their explicit connection to particular features of form.
3 It should be possible to connect these characteristics to the important goals and values of any culture, at least through a chain of reasonable assumptions.
4 The set should cover all the features of settlement form which are relevant, in some important way, to those basic values.
5 These characteristics should be in the form of dimensions of performance, along which various groups in various situations will be free to choose optimum points or "satisficing" thresholds. In other words, the dimensions will be usable where values differ or are evolving.
6 Locations along these dimensions should be identifiable and measurable, at least in the sense of "more or less," using available data. They may be complex dimensions, however, so that locations on them need not be single points. Moreover, the data, while conceivably available, may for the present escape us.
7 The characteristics should be at the same level of generality.
8 If possible, they should be independent of one another. That is, setting a level of attainment along one dimension should not imply a particular setting on some other dimension. If we

are unable to produce uncontaminated dimensions of this kind, we can settle for less, if the cross-connections are explicit. Testing for independence will require detailed analysis.

9 Ideally, measurements on these dimensions should be able to deal with qualities which change over time, forming an extended pattern which can be valued in the present. More likely, however, the measurements will deal with present conditions, but may include the drift of events toward the future.

There have been many previous attempts to outline a set of criteria for a "good city." The dimensions I propose below are not original inventions. Appendix C [not included here] indicates some of my sources. Previous sets have always broken at least some of the rules above. They have at times been so general as to go far beyond settlement form and to require a complex (and usually impossible) calculation which involves culture, political economy, and many other nonformal features. Or they refer to some particular physical solution that is appropriate only in a particular situation. They may mix spatial and nonspatial features, or mix levels of generality, or mix the scale of application. Frequently, they are bound to a single culture. They do not include all the features of city form which are important to human values. They are often given as absolute standards, or they call for minimizing or maximizing, instead of being dimensions. The qualities are sometimes not measurable, or even identifiable, in any clear way. They frequently overlap each other.

The list that follows is an attempt to rework and reorder the material in a way that escapes those difficulties. The presumed generality of this list lies in certain regularities: the physical nature of the universe, the constants of human biology and culture, and some features which commonly appear in contemporary large-scale settlements, including the processes by which they are maintained and changed.

But some view of the nature of human settlements, however unclear or general, is necessarily assumed in making any list. Unfortunately, it is much easier to say what a city is not: not a crystal, not an organism, not a complex machine, not even an intricate network of communications – like a computer or a nervous system – which can learn by reorganizing its own patterns of response, but whose primitive elements are forever the same. True, somewhat like the latter, the city is interconnected to an important degree by signals, rather than by place order or mechanical linkages or organic cohesion. It is indeed something changing and developing, rather than an eternal form, or a mechanical repetition which in time wears out, even a permanent recurrent cycling which feeds on the degradation of energy, which is the concept of ecology.

Yet the idea of ecology seems close to an explanation, since an ecosystem is a set of organisms in a habitat, where each organism is in some relation to others of its own kind, as well as to other species and the inorganic setting. This system of relations can be considered as a whole, and has certain characteristic features of fluctuation and development, of species diversity, of intercommunication, of the cycling of nutrients, and the pass-through of energy. The concept deals with very complex systems, with change, with organic and inorganic elements together, and with a profusion of actors and of forms.

Moreover, an ecosystem seems to be close to what a settlement is. Complicated things must in the end be understood in their own terms. An image will fail to stick if it is only a borrowing from some other area, although metaphorical borrowings are essential first steps in understanding.

Apt as it is, the concept of ecology has its drawbacks, for our purpose. Ecological systems are made up of "unthinking" organisms, not conscious of their fatal involvement in the system and its consequences, unable to modify it in any fundamental way. The ecosystem, if undisturbed, moves to its stable climax of maturity, where the diversity of species and the efficiency of the use of energy passing through are both at the maximum, given the fixed limits of the inorganic setting. Nutrients recycle but may gradually be lost to sinks, while energy inevitably escapes the system or becomes unavailable. Nothing is learned; no progressive developments ensue. The inner experiences of the organisms – their purposes and images – are irrelevant; only their outward behavior matters.

An evolving "learning ecology" might be a more appropriate concept for the human settlement, some of whose actors, at least, are conscious, and capable of modifying themselves and thus of changing the rules of the game. The dominant animal consciously restructures materials and switches the paths of energy flow. To the familiar ecosystem characteristics of diversity, interdependence, context, history, feedback, dynamic stability, and cyclic processing, we must add such features as values, culture, consciousness, progressive (or regressive) change, invention, the ability to learn, and the connection of inner experience and outer action. Images, values, and the creation and flow of information play an important role. Leaps, revolutions, and catastrophes can happen, new paths can be taken. Human learning and culture have destabilized the system, and perhaps, some day, other species will join the uncertainty game. The system does not inevitably move toward some fixed climax state, nor toward maximum entropy. A settlement is a valued arrangement, consciously changed and stabilized. Its elements are connected through an immense and intricate network, which can be understood only as a series of overlapping local systems, never rigidly or instantaneously linked, and yet part of a fabric without edges. Each part has a history and a context, and that history and context shift as we move from part to part. In a peculiar way, each part contains information about its local context, and thus, by extension, about the whole.

Values are implicit in that viewpoint, of course. The good city is one in which the continuity of this complex ecology is maintained while progressive change is permitted. The fundamental good is the continuous development of the individual or the small group and their culture: a process of becoming more complex, more richly connected, more competent, acquiring and realizing new powers – intellectual, emotional, social, and physical. If human life is a continued state of becoming, then its continuity is founded on growth and development (and its development on continuity: the statement is circular). If development is a process of becoming more competent and more richly connected, then an increasing sense of connection to one's environment in space and in time is one aspect of growth. So that settlement is good which enhances the continuity of a culture and the survival of its people, increases a sense of connection in time and space, and permits or spurs individual growth: development, within continuity, via openness and connection. These values could, of course, be applied to judging a culture as well as a place. In either case, there is an inherent tension as well as a circularity between continuity and development – between the stabilities and connections needed for coherence and the ability to change and grow. Those cultures whose organizing ideas and institutions deal successfully with that tension and circularity are presumably more desirable, in this view. Similarly, a good settlement is also an *open* one: accessible, decentralized, diverse, adaptable, and tolerant to experiment. This emphasis on dynamic openness is distinct from the insistence of environmentalists (and most utopians) on recurrence and stability. The blue riband goes to development, as long as it keeps within the constraints of continuity in time and space. Since an unstable ecology risks disaster as well as enrichment, flexibility is important, and also the ability to learn and adapt rapidly. Conflict, stress, and uncertainty are not excluded, nor are those very human emotions of hate and fear, which accompany stress. But love and caring would certainly be there.

Any new model of the city must integrate statements of value with statements of objective relationships. The model I have sketched is neither a developed nor an explicit one, and I retreat to my more narrow concern with normative theory. But the surviving reader will see that these general preferences – for continuity, connection, and open ness – underlie all the succeeding pages, even while the theory makes an effort to see that it is applicable to any context.

Given that general view and the task of constructing a limited set of performance dimensions for the spatial form of cities, I suggest the following ones. None of them are single dimensions; all refer to a cluster of qualities. Yet each cluster has a common basis and may be measured in some common way. I simply name the dimensions at this point. . . .

There are five basic dimensions:

1 *Vitality:* the degree to which the form of the settlement supports the vital functions, the biological requirements and capabilities of human beings – above all, how it protects the survival of the species. This is an anthropocentric criterion, although we may some day consider the way in which the environment

supports the life of other species, even where that does not contribute to our own survival.

2 *Sense:* the degree to which the settlement can be clearly perceived and mentally differentiated and structured in time and space by its residents and the degree to which that mental structure connects with their values and concepts – the match between environment, our sensory and mental capabilities, and our cultural constructs.
3 *Fit:* the degree to which the form and capacity of spaces, channels, and equipment in a settlement match the pattern and quantity of actions that people customarily engage in, or want to engage in – that is, the adequacy of the behavior settings, including their adaptability to future action.
4 *Access:* the ability to reach other persons, activities, resources, services, information, or places, including the quantity and diversity of the elements which can he reached.
5 *Control:* the degree to which the use and access to spaces and activities, and their creation, repair, modification, and management are controlled by those who use, work, or reside in them.

If these five dimensions comprise all the principal dimensions of settlement quality, I must of course add two meta-criteria, which are always appended to any list of good things:

6 *Efficiency:* the cost, in terms of other valued things, of creating and maintaining the settlement, for any given level of attainment of the environmental dimensions listed above.
7 *Justice:* the way in which environmental benefits and costs are distributed among persons, according to some particular principle such as equity, need, intrinsic worth, ability to pay, effort expended, potential contribution, or power. Justice is the criterion which balances the gains among persons, while efficiency balances the gains among different values.

These metacriteria are distinct from the five criteria that precede them. First, they are meaningless until costs and benefits have been defined by specifying the prior basic values. Second, the two metacriteria are involved in each one of the basic dimensions, and thus they are by no means independent of them. They are repetitive subdimensions of each of the five. In each case, one asks: (1) What is the cost (in terms of anything else we choose to value) of achieving this degree of vitality, sense, fit, access, or control? and (2) Who is getting how much of it?

I propose that these five dimensions and two metacriteria are the inclusive measures of settlement quality. Groups and persons will value different aspects of them and assign different priorities to them. But, having measured them, a particular group in a real situation would be able to judge the relative goodness of their place, and would have the clues necessary to improve or maintain that goodness. All five can be defined, identified, and applied to some degree, and this application can be improved.

Now, is this really so? Do the dimensions really meet all the criteria which were given at the beginning of this section? Do they in fact illuminate the "goodness" of a city, or are they only a verbal checklist? Can locations on these dimensions be identified and measured in a concrete way? Are they useful guidelines for research? Do they apply to varied cultures and in varied situations? Can general propositions be made about how optima vary according to variations in resource, power, or values? Can degrees of achievement on these dimensions be related to particular spatial patterns, so that the benefits of proposed solutions can be predicted? Do our preferences about places indeed vary significantly as performance changes? All that remains to be seen.

[...]

15

Downtown is for People

From *The Exploding Metropolis* (1958) by the editors of *Fortune*

Jane Jacobs

Editor's introduction

Three challenging books that would influence the theory and practice of urban and regional planning appeared in the early 1960s. They were *The Death and Life of Great American Cities* (New York: Random House, 1961) by Jane Jacobs; *Silent Spring* (Boston, MA: Houghton Mifflin, 1962) by Rachel Carson; and *The Feminine Mystique* (New York: Norton, 1963) by Betty Friedan. Jacobs's book celebrated the organic, unplanned city; Carson's raised environmental consciousness; and Friedan anticipated the women's movement and its effects on the labor force and household composition. They informed urban design, environmental planning, housing, and demography.

Of the three authors, Jacobs probably had the greatest impact on urban and regional planning. Her wholesale condemnation of the field initially caused a good deal of angst, exemplified by the following comments from the executive director of one of the professional organizations: "Jane Jacobs' book is going to do a lot of harm . . . [but] we are going to have to live with it. So batten down the hatches." In time, however, her recommendations became deeply ingrained as the norm for urban life in contemporary urban and regional planning teachings.

Jacobs highlighted the benefits of city living, comparing her own New York neighborhood, Greenwich Village, with the public housing and urban renewal projects being built in surrounding districts. She argued that her neighborhood's complex, unplanned physical environment fostered an urbane, safe community. She believed its success was due to four attributes: the neighborhood had mixed uses, short blocks and narrow streets lined with continuous commercial use, a dense population (75,000 to 100,000 people), and structures built over time. These components, she asserted, encouraged different kinds of people to walk around at all times of day. Their presence, as well as the surveillance of the area's many shopkeepers and residents, provided "eyes on the street" that discouraged crime and encouraged economic vitality. In contrast, federal urban renewal and housing programs produced sterile, high-rise tower complexes with long, unwalkable blocks lacking ground-floor stores. These designs, Jacobs claimed, stifled or discouraged the busy, urban scene that made cities lively, secure, and attractive.

When *The Death and Life of Great American Cities* appeared, it was an immediate success. The publisher had released excerpts to several popular magazines, including *Harper's*, *The Saturday Evening Post*, and *Vogue*. The *New York Times* reviewed the book twice, in the daily paper and in its prestigious Sunday *Book Review*. Jacobs, a first-time book author, became an instant celebrity. Her message appealed to both the Left, and the Right, both of which had issues with federal urban policy, the former objecting to its inattention to the poor, the latter its exercise of power.

Indeed, Jacobs struck the perfect tone for the times. The public was becoming wary of the clearance practices of urban renewal and public housing: the costly, disruptive, and slow-to-complete projects dislocated so many people and made whole neighborhoods disappear without stemming urban decline. Jacobs exercised sharp observational power in her analyses of city life, describing it in clear, simple, and sometimes indignant language. She added a qualitative dimension to urbanism that urban planners and other professionals had rejected in favor of boring quantitative analysis and dull maps. Furthermore, she had picked a tangible scape-goat: urban planners. Demonizing these professionals was more to her style and taste than writing about the complex economic, social, and political reasons for the massive metropolitan transformations then occurring, and certainly more interesting to her readers.

While Jacobs made a major contribution to the field, her work had several weaknesses. First, she exhibited no sensitivity to the historical conditions that had inspired the urban programs of her era. Nor did she appre-ciate the hard-fought campaigns waged by previous reformers to achieve the smallest improvements, like regula-tions to require running water and toilets in every dwelling, much less their part in securing a federal role in addressing urban problems. She failed to distinguish among the serious contributions to urban planning theory that emerged in the early twentieth century, dismissing all with a sweep of her pen. With an opening sentence throwing down the gauntlet – "This book is an attack on current city planning and rebuilding" – she derisively skewered every idea but her own. Second, while she was brilliantly alert to the daily choreography of different individuals in her neighborhood – children, mothers, shopkeepers, local policeman – she had little awareness of other urban complexities related to race, class, and gender that created large dysfunctions in the metropolitan arena. Third, in elevating the amateur-citizen as the arbiter of city rebuilding, she helped ignite NIMBYism, the not-in-my-backyard syndrome that was a disastrous, unintended consequence of such thinking.

Finally, powerful as Jacobs's message was, it did not immediately change the course of U.S. metropolitan development. The steady movement of people and jobs to the suburbs and the depletion of the traditional cities she extolled continued throughout the 1960s and accelerated in the 1970s. While her readers univer-sally admired the book's prescriptions for successful neighborhoods, mainstream developers and consumers rushed to suburban, automobile-oriented, low-density, single-use areas.

This selection originally appeared in *Fortune Magazine* (April 1958) and later was reprinted in *The Explod-ing Metropolis* (New York: Doubleday, 1958). *Fortune* and *Architectural Forum* were sister publications, both part of Henry Luce's Time-Life corporation. In putting together a series on the metropolis, *Fortune* senior editor William H. Whyte (1917–99) tapped *Architectural Forum* associate editor Jacobs for one essay over the objections of his colleagues. He was aware of her extracurricular activism in Greenwich Village fend-ing off Robert Moses' plans for urban renewal and, as he later related in the forward to *The Exploding Metropolis*, "she was a female, she was untried, having never written anything longer than a few paragraphs. She lived in the West Village and commuted to work on a bicycle." Through conversation, Whyte, himself a student of metropolitan life – he had already authored a bestselling portrait of suburban life, *The Organization Man* (New York: Doubleday, 1956) and would go on to make substantial contributions in suburban open space protection in *Conservation Easements* (Washington, DC: Urban Land Institute, 1959) and in the design of urban public space in *The Social Life of Small Urban Spaces* (New York: Project for Public Spaces, 1980) – recognized her observational skills. The resulting essay, "Downtowns are for People," attracted the favorable attention of the Rockefeller Foundation, which supported Jacobs for two years, allowing her to expand the piece into *The Death and Life of Great American Cities.*

"Downtowns are for People" contains not only the seminal ideas of *The Death and Life of Great American Cities* but also a message about downtowns that has contemporary resonance. Jacobs reminds the reader that two important characteristics make downtowns special: individuality (drawn from the district's particular history and natural resources) and people (attracted to the place by its centrality and clustered activities). Her suggestions for improvement are just now being implemented in many downtowns under the rubric of creating a "24/7" place. Twenty-first-century downtowns – whether traditional like Philadelphia's or brand-new like Plano, Texas's – are enhancing their amenities (open space, cultural, and entertainment facilities) and seeking residents in addition to pursuing their older strategies of courting employment and retail.

Jacobs (1916–2006), daughter of a physician and schoolteacher, grew up in the Northeastern coal town of Scranton, Pennsylvania. She migrated to New York shortly after graduating from high school. There, as an aspiring writer, she secured freelance assignments that sharpened her knowledge of the details of city life –

she sold four articles to *Vogue* on New York's fur, diamond, leather, and flower districts and another one on manhole covers. Her big breakthrough was the invitation to write the downtown article and the subsequent book. After living in New York for thirty-four years, she moved with her husband and three children to Toronto to prevent her sons from being drafted into the Vietnam War. In Toronto, as in New York City, she was a vociferous activist, leading citizen opposition to highway and urban renewal projects. In addition to *The Death and Life of Great American Cities*, she wrote eight other books, including *The Economy of Cities* (New York: Random House, 1969), *Cities and the Wealth of Nations* (New York: Random House, 1984), *The Nature of Economies* (New York: Random House, 2000), and *Dark Age Ahead* (New York: Random House, 2004).

For more on Jane Jacobs, see Alice Sparberg Alexiou's biography *Jane Jacobs: Urban Visionary* (New Brunswick, NJ: Rutgers University Press, 2006). For a lively account of Jacobs fighting urban renewal in her own New York neighborhood, see Christopher Klemek, *Urbanism as Reform: Modernist Planning and the Crisis of Urban Liberalism in Europe and North America, 1945–1975* (Chicago: University of Chicago Press, forthcoming).

This is a critical time for the future of the city. All over the country civic leaders and planners are preparing a series of redevelopment projects that will set the character of the center of our cities for generations to come. Great tracts, many blocks wide, are being razed; only a few cities have their new downtown projects already under construction; but almost every big city is getting ready to build, and the plans will soon be set.

What will the projects look like? They will be spacious, park-like, and uncrowded. They will feature long green vistas. They will be stable and symmetrical and orderly. They will be clean, impressive, and monumental. They will have all the attributes of a well kept, dignified cemetery.

And each project will look very much like the next one: the Golden Gateway office and apartment center planned for San Francisco; the Civic Center for New Orleans; the Lower Hill auditorium and apartment project for Pittsburgh; the Convention Center for Cleveland; the Quality Hill offices and apartments for Kansas City; the Capitol Hill project for Nashville. From city to city the architects' sketches conjure up the same dreary scene; here is no hint of individuality or whim or surprise, no hint that here is a city with a tradition and flavor all its own.

These projects will not revitalize downtown; they will deaden it. For they work at cross-purposes to the city. They banish the street. They banish its function. They banish its variety . . .

There are, certainly, ample reasons for redoing downtown – falling retail sales, tax bases in jeopardy, stagnant real estate values, impossible traffic and parking conditions, failing mass transit, encirclement by slums. But with no intent to minimize these serious matters, it is more to the point to consider what makes a city center magnetic, what can inject the gaiety, the wonder, the cheerful hurly-burly that make people want to come into the city and to linger there. For magnetism is the crux of the problem. All downtown's values are its byproducts. To create in it an atmosphere of urbanity and exuberance is not a frivolous aim.

We are becoming too solemn about downtown. The architects, planners – and businessmen – are seized with dreams of order, and they have become fascinated with scale models and bird's-eye views. This is a vicarious way to deal with reality, and it is, unhappily, symptomatic of a design philosophy now dominant: buildings come first, for the goal is to remake the city to fit an abstract concept of what, logically, it should be. But whose logic? The logic of the projects is the logic of egocentric children, playing with pretty blocks and shouting "See what I made!" – a viewpoint much cultivated in our schools of architecture and design. And citizens who should know better are so fascinated by the sheer process of rebuilding that the end results are secondary to them.

With such an approach, the end results will be about as helpful to the city as the dated relics of the City Beautiful movement, which in the early years of this century was going to rejuvenate the city by making it park-like, spacious, and monumental. For the underlying intricacy, and the life that makes downtown worth fixing at all, can never be fostered synthetically. No one can find what will work for our cities by looking at the boulevards of Paris, as the City Beautiful people did; and they can't find it by looking at suburban Garden Cities, manipulating scale models, or inventing dream cities.

You've got to get out and walk. Walk, and you will see that many of the assumptions on which the projects depend are visibly wrong. . . . If you get out and walk, you see all sorts of other clues. Why is the hub of downtown such a mixture of things? . . . Why is a good steak house usually in an old building? Why are short blocks apt to be busier than long ones?

It is the premise of this critique that the best way to plan for downtown is to see how people use it today; to look for its strengths and to exploit and reinforce them. There is no logic that can be superimposed on the city; people make it, and it is to them, not buildings, that we must fit our plans. This does not mean accepting the present; downtown does need an overhaul: it is dirty, it is congested. But there are things that are right about it too, and by simple old-fashioned observation we can see what they are. We can see what *people* like.

HOW HARD CAN A STREET WORK?

The best place to look at first is the street. . . . [T]he street works harder than any other part of downtown. It is the nervous system; it communicates the flavor, the feel, the sights. It is the major point of transaction and communication. Users of downtown know very well that downtown needs not fewer streets, but more, especially for pedestrians. They are constantly making new, extra paths for themselves, through mid-block lobbies of buildings, block-through stores and banks, even parking lots and alleys. Some of the builders of downtown know this too, and rent space along their hidden streets. . . .

The animated alley

The real potential is in the street, and there are far more opportunities for exploiting it than are realized. Consider, for example, Maiden Lane, an odd two-block-long, narrow, back-door alley in San Francisco. Starting with nothing more remarkable than the dirty, neglected back sides of department stores and nondescript buildings, a group of merchants made this alley into one of the finest shopping streets in America. Maiden Lane has trees along its sidewalks, redwood benches to invite the sightseer or window shopper or buyer to linger, sidewalks of colored paving, sidewalk umbrellas when the sun gets hot. All the merchants do things differently: some put out tables with their wares, some hang out window boxes and grow vines. All the buildings, old and new, look individual; the most celebrated is an expanse of tan brick with a curved doorway, by architect Frank Lloyd Wright. The pedestrian's welfare is supreme; during the rush of the day, he has the street. Maiden Lane is an oasis with an irresistible sense of intimacy, cheerfulness, and spontaneity. It is one of San Francisco's most powerful downtown magnets.

All of downtown can't be remade into a bunch of Maiden Lanes, and would be insufferably quaint if it were. But the basic principles illustrated can be realized by any city and in its own particular way. . . . Think of any city street that people enjoy and you will see that characteristically it has old buildings mixed with the new. This mixture is one of downtown's greatest advantages, for downtown streets need high-yield, middling-yield, low-yield, and no-yield enterprises. The intimate restaurant or good steak house, the art store, the university club, the fine tailor, even the bookstores and antique stores – it is these kinds of enterprises for which old buildings are so congenial. Downtown streets should play up their mixture of buildings with all its unspoken – but well understood – implications of choice. . . .

The pedestrian's level

Let's look for a moment at the physical dimensions of the street. The user of downtown is mostly on foot, and to enjoy himself he needs to see plenty of contrast on the streets. He needs assurance that the street is neither interminable nor boring, so he does not get weary just looking down it. Thus streets that have an end in sight are often pleasing; so are streets that have the punctuation of contrast at frequent intervals. . . . Narrow streets, if they are not *too* narrow . . . and are not choked with cars, can also cheer a walker by giving him a continual choice of this side of the street or that, and twice as much to see. The differences are something anyone can try out for himself by walking a selection of downtown streets.

This does not mean all downtown streets should be narrow and short. Variety is wanted in this respect too. But it does mean that narrow streets or reasonably wide alleys have a unique value that revitalizers of downtown ought to use to the hilt instead of wasting. It also means that if pedestrian and automobile traffic is separated out on different streets, planners would do better to choose the narrower streets for pedestrians, rather than the most wide and impressive. Where monotonously wide and long streets are turned over to exclusive pedestrian use, they are going to be a problem. They will come much more alive and persuasive if they are broken into varying parts. . . .

Maps of reality

But the street, not the block, is the significant unity. When a merchant takes a lease he ponders what is across and up and down the street, rather than what is on the other side of the block. When blight or improvement spreads, it comes along the street. Entire complexes of city life take their names, not from blocks, but from streets – Wall Street, Fifth Avenue, State Street, Canal Street, Beacon Street.

Why do planners fix on the block and ignore the street? The answer lies in a shortcut in their analytical techniques. After planners have mapped building conditions, uses, vacancies, and assessed valuations, block by block, they combine the data for each block, because this is the simplest way to summarize it, and characterize the block by appropriate legends. No matter how individual the street, the data for each side of the street in each block is combined with data for the other three sides of its block. The street is statistically sunk without a trace. The planner has a graphic picture of downtown that tells him little of significance and much that is misleading.

Believing their block maps instead of their eyes, developers think of downtown streets as dividers of areas, not as the unifiers they are. Weighty decisions about redevelopment are made on the basis of what is a "good" or "poor" block, and this leads to worse incongruities than the most unenlightened laissez-faire. . . .

If redevelopers of downtown must depend so heavily on maps instead of simple observation, they should draw a map that looks like a network, and then analyze their data strand by strand of the net, not by the holes in the net. This would give a picture of downtown that would show Fifth Avenue or State Street or Skid Row quite clearly. In the rare cases where a downtown street actually is a divider, this can be shown too, but there is no way to find this out except by walking and looking.

The customer is right

In this dependence on maps as some sort of higher reality, project planners and urban designers assume they can create a promenade simply by mapping one in where they want it, then having it built. But a promenade needs promenaders. People have very concrete reasons for where they walk downtown, and whoever would beguile them had better provide those reasons.

The handsome, glittering stretch of newly rebuilt Park Avenue in New York is an illustration of this stubborn point. People simply do not walk there in the crowds they should to justify this elegant asset to the city with its extraordinary crown jewels, Lever House and the new bronze Seagram Building. The office workers and visitors who pour from these buildings turn off, far more often than not, to Lexington Avenue on the east or Madison Avenue on the west. Assuming that the customer is right, an assumption that must be made about the users of downtown, it is obvious that Lexington and Madison have something that Park doesn't. . . .

The deliberately planned promenade minus promenaders can be seen in the first of the "greenway" streets developed in Philadelphia. Here are the trees, broad sidewalks, and planned vistas – and there are no strollers. Parallel, just a few hundred feet away, is a messy street bordered with stores and activities – jammed with people. This paradox has not been lost on Philadelphia's planners: along the next greenways they intend to include at last a few commercial establishments. . . .

FOCUS

No matter how interesting, raffish, or elegant downtown's streets may be, something else is needed: focal points. A focal point can be a fountain, or a square, or a building – whatever its form, the focal point is a landmark, and if it is surprising and delightful, a whole district will get a magic spillover. All the truly great downtown focal points carry a surprise that does not stale. No matter how many times you see Times Square, with its illuminated soda-pop waterfalls, animated facial tissues, and steaming neon coffee cups, alive with its crowds, it always makes your eyes pop. No matter how many times you look along Boston's Newbury Street, the steeple of the Arlington Street Church always comes as a delight to the eye.

Focal points are too often lacking where they would count most, at places where crowds and activities converge. Chicago, for instance, lacks any focal point within the Loop. In other cities perfectly placed points in the midst of great pedestrian traffic have too little made of them – Cleveland's drab public square, for example, so full of possibilities, or the neglected old Diamond Market in Pittsburgh, which, with just a little showmanship, could be a fine threshold to Gateway Center. . . .

THE ECHO

Backers of the project approach often argue that giant superblock projects are the only feasible means of rebuilding downtown. Projects, they point out, can get government redevelopment funds to help pay for land and the high cost of clearing it. Projects afford a means of getting open spaces in the city with no direct charge on the municipal budget for buying or maintaining them. Projects are preferred by big developers, as more profitable to put up than single buildings. Projects are liked by the lending departments of insurance companies, because a big loan requires less investigation and fewer decisions than a collection of small loans; the larger the project and the more separated from its environs, moreover, the less the lender thinks he need worry about contamination from the rest of the city. And projects can tap the public powers of eminent domain; they don't have to be huge for this tool to be used, but they can be, and so they are. . . .

WHERE IS THIS PLACE?

The project approach thus adds nothing to the individuality of a city; quite the opposite – most of the projects reflect a positive mania for obliterating a city's individuality. They obliterate it even when great gifts of nature are involved. For example, Cleveland, wishing to do something impressive on the shore of Lake Erie, is planning to build an isolated convention center, and the whole thing is to be put on and under a vast, level concrete platform. You will never know you are on a lake shore, except for the distant view of water.

But every downtown can capitalize on its own peculiar combinations of past and present, climate and topography, or accidents of growth. Pittsburgh is on the right track at Mellon Square (an ideally located focal point), where the sidewalk gives way to tall stairways, animated by a cascade. This is a fine dramatization of Pittsburgh's hilliness, and it is used naturally where the street slopes steeply.

Waterfronts are a great asset, but few cities are doing anything with them. Of the dozens of our cities that have river fronts downtown, only one, San Antonio, has made of this feature a unique amenity. Go to New Orleans and you find that the only way to discover the Mississippi is through an uninviting, enclosed runway leading to a ferry. The view is worth the trip, yet there is not a restaurant on the river frontage, nor any rooftop restaurants from which to view the steamers, no place from which to see the bananas unloaded or watch the drilling rigs and dredges operating. New Orleans found a character in the charming past of the Vieux Carré, but the character of the past is not enough for any city, even New Orleans.

A sense of place is built up, in the end, from many little things too, some so small people take them for granted, and yet the lack of them takes the flavor out of the city: irregularities in level, so often bulldozed away; different kinds of paving, signs and fireplugs and street lights, white marble stoops.

THE TWO-SHIFT CITY

It should be unnecessary to observe that the parts of downtown we have been discussing make up a whole. Unfortunately, it is necessary; the project approach that now dominates most thinking assumes that it is desirable to single out activities and redistribute them in an orderly fashion – a civic center here, a cultural center there.

But this notion of order is irreconcilably opposed to the way in which a downtown actually works; what makes it lively is the way so many different kinds of activity tend to support each other. We are accustomed to thinking of downtowns as divided into functional districts – financial, shopping, theatre – and so they are, but only to a degree. As soon as an area gets too exclusively devoted to one type of activity and its direct convenience services, it gets into trouble; it loses its appeal to the users of downtown and it is in danger of becoming a has-been. In New York the area with the most luxuriant mixture of basic activities, midtown, has demonstrated an overwhelmingly greater attractive power for new building than lower Manhattan, even for managerial headquarters, which, in lower Manhattan, would be close to all the big financial houses and law firms – and far away from almost everything else.

Where you find the liveliest downtown you will find one with the basic activities to support two shifts of foot traffic. By night it is just as busy as it is by day. New York's Fifty-seventh Street is a good example: it works by night because of the apartments and residential hotels near by; because of Carnegie Hall; because of the music, dance, and drama studios and special motion-picture theatres that have been generated by Carnegie Hall. It works by day because of small office buildings on the street and very large office buildings to the east and west. A two-shift operation like this is very stimulating to restaurants, because they get both lunch and dinner trade. But it also encourages every kind of shop or service that is specialized, and needs a clientele sifted from all sorts of population.

It is folly for a downtown to frustrate two-shift operation, as Pittsburgh, for one, is about to do. Pittsburgh is a one-shift downtown but theoretically this could be partly remedied by its new civic auditorium project, to which, later, a symphony hall and apartments are to be added. The site immediately adjoins Pittsburgh's downtown, and the new facilities could have been tied into the older downtown streets. Open space of urban – not suburban – dimensions could have created a focal point or pleasure grounds, a close, magnetic juncture between the old and the new, not a barrier. However, Pittsburgh's plans miss the whole point. Every conceivable device – arterial highways, a wide belt of park, parking lots – separates the new project from downtown. The only thing missing is an unscalable wall.

The project will make an impressive sight from the downtown office towers, but for all it can do to revitalize downtown it might as well be miles away. . . .

WANTED: CAREFUL SEEDING

When it comes to locating cultural activities, planners could learn a lesson from the New York Public Library; it chooses locations as any good merchant would. It is no accident that its main building sits on one of the best corners in New York, Forty-second Street and Fifth Avenue, a noble focal point. Back in 1895, the newly formed library committee debated what sort of institution it should form. Deciding to serve as many people as possible, it chose what looked like the central spot in the northward-growing city, asked for and got it.

Today the library locates branches by tentatively picking a spot where foot traffic is heavy. It tries out the spot with a parked bookmobile, and if results are up to expectations it may rent a store for a temporary trial library. Only after it is sure it has the right place to reach the most customers does it build. Recently the Library has put up a fine new main circulation branch right off Fifth Avenue on Fifty-third Street, in the heart of the most active office-building area, and increased its daily circulations by 5,000 at one crack.

The point, to repeat, is to work *with* the city. Bedraggled and abused as they are, our downtowns do work. They need help, not wholesale razing. . . .

THE CITIZEN

The remarkable intricacy and liveliness of downtown can never be created by the abstract logic of a few men. Downtown has had the capability of providing something for everybody only because it has been created by everybody. So it should be in the future; planners and architects have a vital contribution to make, but the citizen has a more vital one. It is his city, after all; his job is not merely to sell plans made by others, it is to get into the thick of the planning job himself. He does not have to be a planner or an architect, or arrogate their functions, to ask the right questions:

- How can new buildings or projects capitalize on the city's unique qualities? Does the city have a waterfront that could be exploited? An unusual topography?
- How can the city tie in its old buildings with its new ones, so that each complements the other and reinforces the quality of continuity the city should have?
- Can the new projects be tied into downtown streets? The best available sites may be outside downtown – but how far outside of downtown? Does the choice of site anticipate normal growth, or is the site so far away that it will gain no support from downtown, and give it none?
- Does new building exploit the strong qualities of the street – or virtually obliterate the street?
- Will the new project mix all kinds of activities together, or does it mistakenly segregate them?

In short, will the city be any fun? The citizen can be the ultimate expert on this; what is needed is an observant eye, curiosity about people, and a willingness to walk. He should walk not only the streets of his own city, but those of every city he visits. When he has the chance, he should insist on an hour's walk in the loveliest park, the finest public square in town, and where there is a handy bench he should sit and watch the people for a while. He will understand his own city the better – and, perhaps, steal a few ideas.

Let the citizens decide what end results they want, and they can adapt the rebuilding machinery to suit them. If new laws are needed, they can agitate to get them. The citizens of Fort Worth, for example, are doing this now; indeed, citizens in every big city planning hefty redevelopment have had to push for special legislation.

What a wonderful challenge there is! Rarely before has the citizen had such a chance to reshape the city, and to make it the kind of city that *he* likes and that others will too. If this means leaving room for the incongruous, or the vulgar or the strange, that is part of the challenge, not the problem.

Designing a dream city is easy; rebuilding a living one takes imagination.

16

Home Remedies for Urban Cancer

From *The New Yorker* (1962)

Lewis Mumford

Editor's introduction

In his "The Sky Line" column in *The New Yorker* – where this review of Jane Jacobs's *The Death and Life of Great American Cities* (New York: Random House, 1961) appeared – critic Lewis Mumford (1895–1990) decried the development excesses of the modern city. He railed against the construction of out-of-scale sky-scrapers and the subsequent loss of the intimacy of urban neighborhoods. He abhorred urban renewal and supported historic preservation. So this born-and-bred New Yorker's criticisms of Jacobs – an author with similar predilections – may come as a bit of a surprise. But a close reading reveals the source of his anger.

Mumford's review reflects his fury and disappointment with Jacobs's attacks on contemporary planning and her ahistorical view of city life. He was particularly annoyed that she did not appreciate the contributions of the Garden City movement, of which he was a strong advocate. And since he had just completed his grand historical survey of urbanization, *The City in History: Its Origins, its Transformation and its Prospects* (New York: Harcourt Brace, 1961), a book that won the National Book Award for nonfiction in 1962, he scorned an analysis that completely ignored the role of nineteenth-century industrialization in the growth of slums.

Perhaps their differences in age and in vocational paths accounted for their contrasting points of view. Both were writers, but Mumford's work was more scholarly and theoretical, while Jacobs's was more journalistic. Jacobs had begun her career as an unpaid assistant to the women's editor of *The Scranton Tribune*, her hometown newspaper. (This was not an unusual entry point for a woman who aspired to be a journalist in those times.) After she moved to New York City in the 1930s, Jacobs sold freelance articles on a wide variety of topics to trade and popular publications. In the early 1940s, she married an architect, raised a family, kept writing, and, in 1952, landed an editor's position at *Architectural Forum.* This job exposed her to contemporary planning activities, notably urban renewal.

In contrast, Mumford grew up in a city that had seen 23 million immigrants pass through its portals between 1880 and 1919. By the 1920s, 40 percent of New York's population of 6 million was foreign-born, the majority living in squalid tenements in teeming ethnic neighborhoods. When Jacobs was born in 1916, Mumford was a twenty-one-year-old, who through long walks through the city's neighborhoods had observed first-hand the misery of overcrowding and environmental pollution. By the time Jacobs was age eighteen, Mumford had already written six books on American architecture, art, and literature, and was turning his intellectual prowess to an extended exploration of the relationship between culture and the machine that would result in *Technics and Civilization* (New York: Harcourt Brace, 1934) and subsequent publications.

In these works, he expanded the ideas of his mentor, Scottish biologist and town planner Patrick Geddes (1854–1932), whom Mumford had discovered as a teenager, encountering Geddes's book *Cities in Evolution: An Introduction to the Town Planning Movement and to the Study of Civics* (London: Williams &

Norgate, 1915) by chance in the library. Geddes's advocacy of a "regional survey" technique (understanding a place through its history, geology, biology, economics, and sociology) and positive attitude towards planning deeply influenced Mumford's thinking. Mumford met his hero face-to-face only once, in 1925, but corresponded with him for several years and, as a mark of his admiration, named his first child Geddes.

Mumford had personal experience with what he considered the ideal form of city living: a low-density and decentralized settlement, an antidote to the wildly overgrown place New York City was becoming. As early as the 1920s, he had joined other socially minded colleagues to form the Regional Planning Association of America, which sought remedies for the era's most pressing urban problems, namely congestion and lack of affordable housing. They studied British town planning concepts involving large-scale (multi-acre, low-building-height) public housing located on the outskirts of London and, later, a regional scheme of self-sufficient satellite towns, or Garden Cities, as first articulated by Ebenezer Howard (1850–1928). (Howard's treatise *To-morrow: A Peaceful Path to Real Reform* (London, 1898), renamed *Garden Cities of To-morrow* when it was formally published in 1902, was reprinted in 1946 with an introductory chapter by Mumford.) Mumford also followed the work of Raymond Unwin (1863–1940) and Barry Parker (1867–1941), who designed England's Letchworth (1904), the first Garden City. He admired the Abercrombie plan for London (1944), which called for population decentralization in satellite towns. He celebrated the British government's Towns program (1946–96), which built thirty-two Garden Cities housing 2.3 million people, later writing the introduction to Frederic J. Osborn and Arnold Whittick's account of the effort in *The New Towns: The Answer to Megalopolis* (New York: McGraw-Hill, 1963).

Applying Garden City ideas in the New York area, Mumford and his associates undertook two experiments: a seventy-seven-acre housing settlement, Sunnyside Gardens, in Queens, New York (1924), where Mumford himself lived for a decade; and the more ambitious undertaking, Radburn, New Jersey (1928), also called "A Town for the Motor Age." Although at Radburn they developed only 149 acres and sold the rest of the site due to Depression-caused financial reverses, they came up with planning ideas that would be later emulated worldwide. Their innovations included setting aside 15 percent of the land for parks; designing residential superblocks blending cul-de-sacs, clustered housing, and open space; and employing the "neighborhood unit," or development block, derived from the number of families needed to support a grade school located within walking distance, or a quarter-mile radius.

A personal friend of Radburn's designers – Clarence Stein (1882–1975), Henry W. Wright (1878–1936), Clarence Perry (1872–1944), and Catherine Bauer (1905–64) – Mumford was totally convinced that these groundbreaking planning concepts held the secret to the ideal twentieth-century community and were workable and achievable improvements to urban living. Jacobs's condemnation of these efforts was anathema to him.

A witty, sophisticated, and entertaining author, Mumford was in his sixties and had already published twenty books on urbanism and architectural criticism when his review of Jacobs's work came out – and he would continue publishing for the next two decades. He wrote from a comfortable perch in Amenia, New York, a small rural town where he fled in 1936 to escape the city's congestion and pollution. His works ranged from highly focused architectural studies like the 128-page extended essay *The Brown Decades: A Study of the Arts in America, 1865–1895* (New York: Harcourt Brace, 1931) to wide-ranging epics like his 657 page *The City in History.*

Sampling Mumford's writings is a must for students of urban and regional planning. *The City in History* is a major but rewarding undertaking, and Mumford's "The Sky Line" columns are splendid, especially those collected by Robert Wojtowicz, Mumford's biographer, in *Sidewalk Critic: Lewis Mumford's Writings on New York* (Princeton, NJ: Princeton Architectural Press, 1998). David A. Johnson's "Lewis Mumford, Critic, Colleague, Philosopher," *Planning Magazine*, April 1983, is a short but excellent biography. For a longer account, see Wojtowicz's *Lewis Mumford and American Modernism: Eutopian Themes for Architecture and Urban Planning* (Cambridge and New York: Cambridge University Press, 1996) or Donald L. Miller's *Lewis Mumford: A Life* (New York: Weidenfeld & Nicolson, 1989). Towards the end of his life, Mumford published three revealing autobiographical works: *Findings and Keepings: Analects for an Autobiography* (New York: Harcourt Brace, 1975), *My Works and Days: A Personal Chronicle* (New York: Harcourt Brace, 1979), and *Sketches from Life: The Autobiography of Lewis Mumford* (New York: Dial Press, 1982).

For an eye-witness description of early U.S. Garden City experiments, see Clarence S. Stein, *Toward New Towns for America* (Cambridge, MA: MIT Press, 1966). To read about the times and people surrounding Radburn's chief designer, see Clarence S. Stein and Kermit C. Parsons, *The Writings of Clarence S. Stein, Architect of the Planning Community* (Baltimore, MD: Johns Hopkins University Press, 1998). Finally, a collection edited by Kermit C. Parsons and David Schuyler, *From Garden City to Green City: The Legacy of Ebenezer Howard* (Baltimore, MD: Johns Hopkins University Press, 2002), tracks the influence of Garden City planning to the present.

Ever since 1948, when the national Urban Renewal Act was passed, the cities of this country have been assaulted by a series of vast federally aided building operations. These large-scale operations have brought only small-scale benefits to the city.

Even in municipal projects designed to rehouse the displaced slum dwellers or people of equivalent low income, the physical improvements have been only partial and the social conditions of the inhabitants have been worsened through further social stratification – segregation, actually – of people by their income levels. The standard form of housing favored by the federal government and big-city administrators is high-rise slabs – bleak structures of ten to twenty stories. Superficially, these new buildings are an immense improvement over . . . [t]enements . . . The latest model buildings are only two rooms deep; all the flats have outside exposure; the structures are widely spaced around small play areas and patches of fenced grass spotted with benches. Not merely are the buildings open to the sun and air on all sides but they are also as bug-proof and vermin-proof as concrete floors and brick walls can make them; they have steam heat, hot and cold water, standard bathroom equipment, and practically everything a well-to-do family could demand. . . .

There is nothing wrong with these buildings except that, humanly speaking, they stink. . . . By the very nature of the high-rise slab, its inhabitants are cut off from the surveillance and protection of neighbors and passers-by, particularly when in elevators. In some housing projects, the possibility of casual violence, rape, even murder, a rising menace in all our big cities, is conspicuously present. . . . In short, though the hygiene of these new structures was incomparably superior to anything the market had offered in the past – and in sunlight, air, and open view definitely superior to the congested superslums of the rich on Park Avenue – most of the other desirable facilities and opportunities had descended to a lower level.

From time to time in *The New Yorker* I have pointed out these deficiencies . . . But the person who has lately followed through on all the dismal results of current public housing and has stirringly presented them is Jane Jacobs, whose book *The Death and Life of Great American Cities* has been an exciting theme for dinner-table conversation all over the country this past year. . . . A few years ago, Mrs Jacobs stepped into prominence at a planners' conference at Harvard. Into the foggy atmosphere of professional jargon that usually envelops such meetings, she blew like a fresh, offshore breeze to present a picture, dramatic but not distorted, of the results of displacing large neighborhood populations to facilitate large-scale rebuilding. She pointed out a fact to which many planners and administrators had been indifferent – that a neighborhood is not just a collection of buildings but a tissue of social relations and a cluster of warm personal sentiments, associated with the familiar faces of the doctor and the priest, the butcher and the baker and the candlestick maker, not least with the idea of "home." Sanitary, steam-heated apartments, she observed, are no substitute for warmhearted neighbors, even if they live in verminous coldwater flats. . . .

Mrs Jacobs gave firm shape to a misgiving that many people had begun to express. . . . This able woman had used her eyes and, even more admirably, her heart to assay the human result of large-scale housing, and she was saying, in effect, that these toplofty barracks that now crowd the city's skyline and overshadow its streets were not fit for human habitation. For her, the new pattern of high-rise urban housing was all one – whether undertaken by municipal authorities to rehouse low-income groups displaced from their destroyed slum quarters, or by insurance companies to house, somewhat more spaciously and elegantly, carefully selected members of the middle classes and provide a safe, reasonably high return – or finally, by speculative investors and builders taking advantage of state aid and state subsidies to feather their private nests.

From a mind so big with fresh insights and pertinent ideas, one naturally expected a book of equally large dimensions. But whereas "Sense and Sensibility" could have been the title of her Harvard discourse, what she sets forth in *The Death and Life of Great American Cities* comes close to deserving the secondary title of "Pride and Prejudice." The shrewd critic of dehumanized housing and faulty design is still evident, and has applied some of her sharp observations and her political experience to the analysis of urban activities as a whole. But this excellent clinical analyst has been joined by a more dubious character who has patched together out of the bits and pieces of her personal observation nothing less than a universal theory about the life and death of our great – by "great," Mrs. Jacobs seems always to mean "big" – American cities. . . .

Before seeking to do justice to Mrs Jacobs' work as a whole, I must say a word about her first chapter, in which she does not do justice to herself. Ironically, this doughty opponent of urban renewal projects turns out to have a huge private urban renewal project of her own. Like a construction gang bulldozing a site clean of all habitations, good or bad, she bulldozes out of existence every desirable innovation in urban planning during the last century, and every competing idea, without even a pretense of critical evaluation. She is sensibly opposed to sterile high-rise projects, but she is even more opposed to the best present examples of urban residential planning, such as Chatham Village, in Pittsburgh, and she seems wholly to misunderstand their nature, their purpose, and their achievement. Her misapprehension of any plans she regards as subversive of her own private concepts of urban planning leads her to astounding statements, and she even attempts to liquidate possible opponents by treating anyone who has attempted to improve the design of cities by another method as if such people were determined enemies of the city. To wipe out her most dangerous rival, she concentrates her attack on Sir Ebenezer Howard, the founder of the New Towns (Garden City) movement in England. Her handling of him is, for those who know anything of his biography, comic. Howard, it happens, devoted the last quarter-century of his life to the improvement of cities, seeking to find by actual experiment the right form and size, and the right balance between urban needs and purposes and those of the rural environment. Under the rubric of the "Garden City," he reintroduced into city building two important ideas: the notion that there was a functional limit to the area and population of a city; and the notion of providing for continued population growth by founding more towns, which would form "town clusters," to perform the more complex functions of a metropolis without wiping out the open recreational spaces and the rural activities of the intervening countryside. Fifteen such communities exist in England today as embodiments of his principle, mostly with populations ranging from sixty [thousand] to ninety thousand people – a group of towns that will eventually hold a vast number of people working not as commuters to London but in their local factories and business enterprises. . . .

Ebenezer Howard, Mrs. Jacobs insists, "set spinning powerful and city-destroying ideas. He conceived that the way to deal with the city's functions was to sort and sift out of the whole certain simple uses, and to arrange each of these in relative self-containment. He focused on the provision of wholesome housing as the central problem to which every thing else was subsidiary." No statement could be further from the truth. Mrs Jacobs' wild characterization contradicts Howard's clearly formulated idea of the Garden City as a balanced, many-sided, urban community. In the same vein, Mrs Jacobs' acute dislike of nearly every improvement in town planning is concentrated in one omnibus epithet expressive of her utmost contempt: "Radiant Garden City Beautiful." Obviously, neither radiance (sunlight), nor gardens, nor spaciousness, nor beauty can have any place in Mrs Jacobs' picture of a great city. I shall say no more of Mrs Jacobs' lack of historical knowledge and scholarly scruple except that her disregard of easily ascertainable facts is all too frequent. . . .

"This book is an attack on current city planning and rebuilding." With these words Mrs Jacobs introduces herself. . . . From her point of view, one of the chief mischiefs of contemporary planning is that it reduces the number of streets by creating superblocks reserved almost exclusively for pedestrian movement, free from through wheeled traffic, with the space once preempted by unnecessary paved streets turned into open areas for play or provided with benches and plantations for the sedentary enjoyment of adults. Such a separation of automobile and pedestrian walks runs counter to her private directives for a safe and animated neighborhood; namely, to multiply the number of cross-streets, to greatly widen the sidewalks, to reduce all other open spaces, and to place many types of shops and services on streets now devoted solely to residences. The street is her patent substitute for the more adequate meeting places which traditional cities have always boasted.

What is behind Mrs Jacobs' idea of assigning exclusively to the street the mixed functions and diverse activities of a well balanced neighborhood unit? The answer . . . is simple: her ideal city is mainly an organization for the prevention of crime. To her, the best way to overcome criminal violence is such a mixture of economic and social activities at every hour of the day that the streets will never be empty of pedestrians, and that each shopkeeper, each householder, compelled to find both his main occupations and his recreations on the street, will serve as watchman and policeman, each knowing who is to be trusted and who not, who is defiant of the law and who upholds it, who can be taken in for a cup of coffee and who must be kept at bay.

This is indeed an "original" theory of the city, and a new order of city planning. It comes pretty close to saying that if the planners had kept blocks as small and irregular as they are in many old quarters of Manhattan below Fourteenth Street, and had made universal the mixture of shops and tenement houses that long characterized the main avenues, the blight and corrosion and violence that have now spread over the whole city could have been avoided. . . . But if this remedy were a sound one, eighteenth-century London, which met all of Mrs Jacobs' planning prescriptions, would not have been the nest of violence and delinquency it actually was.

In judging Mrs Jacobs' interpretations and her planning prescriptions I speak as a born and bred New Yorker, who in his time has walked over almost every street in Manhattan, and who has lived in every kind of neighborhood and in every type of housing, from a private row house on the West Side to an Old Law dumbbell railroad flat, from a grim walkup apartment off Washington Square to the thirtieth floor of an East Side hotel, from a block of row houses with no shops on Hicks Street in Brooklyn Heights to a two-room flat over a lunchroom in the same general neighborhood, with the odor of stale fat filtering through the windows, and with a tailor, a laundry, a florist, grocery stores, and restaurants – Mrs Jacobs' favorite constellation for "urban liveliness" – immediately at hand. Like a majority of my fellow citizens, I am still unregenerate enough to prefer the quiet flat with a back garden and a handsome church beyond it on Hicks Street to all the dingy "liveliness" of Clinton Street as it was back in the 1920s. Finally, for ten years I lived in Sunnyside Gardens, the kind of well planned neighborhood Mrs Jacobs despises: modestly conceived for people with low incomes, but composed of one-, two-, and three-family houses and flats, with private gardens and public open spaces, plus playgrounds, meeting rooms, and an infants' school. Not utopia, but better than any existing New York neighborhood, even Mrs Jacobs' backwater in Greenwich Village. As one who has spent more than fifty years in New York, speaking to a native of Scranton who has not, I must remind Mrs Jacobs that many parts of the city she denounces because they do not conform to her peculiar standards – and therefore, she reasons, are a prey to violence – were for over the better part of a century both economically quite sound and humanly secure. . . .

Mrs Jacobs is at her best in dealing with small, intimate urban areas. She understands that the very life of a neighborhood depends upon the maintenance of the human scale, for it fosters relations between visible people sharing a common environment, who meet face-to-face without intermediaries, who are aware of their personal identity and their common interests even though they may not exchange a word. This sense of belonging rests, however, not on a metropolitan dynamism but on continuity and stability, the special virtues of the village. . . .

The contradiction between Mrs Jacobs' perceptions of the intimate values of neighborhood life and her unqualified adoration of metropolitan bigness and activism remains unreconciled, largely because she rejects the principles of urban design that would unite these complementary qualities. Her ultimate criteria of sound metropolitan planning are dynamism, density, and diversity, but she never allows herself to contemplate the unfortunate last term in the present series – disintegration. Yet her concern for local habits and conventions points her in the right direction for overcoming this ultimate disintegration: the recognition of the neighborhood as a vital urban entity, with an inner balance and an inner life whose stability and continuity are necessary for rebuilding the kind of community that the metropolis, in all its cataclysmic economic voracity – "cataclysmic" is Mrs Jacobs' happy epithet – has destroyed.

She recognizes that a city is more than buildings, but she fails to perceive that a neighborhood is more than its streets and street activities. The new street system she proposes, with twice the number of intersecting north-and-south streets, would do nothing to give visible reality to the social functions of a neighborhood – those performed by school, church, market, clinic, park, library, tavern, eating house, theater. Mrs Jacobs has no use for the orderly distribution of these activities or the handsome design of their necessary structures; she prefers the hit-and-miss distribution of the present city. No wonder she opposes the admirable work of Clarence Stein and Henry Wright. These pioneer planners have repeatedly demonstrated – in Sunny-side Gardens, on Long Island; in Radburn, New Jersey; in Chatham Village, Pittsburgh – how much superior a well planned, visibly homogeneous neighborhood can be to the sort of random community she advocates.

In the multidimensional order of the city Mrs Jacobs favors, beauty does not have a place. Yet it is the beauty of great urban cathedrals and palaces, the order of great monastic structures or the university precincts of Oxford and Cambridge, the serenity and spaciousness of the great squares of Paris, London, Rome, Edinburgh, that have preserved intact the urban cores of truly great cities over many centuries. Meanwhile, the sordid dynamism of the dingier parts of these same cities has constantly proved uneconomic, inefficient, and self-destructive. . . .

What has happened is that Mrs Jacobs has jumped from the quite defensible position that good physical structures and handsome design are not everything in city planning to the callow notion that they do not matter at all. That beauty, order, spaciousness, clarity of purpose may be worth having for their direct effect on the human spirit even if they do not promote dynamism, increase the turnover of goods, or reduce criminal violence seems not to occur to Mrs Jacobs. This is esthetic philistinism with a vengeance.

Mrs Jacobs' most original proposal, then, as a theorist of metropolitan development, is to turn its chronic symptom of disorganization – excessive congestion – into a remedy, by deliberately enlarging the scope of the disease. It is her belief, unshaken by irrefutable counterevidence, that congestion and disorder are the normal, indeed the most desirable, conditions of life in cities. . . .

Yet despite blind spots and omissions, this book at times offers valuable insights into the complex activities of the city – especially those urban functions that flourish precisely because of all the interchanges that take place, by chance no less than by plan, most frequently in cities that have reached a certain order of bigness and complexity. Mrs Jacobs recognizes how much of value they will leave behind, unlike the big corporations and research laboratories that are stampeding into suburbia, in exchange for temporary access to a golf course, a private airfield, or a few domestic acres. She also recognizes, by observation and experience, the communal nucleus of the city – the spontaneous "primary" association of families and neighbors, upon which all the later complexities of urban life are based. And though she dislikes the notion of a planned "neighborhood unit," she chooses for her normal neighborhood the size that Clarence Perry, in his studies for the Regional Plan of New York back in the 1920s, hit upon as roughly the proper size for such a unit – about 5,000 people. . . .

Venice was one of the few cities that, from the Middle Ages onward, were deliberately planned and practically organized on the neighborhood principle, each parish with the little campo at the center – occupied by a café, and shops, and fountain – and its guildhall and its church, a building that might boast as fine a Tintoretto as the Ducal Palace. There is still plenty of variety and domestic vitality in such neighborhoods despite their long decay, but they do not follow Mrs Jacobs' formula of shops and factories strewn all over the quarter.

Her overvaluation of the street as a social rendezvous leads to her naive remedy for combating random violence. And her prescription ("eyes on the street") is a result of wishful thinking. Since when has the idea of shopkeepers as substitute policemen kept even themselves from being held up and knifed? And what makes Mrs Jacobs think that policemen are immune to murderous attack?

But about the long-term remoralization of this demoralized metropolitan community she is emphatically right: the stabilities of the family and the neighborhood are the basic sources of all higher forms of morality, and when they are lacking, the whole edifice of civilization is threatened. When no one cares for anyone else, because we have all become mere computer digits or Social Security numbers, the elaborate fabric of urban life breaks down. Out of this rejection and isolation and emptiness comes, probably, the boiling hostility of both juvenile and adult delinquent.

Mrs Jacobs' concern for the smallest unit of urban life is, then, pertinent and well directed. . . . Mrs Jacobs innocently believes that complexity and diversity are impossible without the kind of intense congestion that has in fact been emptying out the big city, hurling masses of people into the vast, curdled Milky Ways of suburbia. In the desire to enjoy amenities impossible at even a quarter of the density of population she considers desirable, millions of people are giving up the delights and stimulations of genuine city life. It is millions of quite ordinary people who cherish such suburban desires, not a few fanatical haters of the city, sunk in bucolic dreams. Now, it is this massive century-old drift to suburbia, not the building of superblocks or Garden Cities, that is mainly responsible for the dilapidation and the near-death of big cities. How could Mrs Jacobs ignore this staring historic fact?

This movement toward the rural periphery in search of things that were the proud possession of every premechanized city has been helped by the most active enemies of the city – the overbudgeted highway programs that have riddled metropolitan areas with their gaping expressways and transformed civic cores into parking lots. Those who leave the city wish to escape its snarling violence and its sickening perversions of life, its traffic in narcotics and its gangster-organized lewdness, which break into the lives even of children. Not least, the suburban exiles seek to find at least nightly surcease from constant bureaucratic regimentation: Punch the time clock! Watch your step! Curb your dog! Do not spit! No parking! Get in line for a ticket! Move on! Keep off the grass! Follow the green line! Wait for the next train! Buy now, pay later! Don't buck the system! Take what you get! The refugees who leave the metropolis may not keep even the fleeting illusion of freedom and security and a normal family life for long: all too soon rising land values and high rents bring high-rise housing, asphalted parking lots, and asphyxiating traffic jams. But their reaction is evidence of their own spontaneous vitality and a quickened desire for autonomy, which most of the rest of their existence as members of a gigantic, overcongested, necessarily impersonal hive defeats. Strangely, the city that so insistently drives its population into the suburbs is the very same city that Mrs Jacobs quaintly describes as "vital." She forgets that in organisms there is no tissue quite so "vital" or "dynamic" as cancer growths.

But if *The Death and Life of Great American Cities*, taken as a critique of modern city planning, is a mingling of sense and sentimentality, of mature judgments and schoolgirl howlers, how does it stand as an interpretation of the larger issues of urban development and urban renewal, which the title itself so boldly points to? Here again Mrs Jacobs heads her argument in the right direction, toward matters that have been insufficiently appreciated or misinterpreted. No one has surpassed her in understanding the reasons for the great metropolis's complexity and the effect of this complexity, with its divisions of labor, its differentiations of occupations and interests, its valuable racial, national, and cultural variety, upon its daily activities. She recognizes that one cannot handle such a multidimensional social organization as one might handle a simple machine, designed for a single function. "A growing number of people have begun, gradually," she notes, "to think of cities as problems in organized complexity – organisms that are replete with unexamined, but obviously intricately interconnected, and surely understandable, relationships."

That is an admirable observation, but the author has forgotten the most essential characteristic of all organic growth – to maintain diversity and balance, the organism must not exceed the norm of its species. Any ecological association eventually reaches the "climax stage," beyond which growth without deterioration is not possible. Despite Mrs Jacobs' recognition of organic complexity in the abstract, she has a very inadequate appreciation of the ecological setting of cities and neighborhoods; she brusquely turns her back to all but the segregated local environment. Yet the overgrowth of our big cities has destroyed those special environmental qualities that made their setting desirable and fostered their growth in the first place. The obvious result of the large-scale metropolitan congestion she advocates she flatly ignores – the poisoning of the human system with carbon monoxide and the 200 known cancer-producing substances usually in the air, the muffling of the vital ultraviolet rays by smog, the befouling of streams and oceanside (once used for fishing and bathing) with human and industrial waste. This is something worse than an oversight; it is willful blindness. . . .

When they have reached a point long ago overpassed by New York, Chicago, London, Tokyo, and Moscow, big cities are under the necessity to expand their operations to a more capacious container – the region. The forces that have formed our cities in the past are now almost automatically, by their insensate dynamism, wrecking them and threatening to destroy whole countries and continents. Against this background, the problem of policing public thoroughfares against violence is minor; violence and vice are symptoms of those far graver forms of disorder that Mrs Jacobs rules out of consideration because they challenge her rosily sentimental picture of the "great American city."

To blame the conditions in the congested, overgrown metropolis of today on the monumental scale and human hollowness of its urban renewal projects is preposterous, for this draws attention from the grim, enveloping realities that our whole metropolitan civilization confronts. The prevailing economic and technological forces in the big city have broken away from the ecological pattern, as well as from the moral inhibitions and the social codes and the religious ideals that once, however imperfectly, kept them under some sort of control, and reduced their destructive potentialities. . . .

Failing to appraise the larger sources of urban disintegration, or to trace the connection between our major adult and our minor juvenile forms of delinquency, Mrs Jacobs mistakenly regards those who may have a better grasp of the situation as enemies of metropolitan life. Now, under more normal circumstances, the special virtue of the great city was that it did, in fact, tend to keep any one idea or institution or group from becoming dominant. . . . "Silent Spring" came to the big city long before it visited the countryside. . . . For Mrs Jacobs to imagine that the horrifying human byproducts of the city's disordered life can be eliminated by a few tricks of planning is as foolish as for her to imagine that a too generous supply of open spaces and superblocks fostered these symptoms.

If our urban civilization is to escape progressive dissolution, we shall have to rebuild it from the ground up. Certainly we shall have to do far more than alter street plans, humanize housing projects, or give wider geographic distribution to economic activities. . . . One cannot control destructive automatisms at the top unless one begins with the smallest units and restores life and initiative to them – to the person as a responsible human being, to the neighborhood as the primary organ not merely of social life but of moral behavior, and finally to the city, as an organic embodiment of the common life, in ecological balance with other cities, big and little, within the larger region in which they lie. A quick, purely local answer to these problems is no better than applying a homemade poultice for the cure of a cancer. And that, I am afraid, is what the more "original" Jacobsean proposals in *The Death and Life of Great American Cities* come to.

17

The Master Plan, an Impermanent Constitution

From *Law and Contemporary Problems* (1955)

Charles M. Haar

In 1955, when Charles M. Haar wrote the article from which this selection is drawn, all but three states allowed their municipalities to undertake urban planning. States had to pass special legislation permitting this function because under Dillon's Rule (named after nineteenth-century Iowa Chief Justice John Forrest Dillon, whose seminal opinion established the principles outlining municipal power) cities are "creatures of the state." They receive their power from the state in three ways: (1) as expressed in words, or (2) by implication from expressed power, or (3) because such powers are indispensable to the municipality's functioning. In the case of urban planning, states passed laws or enabling legislation outlining its administration and functions.

In authorizing urban planning, most states followed a model adopted from the Department of Commerce's widely circulated Standard City Planning Enabling Act (SCPEA), crafted by a volunteer group of lawyers and planners in the 1920s. Herbert Hoover, then secretary of the department, commissioned a model law after successfully issuing an earlier one for zoning, the Standard Zoning Enabling Act (SZEA, 1924). As the author of this article notes, the language on planning in the two models is ambiguous, which would have far-reaching implications.

SCPEA-based enabling legislation permits but does not mandate urban planning. Municipalities elect to engage in planning and then follow the rules set out in the enabling legislation. Such legislation specifies the creation of a planning commission, a body of citizen volunteers usually appointed by the mayor. It outlines the commission's powers as zoning-ordinance oversight; subdivision-regulation supervision; and "master plan" preparation, adoption and implementation – three broad functions aimed at guiding the "coordinated, adjusted, and harmonious development of the municipality and its environs." This format shaped municipal planning through the 1970s and still pervades contemporary practice.

Of the three functions, the first two – having to do with overseeing zoning and subdivision development – are the most clear-cut. Zoning ordinances (rules governing the use and intensity of use of land) and subdivision regulations (rules outlining how to lay out housing developments) are municipal laws. No development can occur without conforming to these laws. In contrast, the "master plan" is advisory, has no force of law, and, while an official document, it is adopted not by the city council but by the planning commission. In development and capital budget matters the city council consults the plan but is not bound by it.

The law specifies physical development as the subject of a plan. It envisions the plan as a written document with general prescriptions for the future, accompanied by maps, plans, charts, and descriptions based on "careful and comprehensive" surveys and an attached zoning ordinance. In order to transcend politics, a plan has a multi-decade time horizon. The master plan covers four areas: (1) streets; (2) public buildings and property, including open space; (3) public utilities and terminals, regardless of ownership; and (4) development of private property as administered through the zoning ordinance.

Since planning was not required, its supporters were hard pressed to justify the activity and the courts were hesitant to insist on a plan when adjudicating zoning cases. By the early 1930s, almost 400 cities had planning commissions carrying out most of their delegated functions under state and local law. However, for the quarter-century between the conception of planning in the SCPEA mode and the 1950s, most municipalities did not produce master plans because of the expense and their ambiguous legal status. In this essay and another published at the same time, "In Accordance with a Comprehensive Plan," 68 *Harvard Law Review* 1154 (1955), Haar made the case for the master plan, showing how it ought to serve as an impermanent constitution, guiding all land use regulation. The five appendices that accompanied the "Master Plan" article included here stand alone as an important catalog of the state of planning in the mid 1950s.

As Haar wrote these articles the suburbs were exploding – between 1950 and 1960 they increased 49 percent – highlighting the need for master plans. In addition, in 1954 the federal government began to require local comprehensive plans as a condition of urban renewal funding. Haar's arguments served as the foundation for a number of subsequent judicial determinations related to master plans. (See Edward J. Sullivan and Matthew J. Michel, "*Ramapo* at Thirty: The Changing Role of the Plan in Land Use Regulation" in Part Two.) They were also the basis of some states' later decisions to make planning mandatory. (See Part Four, "The Plan: Its Origins and Contemporary Uses.")

After the appearance of these articles, other scholars weighed in, referring to the plan using varying titles – "master plan," "general plan," and "comprehensive plan" – each having a slightly different meaning. Building on Haar's framework, they elaborated on the substance and use of the master plan. For example, University of California, Berkeley, professor T. J. Kent, Jr's now classic *The Urban General Plan* (San Francisco: Chandler, 1964; reprinted by the American Planning Association in 1990) discussed the plan's contents. Washington University School of Law professor Daniel R. Mandelker's encyclopedic "The Role of the Local Comprehensive Plan in Land Use Regulation," 74 *Michigan Law Review* 899 (1975), tracked its course through the courts. Mandelker edited *Planning Reform in the New Century* (Chicago: APA Planners Press, 2005), the proceedings of a conference focused on improved state planning-enabling legislation and new ideas for strengthening the master plan.

Charles M. Haar, Louis D. Brandeis Professor of Law emeritus, Harvard University, and Visiting Professor of Law, University of Miami, was the first legal scholar to bring serious attention to planning law. Generations of planning students used his textbook *Land Use Planning: A Casebook in the Use, Misuse and Re-use of Urban Land* (Boston, MA: Little Brown, 1959, 1971, 1976). He served as an advisor on urban matters to Presidents Kennedy, Johnson, and Carter and as the first Assistant Secretary for Metropolitan Development in the newly formed U.S. Department of Housing and Urban Development. In this latter post, he helped draft key legislation including the Demonstration and Model Cities Act (1966), the Safe Streets and Crime Control Act (1968), and Title IV (New Communities) and Section 236 (Affordable Housing Guarantee Program) of the Housing and Urban Development Act of 1968. He has written and cowritten more than a dozen books, ranging from his earliest, *Land Planning in a Free Society* (Cambridge, MA: Harvard University Press, 1951), to *Suburbs under Siege: Race, Space and Audacious Judges* (Princeton, NJ: Princeton University Press, 1998), winner of the Gustavus Myers award for the best book on human rights, to *Mastering the Boston Harbor: Courts, Dolphins and Imperiled Waters* (Cambridge, MA: Harvard University Press, 2005).

For more on the history of the SCPEA and the SZEA, see Ruth Knack, Stuart Meck, and Israel Stollman, "The Real Story behind the Standard Planning and Zoning Acts of the 1920s," *Land Use Law*, February 1996. To learn about Charles M. Haar's role in urban affairs, see Bruce Stave, "A Conversation with Charles M. Haar: Urban History and the Great Society," *Journal of Urban History*, 25: 1 (1998). And to understand the history of legal scholarship on the master plan, see Edward J. Sullivan, "The Rise of Reason in Planning Law: Daniel R. Mandelker and the Relationship of the Comprehensive Plan in Land Use Regulation," *Journal of Law and Policy*, 3 (2000).

> General ideas are no proof of the strength, but rather of the insufficiency of the human intellect; for there are in nature no beings exactly alike, no things precisely identical, no rules indiscriminately and alike applicable to several objects at once. The chief merit of general ideas is that they enable the human mind to pass a rapid judgment on a great many objects at once; but, on the other hand, the notions they convey are never other than incomplete, and they always cause the mind to lose as much in accuracy as it gains in comprehensiveness.
>
> Alexis De Tocqueville, *Democracy in America* (ed. Bradley, 1946), 13

> The plan shall be made with the general purpose of guiding and accomplishing a coordinated, adjusted, and harmonious development of the municipality and its environs which will, in accordance with present and future needs, best promote health, safety, morals, order, convenience, prosperity, and general welfare, as well as efficiency and economy in the process of development; including, among other things, adequate provision for traffic, the promotion of safety from fire and other dangers, adequate provision for light and air, the promotion of the healthful and convenient distribution of population, the promotion of good civic design and arrangement, wise and efficient expenditure of public funds, and the adequate provision of public utilities and other public requirements.
>
> U.S. Department of Commerce, *A Standard City Planning Enabling Act*, 7 (1928)

City planning in this country has witnessed a combination of professions and talents rare in any reform movement. From the outset sound legislation was recognized as essential for the development programs of the "new city." Great impetus was lent to erecting a legal framework for land planning by the United States Department of Commerce, which, through its Advisory Committees, promulgated and popularized standard enabling legislation for city planning and zoning. Consequently, the theory of city planning has had a decisive imprint in at least one area – state enabling legislation permitting municipalities to plan for and control the uses of land within their corporate areas.

Today enabling legislation for urban planning exists in all states but three. Within this legal matrix the master plan concept is an established element. Since it has arrived at such status, one may assume that it is legislative policy to encourage, or enjoin, recognition of the master plan's significance in the process leading from planning to reality. But, as is the case with most statutes . . . planning enabling laws are cast in broad, amorphous terms. Hence, the enabling Acts indicate a general area of purpose which forms the basis for more detailed elaboration, initially by local legislatures and administrators, and finally by courts. To perform this task of elaboration it is necessary to grasp the motivations and uses of city planning . . .

THE CONCEPT OF THE MASTER PLAN

[The] "master plan" has a variety of meanings, dependent both upon the context in which it is employed, and the purposes for which it is invoked. . . . Master plan may mean one thing when used to advise on the timing of construction of New York City schools, and quite another in the allocation of lands for recreational uses in a rural setting. Again, when utilized by a federal agency for ensuring that a locality is beginning a serious and coordinated attack on slums so that the granting of federal funds is warranted, its contents and scope must differ from the case of a court scrutinizing a zoning ordinance under attack as not consonant with the master plan. And, of course, whether viewed historically over time within one nation, or across-the-board between a country dominated by the institution of private property and one where nationalization of development rights – or of land itself – has become the accepted way of dealing with land use problems, the master plan concept cannot be said to have a universal meaning. Nor is there any one way of formulating or administering it.

This is not to deny the necessity of paying close attention to the master plan concept in and of itself. Even though there are disagreements at the periphery, there is a core meaning that is fairly well agreed upon. Moreover, the empiric situation remains: the concept is constantly used. If its employment were confined to the planning profession alone, there would be small incentive to join in the fray, painful and so often fruitless are the awards of dictionary definition disputes. But the inescapable dilemma persists: the word "master plan" is used in enabling legislation, local ordinances, and judicial decisions.

[. . .]

The larger share of the typical enabling Act concerns itself with the making of plans. The uses to society of this mechanism are envisioned as six broad types: (1) a source of information; (2) a program for correction; (3) a guide to effectuating procedures and measures; (4) an ordinance regulating the use of land; and (5) a guard against the arbitrary. . . .

WHAT THE MASTER PLAN MEANS TO THE PLANNER

The following are conceived to be the uses of the master plan relating primarily to the formulation of plans.

1 Source of information

The acknowledged initial step of the master plan procedure is . . . the "survey." It furnishes a picture of the present state of conditions in the city. . . . These include . . . studies of economic activity, population composition and growth, land uses, channels of movement, systems of public facilities, and physical resources and liabilities.

Gathering and analysis of information is essential; it is the explanation and the buttress of the various conclusions embodied in the master plan. Further, the inventory process has value in itself. For even if the plan becomes a dust gatherer after it is set on its way, this information can prove of use in injecting some light into the operations of such haphazard physical developments as do occur in the future. Thus, a formal attempt to abide by the master plan idea leaves at least this trace.

Of course, if it is to have meaning, the plan itself sets goals, embodies decisions; if it is to have practical effect, procedures must be established to see that these decisions have effect on land. The data itself is not necessarily part of the plan, but a necessary antecedent and, occasionally, a supporting reference. Thus, this category relates to the value of making a plan, and not properly to the plan itself. This thrusts back to a basic precept – planning as a process rather than a rigid blueprint, so that for purposes of analysis master-plan-in-progress would be a more accurate though cumbersome title. . . .

2 A program for correction

By hypothesis the plan serves to indicate the area's sore spots and functional deficiencies. . . . By comparing these sore spots in relation to their effects upon other aspects of the area's physical development and the magnitude of their repercussion upon the people, as well as inadequacies in the rendering of any municipal services to which people aspire, some priority of action can be recommended. In making determination of priority, of the city's "resources, possibilities and needs," the planners have stressed that financial ability and community predilections must be weighed.

3 An estimate of the future

. . . In determining goals some attempt must be made to grapple with the changes of the morrow, for obviously, as the term "planning" readily implies, the planner should be concerned with emerging conditions. Thus . . . plans must be premised upon estimates of industrial growth, of future age and group compositions of the population, and the other variables affecting the physical development of the community.

Thereby master planning puts a brake on the natural tendency to plan only for the immediate. It is the long-range point of view that is put forth as a unique contribution of the planning perspective. . . . Of course, only approximations of the city's future can be made. Consequently, to realize the full potentialities of the use of the plan, periodic modifications of the general plan should be required. . . .

4 An indicator of goals

The master plan should not merely incorporate ascertained or probable trends of development. Otherwise, only an incomplete job would ensue. Objectives should be set in terms of what kind of city the community wants. After the alternative courses of conduct have been presented, debated, and a selection made, the plan represents the decisions and judgments of a community concerning its desirable physical form and character. In this respect it is a blueprint of values – although once more its evolving nature must be emphasized. The plan can never be a total solution, for it exists over time, just as it is a statement of values at one moment in time. In providing this value scheme it brings to bear upon debate of current physical development long-term considerations founded on basic assumptions. And while predicating goals, the problems that may impede their achievement, as well as the means for circumventing the obstacles, thrust themselves forward for analyses and solution. Hence, its educative force on the planners and the planned is again apparent – and its potentiality in the sphere of land development if these goals are allotted a role in the land use field. . . .

5 A technique for coordination

The planning commission . . . is directed to study and crystallize the interrelationships of the various land uses and structures within the city. With different bodies concentrating on streets, parks, school sites, zoning, etc. (and with the increasing tendency to delegate new measures such as public housing or urban redevelopment to newly created authorities), there is a danger that each specific activity affecting the physical environment will lack coordination with the others, and that maladjustments, inefficiencies, and waste will ensue. It is the special task of city planning – comprehensive planning – to supply this coordination and mutual adjustment. . . . The various land uses and physical installations – the physical expression of the myriad of human activities in the city – are combined into a coordinated system. In so far as possible, each piece of property is to be in the right location for its particular use. This will guide the planning activities to achieve greatest efficiency of the whole.

By embodying information and standards concerning these interrelationships, the plan can provide a pattern against which specific proposals for use or building may be viewed. . . . Through its use as a checklist, a more accurate realization of the consequences of any specific planning action may be acquired. And to the degree that the plan carries weight, a touchstone upon which to judge the merit of a proposed action is provided.

The coordination is not only horizontally with other activities affecting the physical environment, but also over time. It is the long-range point of view and the phasing of the program for reaching the ultimate objectives that emphasize the potential contribution of the master plan.

6 A device for stimulating public interest and responsibility

What the previous categories of the values served by the master plan may very well add up to is simply this: the chief purpose of the master plan is that of mutual education. In the process of making a master plan, the planner may learn which issues are the relevant ones so far as the people are concerned, what terms are meaningful to them, and which alternatives make sense as they view them. This education of the planning board and staff is crucial for any plan to survive. Concomitantly, mustering public interest and participation in city planning is one of the most serious problems faced by the profession: preparing the plan can be an effective channel of communication. It is generally understood that today full use must be made of the democratic process to achieve understanding and acceptance by the people who are affected by planning, and who must undertake the responsibility of enacting and maintaining it. . . .

EVALUATION: THE MASTER PLAN IS HORTATORY

It should be evident from the foregoing uses of the master plan – if these were the sole values derived therefrom – that the statutory mandate to make and adopt a master plan is really synonymous with a mandate to plan. The master plan embodies recommendations for an area's development based on predictions of needs and resources for an estimated period of time. Comprehensiveness (a concern with the interaction of the elements of physical development), projection (a concern with the indicia of change), and policy (a commitment to desired goals) are its major premises.

Considered in this light, the generalized statutory emphasis of the master plan concept as it has thus far been limitedly articulated, is purely hortatory. So perceived, the property owner – and the lawyer in his professional capacity – can remain indifferent to this intellectual exercise of the planning profession. Hence, also, the difficulty of defining more precisely what the master plan is – for it becomes another way of asking what is physical planning.

WHAT THE MASTER PLAN MEANS TO PROPERTY INTERESTS

Planning law is directed towards (a) having a plan made; and (b) having it influence development. Given the requisite skill and energy, goal (a) may present little difficulty. The core problem, however, rests in achieving goal (b): how to get the plan, a process of ideas, to touch and concern controls, the process of doing. Here is the area where the property owner's interest, and that of the lawyer he hired to represent it, comes into play; it is also the sphere of activity which concerns the other agencies of local and state government.

Thus far there have developed four primary ways in which local governments exert impact on physical development – public works, zoning, subdivision controls, and protection of mapped streets. To the city planner, the relation of the master plan to such regulatory ordinances is simple and clear. The plan is a long-term general guide for the development of the city; the regulatory laws are tools to bring the plan's goals into realization. . . .

In this translation into results in the physical form and character of the community, what are the advantages attributable to the existence of a master plan? More specifically, what has led to the theoretical desideratum of a two-step process – first the master plan, second the implementary legislation – which is to be found in planning literature and in most planning acts? And what is the bridge between them?

1 A prophesy of public reaction

The first use that may be listed, viewing the master plan from the vantage of the impact it has on men's affairs, flows from the previous categories of Part II, especially No. 6. At least one practical event of great importance emerges for the perspicacious developer of land. The master plan is at the very minimum an intelligent prophesy as to the probable reaction of the local governmental authorities to a given proposal for development. . . . In the light of the master plan, the private landowner may shape his own plans in the plastic stage when they have not yet crystallized; collision with the public interest can in some instances be deflected. Hence, the inclusion of the public interest in programs of land development may be effected without controversy.

2 A tool for the planning commission in making reports

The previous category of prevision of the future on behalf of the private land developer merges into this one – a basis for internal coordination of government actions and programs. Public action – streets, schools, public buildings, housing – vitally affects community development. Yet different programs may vary widely in objective and timing. . . .

This potentially vital review function has been assigned to the planning commission. The usual procedure requires that before taking action necessitating expenditure of public funds, incidental to the location, character, or extent of a government building, the proposal shall be referred to the planning commission for review and recommendations. The effect of such recommendation varies widely among the states. In some instances it has no consequence; only the moral and publicity preventives are available. In others, an overriding vote by the local legislature is necessary. Sometimes this is a unique veto power, where more than a majority – as much as three-fourths – is required to override the commission's disapproval. And, in some instances, two steps are required: it must be overruled by the sponsoring municipal agency, and then by the local legislature.

This coordination not only is between various governmental agencies, but may also be extended by the enabling Act to include these activities and those of private developers. An example is the recent spate of legislation setting up public housing and urban redevelopment authorities. Nearly all of these require the new social welfare programs to accord with a master plan of land use for the community. Again, where referral of subdivision applications for a report by the commission is required before the plat may be filed, the master plan may influence the commission's decision. In some instances it is conclusive. . . .

3 A guide to effectuating procedures and measures

The key role of the master plan in the coordination of diverse activities affecting the city's land has been noted. This, too, is the role that it can provide for the whole series of legislative Acts dealing with the whole series of such activities. The master plan can be most useful in establishing the framework within which to set the legal regulatory devices. Without such coordination, one regulatory device affecting one parcel of land, like zoning regulations, may undo the efforts of other controls over the same parcel, like subdivision regulations. Special regulation of tenement buildings may be rendered wholly ineffective by other laws taking a different approach to the control of the general environment. . . .

4 An ordinance regulating the use of land

The guide may become the ruler. Sometimes, enabling Acts lend immediate binding effect to certain aspects of the master plan. . . .

It should be noted that this is contrary to the theory of the master plan – at least as understood by many planners. "It [master plan] is in no way legally binding upon private property," to select one example of this thought, "until or unless its recommendations are translated into official changes of the zoning map." But in many spheres this binding effect is accorded the master plan by the enabling Acts. Here, its function becomes the simple and familiar one of a government control on private activity. Consequently planning and enforcement may become undesirably confused.

5 A guard against the arbitrary

A basic legal consequence of the master plan follows from its "comprehensiveness." This can be broken down into two aspects: by its requirement of information gathering and analysis, controls are based on facts, not haphazard surmises – hence their moral and consequent legal basis; by its comprehensiveness, diminished are the problems of discrimination, granting of special privileges, and the denial of equal protection of the laws. Hence, the two most frequent sorts of attack upon government regulation become less available to the private landowner. If the local community has gone to the point of preparing a master plan, his chances of success in attacking an ordinance, based on the plan, are considerably diminished.

[. . .]

THE WRITTEN MASTER PLAN

The master plan is an ever changing recordation of the city planner's *end-result thinking*, embodied in a series of diagrams, charts, standards, and policies. Theoretically there is no need for the recordation of these results. The fact that a planning jurisdiction has not deliberately produced a "master plan" in progress does not conclusively indicate its absence. Given an individual who (1) is engaged in city planning and (2) has the capacity to retain mentally all the ingredients that make up that process, there would be no need for that body of materials called the master plan. The improbability of such a mnemonic freak indicates, however, that the failure to engage in the task of producing a tangible master plan shows a failure to engage in city planning. In short, the need for the master plan manifests nothing more than the need for city planning itself.

Nor is it a self-proving proposition that the existence of a master plan affects the constitutional validity of specific land use controls. The injury alleged in each case of land use regulation must be pitted against the measure's relation to the health, safety, morals, and welfare of the community. Production of facts and arguments to substantiate the relation would not seem to be dependent upon evidence of a master plan: facts and arguments do or do not have strength independent of their embodiment in a tangible plan. . . . But the fact nevertheless remains that zoning and subdivision litigation gives rise to questions with respect to which courts might well feel the lack of touchstone of decision lends considerable psychological sway to guiding policies which are presented in tangible form. Courts inevitably do lend weight to expertise. . . . Yet the existence of a master plan (assuming the measure in issue conforms thereto) indicates in a more satisfactory fashion that this expertise has really been put to work on the particular problem before the court. Thus the deference it commands is more likely to be brought into play. In this oblique manner the master plan principle may affect planning litigation. Hence, the value of making a master plan both as a basis for winning community consent to any proposed regulations, as well as enhancing the chances for judicial approval of a particular regulation. Hence, also, its importance for the property owner.

This may be the major significance of the master plan today in terms of impact on the property owner, as listed before in Part III. If the plan is regarded not as the vest-pocket tool of the planning commission, but as a broad statement to be adopted by the most representative municipal body – the local legislature – then the plan becomes a law through such adoption. A unique type of law, it should be noted, in that it purports to bind future legislatures when they enact implementary materials. So far as impact is concerned, the law purports to control the enactment of other laws (the so-called implementary legislation) solely. It thus has the cardinal characteristic of a constitution. But unlike that legal form it is subject to amendatory procedures not significantly different from the course followed in enacting ordinary legislation. To enact a nonconforming measure amounts merely to passing the law twice.

At the present stage of development, however, it is on so slender a reed that the touchstone values of the master plan must hang. This may prove disappointing to planners. . . . The yearning for an absolute principle, and a master plan that truly answers all questions is understandable.

Yet this seems the limited function to which the master plan can withdraw in order to perform most effectively in the grand effort to improve American cities: a reminder of the myriad of activities affecting land, their interrelation, their long-run effects which the day-to-day administrator is too busy to consider. The implementing legislation, on pain of being outside the statute, must conform to its generalized propositions. True, to remove any conflict, the local legislature need but repass the master plan, changed so as to permit the regulation presently desired. But the need of the formal step of amending the plan insures to some degree that the expert's long-range and coordinative contributions are given play in the real world. It may also be desirable – along the lines of the greater than majority vote required by some statutes if the local legislature desires to reverse the planning commission's recommendation concerning a proposed municipal construction – to require that for this purpose the legislature can amend the plan only by a two-thirds or three-quarters vote. This will highlight the master plan's primary role as a constitution. It is a point of view which should be introduced in a courtroom when a particular measure is being assayed.

[. . .]

18

Advocacy and Pluralism in Planning

Journal of the American Institute of Planners (1965)

Paul Davidoff

Editor's introduction

In August 1964, a brash thirty-year-old University of Pennsylvania professor of city and regional planning, Paul Davidoff (1930–84), electrified the attendees of the annual meeting of the American Institute of Planners (AIP) with a stirring plenary address. Despite its dull title, "The Role of the City Planner in Social Policy," Davidoff's speech was a full-scale attack on contemporary planning practices. Striking at the field's basic beliefs, Davidoff argued that urban and regional planners were not value-free, that a plan prepared by a public agency did not represent the interests of all, and that the focus on physical planning in comprehensive plans was a disaster.

Davidoff then proposed a revolutionary approach to practice, encompassing three big ideas. First, develop a cadre of "advocate planners" who, like lawyers, would represent the interests of clients, primarily disenfranchised, low-income people. Not only would the advocate planner educate the clients about planning, he or she would also help craft plans to present to public decision makers to adopt or adapt in developing comprehensive municipal plans.

Second, drop the pretense of being a neutral, apolitical technician, and instead make your values transparent. Depending on your inclinations, formulate a Republican or Democratic point of view and let choices about planning be determined through the democratic process.

Third, shift planning from relying solely on physical considerations – include social and economic factors as well. In making plans, raise and respond to basic political questions: "Who gets what, when, where, why and how?"

The speech hit the conference like a bolt of lightning. It offered a response to the flurry of antiplanning literature that had appeared at about the same time: Jane Jacobs's *The Death and Life of Great American Cities* (New York: Random House, 1961), Herbert Gans's *The Urban Villagers: Group and Class in the Life of Italian-Americans* (New York: Free Press, 1963), and Martin Anderson's *The Federal Bulldozer: A Critical Analysis of Urban Renewal, 1942–1962* (Cambridge, MA: MIT Press, 1964). It provided a refreshing way to reinvent practice.

This selection, a reworking of the AIP plenary address under a new title, "Advocacy and Pluralism in Planning," appeared in the *Journal of the American Institute of Planners* in 1965 and has remained required reading for urban and regional planners ever since. It prompted a number of changes in the theory and practice of the field.

In the theoretical arena, Davidoff's proposition launched intense, largely unresolved, debates. Was advocacy planning just another form of social control exercised by elites to divert low-income populations from more aggressive means of gaining power? Or did it represent true empowerment? Were advocate planners really able to represent their clients, or did they impose their own values? The questions poured out in the writings of anthropologist Lisa Peattie ("Reflections on Advocacy Planning," *Journal of the American Institute of Planners*, 34: 2, 1968); political scientist Frances Fox Piven ("Whom does the Advocate Planner Serve?"

Social Policy, 1: 1, 1978, who with social work professor Richard Cloward also wrote *Regulating the Poor: The Functions of Public Welfare*, New York: Random House, 1971), and city planners Langley C. Keyes, Jr, and Edward Teitcher ("Limitations of Advocacy Planning: A View from the Establishment," *Journal of the American Institute of Planners*, 36: 4, 1978). Educational programs integrated advocacy planning into their curriculum. Hunter College in New York devoted an entire course of study to training advocate planners, hiring Davidoff to lead the effort.

"Advocacy and Pluralism in Planning" also energized planning practice. It stimulated the emergence of "equity planning," an effort that aims to redistribute power, resources, and participation from the rich to the poor. Cleveland city planning director Norman Krumholz became its spokesman with the city's publication of the *Cleveland Policy Plan* (1975), which focused on the needs of the city's low-income population. Krumholz later documented this work in two books – the first, with John Forester, *Making Equity Planning Work: Leadership in the Public Sector* (Philadelphia: Temple University Press, 1990), and, the second, with Pierre Clavel, *Re-inventing Cities: Equity Planners tell their Stories* (Philadelphia: Temple University Press, 1994).

Advocacy planning spread rapidly into the public sector. Municipal departments of city planning focused on subdistricts (often neglecting comprehensive planning in the process), hiring "neighborhood" planners to work with grassroots groups to develop small-area plans. Some cities gave legal standing and financial support to neighborhood planning. In New York City, for example, a 1975 city charter revision permitted local groups to submit their own plans (known as 197a plans, after the charter section) for Planning Commission approval. In 1990, Minneapolis started a ten-year, $20 million-per-year Neighborhood Revitalization Program (NRP) centered on planning. In 2001 Philadelphia launched a $1.6 billion Neighborhood Transformation Initiative focused on neighborhood planning.

And what of the man who set all these changes in motion? A year after the plenary address, Davidoff moved back to New York, his hometown – in his youth he lived in Sunnyside Gardens, the model housing development – to direct Hunter College's new urban planning program. In 1969, he founded the Suburban Action Institute, a legal research group focused on challenging exclusionary zoning and promoting fair housing. The Suburban Action Institute was involved in several zoning suits, notably the Mount Laurel I case (*Southern Burlington County NAACP v. Mt Laurel*, 67 N.Y. 151 [1975]), which established that New Jersey municipalities could not use zoning to exclude low- and moderate-income families, and the successful suit against Starrett Housing (*Arthur v. Starrett City Associates* 98 F.R.D. 500 [E.D.N.Y. 1983]), which struck down the use of racial quotas in tenant selection.

Davidoff, who earned degrees in planning (University of Pennsylvania, 1956) and law (University of Pennsylvania, 1961), wrote from the perspective of a practitioner and academic. He served as a planner in the public and private sectors (Delaware County Planning Commission; New Canaan Planning Commission; New York City Department of City Planning; and Voorhees Walker Smith & Smith) and on the faculties of the University of Pennsylvania (1958–65), Hunter College (1965–69), and Queens College (1969–82). As a member of the board of directors of the American Institute of Planners he successfully campaigned for the insertion of wording into the profession's code of ethics to recognize "a special responsibility to plan for the needs of disadvantaged groups and persons."

For additional writings by this firebrand, see Paul Davidoff and Thomas Reiner, "A Choice Theory of Planning," *Journal of the American Institute of Planners*, 28: 2 (1962); Paul and Linda Davidoff, "Suburban Action: Advocate Planning for an Open Society," *Journal of the American Institute of Planners*, 36: 1 (1969); and Paul Davidoff, Linda Davidoff, and Neil Gold, "Suburbs have to Open up their Gates," *New York Times Magazine*, November 7, 1971.

Thirty years after Davidoff delivered his landmark plenary address (and ten years after his untimely death from cancer), University of Michigan professor Barry Checkoway edited a special section of the *Journal of the American Planning Association* (60: 2, 1994) entitled "Paul Davidoff and Advocacy Planning in Retrospect," with contributions from recipients of the Paul Davidoff Award, an honor initiated by the Association of Collegiate Schools of Planning for a book that most represents his values. This retrospective offers some measure of Davidoff's influence on the field.

[T]he planning process . . . must operate so as to include rather than exclude citizens from participating in the process. "Inclusion" means not only permitting the citizen to be heard. It also means that he be able to become well informed about the underlying reasons for planning proposals, and be able to respond to them in the technical language of professional planners.

A practice that has discouraged full participation by citizens in plan making in the past has been based on what might be called the "unitary plan." This is the idea that only one agency in a community should prepare a comprehensive plan; that agency is the city planning commission or department. Why is it that no other organization within a community prepares a plan? Why is only one agency concerned with establishing both general and specific goals for community development, and with proposing the strategies and costs required to effect the goals? Why are there not plural plans?

If the social, economic, and political ramifications of a plan are politically contentious, then why is it that those in opposition to the agency plan do not prepare one of their own? It is interesting to observe that "rational" theories of planning have called for consideration of alternative courses of action by planning agencies. As a matter of rationality it has been argued that all of the alternative choices open as means to the ends sought be examined. But those, including myself, who have recommended agency consideration of alternatives have placed upon the agency planner the burden of inventing "a few representative alternatives". The agency planner has been given the duty of constructing a model of the political spectrum, and charged with sorting out what he conceives to be worthy alternatives. This duty has placed too great a burden on the agency planner and has failed to provide for the formulation of alternatives by the interest groups who will eventually be affected by the completed plans.

Whereas in a large part of our national and local political practice contention is viewed as healthy, in city planning, where a large proportion of the professionals are public employees, contentious criticism has not always been viewed as legitimate. Further, where only government prepares plans, and no minority plans are developed, pressure is often applied to bring all professionals to work for the ends espoused by a public agency. . . .

In presenting a plea for plural planning, I do not mean to minimize the importance of the obligation of the public planning agency. It must decide upon appropriate future courses of action for the community. But being isolated as the only plan maker in the community, public agencies as well as the public itself may have suffered from incomplete and shallow analysis of potential directions. Lively political dispute aided by plural plans could do much to improve the level of rationality in the process of preparing the public plan.

The advocacy of alternative plans by interest groups outside of government would stimulate city planning in a number of ways. First, it would serve as a means of better informing the public of the alternative choices open, *alternatives strongly supported by their proponents.* In current practice those few agencies, which have portrayed alternatives, have not been equally enthusiastic about each. A standard reaction to rationalists' prescription for consideration of alternative courses of action has been "It can't be done; how can you expect planners to present alternatives which they don't approve?" The appropriate answer to that question has been that planners like lawyers may have a professional obligation to defend positions they oppose. However, in a system of plural planning, the public agency would be relieved of at least some of the burden of presenting alternatives. In plural planning the alternatives would be presented by interest groups differing with the public agency's plan. Such alternatives would represent the deep-seated convictions of their proponents and not just the mental exercises of rational planners seeking to portray the range of choice.

A second way in which advocacy and plural planning would improve planning practice would be in forcing the public agency to compete with other planning groups to win political support. In the absence of opposition or alternative plans presented by interest groups the public agencies have had little incentive to improve the quality of their work or the rate of production of plans. The political consumer has been offered a yes–no ballot in regard to the comprehensive plan; either the public agency's plan was to be adopted or no plan would be adopted.

A third improvement in planning practice which might follow from plural planning would be to force those who have been critical of "establishment" plans to produce superior plans, rather than only to carry out the very essential obligation of criticizing plans deemed improper.

THE PLANNER AS ADVOCATE

Where plural planning is practiced, advocacy becomes the means of professional support for competing claims about how the community should develop. Pluralism in support of political contention describes the process; advocacy describes the role performed by the professional in the process. Where unitary planning prevails, advocacy is not of paramount importance, for there is little or no competition for the plan prepared by the public agency. The concept of advocacy as taken from legal practice implies the opposition of at least two contending viewpoints in an adversary proceeding.

The legal advocate must plead for his own and his client's sense of legal propriety or justice. The planner as advocate would plead for his own and his client's view of the good society. The advocate planner would be more than a provider of information, an analyst of current trends, a simulator of future conditions, and a detailer of means. In addition to carrying out these necessary parts of planning, he would be a proponent of specific substantive solutions.

The advocate planner would be responsible to his client and would seek to express his client's views. This does not mean that the planner could not seek to persuade his client. In some situations persuasion might not be necessary, for the planner would have sought out an employer with whom he shared common views about desired social conditions and the means toward them. In fact one of the benefits of advocate planning is the possibility it creates for a planner to find employment with agencies holding values close to his own. Today the agency planner may be dismayed by the positions affirmed by his agency, but there may be no alternative employer.

The advocate planner would be above all a planner. He would be responsible to his client for preparing plans for all of the other elements comprising the planning process. Whether working for the public agency or for some private organization, the planner would have to prepare plans that take account of the arguments made in other plans. Thus the advocate's plan might have some of the characteristics of a legal brief. It would be a document presenting the facts and reasons for supporting one set of proposals, and facts and reasons indicating the inferiority of counterproposals. The adversary nature of plural planning might, then, have the beneficial effect of upsetting the tradition of writing plan proposals in terminology which makes them appear self-evident.

A troublesome issue in contemporary planning is that of finding techniques for evaluating alternative plans. Technical devices such as cost–benefit analysis by themselves are of little assistance without the use of means for appraising the values underlying plans. Advocate planning, by making more apparent the values underlying plans, and by making definitions of social costs and benefits more explicit, should greatly assist the process of plan evaluation. Further, it would become clear (as it is not at present) that there are no neutral grounds for evaluating a plan; there are as many evaluative systems as there are value systems.

The adversary nature of plural planning might also have a good effect on the uses of information and research planning. One of the tasks of the advocate planner in discussing the plans prepared in opposition to this would be to point out the nature of the bias underlying information presented in other plans. In this way, as critic of opposition plans, he would be performing a task similar to the legal technique of cross-examination. While painful to the planner whose bias is exposed (and no planner can be entirely free of bias) the net effect of confrontation between advocates of alternative plans would be more careful and precise research.

Not all the work of an advocate planner would be of an adversary nature. Much of it would be educational. The advocate would have the job of informing other groups, including public agencies, of the conditions, problems, and outlook of the group he represented. Another major educational job would be that of informing his clients of their rights under planning and renewal laws, about the general operations of city government, and of particular programs likely to affect them.

The advocate planner would devote much attention to assisting the client organization to clarify its ideas and to give expression to them. In order to make his client more powerful politically the advocate might also become engaged in expanding the size and scope of his client organization. But the advocate's most important function would be to carry out the planning process for the organization and to argue persuasively in favor of its planning proposals.

Advocacy in planning has already begun to emerge as planning and renewal affect the lives of more and more people. The critics of urban renewal have forced response from the renewal agencies, and the ongoing debate has stimulated needed self-evaluation by public agencies. Much work along the lines of advocate planning has already taken place, but little of it by professional planners. More often the work has been conducted by trained community organizers or by student groups. . . .

Pluralism and advocacy are means for stimulating consideration of future conditions by all groups in society. But there is one social group which at present is particularly in need of the assistance of planners. This group includes organizations representing low-income families. At a time when concern for the condition of the poor finds institutionalization in community action programs, it would be appropriate for planners concerned with such groups to find means to plan with them. The plans prepared for these groups would seek to combat poverty and would propose programs affording new and better opportunities to the members of the organization and to families similarly situated.

The difficulty in providing adequate planning assistance to organizations representing low-income families may in part be overcome by funds allocated to local anti-poverty councils. But these councils are not the only representatives of the poor; other organizations exist and seek help. How can this type of assistance be financial? This question will be examined below, when attention is turned to the means for institutionalizing plural planning.

THE STRUCTURE OF PLANNING

[. . .]

What organizations might be expected to engage in the plural planning process? The first type that comes to mind are the political parties; but this is clearly an aspirational thought. There is very little evidence that local political organizations have the interest, ability, or concern to establish well developed programs for their communities. Not all the fault, though, should be placed upon the professional politicians, for the registered members of political parties have not demanded very much, if anything, from them as agents.

[. . .]

The second set of organizations which might be interested in preparing plans for community development are those that represent special interest groups having established views in regard to proper public policy. Such organizations as chambers of commerce, real estate boards, labor organizations, pro- and anti-civil rights groups, and anti-poverty councils come to mind. Groups of this nature have often played parts in the development of community plans, but only in a very few instances have they proposed their own plans.

[. . .]

There is a third set of organizations that might be looked to as proponents of plans and to whom the foregoing comments might not apply. These are the *ad hoc* protest associations which may form in opposition to some proposed policy. An example of such a group is a neighborhood association formed to combat a renewal plan, a zoning change, or the proposed location of a public facility. Such organizations may seek to develop alternative plans, plans which would, if effected, better serve their interests.

From the point of view of effective and rational planning it might be desirable to commence plural planning at the level of citywide organizations, but a more realistic view is that it will start at the neighborhood level. Certain advantages of this outcome should be noted. Mention was made earlier of tension in government between centralizing and decentralizing forces. The contention aroused by conflict between the central planning agency and the neighborhood organization may indeed be healthy, leading to clearer definition of welfare policies and their relation to the rights of individuals or minority groups.

Who will pay for plural planning? Some organizations have the resources to sponsor the development of a plan. Many groups lack the means. The plight of the indigent association seeking to propose a plan might be analogous to that of the indigent client in search of legal aid. If the idea of plural planning makes sense, then support may be found from foundations or from government. In the beginning it is more likely that some foundation might be willing to experiment with plural planning as a means of making city planning more effective and more democratic. Or the federal government might see plural planning, if carried out by local antipoverty councils, as a strong means of generating local interest in community affairs.

[. . .]

THE PUBLIC PLANNING AGENCY

A major drawback to effective democratic planning practice is the continuation of that nonresponsible vestigial institution, the planning commission. If it is agreed that the establishment of both general policies and implementation policies are questions affecting the public interest and that public interest questions should be decided in accord with established democratic practices for decision making, then it is indeed difficult to find convincing reasons for continuing to permit independent commissions to make planning decisions. . . .

Aside from important questions regarding the propriety of independent agencies, which are far removed from public control determining public policy, the failure to place planning decision choices in the hands of elected officials has weakened the ability of professional planners to have their proposals effected. Separating planning from local politics has made it difficult for independent commissions to garner influential political support. The commissions are not responsible directly to the electorate and in turn the electorate is, at best, often indifferent to the planning commission.

During the last decade in many cities power to alter community development has slipped out of the hands of city planning commissions, assuming they ever held it, and has been transferred to development coordinators. This has weakened the professional planner. Perhaps planners unknowingly contributed to this by their refusal to take concerted action in opposition to the perpetuation of commissions.

Planning commissions are products of the conservative reform movement of the early part of this century. The movement was essentially antipopulist and pro-aristocracy. Politics was viewed as dirty business. The commissions are relics of a not too distant past when it was believed that if men of goodwill discussed a problem thoroughly, certainly the right solution would be forthcoming. We know today, and perhaps it was always known, that there are no right solutions. Proper policy is that which the decision-making unit declares to be proper.

Planning commissions are responsible to no constituency. The members of the commissions, except for their chairman, are seldom known to the public. In general the individual members fail to expose their personal views about policy and prefer to immerse them in group decisions. If the members wrote concurring and dissenting opinions, then at least the commissions might stimulate thought about planning issues. It is difficult to comprehend why this aristocratic and undemocratic form of decision making should be continued. The public planning function should be carried out in the executive or legislative office and perhaps in both. There has been some question about which of these branches of government would provide the best home, but there is much reason to believe that both branches would be made more cognizant of planning issues if they were each informed by their own planning staffs. To carry this division further, it would probably be advisable to establish minority and majority planning staffs in the legislative branch.

At the root of my last suggestion is the belief that there is or should be a Republican and Democratic way of viewing city development; that there should be conservative and liberal plans, plans to support the private market and plans to support greater government control. There are many possible roads for a community to travel and many plans should show them. Explication is required of many alternative futures presented by those sympathetic to the construction of each such future. As indicated earlier, such alternatives are not presented to the public now. Those few reports which do include alternative futures do not speak in terms of interest to the average citizen. They are filled with professional jargon and present sham alternatives. These plans have expressed technical land use alternatives rather than social, economic, or political value alternatives. Both the traditional unitary plans and the new ones that present technical alternatives have limited the public's exposure to the future states that might be achieved. Instead of arousing healthy political contention as diverse comprehensive plans might, these plans have deflated interest.

The independent planning commission and unitary plan practice certainly should not coexist. Separately they dull the possibility for enlightened political debate; in combination they have made it yet more difficult. But when still another hoary concept of city planning is added to them, such debate becomes practically impossible. This third of a trinity of worn-out notions is that city planning should focus only upon the physical aspects of city development.

AN INCLUSIVE DEFINITION OF THE SCOPE OF PLANNING

The view that equates physical planning with city planning is myopic. It may have had some historic justification, but it is clearly out of place at a time when it is necessary to integrate knowledge and techniques in order to wrestle effectively with the myriad of problems afflicting urban populations.

The city planning profession's historic concern with the physical environment has warped its ability to see physical structures and land as servants to those who use them. Physical relations and conditions have no meaning or quality apart from the way they serve their users. But this is forgotten every time a physical condition is described as good or bad without relation to a specified group of users. High density, low density, green belts, mixed uses, cluster developments, centralized or decentralized business centers are *per se* neither good nor bad. They decide physical relations or conditions, but take on value only when seen in terms of their social, economic, psychological, physiological, or aesthetic effects upon different users.

The profession's experience with renewal over the past decade has shown the high costs of exclusive concern with physical conditions. It has been found that the allocation of funds for removal of physical blight may not necessarily improve the overall physical condition of a community and may engender such harsh social repercussions as to severely damage both social and economic institutions. Another example of the deficiencies of the physical bias is the assumption of city planners that they could deal with the capital budget as if the physical attributes of a facility could be understood apart from the philosophy and practice of the service conducted within the physical structure. This assumption is open to question. The size, shape, and location of a facility greatly interact with the purpose of the activity the facility houses. Clear examples of this can be seen in public education and in the provision of low-cost housing. The racial and other socioeconomic consequences of "physical decisions" such as location of schools and housing projects have been immense, but city planners, while acknowledging the existence of such consequences, have not sought or trained themselves to understand socioeconomic problems, their causes or solutions.

The city planning profession's limited scope has tended to bias strongly many of its recommendations toward perpetuation of existing social and economic practices. Here I am not opposing the outcomes, but the way in which they are developed. Relative ignorance of social and economic methods of analysis [has] caused planners to propose solutions in the absence of sufficient knowledge of the costs and benefits of proposals upon different sections of the population.

Large expenditures have been made on planning studies of regional transportation needs, for example, but these studies have been conducted in a manner suggesting that different social and economic classes of the population did not have different needs and different abilities to meet them. In the field of housing, to take another example, planners have been hesitant to question the consequences of locating public housing in slum areas. In the field of industrial development, planners have seldom examined the types of jobs the community needed; it has been assumed that one job was about as useful as another. But this may not be the case where a significant sector of the population finds it difficult to get employment.

Who gets what, when, where, why, and how are the basic political questions which need to be raised about every allocation of public resources. The questions cannot be answered adequately if land use criteria are the sole or major standards for judgment.

The need to see an element of city development, land use, in broad perspective applies equally well to every other element, such as health, welfare, and recreation. The governing of a city requires an adequate plan for its future. Such a plan loses guiding force and rational basis to the degree that it deals with less than the whole that is of concern to the public.

[. . .]

The new city planner will be concerned with physical planning, economic planning, and social planning. The scope of his work will be no wider than that presently demanded of a mayor or a city councilman. Thus, we cannot argue against an enlarged planning function on grounds that it is too large to handle. The mayor needs assistance; in particular he needs the assistance of a planner, one trained to examine needs and aspirations in terms of both short and long-term perspectives. . . . [I]t is apparent that our cities are in desperate need of the type of assistance trained planners could offer. Our cities require for their social and economic programs the type of long-range thought and information that have been brought forward in the realm of physical planning. Potential resources must be examined and priorities set.

What I have just proposed does not imply the termination of physical planning, but it does mean that physical planning be seen as part of city planning. Uninhibited by limitations on his work, the city planner will be able to add his expertise to the task of coordinating the operating and capital budgets and to the job of relating effects of each city program upon the others and upon the social, political, and economic resources of the community.

An expanded scope reaching all matters of public concern will make planning not only a more effective administrative tool of local government but it will also bring planning practice closer to the issues of real concern to the citizens. A system of plural city planning probably has a much greater chance for operational success where the focus is on live social and economic questions instead of rather esoteric issues relating to physical norms.

[. . .]

19

The Metropolitan Region

From *Design with Nature* (1969)

Ian McHarg

Editor's introduction

In the 1960s, Americans became increasingly conscious of environmental issues. Post-war suburbanization had transformed rural land at an unprecedented rate, something noted by the French geographer Jean Gottmann, who, in his massive study *Megalopolis: The Urbanized Northeastern Seaboard of the United States* (Cambridge, MA: MIT Press, 1961), predicted that the area from Boston to Washington, DC, would become one large city. Other observers documented the pernicious results of the carelessness with which humans were treating their natural surroundings. Rachel Carson's best-selling *Silent Spring* (Boston, MA: Houghton Mifflin, 1962) called attention to the damage caused by insecticides, and William H. Whyte mourned the loss of metropolitan open space in *The Last Landscape* (New York: Doubleday, 1968). Pollution-induced ecological disasters – Cleveland's Cuyahoga river on fire, captured in photographs by *Time Magazine*, August 19, 1969, and Lake Erie, the world's eleventh largest lake, being declared "dead," unable to support animal and plant life – galvanized people. So did biologist Paul Ehrlich's *The Population Bomb* (New York: Ballantine Books, 1968), which predicted imminent global famine.

Symbolizing these growing concerns was the nation's first Earth Day, April 22, 1970. Conceived by Wisconsin Senator Gaylord Nelson as a national teach-in modeled on the antiwar demonstrations occurring at the time, this well-publicized grassroots effort mobilized an estimated 20 million Americans, including students from 2,000 colleges and universities, in massive gatherings across the nation. (The author of this selection, Ian McHarg, was an organizer of the 30,000-person turnout in Philadelphia.)

Tapping into this sentiment, President Richard M. Nixon in 1969 created a presidential advisory body, the Council on Environmental Quality (CEQ), appointing Russell E. Train, president of the Conservation Foundation, as chair. Congress soon passed the National Environmental Protection Act (NEPA, 1969), later called the Magna Carta of environmental policy. It required government agencies to undertake environmental impact assessments prior to the construction of federal facilities and mandated CEQ to issue annual "state of the environment" reports.

To coordinate the NEPA work, Nixon formed the cabinet-level U.S. Environmental Protection Agency (EPA) in 1970, while Congress voted for a succession of protective laws. They included the Clean Air Act (1970) and Clean Water Act (1972) to attack pollution, the Coastal Zone Management Act (1972) to protect waterfronts, the Endangered Species Act (1973) to identify and conserve at-risk plants and animals on public and private land, and the Comprehensive Environmental Response, Compensation and Liability Act (CERCLA, or the Superfund Act, 1980) to deal with brownfields.

Over time, the United States collected a vast amount of environmental information. For example, in 1972 the National Aeronautics and Space Administration (NASA) launched the first Earth Resources Technology Satellite to create Landsat, imagery monitoring the earth's geological and environmental conditions. CEQ's State of the Environment report, issued from 1970 to 1997, was a useful resource – until the Federal Reports

Elimination and Sunset Act (Public Law 104-66, 1995) shut it down. Since that time the CEQ maintains environmental statistics at http://ceq.eh.doe.gov/NEPA/reports/statistics.

EPA developed best practices, case studies, and scientific measures for different types of pollution – and evolved into an 18,000 employee agency with ten regional offices that supervises federal initiatives dealing with air, water, hazardous waste, Superfund, coastal zone management, wetland protection, and other environmental concerns. Finally, the U.S. and state supreme courts handed down judicial decisions that frame today's environmental practices. The cases range from *Citizens to Preserve Overton Park Inc. v. Volpe* (401 U.S. 402 [1971]), a ruling strengthening the environmental impact assessment process, to *Rapanos et ux et al.* (*Nos. 04-1034*) *v. United States* (*Nos 04-1384*), June 19, 2006, which defines wetland.

The environmental movement had a huge effect on urban and regional planning. Practitioners developed new, enlightened methods of plotting land use that would take environmental issues into account. Fortuitously, McHarg, University of Pennsylvania professor of landscape architecture, had just published a 197-page treatise, *Design with Nature* (New York: Ballantine Books, 1969), written at the behest of then Conservation Foundation president Russell Train. The book not only captured the ecological sensitivities related to land use, but also provided a simple but brilliant methodology to incorporate environmental considerations into planning. Lewis Mumford, the renowned urbanist, wrote an introduction to the book, highlighting its systematic, multidisciplinary approach, one that carried on the traditions of his own mentor, another Scot, Patrick Geddes.

Design with Nature was an instant hit, going through four printings in its first year. More than 300,000 copies have been sold to date, with sales rebounding after the publication of a twenty-fifth anniversary edition (New York: Wiley, 1992). The book's enduring success is due to McHarg's elegant prose and persuasive graphics. The cover of the first edition featured an enormous moon accompanied by McHarg's introductory passage:

> This book is a personal testament to the power and importance of the sun, moon and stars . . . the creatures of the earth and the herbs. They are with us as co-tenants of the phenomenal universe . . . The world is abundant . . . Man is that uniquely conscious creature . . . who must become the steward of the biosphere. To do this he must design with nature.

Using the McHarg method, a planner produced a series of maps – at the time of the book's first publication, working with clear Mylar and magic markers – for an area, each summarizing its key ecological, geologic, social, or other characteristics. The planner then laid the individual maps over each other, to analyze suitable patterns of land development to be used in the formulation of a plan. McHarg himself applied this technique in several projects. The first was the 1962 *Plan for the Valleys*, an assessment and recommendations for a seventy-square-mile rural area north of Baltimore, followed by Amelia Island, Florida (1971), noted for its wide buffer at the waterfront and protected inland forests, and, later, The Woodlands, Texas (1974), whose design blended forest preservation and aquifer protection with relatively dense development.

McHarg's new mode of thinking – his layering of concerns – became the basis of today's geographic information system (GIS) analysis. It inspired Harvard graduate student Jack Dangerman, who later founded Environmental Systems Research Institute (ESRI), now a significant GIS developer. Many other designers also claim his legacy, including Andres Duany of Duany Plater-Zyberk & Co., the principals of the Philadelphia landscape firm Andropogen, and, most recently, the National Committee to Map Ecological Constraints/University of Texas project with EDAW to provide a McHarg-inspired analysis of the Gulf Coast region.

Born in Clydebank, near Glasgow and educated at Harvard University, where he earned degrees in landscape architecture (BLA, 1949; MLA, 1950) and city planning (MCP, 1951), Ian McHarg (1920–2001) founded the Department of Landscape Architecture and Regional Planning at the University of Pennsylvania and chaired the department from 1954 to 1986. In 1962, along with Penn colleague David Wallace, he founded Wallace McHarg Associates, later joining with William H. Roberts and Thomas A. Todd, an association that lasted until 1979. (Today the firm is Wallace Roberts & Todd [WRT]). Here he employed his ecological planning approach in several regional plans in addition to those cited above: Potomac river valley, the Twin Cities,

Washington, DC, and Denver. McHarg is the recipient of numerous awards, including Planning Pioneer, American Institute of Certified Planners (1997), and the Japan Prize, in recognition of his lifetime achievements (2000).

To learn more about McHarg, read his autobiography, *A Quest for Life* (New York: Wiley, 1996), or his collected writings, *To Heal the Earth* (Washington, DC: Island Press 1998) with Frederick Steiner. Andres Duany's "Introduction to the Special Issue: The Transect" in the *Journal of Urban Design*, 7: 3 (2002), describes his use of McHarg's work in developing New Urbanist thinking. Ann Whiston Spirn's "Ian McHarg, Landscape Architecture and Environmentalism: Ideas and Methods in Context," in Michael Conan (ed.), *Environmentalism in Landscape Architecture* (Washington, DC: Dumbarton Oaks Research Library and Collection, 2000), is a tribute from a former student and colleague. For an analysis of The Woodlands, see Ann Forsyth, *Reforming Suburbia: The Planned Communities of Irvine, Columbia and the Woodlands* (Berkeley, CA: University of California Press, 2005). Two works capture aspects of the early environmental movement: J. Brooks Flippen, *Conservative Conservationist: Russell E. Train and the Emergence of American Environmentalism* (Baton Rouge, LA: Louisiana State University Press, 2006), and William McGucken, *Lake Erie Rehabilitated: Controlling Cultural Eutrophication, 1960s–1990s* (Akron, OH: University of Akron Press, 2000). For a solid introduction to environmental planning, see Thomas L. and Katherine Daniels, *The Environmental Planning Handbook for Sustainable Communities and Regions* (Chicago: APA Planners Press, 2003).

A city occupies an area of land and operates a form of government; the metropolitan area also occupies an area of land but constitutes the sum of many levels and forms of government. It is united neither by government, planning, nor the expression of these. While the name has been coined to describe the enlargement of the older city, it is appropriate to observe that this is more a convenience for cartographers than a social organism. . . .

The American dream envisioned only the single-family house, the smiling wife and healthy children, the two-car garage, eye-level oven, foundation planting and lawn, the school near by and the church of your choice. It did not see that a subdivision is not a community, that the sum of subdivisions that make a suburb is not a community, that the sum of suburbs that compose the metropolitan fringe of the city does not constitute community, nor does a metropolitan region. It did not see that the nature that awaited the subdivider was vastly different from the pockmarked landscape of ranch and split-level houses.

And so the transformation from city to metropolitan area contains all the thwarted hopes of those who fled the old city in search of clean government, better schools, a more beneficent, healthy and safe environment, those who sought to escape slums, congestion, crime, violence and disease.

There are many problems caused by the form of metropolitan growth – the lack of institution, which diminishes the power to effect even local decisions, the trauma that is the journey to work, the increasingly difficult problem of providing community facilities. Perhaps the most serious is the degree to which the subdivision, the suburb, and the metropolitan area deny the dream and have failed to provide the smiling image of the advertisements. The hucksters made the dream into a cheap thing, subdivided we fell, and the instinct to find more natural environments became the impulse that destroyed nature, an important ingredient in the social objective of this greatest of all population migrations.

Let us address ourselves to this problem. In earlier studies we saw that certain types of land are of such intrinsic value, or perform work for man best in a natural condition or, finally, contain such hazards to development that they should not be urbanized. Similarly, there are other areas that, for perfectly specific reasons, are intrinsically suitable for urban uses. This method has been applied to the Potomac river basin, its constituent physiographic regions: there is no good reason why it should not be applicable to the metropolitan region of Washington.

Thus we can state as a proposition that certain lands are unsuitable for urbanization and others are intrinsically suitable. If our hearts are pure and our instincts good, then the lands that best perform work for man in a natural condition will not be those that are most suitable for urbanization. And because we are not necessarily pure or good, but lucky, it transpires, as we have seen before, that if one selects eight natural features, and ranks them in order of value to the operation of natural process, then that group reversed will constitute a gross order of suitability for urbanization. These are: surface water, floodplains, marshes, aquifer recharge areas, aquifers, steep slopes, forests and woodlands, unforested land. As was discussed in the study of metropolitan open space, natural features can absorb degrees of development – ports, harbors, marinas, water-related and water-using industries must be on riparian land and may occupy floodplains. Surface water, floodplains and marshes may be used for recreation, agriculture, and forestry. The aquifer recharge areas may absorb development in a way that does not seriously diminish percolation or pollute groundwater resources. Steep slopes, when forested, may absorb housing of not more than one house per three acres, while forests on relatively flat land may support a density of development up to one-acre clusters.

We can expect that there will be regional and subregional variation in intrinsic suitabilities. After all, this region includes parts of the coastal plain and the piedmont; within the latter there is the important Triassic subprovince. Indeed, topographic variation is least in the Triassic area and greatest in the crystalline piedmont. Stream dissection, and thus steep slopes, are greatest in the piedmont, followed by deposits of the Lower Cretaceous. Streams are all but absent in the Triassic area, but abundant in the remainder of the piedmont – less so in the coastal plain. Aquifers are concentrated in the Triassic area and coastal plain, but absent in the remaining piedmont.

This being so, there are revealed areas that are intrinsically unsuitable for urbanization, and these are shown. We can now plot the reverse, and indicate the areas that are suitable for urban uses. Here is the obverse of the first, and what is revealed is the regionality of urban suitability: a broad swath of land running parallel to the Triassic formation is shown to be the most suitable area, with a greater opportunity north of the Potomac than south of it. Lesser areas are visible in the coastal plain.

When an uncontrolled growth model is projected, it is seen that development bears no relation either to definitions of natural process values or to intrinsic suitability. Indeed, when the Year 2000 Plan developed for Washington is examined against these factors, it is seen that it is almost as oblivious as is unplanned growth.

It is most disconcerting to conclude that not only does uncontrolled growth fail to recognize intrinsic suitabilities and unsuitabilities for urban growth, but that the formal planning process is almost as culpable.

We require more precise information on which to base our decisions. It is not enough to describe land as unforested: one must examine its agricultural value, factors of foundations, suitability of soils for septic tanks and their susceptibility to erosion and the relative values of groundwater resources. To this end, a sector of the metropolitan region has been examined more specifically. It extends north and west from the Capitol to enclose an area of almost 400 square miles, reaching out to the rural perimeter, including both Dulles Airport and the new town of Reston.

THE QUADRANT

The method has now been used repeatedly and follows the familiar litany of historical geology, physiography, hydrology, and so on, and thereafter interprets these data to reveal intrinsic suitabilities. In this exercise the major prospective land use is urban; it is this that the method seeks to select. There is also an effort made to relate the density of prospective development, not only to the characteristics of the land, but also to its carrying capacity.

The area under study reveals both its characteristics and its variability when geology, physiography, hydrology, soils and slopes are examined. The major divisions of the Triassic piedmont, the crystalline piedmont and the coastal plain are immediately evident. In the Triassic formation the area is quite flat, streams are few and show little dissection, floodplains are absent, the substructure is limestone – it is associated with an important deep aquifer – and the soils are rich.

In the crystalline piedmont there are two subdivisions, which are roughly divided by the

Potomac. In the southern section the slope map most clearly reveals the fissures of streams and the marked dissection that accompanies them. The study area reveals the greatest topographic variety, which is displayed by the entire piedmont. The landscape is folded, small in scale, and soils are variable in response to topographic change and conditions of slope and exposure. There is little groundwater.

The coastal plain is generally 200 ft lower than the adjacent piedmont and the change occurs in the fall zone. The topography of the coastal plain is flat, but it is distinguishable from the Triassic formation by the pattern of streams, their floodplains, and the associated dissection. There are several elements unique to this region – escarpments and terraces, the marshes and bays and the estuarine aspect, which the Potomac assumes in this region.

When this area is considered in terms of the incidence of factors that render it unsuitable for urbanization, it is seen that the major aquifer and the productive soils of the Triassic formation represent an important value and thus a constraint. Dulles Airport now withdraws 1,000 gallons per minute from this groundwater. The southern section of the piedmont is broadly unsuitable because of the abundance of steep slopes and the absence of large areas of relatively flat land. The northern section of this region has the fewest constraints and offers the greatest opportunity for urbanization. The coastal plain does contain aquifers, floodplains and marshes, but also areas of land that impose few constraints.

Existing woodlands persist as residues of earlier and larger forests or as areas of farmland that have been abandoned and have returned to forest. Because of their value in diminishing runoff, reducing erosion and sedimentation and sustaining wildlife – in addition to their scenic and recreational uses – it was decided that, in this study, such woodlands should be considered a value and thus only marginally capable of accepting urbanization.

The presence of Dulles International Airport exerts a significant influence upon land uses in the area – mainly detrimental. The zones of 90 dB and 80 dB have been identified – this sound level is equivalent to an average machine shop or the noisiest street corner in New York City. For this reason, the FHA has refused to insure loans on residential construction within these zones. They are, therefore, considered unsuitable for urban development.

In this study it is assumed that, given the possibility of choice, prime agricultural land should not be employed for urban land uses on the grounds that this sterilizes an irreplaceable resource, all but irreversibly. For this reason, soils were identified in terms of four grades of agricultural potential product ion – row-cropland, cropland, pasture and permanent woodland – to diminish erosion on susceptible soils or excessive slopes.

From the foregoing information, certain lands were selected as unsuitable for urbanization. These included row-cropland and cropland, floodplains, slopes in excess of 15 percent, areas which (for reasons of erosion due to slope or soils) should be in forest cover, major aquifer recharges, forests and noise zones. The areas suitable for urbanization in this initial examination were the least productive agricultural soils – mainly pasture land. Examination quickly revealed that this provided inadequate space for future urban growth and that other lands would have to be utilized.

Clearly some agricultural land will be absorbed by urbanization. It was decided that both cropland and woodland should be investigated for urban suitability, but that this should be based on the characteristics of soils for the provision of foundations and for their usability for septic tanks, and that these qualifications would determine which of the cropland should be designated as suitable for urban development. The same analysis was accorded to forest.

The highest category of suitability in this analysis is noncropland with a capacity to bear foundations of high-density construction. The second category is identical to the first, save that it is cropland. The next category consists of low-bearing-capacity soils that are incapable of supporting septic tanks. These, it was thought, would be suitable for medium-density development, served by sewers. The next category was identical to the preceding, save that it involved the employment of cropland. A further category included poor-bearing-capacity soils that were capable of supporting septic tanks and thus could be used for low-density development. This condition, where cropland is lost, constitutes the next category.

The aggregate of these conclusions is a discrimination system whereby the propensity of the region to support urban land uses is disclosed. It selects surface water and floodplains, steep slopes (over 15 percent), major aquifer recharges, noise zones, and soils susceptible to erosion from which it is recommended that development be excluded.

Agricultural land is examined in detail, pasture land is identified as the least sacrifice of agricultural land for urban use, cropland is divided into categories of suitability, for foundations and septic tanks. Those forests able to support septic tanks are indicated for low-density development.

As cities are not composed entirely of buildings, and countryside is not entirely without them, the entire area has been examined to find large blocks of land that are preponderantly suitable for urban uses. The study of the enlarged metropolitan region and that of the Quadrant revealed the propensity of the land itself to support urbanization and showed the resulting system of both open space and urban structure that would result if this method were employed. The most arresting fact is the regional variability and the abundance of land available for prospective metropolitan growth. It is clear that many alternative patterns could be employed within any conscious plan.

This, of course, is not a plan. It merely shows the implications that the land and its processes display for prospective development and its form. The plan can be developed only when there is adequate information on the nature of demand, its locational and resource characteristics, the capacities to realize objectives and, indeed, the social goals of the community. It is enough to say here that – whatever the characteristics of demand – the formulation of a plan for the metropolitan region should respond to an understanding of natural processes. It must plan with nature.

Finally, this investigation is concerned with the matter of form. If growth responds to natural processes, it will be clearly visible in the pattern and distribution of development – and, indeed, in its density. But the formal investigation, it must be admitted, is not at a very high level. By responding to nature, one only avoids the allegation of ignorance, stupidity and carelessness. When we can demonstrate this elementary level of intelligence and perception, then we may aspire to more elevated objectives; but that is clearly premature.

At the moment, much of the area is unaffected by planning. Where planning does occur, its single instrument is zoning and by this device political subdivisions are allocated densities irrespective of geology, physiography, hydrology, soils, vegetation, scenery or historic beauty. The adoption of the ecological method would at least produce the negative value of a structure of open space wherein nature performed work for man, or wherein development was dangerous. It would canalize development to areas that were propitious. Positively, it can be employed to find the morphology of man–nature–Washington.

20

Paying for the Plan

From *A Region at Risk: The Third Regional Plan for the New York–New Jersey–Connecticut Metropolitan Area* (1996)

Robert D. Yaro and Tony Hiss

Editor's introduction

Urban planning reports rarely make front-page news. But in 1996 when the Regional Plan Association – the prominent nonprofit organization representing New York, New Jersey, and Connecticut – issued its third regional plan, that's exactly what it got, from the *New York Times*, no less. The plan's title, "A Region at Risk," likely captured an editor's eye, but its contents were also riveting. *Times* reporter Kirk Johnson bought the RPA line that the "nation's pre-eminent urban center and worldwide symbol of opportunity" could decline. According to RPA, he noted, the thirty-one-county region would be a "bleak landscape of crumbling economic core, increasingly isolated wealthy enclaves, transportation paralysis, deteriorating air and water quality and feeble economic growth."

However, if the area's leaders would pursue RPA's recommendations – grouped in five campaigns and focused on open space, centers, transportation, workforce development, and governance – it could have a far brighter future. This future would be one in which "residents zip about in high-speed trains on a fully interconnected regional rail system patterned after the Paris Metro, the aging urban centers have been recaptured as places to live, work and play, and lifelong adult education and job banks connect the population to new opportunities in the global economy." And it would cost only about $75 billion over the next twenty-five years. But Johnson was skeptical about the region marshalling such resources, observing that the governors of the region's three states were verbally supportive but had not backed up their support with the required dollars.

As is customary with regional plans, "A Region at Risk" focuses on traditional projects related to metropolitan spatial organization, transportation, and greening. Unlike earlier regional plans, it also incorporates suggestions for workforce development and governance. Overall, it envisions a polycentric region, ample open space, and efficient transportation. Within this vision, it calls for curbing sprawl, creating green networks, redeveloping cleaned-up brownfields, and enhancing mass transit.

Achieving this vision entails specific public policy and fiscal interventions. For example, the Greensward campaign identifies eleven existing regional reserves, region-shaping landscapes, and coastal waterways, whose preservation will protect the area's water sources, add to the supply of fresh produce, promote clean air, and provide recreation. They encompass estuarine systems, rivers, pine barrens, mountain chains, and agricultural lands, and range in size from 100,000 acres to 3 million acres, with ownership in many small parcels, a fact that complicates but does not prevent conservation efforts. And the Mobility campaign differentiates commuter/personal travel and freight, focuses on hard infrastructure investments (rail, light rail, subway, ferry service, highway and bike, and pedestrian ways) and on policies such as congestion pricing that affect vehicular travel. Its most powerful recommendation, the "Regional Rx," centers on the creation of a seamless mass-transit system connecting the Tri-state metropolitan region and enhancing travel within it. This goal could be achieved by adding only twenty-five miles of new transit lines to the existing 1,250 mile network – a mere 2 percent increase. The three remaining campaigns have equally compelling projects, all oriented to fulfilling the overall vision, arresting economic and physical decline, and maintaining the region's position as a global leader.

The plan's strength lies in its unity, internal logic, and comprehensiveness. Although this plan represents the views of the Regional Plan Association, a nonpartisan coalition of business and civic leaders dating from the 1920s, and has no legal standing, it is influential due to RPA's longstanding reputation for solid research and policy analysis.

In a little more than a decade after the plan's publication, an impressive number of the recommendations have been realized. These successes stem from the genius of RPA's incorporating ideas from many specialized groups. By helping to establish common cause among separate groups, RPA has empowered and given identity to alliances and shared goals. For example, in 1999 Robert Yaro, one of the authors of this selection, helped create the Empire State Transportation Alliance (ESTA) – composed of civic, academic, business, environmental, and transportation user groups – to lobby the Metropolitan Transportation Authority and other regional transportation agencies for capital budget allocations in support of the plan.

Land acquisition and conservation protection for seven of the eleven regional reserves is now under way. The transportation authorities in the three states are spending about $75 billion in today's dollars on some of the big-ticket items called for in the plan – a new subway line, a tunnel under the Hudson river, airport connections, and other critical links between the commuter-rail or light-rail lines and centrally located terminals.

This selection, written by Robert Yaro, then research director, now president, of the Regional Plan Association, and author Tony Hiss, is an example of how to think conceptually about funding regional-scale initiatives. Yaro and Hiss match the expense of the individual campaigns to revenue sources and break the total bill into understandable and manageable increments.

Yaro and his associates have extended regional thinking beyond the New York metropolitan area to encompass a larger geographic entity called a megaregion. (See Plate 1.10.) A megaregion is an urban land mass linked by economic or trade relationships, a transportation network, environmental systems, common culture and growth concerns. Working with the University of Pennsylvania, Lincoln Institute of Land Policy, and several foundations, including Ford and Rockefeller, they have identified ten U.S. megaregions that they forecast will accommodate 70 percent of the nation's population growth by 2050. This likely growth, they argue, calls for new planning approaches. For more information, see Armando Carbonell and Robert Yaro, "American Spatial Development and the New Megalopolis," in *Land Lines*, 17: 2 (2005).

In addition to being RPA president, Robert Yaro is Professor of Practice in the Department of City and Regional Planning at the University of Pennsylvania. He is author of the prize-winning *Dealing with Change in the Connecticut River Valley: A Design Manual for Conservation and Development* (Cambridge, MA: Lincoln Institute of Land Policy, 1990), co-author of *Smart Growth in a Changing World* (Chicago: Planners Press, 2007) with Jonathan Barnett, F. Kaid Benfield, Paul Farmer, Armando Carbonell, and Shelley Poticha, and co-author of *America 2050: The Next Exploration* (Philadelphia: University of Pennsylvania Press, forthcoming) with Armando Carbonell. Tony Hiss is a staff writer for the *New Yorker*, a contributor to *Preservation* and other magazines, and a visiting scholar at New York University. He has written widely about the conservation and restoration of urban places and landscapes. His book *The Experience of Place: A New Way of Looking at and Dealing with our Radically Changing Cities and Countryside* (New York: Vintage Books, 1991) analyzes the public realm in cities around the world. He is currently working on *From Place to Place*, dealing with transportation and sprawl.

To read more about the Regional Plan Association's contemporary work go to http://rpa.org. David A. Johnson's *Planning the Great Metropolis: The 1929 Regional Plan of New York and its Environs* (New York: Chapman & Hall, 1996) is a history of the RPA's first regional plan and its impact on the development of the New York area. Another example of private sector-led regional planning is the work of Chicago Metropolis 2020, which was founded in 1996, and in 2000 published its plan, *Chicago Metropolis 2020: The Chicago Plan for the Twenty-first Century* (Chicago: University of Chicago Press, 2000). Like the RPA, Chicago Metropolis 2020 has ongoing advocacy and research programs (see www.chicagometropolis2020.org/). For a general textbook on regional planning, see David A. Plane, Lawrence D. Mann, Kenneth Button, and Peter Nijkamp,

Regional Planning (Northampton, MA: Elgar, 2007), containing classic writing in the field from the 1970s to the 1990s. For a view of government-sponsored regional planning, read about European practices in Sir Peter Hall's *Urban and Regional Planning*, fourth edition (London: Routledge, 2002). To follow the seminal research on megaregions, see Peter Hall and Kathy Pain, *The Polycentric Metropolis: Learning from Mega-city Regions in Europe* (London: Earthscan, 2006); Petra Todorovich (ed.), *The Healdsburg Research Seminar on Megaregions* (New York: Regional Plan Association and Lincoln Institute of Land Policy, 2007); and the numerous local megaregion studies now emerging. Especially noteworthy are the Brookings Institution's Metropolitan Policy Program, the Great Lakes Economic Initiative, and Georgia Tech's Center for Quality Growth and Regional Development's work on the Piedmont Atlantic MegaRegion, all found on the America 2050 website: www.america2050.org.

If fully implemented, the plan's five campaigns will yield a range of benefits that include stronger, more sustainable economic growth, a more cohesive society with a more equitable distribution of the region's prosperity, cleaner air and water, less congestion, more attractive communities, and a rich legacy for future generations.

Many of these benefits cannot be quantified, either because they represent qualitative improvements in the lives of the region's citizens or because we lack the tools to measure them effectively. For example, we have only gross measures of the contributions of health and education in a productive society and no clear understanding of the role that informal businesses, volunteers, and family workers have in our economy. The true costs of production, investments, and public services are even more difficult to ascertain because we do not incorporate the costs of depleted natural or public resources in our economic accounting systems, nor do we determine the full costs of highways, real estate development, and public facilities and services in making public investment choices.

Even for traditional measures of economic growth, benefits estimates are constrained by a lack of data at the regional level. Within these constraints, however, a plausible scenario can be depicted for the difference between current trends and the result of a proactive strategy to reinvest in the region's human, physical, and natural resources. . . .

[. . .]

Achieving . . . the . . . benefits of the plan will require an investment of financial resources, political will, and civic energy. Of the five campaigns outlined in this plan, four – Greensward, Centers, Mobility, and Workforce – call for additional public investment in systems to restore the region's competitiveness, social equity, and quality of life. The final campaign, Governance, illustrates ways that new funds could be raised and public services reorganized. The investments carry significant price tags that must be put in the context of current spending, potential revenue sources, and estimated return. With the movement in Washington toward devolution of government responsibilities – a reality, at least in the short run – the region needs to develop its own investment strategy and the means to finance it with declining guidance and support from the federal government. Currently, public sector expenditures in the region from state and local government, excluding infrastructure spending, total approximately $80 billion. In addition, RPA estimates from available agency data that the region spends an additional $12 billion annually on infrastructure investments. Transportation expenditures – which are weighted heavily in favor of highway maintenance and construction – dominate the region's capital budget.

RPA estimates that the region will need to spend $75 billion in capital expenditures above current levels over the next twenty-five years to implement the recommendations of the Greensward, Centers, and Mobility campaigns. Some of the larger region-shaping investments, such as Regional Express Rail (Rx), regional reserves and revitalized downtowns can be completed within a fifteen-year time period, increasing capital investments by approximately $4 billion per year during that period.

The total cost of the human capital investments identified in the Workforce campaign have not been estimated, but fully addressing the needs of public education, early childhood development, and adult education and training could require resource commitments approaching those of the other three campaigns combined. With the region's school-age population expected to surge by 700,000 over the next fifteen years, additional enrollment alone will require a significant increase in both capital and operating expenditures in public schools. The capital requirements to bring schools to a state of good repair and meet expansion needs could be as high as $25 billion for New York City alone. Actual improvements in instruction – reductions in class size; expanded programs in science, the arts, and foreign language instruction; and implementation of school-to-work initiatives – could require region-wide expenditures of similar magnitude even with aggressive efficiency measures. Investments in early childhood development, daycare, and adult literacy are also essential components of workforce development that carry significant price tags.

Specific investments for the campaigns . . . include:

- Establishing the eleven reserves, greenways, and urban open spaces envisioned in the Greensward campaign will cost an estimated $11 billion, with the largest expenditures for wetland restoration, improved access, and waterfront redevelopment for Long Island Sound and the New York–New Jersey Harbor Estuary.
- The Centers campaign calls for $17 billion to attract jobs and development to the region's downtowns and to make significant progress in building affordable housing. At $10 billion the housing recommendations represent the largest expenditure in this category.
- Building a world-class express rail system and improving both highway and freight movement, as outlined in the Mobility campaign, will cost an estimated $47 billion. Regional Express Rail, the largest single capital budget item in the plan, has a price tag of $21 billion.

The plan's capital investment recommendations do not include funding for essential maintenance needed to keep transportation and environmental systems in a state of good repair, for required new or upgraded wastewater collection, and for treatment plants. These costs, along with currently planned projects and plan recommendations, should be included in a regional capital investment plan that sets priorities based on a comprehensive cost–benefit analysis.

RPA proposes that these investments be funded through a combination of strategies that increase government efficiency or charge users for services, rather than through increases in general taxes. RPA's strategy is for the region to make better use of public funds by streamlining and restructuring government, in much the same way that every other sector of the economy has restructured over the last decade, to take advantage of new approaches to organization, communications, and data management. In so doing, government can purchase the same services and meet the same objectives with less money. All three states and New York City have already initiated restructuring efforts of the type envisioned here. New or restructured revenues for investments should be raised through fee systems that charge those who either create the need or would benefit most from the new investments.

INCREASING PUBLIC EFFICIENCY

Just as sustainable business principles call for private enterprise to invest more efficiently by focusing on pollution prevention rather than environmental cleanup, government should use public funds with much greater cost-effectiveness. Four approaches are proposed that could fund a substantial portion of the plan's recommendations: smart infrastructure, right-sizing government (both described in Governance), redirecting inefficient economic development incentives, and reducing the costs of sprawl.

Smart infrastructure

Existing examples of smart infrastructure in New York City and other places have demonstrated the potential to save half or more of the cost of some projects compared with conventional approaches. Three examples . . . include:

- The restoration of Jamaica Bay by using an ecosystem restoration strategy is expected to be $1.1 billion less expensive than the originally proposed $2.3 billion treatment facility.

- The region's ongoing shift toward an integrated solid waste management system, including waste prevention, will save public agencies an estimated $475 million per year in waste management costs by 2015 over disposal-based strategies.
- Expenditures of about $1.5 billion for a comprehensive New York City watershed protection program should avoid the estimated $6 billion otherwise required to build the world's largest water filtration system, while achieving at least the same water quality standards.

Less dramatic savings may be available from transportation and other types of infrastructure than from the environmental examples cited above. Studies also indicate substantial savings potential in procurement reform, the elimination of duplicative oversight, and privatization of some systems and facilities. If the region were to save just 10 percent overall through smart infrastructure reforms and strategies, it would translate into a $1.2 billion saving annually, or $30 billion over twenty-five years. The trend toward privatization, in particular, is only now beginning in a series of controversial and experimental programs that go beyond private service contracting to the actual investment of private capital – and subsequent private ownership of facilities – for public services. These efforts need to be examined and, where promising, nurtured and developed. However, this must not mean an abrogation of public oversight but rather a new role for public agencies in setting appropriate standards for construction and operation of privatized infrastructure.

Right-sizing governance

Right-sizing governance also presents the opportunity for dramatic savings in public spending, particularly in this region. State and local government expenditures, excluding public authorities, have more than doubled in real terms since 1967. Much of this increase represents expansion in medical, social welfare, and environmental protection costs, in addition to higher service levels in much of the region. Other regions have experienced tremendous growth in government expenditures over this period, and many are also now implementing aggressive efficiency and privatization measures. Excluding capital outlays, but including the operating expenditures of public authorities, current expenditures total approximately $80 billion. If institutional reforms, governmental consolidation, and better inter-regional cooperation could produce a 5 percent reduction in these expenditures, then an additional $100 billion in savings would be produced over twenty-five years.

Currently, the political leadership of the region recognizes the need to improve government efficiency but is coupling efficiencies primarily with reductions in taxes. While tax reductions may make the region marginally more competitive in the short run, some of these savings should be invested wisely in human capital that both increase long-term economic growth and address the widening divide between well educated and low-skill workers. Particularly since low-skill workers and minorities are likely to bear the brunt of reduced government employment, the need to provide these workers with the means to obtain other employment becomes more compelling.

The cost of sprawl

Successful implementation of the plan's Greensward, Centers, and Mobility campaigns would result in a third category of investment savings and increased efficiency: a reduction in the costs of sprawl development patterns. Various studies have demonstrated through case study analysis that planned forms of urban development cost local governments less in services, particularly in roads and utilities. For example, with compact development patterns, fewer miles of roads and utilities have to be built and school districts can more efficiently operate at full capacity. Savings of around 25 percent in road construction and maintenance, 15 percent in utilities, and smaller savings in public education offer potentially large savings overall in service costs. A comprehensive study of the New Jersey State Plan has indicated potential savings of up to $400 million annually to municipalities and school districts statewide if new development proceeded in accordance with State Plan guidelines. Region-wide, these savings could be twice the estimate for New Jersey alone. Assuming that it will take a decade before significant savings from reduced sprawl could begin to accrue, the cost of government could be reduced by as much as $1 billion per year by the end of the first decade of the next century.

Eliminating border warfare

A fourth approach that would more efficiently use government resources would be to end "border warfare" economic development strategies in which different sections of the region compete against each other to lure businesses with tax breaks. Tax waivers and economic development incentives represent a substantial source of forgone revenue for investment. New York City alone exempts commercial and industrial taxpayers from some $800 million each year in real property, business, and excise taxes to induce business to remain in or relocate to the city.

Some of these waivers are included in specific incentive packages designed to lure or retain particular companies, such as deals in 1995 to attract Swiss Bank to Stamford or to keep CS First Boston in New York City. These transactions add nothing to the region's economy and deplete revenues that could be used for general tax reductions, service improvements, or capital investments. The bulk of these expenditures, however, are "as of right" programs (available to any firm moving to a designated area), such as New York City's $96 million Industrial and Commercial Incentive Program for companies locating outside Manhattan's central business district; industry-wide programs; or exemptions for property owned by public authorities, such as Battery Park City. . . . Even for as-of-right programs, intraregional competition for jobs and tax base is often a strong motivating force, as municipalities respond to relocations in particular industries or areas.

As levers for economic development, it is highly questionable that targeted incentives are more effective than general tax reductions or equivalent investments in infrastructure or education. While they reduce business costs for selected recipients, they do nothing to improve overall productivity or efficiency, or to improve the region's attractiveness as a place to live and work. They also tend to ratchet up destructive border warfare.

Complete elimination of tax waivers and targeted incentive programs is not feasible in the short run, or completely desirable. Many transactions are legally binding for years to come, and some as-of-right programs may be important tools if they are part of a comprehensive strategy to redirect employment to centers. As a general policy direction, however, the thrust should be to reduce reliance on incentives in favor of investments in workforce, transportation, and the environment. Assuming that economic development incentives currently total well over $1 billion region-wide, a reasonable goal would be to gradually redirect them to investments in centers and communities. Assuming an immediate moratorium on new incentive agreements and a gradual reduction of $50 million per year, the size of incentive programs could be cut in half over a ten-year period.

NEW REVENUES FROM USER FEES AND DEDICATED TAXES

Even with the savings outlined above, this plan calls for investments that will require new sources of public funding. But rather than looking simply for more money to go to public expenditures, RPA recommends a transformation in the manner by which new public funds are procured and allocated. Rather than increases in general taxation rates, greater emphasis should be given to methods of finance that create visible links between fees collected and services provided. Relating revenues directly to services provides the public with greater assurance that their taxes and fees fund specific service improvements, and it lets government rely more on market forces to achieve public purposes. For example, pricing tolls differently during peak and off-peak periods can reduce congestion and lets users decide whether time or price is a higher priority. Dedicated taxes, such as a cigarette tax to fund health care or a gasoline tax to fund transportation improvements, do not have as direct a link between a user and a specific service, but they affect consumption rates of activities that have external costs to society – such as medical costs and highway maintenance.

To produce new revenues for public investment, the plan proposes user fees or dedicated taxes for transportation and other systems. Research has demonstrated both that certain user fees can efficiently achieve public goals, such as reduced congestion, and make political sense, as RPA's 1995 Quality of Life Poll demonstrated. The significant finding of the poll was that, in an era of cost cutting and government cutbacks, people would be willing to pay for increased services as long as the new fees are directly and visibly tied to the production of improved public service.

To finance a Tri-state Infrastructure Bank, RPA proposes various user fees and dedicated taxes on automobile travel. A 20c per gallon regional gasoline tax would produce approximately $1.2 billion a year, and a 1.5c vehicle-per-mile (VMT) charge, to be collected when vehicles are registered, would create a $1.5 billion revenue stream. By combining these two – for example, a 10c per gallon gas tax and a 1c per mile VMT charge – around $1.6 billion would be raised annually for the Tri-state Infrastructure Bank. This strategy would introduce long overdue market forces to the region's transit funding policies. In addition to raising capital for infrastructure improvements, congestion on the region's highway system would be reduced, particularly as new or improved transit alternatives come on line.

Over time, the goal should be to replace as much of this dedicated tax revenue as possible with user fees. New technology, such as automated toll and fare collection, represents significant potential to charge users according to how much they use the transportation system with both distance and congestion-based pricing. Once the technology is in place, revenue from differential tolls and fares could allow dedicated taxes to be reduced. Until such systems are in use, however, it is difficult to determine how much revenue would actually be generated.

While these vehicle taxes and fees may seem excessive to some, they represent a fraction of the current direct subsidy provided to automobile and truck users by taxpayers. In New York State alone, these subsidies were estimated at $1.9 billion in 1990. They also represent an even smaller fraction of vehicle taxes paid in competing European and Asian countries.

Similarly, new fee systems and dedicated taxes should be implemented to finance acquisition and protection of the region's natural resources and open spaces. As experience in the New Jersey Pinelands has demonstrated, land values increase significantly in areas where open space and ecological systems are protected. A portion of this increased value should be used to pay for conservation measures. Specific options, outlined in Greensward, include a dedicated real estate tax tied to development activity of large landowners or short-term transactions and a property tax surcharge that would capture a portion of the rise in property values associated with adjacent conservation lands. In addition, a surcharge on water bills for areas that draw water from the Regional Reserves could be used to finance land conservation, water quality infrastructure, and pollution prevention programs in the Appalachian Highlands and other water-producing areas. These incremental fees, tied to those who directly benefit from improved environmental quality, could raise in excess of $300 million a year for environmental investments.

REVENUE FROM EXPANDED GROWTH

A final source of revenue is the potential for increased tax revenues if the region is able to increase its rate of economic growth. If the implementation of the plan is successful in enabling the region to stabilize its share of national growth, then tax revenues would increase substantially.

As described above, stabilization of the region's share of national economic activity would result in an additional $200 billion per year in gross regional product (GRP) by 2020. Currently, state and local tax revenues approach 13 percent of GRP. Revenues equivalent to 10 percent of GRP increases would yield an additional $20 billion per year in tax revenues by 2020, if this rate of growth is achieved. Over the next twenty-five years, $180 billion in revenue could be raised without an increase in tax rates, with most of the increase coming in the later years.

A TWENTY-FIVE YEAR INVESTMENT STRATEGY

The potential revenue sources described above are a combination of potential savings from government efficiencies, fees, and dedicated taxes that could be adopted by state and city governments, reallocations of existing revenues in the case of economic development incentives, and revenue potential from faster economic growth. All will require considerable effort and ingenuity to implement but are achievable if the political will exists.

To provide public assurance that efficiency efforts and new revenue streams result in improved services and a stronger economy, the plan recommends that specific funding mechanisms and protocols be instituted, including:

1 A Tri-state Infrastructure Bank to administer a coordinated long-range capital budget for transportation and environmental improvements.
2 Along with state assumption of education financing, a requirement that productivity savings in the education system be reinvested in schools. In recent years, cost-saving efficiencies implemented in the schools have been used to balance municipal budgets, reduce taxes, or fund other purposes. At a minimum, the increasingly heard refrain in all levels of government that children are the highest priority should result in the reinvestment of these savings, if not savings achieved from right-sizing government in other areas.
3 Identification of dedicated revenue streams for specific workforce-related objectives, combined with the implementation of strong management reforms to insure that objectives are achieved. In this regard, implementation of the investment plan advocated by the Commission on School Facilities and Maintenance Reform can be a first step, as well as serve as a model for other jurisdictions and issues.

The Infrastructure Bank would receive funds from environment- and transportation-related fees and taxes. A portion of this revenue stream should be devoted to training and employment services to provide low-income communities with access to the increased job opportunities, particularly in construction jobs directly tied to the infrastructure investments. These expenditures could also help reduce the total construction costs by easing labor shortages during peak construction periods.

Capital costs funded by the Infrastructure Bank could be paid directly out of government receipts ("pay as you go") or through government bonds with debt service paid from the above mix of revenues. The "pay as you go" approach has the advantage of reducing government debt ratios and relying more on committed rather than expected revenues. It has the disadvantage of increasing upfront costs and reducing the early stimulative effect of the capital program. The scenario below assumes that the entire program is funded through the more traditional approach of government bonding. However, a "pay as you go" approach for at least part of the expenditures might be appropriate for a program of this size, even though it might slow the initial pace of implementation.

... [O]ne scenario for funding the $75 billion in recommended capital expenditures over twenty-five years ... assumes that all dedicated revenues are used to support debt service for bonded projects. If smart infrastructure is able to achieve a 10 percent efficiency in the plan's infrastructure recommendations, then the cost of the capital plan would be reduced by $7.5 billion. Debt service for the remainder would total approximately $100 billion during the twenty-five-year plan period.

Capital expenditures are somewhat less during the first five "startup" years than the following ten years, but they then decline again on the assumption that Rx and other large systems are completed by 2010. If user fees and taxes are fully implemented from the first year of the program, then dedicated revenue streams from these fees can fully pay for debt service in the first part of the program. As debt levels rise in later years, accelerated growth should be more than sufficient to cover increased costs.

The auto use fees and taxes would yield $1.6 billion per year, or $40 billion total, enough to finance the debt service on the Mobility investments until approximately 2008. The $300 million in annual conservation fees, $7.5 billion total, would finance Greensward investments until approximately 2005. Likewise, redirected economic development incentives would fund more productive investments in Centers until around 2000.

Combined, fees and taxes cover more than half of the debt service payments. The remaining $44 billion needed from additional revenues is less than one-quarter of what the region might achieve if it can only maintain its current share of national growth. For the years beyond 2020, debt service will stabilize at approximately $7 billion per year, but revenues should continue to rise as the productivity and quality of life improvements of the plan generate additional economic growth.

Growth could be well below full potential and still be more than sufficient to fund increasing debt service.

Clearly, implementing the efficiency savings and user fees recommended here represents an imposing challenge for a region with more than its share of fiscal difficulties. However, the means are available if the civic and political will exist. In many ways, today's challenge is no less than the one faced by the region at the time of RPA's first regional plan. In spite of an infrastructure system with needs even greater than those we face today, and in spite of a depression even more severe than our recent downturn, the region was able to build the foundations for two generations of growth. Building on past successes to create more equitable and sustainable growth will require a similar level of commitment.

PART FOUR

The Plan: Its Origins and Contemporary Use

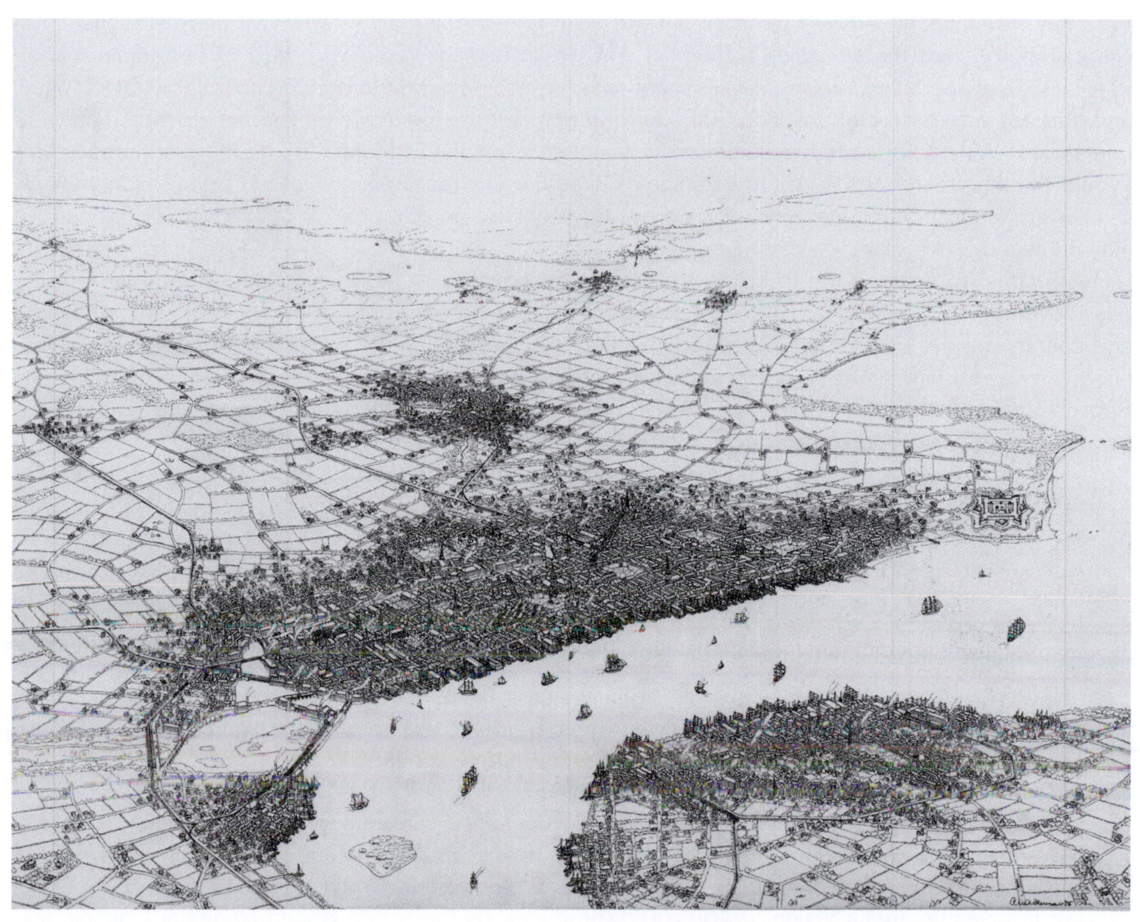

Growth of a typical Northeastern city: 1860. (Drawing by Andrew Whittemore)

INTRODUCTION TO PART FOUR

In April 2007 New York City Mayor Michael R. Bloomberg stood before a huge crowd assembled in the auditorium of the American Museum of Natural History. The audience – which included the city's political elite, civic leaders, members of the press, and as many others as could fit in – had passed through the museum's dinosaur halls on the way to the meeting. But on this afternoon they were all gathered to hear not about the past, but about the future. The mayor approached the podium. Suddenly a stirring video played on a giant screen behind him. Then Bloomberg unveiled a strategic plan entitled "PlaNYC 2030," announcing that over the next two and a half decades the city would absorb a million more people, take control of its growth, and transform itself into one of the greenest cities in the nation. Was this New York hyperbole? No. This businessman-turned-politician – known for understatement, not exaggeration – had thrown down a gauntlet. He would use the remaining days of his administration to enhance transportation, provide development sites for high-density housing, add accessible open space, and address environmental pollution. Although the plan set strategic directions, many details remained unresolved: Would the state legislature allow congestion pricing in Manhattan as the plan suggested? Would the neighborhoods agree to rezoning? Could the city really plant a million trees? Nonetheless, Bloomberg made a political as well as a policy statement like no other mayor before him. He pinned his mayoral legacy on a plan.

From coast to coast, north to south, the plan is becoming ever more important. The media's drumbeat coverage of climate change and peak oil, the rising concern about open space losses when city and country collide, New Urbanist-inspired nostalgia – all are among the societal trends that have stimulated a renewed interest in planning. By winter 2008 interest had risen to such a pitch that National Public Radio ran a multi-week series, "Shifting Ground," that captured the mood of the country and helped explain why the plan has become so important.

The results vary. Legislative mandates drive some planning endeavors; expanding the potential for sustainable development inspires others. Some plans follow decades-old methods, emerging with detailed socioeconomic analyses and associated land use and transportation schemes. Others blend geographic information system and urban design analyses to map growth under different conditions like "business as usual" or with transit-oriented development; they show 3-D sketches of what places would look like according to the various alternatives. Of course, as will be seen in the readings in this section, there are many different kinds of plans.

In New York City, where municipal government has a weak mandate to produce a strategic plan, Michael R. Bloomberg's PlaNYC 2030 serves as one type of example. In the west, where municipal planning is mandated, cities have traditional plans. But here *regional* planning has emerged with new energy. In the past five years, the west's most populous metropolitan areas have voluntarily produced a raft of vision plans that, like PlaNYC 2030, aim to harness the current and future population increases that have resulted in sprawl, threatening environmental resources and a fabled quality of life. These plans – all of which trade higher density for more open space – include Portland's 2040 Growth Concept (2005); the Seattle area's Vision 2040 (2007), produced by the Puget Sound Regional Council; and San Francisco Bay Area's Smart Growth/Livability Footprint Project (2003). Also produced are plans for San Diego, Los Angeles, and Sacramento. Meanwhile, in Florida, frightening growth projections, traffic congestion, and a threatened environment have stimulated a civic coalition – including the University of Central Florida's Metropolitan Center for Regional Studies, my region.org, and 1,000 Friends of Florida – to sponsor an ambitious state plan, "An Alternative Future: Florida in the Twenty-first Century 2020, 2040 and 2060" (2007).

Against the backdrop of widespread attention to the plan in its varied forms, this section explores the plan's evolution, uses, and power. In "Twentieth Century Land Use Planning: A Stalwart Family Tree," Edward J. Kaiser and David R. Godschalk focus on the bread-and-butter of the profession, the land use plan. Their apt family-tree analogy allows a discussion of the general, master, or comprehensive plan as well as new forms called "land use design" plans and "development management plans."

In "Regionalism," Oliver Gillham highlights state-legislated growth management efforts in Portland, Oregon, a story that continues to unfold. Since Gillham wrote his essay, Oregon voters have passed two referenda – Measure 37 (2004) that limited the growth management law and Measure 49 (2007) that restored some elements of the law. Unusual alliances have arisen (the homebuilders and the planners banded together on Measure 49), the courts have been active, and the planning community and property rights advocates at odds. Monitor this important case on the internet to stay up to date. And remember, no one can accuse land use planning of being for the faint at heart – there is too much at stake.

"Does Planning Need the Plan?" Michael Neuman asks in the next selection. Yes, is his response in his piece, written to remind his academic colleagues that the plan – not discourse about a plan or the planning process – is the central product of the field. People are really interested in the design of places, he asserts; they want to have a say about the physical appearance and functioning of their cities. As a result, he insists, they are clamoring for plans they can discuss, kick around, and settle on.

Lewis D. Hopkins not only agrees with Neuman in "How Plans Work," but goes well beyond him with his explanation of the purposes of plans. He sees layers of usefulness – a guide to the future, an agenda, and a whole lot more – but challenges planners to test whether plans work as well as they should.

Edward J. Kaiser and David R. Goldschalk in the "The Land Planning Arena" add other dimensions to thinking about plans. A plan does not appear in a vacuum, they observe; it represents the views of many players – public and private sector representatives, special interest and advocacy groups. Understanding this simple principle is critical to the success of planning in a pluralistic society such as the United States. (See Plate I.13.) Knowing federal, state, and local law as it applies to planning is also important because their decisions, mandates, and those laws form the environment in which a plan evolves.

Finally, in "Shaping Cities through Development Regulations," Jonathan Barnett reminds the reader of the other side of the coin – a plan is only as strong as the implementation tools that accompany it. Too often planners (and their bosses) dissociate the plan and the regulations, and that is when they get in trouble. Different sets of regulations can be on collision courses – a historic district located in a high-density zoning area puts development pressure on landmarks. Barnett's inventory and assessment of the full range of laws, from zoning to design review, not only illustrates the power of the different types but also argues for using them to reinforce, not weaken, plans.

21

Twentieth Century Land Use Planning: a Stalwart Family Tree

Journal of the American Planning Association (1995)

Edward J. Kaiser and David R. Godschalk

Editor's introduction

Since urban and regional planners aim to guide development, the comprehensive plan is a critical tool. In its ideal form, a comprehensive plan is a public advisory document encompassing a long-term vision of a community's future physical growth. It is crafted by technical staff with input from the place's residents, adopted by the city planning commission, and consulted by municipal officials. It has several elements; primary among them is a land use plan.

Land use plans have evolved over time. Starting as inventories of public facilities (buildings, street systems, and open space), utilities (power plants), and zoning districts, they became more detailed as planners with increasing professional competence advanced new methods of analysis and means of discerning and articulating community aspirations.

This selection shows that while modern comprehensive plans and their land use plan components date from the early twentieth century, their proliferation is a post-World War II phenomenon. They gained momentum when the national Housing Act of 1954 mandated plans as a condition of funding, spawning hundreds of plans over the next twenty-seven years. And in the 1970s, high-growth-rate states, particularly in the Sunbelt and coastal areas, fueled the effort by requiring local adoption of comprehensive plans. As of 2002, twenty-four states had some type of compulsory planning.

California, Oregon, and Florida were among the first states to insist upon planning; a brief survey of their requirements highlights the integral role of the land use plan.

In 1971, the California Government Code Section 65301 required municipalities to make land use regulations consistent with a general plan; in this roundabout fashion, the state made comprehensive plans mandatory. The code specified that the plans should include a statement of development policies with objectives, principles, and standards, along with a plan proposal with seven elements: circulation, conservation, housing, land use, noise, open space, safety and seismic safety. It allowed the addition of such optional elements as air quality, economic development, hazardous waste, and parks and recreation. For an example of a California-type comprehensive plan, go to the San Francisco's planning department website (www.sfgov.org/site/planning_index.asp?id=41423).

In 1973, Oregon adopted Senate Bill 100, a state law that required localities to adopt a comprehensive plan and development regulations consistent with the state's nineteen planning goals. The state Department of Land Conservation and Development, along with the Land Conservation and Development Commission, a seven-member volunteer citizens' board, oversees the process. The state goals – the equivalent of California's plan elements – span topics like land use, transportation, and public facilities, but detail extensive natural resources and environmental protection assessments. These assessments resulted in more extensive land use plans. As illustrated in the guidelines for Goal 2 Land Use, compliance follows the rational decision-making model, calling for a written plan that identifies issues and problems, contains inventories and factual information related to each state goal, provides alternative courses of action, and selects one. For an example of an Oregon comprehensive plan, see the Salem, Oregon, website: www.cityofsalem.net.

In 1975, the Florida state legislature passed the Local Government Comprehensive Planning and Land Development Act, updating it in 1985. This law mandates municipal adoption of comprehensive plans, and requires that these plans be consistent with the state's twenty-five planning goals and with local development regulations. Like the California and Oregon legislation, the Florida law prescribes basic plan elements (land use, utilities, transportation, conservation, and housing), but it demands more detail than the others. It has additional requirements for school sites related to residential development and for intergovernmental coordination to demonstrate implementation arrangements. The law also views planning as continuous, mandating a revision process that is completed at eight- to nine-year intervals. To insure compliance, the state's Department of Community Affairs (DCA) reviews the plans. The DCA maintains a web page with several comprehensive plans: www.dca.state.fl.us/fdcp/DCP/compplanning/online.cfm.

The impetus for comprehensive planning in the past came from good-government advocates who envisioned the document as a means to site public facilities, provide a basis for regulating land use through zoning, and coordinate new subdivisions, thus managing local capital expenditures. While these motivations still exist, a key force behind today's comprehensive planning is a desire to control suburban expansion. From the mid 1980s and onward, nine states (Washington, New Jersey, Vermont, Rhode Island, Georgia, Maine, Tennessee, Delaware, and Maryland) passed legislation strengthening comprehensive planning under the rubric of growth management. (See Plates II.2–II.3.)

The growth management movement did more than encourage comprehensive planning; under a title coined in 1990s, "Smart Growth" – defined in general terms as joining economic growth with environmental protection – it fostered a new canon for land use planning. Publicized by the U.S. Environmental Protection Agency (EPA) and the Smart Growth Network (founded in 1996 and funded by the EPA), the Smart Growth movement promotes ten land use principles. They are: "mix land uses, take advantage of compact building design, create a range of housing opportunities and choices, create walkable neighborhoods, foster distinctive attractive communities with a strong sense of place, preserve open space, farmland, natural beauty and critical environmental areas, strengthen and direct development towards existing communities, provide a variety of transportation choices, make development decisions predictable, fair and cost effective, encourage community and stakeholder collaboration in development decisions." When the authors wrote this selection, the Smart Growth movement was just taking shape. However, they capture its incipient efforts in their descriptions of contemporary variations of traditional land use planning.

In 1997, the American Planning Association undertook a multi-year initiative to update a model of state planning legislation. Entitled *Growing Smart Legislative Guidebook: Model Statutes for Planning and the Management of Change* (Chicago: APA Planners Press, 2002), the project reasserted the importance of the land use plan in the comprehensive plan, specifying required sections on land use, transportation, community facilities, housing, economic development, critical or sensitive areas/natural hazards, and implementation.

As planners worked with land use plans, they developed a graphic vocabulary for mapping. By the 1950s, when public documents were increasingly printed in color, a universal color code was in place: residential use was yellow to brown, depending on density; commercial, red; industrial, purple; and institutional, blue. However, by 1965, computers brought about an updating of map coding. The U.S. Urban Renewal Administration and the Bureau of Public Roads collaborated to create a digital system published in the *Standard Land Use Coding Manual* (Washington, DC: U.S. Government Printing Office, 1965). It specified nine one-digit categories for basic land uses (e.g. residential = 1), sixty-seven two-digit classes (e.g. group quarters = 12), 294 three-digit designations (e.g. dormitory = 123), and 772 four-digit groupings (e.g. college dormitory = 1232). About thirty years later, the American Planning Association updated this system to increase the ability to convey more information, crafting the Land-Based Classification Standards (LBCS), a digital coding scheme incorporating five descriptive characteristics: the activity, economic function, structure type, development character, and ownership. Adapted to modern land use practice, this coding system can show mixed as well as single uses and is compatible with GIS technologies.

The authors of this selection, Edward J. Kaiser and David R. Godschalk, are professors emeriti, Department of City and Regional Planning, University of North Carolina, Chapel Hill. They have written extensively on land use, growth management, and hazards mitigation. They coedited the *Journal of the American Institute of Planners* from 1967 to 1971 and more recently cowrote the fourth and fifth editions of *Urban Land Use Planning* with F. Stuart Chapin, Jr, and Philip Berke (Urbana, IL: University of Illinois Press, 1995 and 2006).

For additional reading on land use planning, see *Urban Land Use Planning*, fifth edition, cited above, the most comprehensive reference in the field. Larz Anderson's *Guidelines for Preparing Urban Plans* (Chicago: APA Planners Press, 1995) and *Planning the Built Environment* (Chicago: APA Planners Press, 2000) provide detailed instructions regarding plan and land use specifications. The complete Land-Based Classification Standards codebook and instructions are on the American Planning Association website: www.planning.org/LBCS. For a theoretical overview of the use and functions of plans, see Lewis Hopkins, *Urban Development: The Logic of Making Plans* (Washington, DC: Island Press, 2001).

THE LAND USE PLANNING FAMILY TREE

We liken the evolution of the physical development plan to a family tree. The early genealogy is represented as the roots of the tree. The general plan, constituting consensus practice at mid-century, is represented by the main trunk. Since the 1970s this traditional "land use design plan" has been joined by several branches – the verbal policy plan, the land classification plan, and the development management plan. . . .

[. . .]

The focus of the article is the plan prepared by a local government – a county, municipality, or urban region – for the long-term development and use of the land.

ROOTS OF THE FAMILY TREE: THE FIRST FIFTY YEARS

New World city plans certainly existed before this century. They included L'Enfant's plan for Washington, William Penn's plan for Philadelphia, and General Oglethorpe's plan for Savannah. These plans, however, were blueprints for undeveloped sites, commissioned by unitary authorities with power to implement them unilaterally.

In this century, perhaps the most influential early city plan was Daniel Burnham's plan for Chicago, published by the Commercial Club of Chicago (a civic, not a government, entity) in 1909. The archetypical plan-as-inspirational-vision, it focuses only on design of public spaces as a City Beautiful effort.

The City Beautiful approach was soon broadened to a more comprehensive view. . . . [An] early influence [was] the federal Standard City Planning Enabling Act of 1928. [Although it] shaped enabling Acts passed by many states, [it] left many planners and public officials confused about the difference between a master plan and a zoning ordinance, so that hundreds of communities adopted "zoning plans" without having created comprehensive plans as the basis for zoning. Because the Act also did not make clear the importance of comprehensiveness or define the essential elements of physical development, no consensus about the essential content of the plan existed.

Ten years later, Edward Bassett's book *The Master Plan* (1938) spelled out the plan's subject matter and format – supplementing the 1928 Act, and consistent with it. He argued that the plan should have seven elements, all relating to land areas (not buildings) and capable of being shown on a map: streets, parks, sites for public buildings, public reservations, routes for public utilities, pierhead and bulkhead lines (all public facilities), and zoning districts for private land. Bassett's views were incorporated in many state-enabling laws.

The physical plans of the first half of the century were drawn by and for independent commissions, reflecting the profession's roots in the Progressive Reform movement, with its distrust of politics. The 1928 Act reinforced that perspective by making the planning commission, not the legislative body, the principal client of the plan, and purposely isolating the commission from politics. . . . The commission, not the plan, was intended to be the advisor to the local legislative body and to city departments.

By the 1940s, both the separation of the planning function from city government and the plan's focus on physical development were being challenged. Robert Walker, in *The Planning Function in Local Government*, argued that the "scope of city planning is properly as broad as the scope of city government." The central planning agency might not necessarily do all the planning, but it would coordinate departmental planning in the light of general policy considerations – creating a comprehensive plan but one without a physical focus. That idea was not widely accepted. Walker also argued that the independent planning commission should be replaced by a department or bureau attached to the office of mayor or city manager. That argument did take hold, and by the 1960s planning in most communities was the responsibility of an agency within local government, though planning boards still advised elected officials on planning matters.

This evolution of ideas over fifty years resulted at mid-century in a consensus concept of a plan as focused on long-term physical development; this focus was a legacy of the physical design professions. Planning staff worked both for the local government executive officer and with an appointed citizen planning board, an arrangement that was a legacy of the Progressive insistence on the public interest as an antidote to governmental corruption. The plan addressed both public and private uses of the land, but did not deal in detail with implementation.

THE PLAN AFTER MID-CENTURY: NEW GROWTH INFLUENCES

Local development planning grew rapidly in the 1950s, for several reasons. First, governments had to contend with the postwar surge of population and urban growth, as well as a need for the capital investment in infrastructure and community facilities that had been postponed during the Depression and war years. Second, municipal legislators and managers became more interested in planning as it shifted from being the responsibility of an independent commission to being a function within local government. Third, and very important, Section 701 of the Housing Act of 1954 required local governments to adopt a long-range general plan in order to qualify for federal grants for urban renewal, housing, and other programs, and it also made money available for such comprehensive planning. The 701 program's double-barreled combination of requirements and financial support led to more urban planning in the United States in the latter half of the 1950s than at any previous time in history.

At the same time, the plan concept was pruned and shaped by two planning educators. T. J. Kent, Jr . . . , [whose] book, *The Urban General Plan* (1964), clarified the policy role of the plan [and] F. Stuart Chapin, Jr . . . , whose contribution was to codify the methodology of land use planning in the various editions of his book *Urban Land Use Planning* (1957, 1965).

What should the plan look like? What should it be about? What is its purpose (besides the cynical purpose of qualifying for federal grants)? The 701 program, Kent, and Chapin all offered answers.

The "701" Program comprehensive plan guidelines

In order to qualify for federal urban renewal aid – and, later, for other grants – a local government had to prepare a general plan that consisted of plans for physical development, programs for redevelopment, and administrative and regulatory measures for controlling and guiding development. The 701 program specified what the content of a comprehensive development plan should include:

- A land use plan, indicating the locations and amounts of land to be used for residential, commercial, industrial, transportation, and public purposes.
- A plan for circulation facilities.
- A plan for public utilities.
- A plan for community facilities.

T. J. Kent's urban general plan

Kent's view of the plan's focus was similar to that of the 701 guidelines: long-range physical development in terms of land use, circulation, and community facilities. In addition, the plan might include sections on civic design and utilities, and special areas, such as historic preservation or redevelopment areas. It covered the entire geographical jurisdiction of the community, and was, in that sense, comprehensive. The plan was a vision of the future, but not a blueprint; a policy statement, but not a program of action; a formulation of goals, but not schedules, priorities, or cost estimates. It was to be inspirational, uninhibited by short-term practical considerations.

Kent believed the plan should emphasize policy, serving the following functions:

- Policy determination – to provide a process by which a community would debate and decide on its policy.
- Policy communication – to inform those concerned with development (officials, developers, citizens, the courts, and others) and educate them about future possibilities.
- Policy effectuation – to serve as a general reference for officials deciding on specific projects.
- Conveyance of advice – to furnish legislators with the counsel of their advisors in a coherent, unified form.

The format of Kent's proposed plan included a unified, comprehensive, but general physical design for the future, covering the whole community and represented by maps. It also contained goals and policies (generalized guides to conduct, and the most important ingredients of the plan), as well as summaries of background conditions, trends, issues, problems, and assumptions. So that the plan would be suitable for public debate, it was to be a complete, comprehensible document, containing factual data, assumptions, statements of issues, and goals, rather than merely conclusions and recommendations. The plan belonged to the legislative body and was intended to be consulted in decision making during council meetings.

F. Stuart Chapin, Jr's, urban land use plan

Chapin's ideas, though focusing more narrowly on the land use plan, were consistent with Kent's in both the 1957 and 1965 editions of *Urban Land Use Planning*, a widely used text and reference work for planners. Chapin's concept of the plan was of a generalized, but scaled, design for the future use of land, covering private land uses and public facilities, including the thoroughfare network.

Chapin conceived of the land use plan as the first step in preparing a general or comprehensive plan. Upon its completion, the land use plan served as a temporary general guide for decisions, until the comprehensive plan was developed. Later, the land use plan would become a cornerstone in the comprehensive plan, which also included plans for transportation, utilities, community facilities, and renewal, only the general rudiments of which are suggested in the land use plan. Purposes of the plan were to guide government decisions on public facilities, zoning, subdivision control, and urban renewal, and to inform private developers about the proposed future pattern of urban development.

The format of Chapin's land use plan included a statement of objectives, a description of existing conditions and future needs for space and services, and finally the mapped proposal for the future development of the community, together with a program for implementing the plan (customarily including zoning, subdivision control, a housing code, a public works expenditure program, an urban renewal program, and other regulations and development measures).

THE TYPICAL GENERAL PLAN OF THE 1950S AND 1960S

Influenced by the 701 program, Kent's policy vision, and Chapin's methods, the plans of the 1950s and 1960s were based on a clear and straightforward concept: The plan's purposes were to determine, communicate, and effectuate comprehensive policy for the private and public physical development and redevelopment of the city. The subject matter was long-range physical development, including private uses of the land, circulation, and community facilities. The standard format included a summary of existing and emerging conditions and needs; general goals; and a long-range urban form in map format, accompanied by consistent development policies. The coverage was comprehensive, in the sense of addressing both public and private development and covering the entire planning jurisdiction, but quite general. The tone was typically neither as "inspirational" as the Burnham plan for Chicago, nor as action-oriented as today's plans.

Such was the well defined trunk of the family tree in the 1950s and 1960s, in which today's contemporary plans have much of their origin.

CONTEMPORARY PLANS: INCORPORATING NEW BRANCHES

Planning concepts and practice have continued to evolve through mid-century, maturing in the process. By the 1970s, a number of new ideas had taken root. Referring back to the family tree . . . , we can see a trunk and several distinct branches: [the land use design, land classification plan, verbal policy, and development management plan]. . . .

[. . .]

The land use design plan

The land use design plan is the most traditional of the four prototypes of contemporary plans and is the most direct descendent of the Kent–Chapin–701 plans of the 1950s and 1960s. It proposes a long-range future urban form as a pattern of retail, office, industrial, residential, and open spaces, and public land uses and a circulation system. Today's version, however, incorporates environmental processes, and sometimes agriculture and forestry, under the "open space" category of land use. Its land uses often include a "mixed use" category, honoring the neotraditional principle of closer mingling of residential, employment, and shopping areas. In addition, it may include a development strategy map, which is designed to bring about the future urban form and to link strategy to the community's financial capacity to provide infrastructure and services. The plans and strategies are often organized around strategic themes or around issues about growth, environment, economic development, transportation, or neighborhood/community scale change.

Like the other types of plans in vogue today, the land use design plan reflects recent societal issues, particularly the environmental crisis, the infrastructure crisis, and stresses on local government finance. Contemporary planners no longer view environmental factors as development constraints, but as valuable resources and processes to be conserved. They also may question assumptions about the desirability and inevitability of urban population and economic growth, particularly as such assumptions stimulate demand for expensive new roads, sewers, and schools. While at mid-century plans unquestioningly accommodated growth, today's plans cast the amount, pace, location, and costs of growth as policy choices to be determined in the planning process.

The 1990 Howard County (Maryland) General Plan, winner of an American Planning Association (APA) award in 1991 for outstanding comprehensive planning, exemplifies contemporary land use design. . . .

THE LAND CLASSIFICATION PLAN

Land classification, or development priorities mapping, is a proactive effort by government to specify where and under what conditions growth will occur. Often, it also regulates the pace or timing of growth. Land classification addresses environmental protection by designating "non-development" areas in especially vulnerable locations. Like the land use design, the land classification plan is spatially specific and map-oriented. However, it is less specific about the pattern of land uses within areas specified for development, which results in a kind of silhouette of urban form. On the other hand, land classification is more specific about development strategy, including timing. Counties, metropolitan areas, and regional planning agencies are more likely than cities to use a land classification plan.

The land classification plan identifies areas where development will be encouraged (called urban, transition, or development areas) and areas where development will be discouraged (open space, rural, conservation, or critical environmental areas). For each designated area, policies about the type, timing, and density of allowable development, extension of infrastructure, and development incentives or constraints apply. The planning principle is to concentrate financial resources, utilities, and services within a limited, prespecified area suitable for development, and to relieve pressure on nondevelopment areas by withholding facilities that accommodate growth.

Ian McHarg's (1969) approach to land planning is an early example of the land classification concept. . . . Vision 2005: A Comprehensive Plan for Forsyth County, North Carolina exemplifies the contemporary approach to land classification plans.

THE VERBAL POLICY PLAN: SHEDDING THE MAPS

The verbal policy plan focuses on written statements of goals and policy, without mapping specific land use patterns or implementation strategy. Sometimes called a policy framework plan, a verbal policy plan is more easily prepared and flexible than other types of plans, particularly for incorporating nonphysical development policy. Some claim that such a plan helps the planner to avoid relying too heavily on maps, which are difficult to keep up to date with the community's changes in policy. The verbal policy plan also avoids falsely representing general policy as applying to specific parcels of property. The skeptics, however, claim that verbal statements in the absence of maps provide too little spatial specificity to guide implementation decisions.

The verbal policy plan may be used at any level of government, but is especially common at the state level, whose scale is unsuited to land use maps. The plan usually contains goals, facts and projections, and general policies corresponding to its purposes – to understand current and emerging conditions and issues, to identify goals to be pursued and issues to be addressed, and to formulate general principles of action. Sometimes communities do a verbal policy plan as an interim plan or a first step in the planning process. Thus, verbal policies are included in most land use design plans, land classification plans, and development management plans. The Calvert County, MD, Comprehensive Plan (Calvert County, 1983), winner of a 1985 APA award, exemplifies the verbal policy plan. . . .

THE DEVELOPMENT MANAGEMENT PLAN

The development management plan features a coordinated program of actions, supported by analyses and goals, for specific agencies of local government to undertake over a three-to-ten-year period. The program of actions usually specifies the content, geographic coverage, timing, assignment of responsibility, and coordination among the parts.

[. . .]

The development management plan is a distinct type, emphasizing a specific course of action, not general policy. At its extreme the management plan actually incorporates implementation measures, so that the plan becomes part of a regulative ordinance. Although the spatial specifications for regulations and other implementation measures are included, a land use map may not be. The *Sanibel, Florida, Comprehensive Land Use Plan* (1981) exemplifies the development management plan. . . .

THE CONTEMPORARY HYBRID PLAN: INTEGRATING DESIGN, POLICY AND MANAGEMENT

The rationality of practice has integrated the useful parts of each of the separate prototypes reviewed here into contemporary hybrid plans that not only map and classify land use in both specific and general ways, but also propose policies and management measures. . . .

The states that manage growth have created new land use governance systems whose influence has broadened the conceptual arsenals of local planners. . . . [T]he common elements of these systems [are]:

- Consistency – intergovernmentally and internally (i.e. between plan and regulations).
- Concurrency – between infrastructure and new development.
- Compactness – of new growth, to limit sprawl.
- Affordability – of new housing.
- Economic development or "managing to grow."
- sustainability – of natural systems.

[. . .]

Another important influence on contemporary plans is the renewed attention to community design. The neotraditional and transit-oriented design movements have inspired a number of proposals for mixed use villages in land use plans. *Toward a Sustainable Seattle: A Plan for Managing Growth* (1994) exemplifies a city approach to the hybrid plan. . . .

[. . .]

SUMMARY OF THE CONTEMPORARY SITUATION

Since mid-century, the nature of the plan has shifted from an elitist, inspirational, long-range vision that was based on fiscally innocent implementation advice, to a framework for community consensus on future growth that is supported by fiscally grounded actions to manage change. Subject matter has expanded to include the natural as well as the built environment. Format has shifted from simple policy statements and a single large-scale map of future land use, circulation, and community facilities, to a more complex combination of text, data, maps, and timetables. In a number of states, plans are required by state law, and their content is specified by state agencies. . . .

Today's prototype land use design continues to emphasize long-range urban form for land uses, community facilities, and transportation systems as shown by a map; but the design is also expressed in general policies. Land use design is still a common form of development plan, especially in municipalities.

The land classification plan also still emphasizes mapping, but of development policy rather than policy about a pattern of urban land uses. Land classification is more specific about development management and environmental protection, but less specific about transportation, community facilities, and the internal arrangement of the future urban form. County and regional governments are more likely than are municipalities to use land classification plans.

The verbal policies plan eschews the spatial specificity of land use design and land classification plans and focuses less on physical development issues. It is more suited to regions and states, or may serve as an interim plan for a city or county while another type of plan is being prepared.

The development management plan represents the greatest shift from the traditional land use plan. It embodies a short-to-intermediate-range program of governmental actions for ongoing growth management rather than for long-range comprehensive planning.

In practice, these four types of plans are not mutually exclusive. Communities often combine aspects of them into a hybrid general plan that has policy sections covering environmental/social/economic/housing/infrastructure concerns, land classification maps defining spatial growth policy, land use design maps specifying locations of particular land uses, and development management programs laying out standards and procedures for guiding and paying for growth. Regardless of the type of plan used, the most progressive planning programs today regard the plan as but one part of a coordinated growth management program, rather than, as in the 1950s, the main planning product. Such a program incorporates a capital improvement program, land use controls, small area plans, functional plans, and other devices, as well as a general plan.

THE ENDURING LAND USE FAMILY TREE AND ITS FUTURE BRANCHES

[. . .]

To return to our analogy of the plan's family tree: Roots for the physical development plan became well established during the first half of this century. By 1950, a sturdy trunk concept had developed. Since then, new roots and branches have appeared – land classification plans, verbal policy plans, and development management plans. Meanwhile development of the main trunk of the tree – the land use design – has continued. Fortunately, the basic gene pool has been able to combine with new genes in order to survive as a more complex organism – the 1990s design–policy–management hybrid plan. The present family tree of planning reflects both its heredity and its environment.

The next generation of physical development plans also should mature and adapt without abandoning their heritage. We expect that by the year 2000 plans will be more participatory, more electronically based, and concerned with increasingly complex issues. An increase in participation seems certain, bolstered by interest groups' as well as governments' use of expert systems and computer databases. A much broader consideration of alternative plans and scenarios, as well as a more flexible and responsive process of plan amendment, will become possible. These changes will call upon planners to use new skills of consensus building and conflict management, as more groups articulate their positions on planning matters, and government plans and interest group plans compete, each backed by experts.

With the advent of the "information highway," plans are more likely to be drafted, communicated, and debated through electronic networks and virtual reality images. The appearance of plans on CD-ROM and cable networks will allow more popular access and input, and better understanding of plans' three-dimensional consequences. It will be more important than ever for planners to compile information accurately and ensure it is fairly communicated. They will need to compile, analyze, and manage complex databases, as well as to translate abstract data into understandable impacts and images.

Plans will continue to be affected by dominant issues of the times: aging infrastructure and limited public capital, central city decline and suburban growth, ethnic and racial diversity, economic and environmental sustainability, global competition and interdependence, and land use/transportation/air quality spillovers. Many of these are unresolved issues from the last thirty years, now grown more complex and interrelated. Some are addressed by new programs like the Intermodal Surface Transportation Efficiency Act (ISTEA) and HUD's Empowerment Zones and Enterprise Communities. To cope with others, planners must develop new concepts and create new techniques.

One of the most troubling new issues is an attempt by conservative politicians (see the Private Property Protection Act of 1995 passed by the U.S. House of Representatives) and "wise use" groups to reverse the precedence of the public interest over individual private property rights. These groups challenge the use of federal, state, and local regulations to implement land use plans and protect environmental resources when the result is any reduction in the economic value of affected private property. Should their challenge succeed and become widely adopted in federal and state law, growth management plans based on regulations could become toothless. Serious thinking by land use lawyers and planners would be urgently needed to create workable new implementation techniques, setting in motion yet another planning evolution.

We are optimistic, however, about the future of land use planning. Like democracy, it is not a perfect institution but works better than its alternatives. Because land use planning has adapted effectively to this century's turbulence and become stronger in the process, we believe that the twenty-first century will see it continuing as a mainstay of strategies to manage community change.

22

Regionalism

From *The Limitless City: A Primer on the Urban Sprawl Debate* (2002)

Oliver Gillham

Editor's introduction

Regionalism is an enduring interest of planners, dating from the late nineteenth century, with the diagnosis of the overcrowded, corrupt city as the root of urban problems and various forms of decentralization proposed as possible solutions. By the late twentieth century, sprawl and the associated decline of central cities emerged as central concerns of urban and regional planners, the result of near-universal use of the automobile, federal highway construction, tax and mortgage insurance policies, and consumer preferences favoring suburbanization; in this era growth management became a solution.

Contemporary regional planners group complex sprawl-related issues into two categories: uneven metropolitan growth and local government fragmentation. (See Plate I.11.) Their plans seek to connect inner-city neighborhoods to suburban job centers and balance jobs and housing to minimize traffic. For example, today the majority of new job formation takes place in the suburbs, and many of these new jobs require high school, if not higher, levels of educational achievement. For low-income central-city populations, these conditions pose difficulties related to residential location, transportation, and school quality. Planners also attempt to unite metropolitan areas around common infrastructure (water, sewer, transportation) and open space and find ways to support region-wide planning and funding initiatives.

Many writers, from the scholarly to the popular, assist planners (and the public) in understanding regional dynamics and the need for planning. Regional scientists and economic geographers contribute mathematical systems and spatial analyses, while a select group of urban observers and journalists offer serious, but less quantitative, assessments. For example, the classic scholarly work of Walter Isard (*Location and Space-Economy: A General Theory Relating to Industrial Location, Market Areas, Land Use, Trade, and Urban Structure*, Cambridge, MA: MIT Press, 1956) and William Alonso (*Location and Land Use*, Cambridge, MA: Harvard University Press, 1964) now guides contemporary thinking on location decisions. See, for example, Stuart S. Rosenthal and William C. Strange, "Evidence on the Nature and Sources of Agglomeration Economies," in J. V. Henderson and J. F. Thisse (eds), *Handbook of Urban and Regional Economics* (Amsterdam: Elsevier, 2004). In the urban observer camp are Anthony Downs (*New Visions for Metropolitan America*, Washington, DC: Brookings Institution Press, 1995), William Fulton (*The Regional City: Planning for the End of Sprawl*, Washington, DC: Island Press, 2001) with Peter Calthorpe, and Neal Pierce (*Boston Unbound*, Boston, MA: Boston Globe, 2004).

This selection falls into the latter category. It provides an overview of regional planning strategies employed to limit sprawl and a case study of Portland, Oregon. Famed for its state legislation (Senate Bill 100, 1973) requiring local governments to adopt comprehensive plans in accord with state planning goals, Oregon has 276 cities and counties, many of which have devised innovative approaches to planning – Portland's elected metro government and urban growth boundaries are particularly exemplary. While the narrative relates the failed attempts in 1976, 1978, 1982, and 2000 to gut the law, it ends in 2002, before voter approval of Measure 37 (November 2004), a law mandating landowner compensation when their properties lose value due to land use regulation, and Measure 49 (November, 2007) to restrict application of Measure 37. To stay apprised of the situation, refer to the Land Conservation and Development Commission website: www.lcd.state.or.us/LCD/goals.shtml.

While statewide planning efforts are varied, in general, states rarely override local government planning functions unless special conditions exist. For example, Hawaii, which has a fragile natural environment, pioneered statewide comprehensive planning. Nine states have legislated such efforts, ranging from Maryland's priority funding approach (focusing state capital investments on specific geographic areas) to Florida's mandate that plans be consistent with state goals.

Many other states have created bodies that target regional geographies for planning. For example, in 1999 the Georgia state legislature created the Georgia Regional Transportation Authority (GRTA), with jurisdiction over thirteen counties in metropolitan Atlanta, after the federal government declared the area in violation of the Clean Air Act and suspended transportation funds. With $2 billion in bonding capacity and the power to supersede local land use regulations, the GRTA quickly brought the area into compliance through the institution of a new regional bus service, major arterial road improvements, and regular air monitoring. The Cape Cod Commission – the only regional government planning and regulatory body in Massachusetts – is the state's response to growth-related environmental threats and the absence of affordable housing in this unique area. Covering the fifteen towns of Barnstable County, Massachusetts, the Commission is charged with developing and regularly updating a regional plan and approving development of regional impact (DRI), or large-scale projects. The resulting *Cape Cod Regional Policy Plan* (2003) contains a growth policy, minimum DRI performance standards, and a framework for local comprehensive plans.

Other exemplary regional planning efforts include the work of nongovernmental groups who operate in large and small metropolitan areas. They range from the eighty-year-old Regional Plan Association (RPA), which covers thirty-one counties in New York, Connecticut, and New Jersey and has come out with three regional plans, to eight-year-old Metropolis 2020 in Chicago, which is developing a coordinating structure to unite the area's 1,246 local governments around the major issues articulated in its regional plan, *Chicago Metropolis 2020: Preparing Metropolitan Chicago for the Twenty-first Century*. By 2006 it successfully lobbied the state legislature to unify the metropolitan area's regional planning and transportation agencies into a single body, the Chicago Metropolitan Agency for Planning (CMAP). Other entities are the Joint Venture Silicon Valley Network, which issued a regional vision, *Silicon Valley 2010* (1998), and myregion.org, an active central Florida group.

The work of the Regional Plan Association is notable for its longevity, focus, breadth, and political acumen. Having no official power, it has offered leadership and proselytizing to further long- and short-term regional goals. While currently pursuing its most recent plan, *A Region at Risk* (1996) – containing recommendations for balanced regional development, open space, and transportation – it has been a nimble player in other roles congruent with its regional mission. For example, it organized the Civic Alliance to provide a unified response to the World Trade Center rebuilding efforts. (This alliance sponsored the "Listening to the City" meeting that drew 4,000 people and shaped the subsequent plans for the area). It has also partnered with several groups to secure passage of the Highlands Conservation Act of 2004 that authorizes $110 million for land conservation in a 3 million-acre watershed area encompassing Pennsylvania, New Jersey, New York, and Connecticut. And it is spearheading the national movement to promote unified thinking on megaregions.

A number of voluntary civic and professional groups support regional planning through public education, research, litigation, and advocacy. They range from the American Planning Association to the Sierra Club and the Trust for Public Land. While they have varying interests, their conferences, publications, and other activities are an important source of best practices and information about regionalism.

The author of this selection, Oliver Gillham, is an architect and planner with an active practice in New England. The book from which it comes, *The Limitless City: A Primer on the Urban Sprawl Debate* (Washington, DC: Island Press, 2002), aims to provide the educated reader with an understanding of the definition and impact of uncontrolled development and of the alternatives to sprawl.

For additional reading on regional issues, see Bruce Katz (ed.), *Reflections on Regionalism* (Washington, DC: Brookings Institution Press, 2000). Curtis Johnson and Neal Peirce have written an excellent summary of the political will to address regional issues, *Risks, Rewards and Unresolved Questions* (Chicago: John D. and Catherine T. MacArthur Foundation, 2004), accessible on the Citistates website: www.citistates.com/assocspeakers/n_peirce.html. John M. DeGrove's *Planning Policy and Politics: Smart Growth and the States* (Cambridge, MA: Lincoln Institute of Land Policy, 2005) provides a comprehensive update on state planning efforts. For excellent references on Oregon and its planning efforts, see Connie P. Ozawa, *The Portland Edge: Challenges and Successes in Growing Communities* (Washington, DC: Island Press, 2004), and Edward J. Sullivan, "Year Zero: The Aftermath of Measure 37," *Environmental Law*, 36: 131 (2006). Various regional groups have their regional plans for review on their websites. See, for example, Metropolis 2020's *Chicago Metropolis: Preparing Metropolitan Chicago for the Twenty-first Century* at www.chicagometropolis2020.org/plan.pdf; *Silicon 2010* at www.jointventure.org/PDF/SV2010.pdf; the *Cape Code Regional Policy Plan* at www.capecodcommission.org/RPP/RPPrev2003illustrated.pdf; and a myriad of preparatory publications for the Central Florida group, myregion.org at www.myregion.org/Home/tabid/36/Default.aspx.

RECURRENT REGIONALISM

Regionalism, or metropolitanism (as it is sometimes also called), has recurred from time to time as a possible solution to the urban ills caused by the rapid explosion of American metropolitan areas. There was considerable debate about regionalism in the 1920s, during the first automotive suburban boom. Then, the debate focused on how regions should be controlled – by dominant central cities or by dispersed settlements. That era saw the birth of the Regional Planning Association of America (RPAA), of which Lewis Mumford was a founding member. The RPAA served as a forum for much of this debate until its demise in the late 1930s.

[. . .]

CONTEMPORARY REGIONALISM

During the past few years, regionalism has once again come into fashion. In some respects, this may be due to the cycle of growth and change, because development has mushroomed during the 1990s, resulting in increased concern about the effects of sprawl. The recognition that our nation's metropolitan areas have reached a whole new level in transcending their boundaries has lent renewed urgency to the subject.

. . . [W]here regional strategies are being attempted, they tend to be organized around one or more of the following concepts:

- Regional growth control, including boundaries limiting the outward extension of growth.
- Regional coordination of transportation and land use.
- Regional sharing of fiscal resources (tax-base sharing).
- Regional growth control.

[. . .]

. . . A regional UGB [urban growth boundary] can cover an entire metropolitan area, including multiple cities, towns, and counties. If properly planned and administered, such a technique can be used to manage and direct growth on a regional basis, effectively limiting sprawl. Because the land both within and outside of the UGB crosses community borders, the chances for local plans canceling each other out with leapfrog development are greatly reduced.

. . . [A] regional UGB cannot be implemented or administered by existing local government unless a group of localities agrees to work together in a formal process. Even that usually requires the cooperation of a broader regional entity, such as the state in which the localities exist. By definition, a regional growth control program is broader than any one city or town and is likely to include multiple counties. Because it is aimed at containing sprawl on a regional basis, the program ultimately must be implemented and administered at the regional level either by state government or by a specially created regional governmental entity.

Thus, although such a UGB may be more effective on a regional basis, it clearly implies a greater order of complexity and an ambitious governmental agenda that may be very difficult to accomplish in some parts of the nation. . . .

REGIONAL COORDINATION OF TRANSPORTATION AND LAND USE

Ultimately, the primary transportation burden will continue to fall upon the nation's roadway system in serving the existing land uses that make up the majority of the nation's urbanized areas. But it is widely recognized that congestion is a very serious issue and that something needs to be done to keep this primary system from seizing up entirely. It also is considered unlikely by many planners that roadway capacity measures alone will be able to fulfill this need. Ultimately, more transit is needed to bring balance to the system and to protect the nation's huge investment in its roadway network.

[. . .]

In most metropolitan areas, there simply is no formal regional structure with any significant powers linking expansion of transit or highway systems to land use planning for their areas. Metropolitan planning organizations (MPOs) were originally established with planning coordination in mind – and they do provide valuable and needed transportation planning and coordination functions – but the MPOs typically have no real power over how land gets developed. That power remains with localities. The result is that planning for most rail extensions and highway improvements [is] not formally coordinated with regional and local land use planning in any enforceable way.

In the future, things may change slightly. It is possible that coordination between land use and transportation agencies may happen more frequently with new federal mandates for transportation planning, such as ISTEA and TEA-21, in which state agencies will be required to assess alternative modes when considering improvements to a given transportation corridor. Ultimately, it will be up to states and regional governments to initiate formal coordination between land use and transportation and to create the mechanisms necessary to implement it.

Land use and transportation must be coordinated at a regional level if there is to be any alternative to auto-dependent sprawl development. Land use plans must be tailored to transit for transit to be effective and vice versa. Without some kind of "top-down" regional action on this issue, there will be no truly significant change to post-World War II sprawl development patterns. Furthermore, without any counterbalancing transit initiatives, it is quite possible that the roadway system that has served the nation's growth will no longer be able to keep up, even after significant capacity improvements have been made.

REGIONAL TAX BASE SHARING

The fiscal measure of tax revenue sharing is, among other things, intended to promote greater equity of resources between both suburban and urban districts. For example, tax revenue sharing can be used to improve education in urban areas, a necessary ingredient to making downtowns more attractive and leveling the playing field between city and suburb. Such measures can also be used to relieve the pressured competition between municipalities to attract commercial projects to offset losses on residential properties. . . .

[. . .]

This type of regional measure is clearly a controversial issue. . . . Many taxpayers believe they should get back what they invest in their own communities. As a rule, those in wealthy communities are paying more money, based on the valuation of their property. Why should their taxes, or the taxes of upscale commercial development that they are able to attract to their community, go to serve someone else in another part of the region? This point of view might hold that those in communities of modest means get the services they can afford. If individuals in those communities want better services, then they can always work to improve themselves and move to a community that provides those services. America's traditional ethos of individual rights, self-determination, and private property ownership tends to support this viewpoint.

. . . Americans also have a strong tradition of community service and reaching out to help others within their communities. The problem is that the local community usually ends at the town boundaries, where local control and local determination also end. Regional tax base sharing takes funds out of local control and directs it to other communities. This can seem very unfair unless a wider perspective is taken and the economic health of the local community is viewed in a broader regional context.

CASE STUDIES IN REGIONALISM

Throughout the nation, only a small number of metropolitan areas have formally established regional entities to carry out any of the above measures. Portland, Oregon, is among the most widely cited, having adopted the most ambitious regional agenda in the nation, but there are other initiatives that have been undertaken or are under way throughout the United States. These include both metropolitan area undertakings as well as state-level programs. In addition to Portland, the efforts of Minneapolis/St Paul, Maryland, Atlanta, and the Regional Plan Association of the New York metropolitan region are worthy of special mention because of what they have tried (or are trying) to do and how they have fared. . . .

Portland, Oregon

Portland is generally considered to be the national model for regionalism in the United States. Like Boulder in Colorado, Portland has an urban growth boundary based on an urban service district, in this case, the Portland Metropolitan Services District. . . . Portland's UGB is administered by a directly elected regional government called Metro, which serves 1.3 million residents and twenty-four cities in the Portland area. Metro is the result of a multi-decade planning and political process that dates back to the 1970s.

Although Metro has chalked up many remarkable achievements, its existence remains threatened even today, with recurring ballot initiatives that would effectively curtail its powers. Until recently, Metro has managed to survive these assaults, but in fall 2000 Oregon voters passed a ballot initiative called Measure 7. Measure 7 requires payment to property owners if their land loses value due to growth boundaries or other land use regulations. Supporters of Oregon's planning programs consider this measure nothing less than a "hydrogen bomb" directed at all the work that has been done to date. As of this writing, Measure 7 was overturned in county court and is now being appealed in the Oregon State Supreme Court.

Metro

In 1973, reacting to the suburban sprawl that was rapidly consuming neighboring California, the Oregon state legislature made it mandatory for cities, towns, and counties in the Portland region to join together in the Columbia Regional Association of Governments (CRAG). At the same time, the state legislature created a Land Conservation and Development Commission (LCDC), whose mandate was to preserve forest and farmland resources by encouraging orderly development adjacent to existing communities. The LCDC in turn required all cities and counties in Oregon to draw up growth management plans together with UGBs to combat sprawl. These moves, like others that followed them, did not go unchallenged. For example, there was a statewide ballot measure to abolish all councils of government (including CRAG) that nearly won in 1976. Despite this, Oregon continued to pursue the regional approach. In another narrow victory in 1978, voters agreed to create Metro, a directly elected regional government for the Portland area that is the only one of its kind in the United States. Metro includes three counties and twenty-four municipalities in the Portland area. It is governed by an executive officer that is elected at large and a thirteen-member council that is directly elected from council districts in the region.

Portland's regional UGB

Following its creation, Metro took up where CRAG had left off in drawing up a UGB for the Portland region. . . . The first UGB drawn for Portland in 1978 contained an area of 234,000 acres, or 365 square miles. It was projected to be enough land to accommodate development for twenty years.

Although the UGB is based to some degree on Metro's service district, its shape and enforcement are largely determined by other factors. The power to define the UGB is derived from state legislation and zoning. Metro has the responsibility for drawing the boundary and for any modifications that may be made to it. The boundary is drawn in part to preserve the best agricultural land. Metro is also charged with requiring consistency of local comprehensive plans with regional planning goals. This requirement is enforced through review powers granted to Metro. Thus, local cities and towns, which retain actual zoning powers, are obliged to implement the boundary using those powers through the consistency requirement. . . .

Success of the UGB

The UGB program in Oregon appears to be quite successful in meeting its original objectives. Between 1979 and 2000, only about 6,000 acres have been added to the UGB – slightly more than a 2 percent increase in overall territory. Nearly 90 percent of the state's population growth during the 1980s occurred inside UGB limits. In Portland, it is estimated that 95 percent of population growth occurred within the UGB. Furthermore, between 1979 and 1997, the average size of a residential lot shrank from 13,000 sq. ft to 7,400 sq. ft. Although this trend clearly is moving in the intended direction, buildout within Portland's UGB is also said to be occurring at only 70 percent of planned density. That fact, combined with rapid development during the 1990s and a projected 65 percent population increase by 2040, has created considerable pressure for expansion of the UGB.

In a 1993 referendum, voters opted for restricting development to the existing growth boundary. The policy preference that won the referendum called for retaining the existing UGB and concentrating on infill development and higher density near transit stations. Even so, development pressures later in the decade caused Metro to set aside 19,000 acres outside the UGB as a thirty-year reserve for future development.

Housing costs

Population increase and development pressure have raised concerns about rising housing costs in the Portland metro area given the restrictions on the amount of land available for growth. Some groups strongly believe that readily available, inexpensive land is an essential ingredient for providing moderately priced housing. This may be true when the housing is produced in the mold of the typical subdivision of single-family homes.

There is a countervailing theory that the price of housing could be kept down by simply increasing the number of units per acre (for example, density or compactness) as land cost goes up. This is part of the UGB s intent: to promote higher density and to conserve land. Although construction costs can also rise with increasing density, this general premise should hold true at moderate levels of compactness (single-family homes on smaller lots and townhouses, for example). As we have already seen, the UGB appears to have resulted in a marked downsizing of lot size, which was one of its goals. Furthermore, Portland has not left housing patterns to market forces alone. The Metropolitan Portland Housing Rule requires that 50 percent or more of new housing developments be built in attached single-family or multifamily structures.

Density theories and housing requirements aside, the fact remains that housing prices have gone up considerably for the region, but this does not entirely resolve the debate. Studies to date have shown that the median price of housing in the Portland region more than doubled from $64,000 in 1989 to almost $140,000 in 1996, an increase of about 25 percent per year over an eight-year period. This is clearly a substantial upsurge for any region. On the other hand, as dramatic as that escalation has been, the resultant median price of housing in the Portland area was actually about 8 percent below the average median price for the western part of the United States in 2001. Moreover, housing price escalation appears to be moderating, having increased only slightly more than 2 percent in 2000, which is about half of the 4 percent national average. Thus, it has been argued that Portland simply is catching up to the rest of the nation.

Furthermore, it is not entirely clear that housing prices went up during the 1990s simply because of limits on developable land. The economic picture in the region was also improving during the same period. During the past decade, Portland has continued its transition from forestry and farm products to newer industries, with high-tech companies investing $13 billion in new plants in the region and with Intel alone adding 1,000 jobs in 1999. The high-tech boom has been so successful that part of the Portland region is now known as the "Silicon Forest." Good jobs and an attractive environment naturally have drawn more people to the area. Portland was growing at the rate of 50,000 new residents per year in 1996, a number sure to put pressure on housing prices regardless of boundaries.

Nor is Portland alone in the nation in experiencing higher housing costs due to economic growth. Other prospering U.S. metropolitan areas that have no growth boundaries have also experienced rapid runups in housing price. In the Boston region, where housing prices are nearly double those of Portland, housing prices increased by nearly 14 percent in 2000. Prices in parts of California increased by more than 20 percent in 2000. In the most expensive home market in the nation, the San Francisco Bay area, where median home prices are nearly triple those of Portland, costs were increasing at more than 15 percent in 2000. In addition to experiencing economic growth, these areas offer special attractions for many people. In fact, it can be argued that part of the increase in housing prices in the Portland area is due to the increased appeal of the region to both companies and individuals. Part of the region's magnetism can be attributed to the regional planning measures undertaken since the 1970s (including the UGB) that have preserved the amount of green space in the region.

Another way of looking at the issue is income and housing value disparity. In many metropolitan areas that have no growth boundaries, there are pronounced disparities in income and housing value between city and suburb. This gap is widely perceived to be one of the major problems generated by sprawl. In Portland, disparities in income level and housing value between central city and suburban areas are said to be about half of those in many other urban areas. Thus, it could be interpreted that Portland's growth management actions have helped to even out income levels and housing costs throughout the region. Portland's efforts essentially have redirected middle- and upper-income home buyers from the outer suburbs back to the center city and inner suburbs.

Transportation initiatives

Portland's efforts at growth management have not been limited to the UGB. During the mid-1970s, Portland vetoed a series of highway projects, including a planned Mount Hood Freeway. The city tore up the six-lane Harbor Drive expressway and replaced it with a waterfront park. They made investments in downtown Portland and its historic buildings while placing a cap on downtown parking spaces and creating a downtown transit mall. Meanwhile, Portlanders used federal funds to build a fifteen-mile light rail line during the 1980s. Within fifteen years after opening, the new light rail and bus lines carried nearly one-third of all commuters to downtown. As a result, downtown's workforce has grown by 67 percent while adding half a million square feet of retail space since 1972.

During the 1990s, Portland put into place the Transportation Planning Rule (TPR), which advocates a direct approach to integrating land use and transportation. Adopted in 1991, the TPR requires local governments to amend their local land use ordinances to encourage higher-density, mixed use development near transit lines. The TPR was modified in 1995, and suggested examples were issued as guidelines. Subsequently, Portland has adopted the 2040 Regional Framework Plan and the Metro 2040 Growth Concept Plan. These plans expand the UGB only slightly while turning attention to the land use–transportation connection. The new plan calls for encouraging the development of regional town centers around light rail stops, thereby maintaining jobs and services in better balance while reducing auto trips. Although controversy over the plan has arisen (as witness Oak Grove), there have also been successes, such as the new development at Orenco Station. In a continuing endorsement of transit in the region, voters in Portland also approved a $472 million bond issue to fund a new $900 million light rail line.

Results

Although Portland's measures have been successful in many ways, so far they have not proven to be a "silver bullet" for sprawl. Specifically, whereas total employment in downtown Portland has gone up, the city's share of regional jobs has trended downward. In 1972, the city held 90 percent of the region's best office space and 70 percent of its jobs. By the 1990s, those shares had dropped to 61 percent and 14 percent, respectively. Meanwhile, even though transit use to and from downtown is relatively high, with up to 35 percent of peak-hour work trips using transit downtown, transit use in the region as a whole actually dropped between 1980 and 1990, with the transit share of the daily region-wide commute standing at a little more than 5 percent of the total in 1990. The automobile remains by far the preferred mode of transportation, with about 87 percent of the total trips in the Portland region. The problem is that sprawl has continued inside the UGB. As mentioned earlier, Metro has only achieved 70 percent of density projected for development within the UGB, even though average lot size has decreased. Additionally, many of the new high-tech jobs are in suburban areas.

Another issue has to do with equity to property owners. The value of land can change radically from one side of the boundary to the other. When land is moved inside the boundary (as it is when the boundary is occasionally adjusted), the value of land can increase tenfold, from $15,000 an acre to $150,000. As Mike Burton, the executive officer of Metro, was recently quoted in the *New York Times*: "we make instant millionaires out of some people." This type of property value disparity may be a contributing factor to the public sentiment behind the passage of Ballot Measure 7 in fall 2000.

On the other hand, Portland has made some bold and positive forward steps. The Metro regional system is still evolving to address new issues as they become evident. The 2040 Plans hold forth particular promise for directing and containing sprawl while shifting a greater share of the region's transportation burden to transit. Making a real effort to integrate land use and transportation at the scale of Portland's TPR and the 2040 Plans is something that is rare in the United States. Portland's approach to metropolitan government and land use controls is equally rare – though they have not fully succeeded in stopping sprawl, what they have achieved is remarkable. One can only imagine what might have occurred had nothing been done.

An evolving process

Programs such as Portland's can take a very long time to leaf out, and no one can know all of the exact consequences before setting out on an untried course of action. Because of this, Portland has been cautious. Portlanders didn't seek to stop growth or even to radically alter it. They sought only to manage it, reducing lot sizes, encouraging slightly greater density, establishing boundaries, and creating alternative transportation modes. If anything, they seem to be moving gradually toward increased control in response to the pattern that has developed. They are fine-tuning their program to improve the regional results. The Portland program is still evolving and will continue to do so, provided it isn't disassembled first.

[. . .]

REGIONALISM IN SUMMARY

The expansion of metropolitan areas over the past fifty years has made them into the nation's primary economic and social units. This same expansion has created local and regional tensions that call for metro-wide solutions to the twenty-first-century problems of spreading urbanization. Major debates are occurring over urban growth control, coordination of land use planning and regional transportation systems, and regional fiscal disparities and social and economic burdens.

Different metro areas have attempted to deal with regional problems in different ways. Some metro areas have been able to address these problems by simply expanding and incorporating their mushrooming suburbs into their municipal boundaries. Others have instituted metropolitan governments, as in Portland . . .

Each solution is appropriate to its place and is not applicable in every situation. Metro governments like Portland's could be a model for some cities in some states but certainly not for all. Portland has made significant progress in trying to manage its growth, and it is often hailed as a national model for smart growth. But even in Portland, there are problems, and growth is spilling over into neighboring Washington State – beyond Metro's control. . . .

[. . .]

Portland would not have Metro or its UGB were it not for state legislation that started during the 1970s. . . . State governments typically are in control of regional transportation funds, and state legislatures are the source of land use control laws. State governments ultimately sign agreements with other states in multistate metro areas. In the quest for metropolitan solutions, state government has generally been the place to start.

23

Does Planning need the Plan?

From *Journal of the American Planning Association* (1998)

Michael Neuman

Editor's introduction

Urban and regional planning, like any dynamic profession, has participants who engage in lively dialogue about its contents, methods, and critical questions. They agree that the field's fundamental concern is a focus on human settlements and their futures, or, in popular parlance, building healthy communities. Their understanding of human settlements comes from the social sciences, environmental sciences, design, and law. In addition to possessing substantive knowledge about urbanism, planners have skills in quantitative analysis, graphic representation, and collaborative teamwork. Their training encourages them to have a holistic approach to the places they are planning. Like other professionals, they adhere to a code of ethics and accredit educational programs overseeing the knowledge, rules, values, and methods that define their expertise.

Yet expertise does not always provide complete answers to all situations. As might be expected in a field that deals with human behavior, planners balance routine and predictable elements with less routine and unpredictable ones as they determine how and where people will live, work, and play far into the future. For example, practitioners employ statistical techniques and mathematical models to forecast demographic patterns, but these methods can't always predict human conduct. Planners understand the effects of land development on the environment, but politics can restrict their ability to control harmful aspects of growth.

Planners have attempted to reconcile these contradictions. Scholars explore the planning process, finding fault with rational decision making and looking for other theories to explain and predict behavior. At the same time, practitioners employ rational decision making in their work but adapt it to accommodate nonroutine elements, thus advancing the form, content, and process of devising and implementing plans. They are all involved in the laborious process of understanding the nonscientific, nontechnical aspects of their work, or what Donald A. Schön, the late philosopher and MIT professor, called the "artistry" of planning. Schön's classic books, *The Reflective Practitioner: How Professionals Think in Action* (New York: Basic Books, 1983) and *Educating the Reflective Practitioner: Toward a New Design for Teaching and Learning in the Professions* (San Francisco: Jossey-Bass, 1987), provide an explanation of the interplay between the art and science of planning.

In this selection, the author is decidedly provocative. He challenges the academics to turn their attention to what he declares is the substance of planning – the plan, not the process of planning – as a central feature of the profession. He defines a plan as "a two-dimensional representation of the layout of the physical form of the city," meaning a comprehensive plan, not a functional or sectoral plan. The pictorial qualities are important for reasons he details. Some may find this position as too restrictive, others may view it as a needed corrective.

Several phenomena have led to this debate. First, the comprehensive plan, as has been described in preceding readings, has always been part of the planners' realm but gained substantial standing primarily through growth management programs emerging in the 1970s onward. However, not all places have subscribed to growth management and not all are interested in developing comprehensive plans, due to a lack of resources, or prioritizing other planning activities. New York City is an example. While it has no comprehensive plan, it has one of the nation's most developed planning systems, with a compulsory land use review process and neighborhood community boards with mandated roles in planning. Its mayor, Michael Bloomberg, issued PlaNYC 2030 in April of 2007, but this is a mayoral policy document, not a comprehensive plan. In contrast, all cities and counties in Oregon have comprehensive plans. These varied conditions are the reality of planning in a decentralized system that favors home rule, as does the United States.

Second, many practitioners – primarily those engaged in local municipal planning, or subfields like economic development or transportation or urban design – may develop plans that are small-scale rather than comprehensive. They might create neighborhood plans dealing with abandoned property, or a vision plan for a downtown. While they employ the same techniques as those involved in comprehensive planning, their products are quite different. This certainly is planning, but it has a more limited scope than the norm expressed by the author.

Third, some planners are actively engaged in bringing various stakeholders to agreement about the future course of action for a place. The results might be a policy document, or a map (as advocated by the author), but they also might be the implementation of a specific government program, the expenditure of capital funds on a project, or the passage of a local law – not a long-term vision for the physical layout of the city. Advocates maintain that this, too, is planning.

Urban and regional planning encompasses all of these approaches, whether focused on producing a comprehensive plan, a small-scale plan, or developing consensus about a project. However, the process of developing these widely different plans varies, and the professional planner takes on several roles. While planning certainly involves extensive citizen participation, it also requires the injection of professional expertise at different stages. For example, when a community is just beginning to develop a plan (any type of plan), the planner acts as a facilitator, mediator, or educator in bringing together the disparate opinions, values, and assertions of community representatives. During successive stages of the planning process, planners lend professional expertise at specific points – for example, formulating different alternatives based on his/her knowledge and the information culled from the community, or suggesting various implementation strategies.

The selection's author, Michael Neuman, is Associate Professor of Urban Planning Program, College of Architecture, Texas A&M University. He has worked on several plans, including the New Jersey State Plan – he was the principal author of the background report framing that state's efforts to guide growth into communities and centers. He has also been an advisor to the Regional Government of Catalonia for the *Pla metropolità de Barcelona* on planning strategy and process, and a consultant to Public Policy Institute of California, authoring *Building California's Future: Current Conditions in California's Infrastructure Planning, Budgeting, and Financing*, and to the City of Oakland, developing the framework for *Oakland: Sharing the Vision's Strategic Plan.*

For more on comprehensive plans and the plan-making process, see Elisabeth M. Hamin, Priscilla Geigis, and Linda Silka (eds), *Preserving and Enhancing Communities: A Guide for Citizens, Planners, and Policymakers* (Amherst, MA: University of Massachusetts Press, 2007); Barbara Becker and Eric Damian Kelly, *Community Planning: An Introduction to the Comprehensive Plan* (Washington, DC: Island Press, 2000); American Planning Association, *Planning and Urban Design Standards* (Hoboken, NJ: Wiley, 2006); and Thomas L. Daniels *et al.*, *The Small Town Planning Handbook*, third edition (Chicago: APA Planners Press, 2006). For a history of plans up to the 1960s, see Mel Scott, *American City Planning since 1890* (Berkeley, CA: University of California Press, 1969). For information on plans since the 1970s, see Alexander Garvin, *The American City: What Works, What Doesn't* (New York: McGraw-Hill, 2002).

THE PLAN IS DEAD?

For a century the Plan was the centerpiece of modern city planning. It maintained that status in Europe and North America from modern planning's inception in the mid nineteenth century. After World War II, however, the plan's fortunes began to ebb. Plans and comprehensive planning were subjected to critiques that led practitioners and scholars to question their value. In spite of elegant plans and eloquent defenses, planning practice and theory shifted from plan to process. Urban planners were advised to perform "middle-range" rather than comprehensive tasks. Thus, since the early 1960s, the plan has not occupied center stage. More attention has been given to process. . . .

[. . .]

REVIVING THE PLAN

In the midst of the sea-change from plans to process, exceptional plans made their presence felt. Perhaps the first to get much attention in the United States were the 1971 Urban Design Plan and the 1983 Downtown Plan (also an urban design plan) for San Francisco. Large "urban pieces" – a term in use in Europe – such as the designs for Battery Park City in New York and the Docklands in London followed. The plan for the new town of Seaside, Florida, became New Urbanism's first emblem. Portland, Oregon, prepared an ambitious plan that took public participation seriously, giving citizen involvement a new dimension. The Regional Plan Association published its third plan for the New York metropolitan region, titled *A Region at Risk*, in 1996. A draft plan for Washington, DC, *Extending the Legacy*, was released in 1996 by the National Capital Planning Commission. . . .

National institutions latched on to physical planning. The American Institute of Architects formed Urban Design Assistance Teams. The American Planning Association followed with Community Planning Teams. Meanwhile the National Endowment for the Arts weighed in with the Mayors Institute on City Design. The 1992 New Jersey State Plan and the 1994 San Diego Regional Growth Management Strategy provided new visions for their jurisdictions and new models for state and regional planning. These two plans also redefined the relations between planning and governance. New Jersey invented a plan preparation process called "cross-acceptance," a collaborative and iterative model for negotiation. In San Diego, local planners wrote their regional plan. These two plans held out the promise that planning, by redesigning governing institutions, could be a path to real democratic reform. San Diego and New Jersey thus set a new agenda for planning and research.

In Europe, map- and design-based plans had yielded to policy-based plans in the 1960s and 1970s, as they had in North America. In the 1980s, strategies and frameworks became popular. In the midst of this shift, a similar blip on the radar screen of physical plans appeared. An example was the plan that guided the renovation of Bologna, Italy, in the 1960s. In renewing Bologna, its planners created new modes of grassroots planning and participative democracy. The Bologna plan was seminal in Italy and Europe, . . . [and] called attention to the roles of architecture in the building of the city, and of physical design in guiding its planning and politics. In Madrid and Barcelona, grassroots movements led to citizen-based city plans. Architects crafted Madrid's 1985 General Plan, which guided the restoration of its historic center and the provision of infrastructure and services in the periphery. The transformation of Barcelona in the 1980s and 1990s was guided by its 1976 *Plan general metropolitano de Barcelona*. The Thames River Gateway Strategy signaled a departure from the rule-based norms common in Britain. Although these plans were the exemplars, after 1980 the entire continent witnessed a resurgence of physical plans and strategies.

Is the revival due to the centering influence of the plan, always and again at planning's heart? In some cases, the plan has proved to be an effective instrument of urban policy and a spark for urban change. It still serves its traditional functions of guiding urban facilities and setting parameters for zoning and other legal controls on real property. It is serving newer purposes as well. Physical plans put forth graphic images of the future that can rally stakeholders to act. Citizens and interest groups like to back a plan that lets them "see" what they will get. Politicians like to back a consensus plan that deals with thorny issues they often find too risky to tackle themselves. By bringing a ready-made consensus to political bodies, planners do political work. Plans serve as "single-text negotiating documents," to use the language of dispute resolution. Around a well written plan diverse interests can negotiate and agree on policy. In these ways, plans have begun to breathe life again into the comprehensive planning ideal. This article explores the new claims put forward for the plan, by comparing plan-based and non-plan-based planning, looking at both practice and theory in historical perspective.

ON THE ORIGINS OF MODERN CITY PLANNING

Planning historians customarily attribute the origins of modern planning to Haussmann's plan for Paris at the middle of the nineteenth century. Alternatively, the beginning is pegged to the late nineteenth century, with the rise of the movements for tenement improvement and civic hygiene in Germany, Britain, and the United States. This chronology relegates Haussmann and his imitators to placing monuments and laying out boulevards. The first *comprehensive* city plan dealing with the concerns that claim contemporary planners' attention – housing, environment, traffic, social and health conditions, urban design, density, infrastructure, etc. – was drafted by the Catalonian civil engineer Ildefons Cerda for Barcelona in 1859. He based his plan on a theory of urbanization.

Other plans ensued, which shaped the fledgling profession. Letchworth Garden City, the 1893 Columbian Exposition's Great White City, Daniel Burnham's and Edward Bennett's 1909 *Plan for Chicago*, Arturo Soria's Lineal City of 1882 for an extension of Madrid, Walter Burley Griffin's 1912 plan for Canberra, and Otto Wagner's 1893 extension plan for Vienna were some of the most prominent. The 1929–30 *Plan for New York and Environs* by the Regional Plan Association and the 1944 *Greater London Plan* by Patrick Abercrombie were landmarks of the plan movement. Postwar plans such as Copenhagen's famous Finger Plan and Holland's Green Heart and Randstad extended the tradition. In the early 1960s, Ed Bacon's plan for Philadelphia got him on the cover of *Time* magazine. Thirty years later its vision has largely been implemented, bringing lasting improvements to the City of Brotherly Love.

Plan-based Utopian treatises also exercised influence. Camillo Sitte's *The Art of Building Cities* (1889); Ebenezer Howard's *Garden Cities* (1898); Tony Garnier's *Industrial City* (1917); Le Corbusier's *Plan Voisin* (1925), *Radiant City* (1933), and *A Contemporary City of 3 Million Inhabitants* (1922); and Frank Lloyd Wright's *Broadacre City* (1935) are a few noteworthy examples.

These seminal thinkers showed in no uncertain terms what they thought the future city should look like. Their plans did not rely on mere words or abstract theories. Even Cerda's theory was empirical, based on over ten years of detailed data gathering and comparative analysis. The pioneers adopted the stance boldly stated by Daniel Burnham in 1907, in what has nearly become a mantra for planners:

> Make no little plans. They have no magic to stir men's blood and probably themselves will not be realized. Make big plans; aim high in hope and work, remembering that a noble, logical diagram once recorded will never die, but long after we are gone will be a living thing, asserting itself with ever-growing insistency.

In this article, the word plan refers to a two-dimensional representation of the layout of the physical form of the city. Two-dimensional diagrams are abetted by three-dimensional models and illustrations portraying ground-level and bird's eye views. . . . My use of the term "plan," unless noted otherwise, refers to a general, comprehensive, master structure, or strategic plan, rather than a sectoral or functional plan such as for transport or housing, or a site-specific plan for an area or project. . . .

The images in those historic plans are etched in our minds. They are icons of the profession. The plan assumed heroic status, and the creators of plans became legends. They played starring roles in building the profession and institution of city planning. The pioneers did not separate practice from theory. "Survey before plan" and "garden cities" were concepts at the core of planning theory a century ago. Practitioners and theorists, process and substance were one and the same.

WORDS AND PLANS

In the 1920s, governments in the United States took what became the first steps in moving away from the plan and towards zoning as a determinant of urban and suburban growth. The 1926 Supreme Court decision in the case of *Village of Euclid v. Ambler Realty Company* and the 1926 United States Department of Commerce Standard State Zoning Enabling Act shifted the emphasis away from plans, designs, arid urban form to zoning, laws, and land use. It was prophetic that the Department of Commerce released the Standard State Zoning document before its Standard City Planning Enabling Act of 1928. Lawyers and planners were to replace designers and engineers as the leading professionals shaping urban growth policy.

As World War II erupted, national planning efforts around the world took a turn at churning out materiel for their armies and navies. As soldiers and sailors returned home after the war, the priorities of national planning and programs continued to supersede those of local planning, as they had during the war and the Depression before it. Service men and women and their baby boom offspring needed homes and schools. As the Marshall Plan was helping wartorn Europe to rebuild its cities and infrastructure, another part of America's production capacity was channeled to building highways and suburbs, and rebuilding "blighted" cities by clearing slums.

National planning, because of its scale, was less sensitive to design and place than was its local-scale kin. It opted for replicable programs and contributed to the specialization of planning. Federal highway and urban renewal programs foreshadowed the hold over cities and their planning that national programs in the United States were to exert from the 1950s onward. Those programs ensured the ascendancy of specialists and the fall of generalists, despite pleas such as Jane Jacobs's *The Death and Life of Great American Cities* (1961) and Ian McHarg's *Design with Nature* (1969). Segmentation occurred in other areas as well. Instead of an ecological view of the environment, in the United States there appeared separate programs (not plans) for clean air, clean water, endangered species, coasts, flood zones, waste, etcetera. Fragmentation of this sort fueled the need for coordination, the coordination innate in synthetic comprehensive plans.

[. . .]

A new player made the planning scene in the decades after World War II, often at center stage. The private real estate developer emerged as a formidable force to which localities could mostly only react. Tract homes in large residential subdivisions formed a patchwork suburban quilt, dotted by shopping malls. Standardized site plan and floor plan layouts, national building codes, and easy financing enabled developers to mass produce a panoply of "ticky-tacky little boxes" that subverted the traditional subdivision, which until then had been craft-made. New tools such as Planned Unit Developments further proliferated large-scale projects. New subdivisions spread over the landscape like an intractable tract home rash. Where developers overmatched localities' powers, control over local destinies was wrested from the cities and towns.

In sum, planning and general plans gave way to developers' site plans, highway engineers' concrete cloverleafs and asphalt ribbons, federal officials' urban renewal, environmental regulations and impact reports, and lawyers' codes. Texts such as Kevin Lynch's *Site Planning* (1962) and Norman Williams's *American Land Planning Law* (1974) influenced entire generations. Kent's *The Urban General Plan* (1964) receded from view. Supreme Court decisions and federal programs stirred planners' interest more than new general plans did.

The plans that were produced stood apart from previous physical plans. They were policy plans replete with goals, objectives, policies, criteria, standards, and programs; graphs, charts, projections, and matrices. They were generally devoid of graphic images or proposals for urban form. Again, the Regional Plan Association produced the archetype, in its 1968 *Second Regional Plan* for New York's metro area. Such plans, more often than not, "gathered dust on the shelves," a now well worn expression that entered our lexicon around that time.

Critiques notwithstanding, it seemed in the 1960s that nothing could stop the bureaucratic machinery . . . well oiled by then-fresh ideas. Planning employed systems paradigms and quantitative methods that treated politics, institutions, and other factors as exogenous. Planning's new technocracy applied principles of hierarchy and conformity (plan–program–budget; plan–regulation–permit;

national–state–regional–local) in a linear sequence. Some theorists followed suit. Other theorists of the time elided the plan. . . . Leading practitioners opted for functional plans or for equity planning at the expense of the master plan.

At about the same time, a "quiet revolution" of state planning gave new roles to planning and state government. State planning reinforced general plans and other local instruments, in addition to inserting the new figure of state plans. State laws often specified plans that were laden with goals, objectives, criteria, and standards with which subordinate levels of government had to comply. Creativity, negotiation, and design-based physical plans, though not explicitly excluded, did not appear in this programmatic approach. It took the innovations of conflict resolution and new urbanism to change planners' thinking, largely by providing new images of how to conduct planning.

Planning theory of this era, though fertile, was in a muddle about just what it and planning were. Dichotomous debates ran through theory: content versus context, rational versus political, comprehensive versus incremental, substance versus process. One thing was clear. Theory and practice were distinct and acknowledged as such. Theory was not linked to practice. Out of these debates emerged empirical theorists who closely examined practice. . . .

Practitioners, for their part, took cues from urban conditions, not theories. One of these conditions was that of government itself. The involvement of more levels of government, the explosion of regulatory permits, the proliferation of programs, the splintering of agencies into professional subdisciplines, and the inclusion of new stakeholders created a planning panorama that would have been unfamiliar just a short time before. Coordination became paramount.

Coordination was not a new idea. As early as 1911 Patrick Abercrombie had signaled the "necessity for cooperation." Early coordinating attempts followed the hierarchical model endemic in federal practice. Later, the "quiet revolution" of state intervention in land use and growth management did not break that mold. Florida epitomized state planning in the 1970s and 1980s. Florida laws mandated "consistency" among local, regional, and state plans. Extensive and detailed rules and regulations were specified. Little room was left for variation or interpretation, which stifled local planning. The laws did not allow for the negotiation of differences. Coordination was procedural, its criteria prescribed from above. In this way American planning resembled earlier efforts in Europe, where administrative practices in highly centralized governments followed hierarchical norms, as in France, Italy, and Spain. However, the consistency doctrine was soon to reach its limits.

Consistency and other rigid approaches clashed with the complexity and pace of change in cities and their administration. Moreover, evolving lifestyles demanded flexibility and mobility. Planners adjusted to these developments by questioning planning's domain. Was coordination enough? More generally, was process enough? Some urged a return to physical planning and urban design. Others urged a fuller accounting of politics. Davidoff's critique (1965), urging representation of underrepresented interests, began to take root in institutional settings. Processes opened up to include interests that had been ignored. Citizens and organized interest groups (neighborhood and civic associations, environmental organizations, and developers' lobbies, for example) were brought in. Planning codes were rewritten to mandate public participation, as were the laws governing environmental impact statements and endangered species protection.

Bringing in new stakeholders had the foreseeable result of increasing conflict. Introducing new interests (and thus conflicts) made politics more relevant, and gave planning a higher profile. Planning was front-page news. Planners occupied (and increasingly occupy) seats in Congress and state legislatures, mayor and council posts, city and county manager positions, and university leadership. These advances were not led by planning theory. Planners responded to local situations with local knowledge. In their confrontations with new realities, planners invented. Their ingenuity paid off.

Coordination moved from mandated top-down consistency and strict compliance to voluntary mutual acceptance of plans via comparison and negotiation, called "cross-acceptance" [as in New Jersey]. The image governing coordination was no longer simply top-down or bottom-up, but back and forth. Iterative back-and-forth multilogues entailed successive rounds of multilateral negotiations and fine tuning. Consensus building was the new watchword. . . . Collaboration, cooperation, and consensus became the new 3-C, replacing "comprehensive, continuous, and coordinated."

The stress placed on process from about 1960 until the advent of neotraditionalism came at the expense of place. How to plan, not what to plan, occupied planners' imaginations. Plans were filled with words and numbers rather than maps and designs. It became difficult to envision the future of the place being planned through the haze of statistical data and quantitative analysis. Implementation surfaced as a serious issue. The plan–implementation dichotomy was born in the depths of quantitative plans. How do you implement a matrix? Policy plans did not fare better. How to implement a vague goal or policy? . . .

"OH, GREAT – ANOTHER PARADIGM SHIFT"

As planning for places becomes salient once again, we can discern several plan archetypes now prominent in practice. One archetype is the traditional physical plan, as is apparent when one compares New Urbanists' plans to those of earlier civic designers. Another is the strategic plan, a vogue in business schools and boardrooms, which has now crossed to the public sector. Public sector strategic plans pursue restructuring, privatization, government as business, and customer service. Spatial strategy in the public sector is often transformed into city marketing. A third archetype comprises environmental and community plans that engage a wide range of interests and stakeholders. . . .

We know from experience that some plans have little effect. Worse, some backfire and cause disasters. The critiques of plans are familiar: Plans become marginal when not connected to power. Plans restrict development and impinge on the "free" market. Plans are too general and future-oriented to deal with daily concerns. Plans take too long to prepare, and by the time they are adopted they have been overtaken by events. Plans attempt to accomplish too much and end up doing little or nothing. These points have had a long history of debate, and the debate has been revived.

Recently, theorists have tended to pick up on the discursive and the communicative aspects of planning and politics. Offshoots pointed to the significance of storytelling and metaphors. Others attributed action in the politico-planning sphere to the influence of civic culture, institutional culture, political culture, and planning culture. At the same time, the popular writers and consultants have bombarded us with "visioning," "reinventing," "rethinking," "reengineering," "restructuring," and "downsizing." As the world changed, behemoths were dismantled and hierarchies became outmoded. In planning, a wide range of techniques and theories have competed to fill the cracks thus created. . . .

[. . .]

THE POWER OF THE DREAM

At the Olympic games, athletes refer to attaining their dreams. At the 1996 Atlanta games a theme song was "The Power of the Dream." In addition to visualizing their performances before they compete, world-class athletes create powerful mental images of winning. These images inspire them to exercise at extraordinary levels and to reach almost superhuman levels of focus, commitment, and discipline. Persuasive plans, too, possess the power of the dream. Images and visions, including those in plans, can stir minds, arouse hopes, and inspire action.

The pictorial nature that images, designs, and maps afford plans endows them with qualities that other instruments of public policy often lack. The plan stands as an important part of our discipline's intellectual heritage precisely because of these qualities. Why has so little been written recently about these advantages of plans? . . .

[. . .]

Taken together, plans and images form a research agenda for analysts and an action agenda for practitioners. A framework for such an agenda appears below. . . .

Plans chart collective hope

Plans help connect people to places by bringing people together to shape a common destiny for their places and themselves. . . . We need to better understand the connection of people to place to plan, if we want plans to respond to the needs of residents rather than to regulations.

Plans use images of place to portray collective hopes

Pictures, metaphors, stories, designs, and maps paint images in the mind's eye. . . . Images enhance a plan's capacity to change people's minds, converting plans into political change agents. Using images also enables planners to exploit powerful media networks. We live in a popular culture where images reign and determine fortunes. The more we know about the way all sorts of images work in plans and planning, the more we will succeed.

Plans are the loci of conflict

Comprehensive plans bring peoples, disciplines, urban functions, problems, interests, and ideas together in institutional settings. . . . Plans can be used to set agendas and resolve conflicts, because they are ideal "single texts" that the participants in plan making rely on to make decisions. If plan making is truly pluralist and participative and not just a "staff prepare – others respond" *pro forma* ritual, then it can build community as it builds upon the social, intellectual, and political capital in a community.

Plans are powerful because they are built into the power structure

Plans derive their greatest authority in current practice by being inextricably entrenched in governing institutions. . . . Plan makers, it is clear, should be savvy about politics and institutions. Moreover, they need to be knowledgeable about and critical of institutional power. . . . Although the profession has long recognized the primacy of politics, practitioners often still grapple on the sidelines in political contests. . . . How can research really aid planners in these undertakings? Can front-line planners be engaged more effectively in research design?

Plans have built the profession and institutions of planning

Plans have also been used to design and redesign institutions of government. This transcendental quality of plans stems from their relation to constitutions. Indeed, in nearly every *Federalist Paper*, the authors Alexander, Jay, and Madison called the Constitution of the United States a plan. A better understanding of this contribution by plans can reap benefits to our governments, our communities, our profession, and ourselves.

Powerlines: lines on maps decide "who gets what, when, and how."

City plans and zoning codes have a distinctive feature: maps with lines that mark boundaries. This characteristic is often overlooked by researchers, despite its importance. . . . Fear of drawing lines that are legally binding and fear of presenting maps in public arenas are skeletons in the planning closet. Yet the conflict occasioned in governing urban development by plans and zoning is an indicator of consequential planning. As any planner or politician knows, conflict comes to a head when the lines are drawn. Lines that have this effect appear not only in plan and zoning maps, but also in site plans, transportation and utility plans, process flow diagrams, and organizational charts. Who wins and who loses, who sits at the table, and how the game is played are gauged by planners' lines. Are planners afraid to draw lines because they are conflict avoiders (consensus seekers) by nature? Or are they afraid that, finally, plans and zoning will gain them status in accord with their true place in the urban realm?

Does planning need the plan? Or can planning go planless, naked and exposed? If the latter, why not call our profession "ning" and leave out "plan" entirely? As it is, planning is blessed with an active verb for its name, a characteristic it shares with other professions that nurture and bring things into being: nursing, engineering, design. City planners bring cities to life and life to cities, and have done so for centuries using plans. The recent diversification of tools used by planners has enriched our profession. To be most effective, and to be used with the soundest legal basis, they need to be linked to a general plan. After all, the plan did give planning its name.

How Plans Work

From *Urban Development: The Logic of Making Plans* (2001)

Lewis D. Hopkins

Editor's introduction

When nine government advisors – including Frederick Law Olmsted, Jr, Alfred Bettman, and Edward M. Bassett – wrote the Standard City Planning Enabling Act (SCPEA) in the 1920s, they provided a simple, straightforward prescription for the form and functions of urban planning that, like the U.S. Constitution, has stood the test of time because of its ability to sustain interpretation. Under the SCPEA rubric, governmental urban and regional planning is undertaken by a citizen advisory commission, assisted by a technical staff, which adopts plans for the physical development of a place and oversees their implementation.

The advisors envisioned these plans as guiding public and private development decisions. When they crafted their model, several cities already had created planning commissions of the sort the advisors recommended (e.g. Hartford, 1907; Milwaukee, 1908; Chicago, 1909; and Newark, 1911), and plans existed for Washington, DC, 1901; Chicago, 1909; St Louis, 1917; Memphis, 1920; Mariemont, 1922; and Cincinnati, 1925. The Cincinnati plan was the first one officially adopted by a city planning commission.

While the drafters of the SCPEA saw the master plan as the key official document, they understood it to be composed of several preliminary, less comprehensive plans. Their clear, minimal language, however, laid the foundation for the growth of different kinds of plans. (See Plate II.5.)

In this selection, the author presents the plan as having five purposes. A plan serves as an agenda (a list of things to do), and is also a set of policies (a framework for fair, consistent, and predictable decision making), a vision (a document that presents beliefs), a design (a coordinator of interdependent actions), and a strategy (a path through a decision tree). Some plans have all of these qualities, others have only some of them.

A quick review of landmark American plans shows how these five purposes come into play. Cincinnati's 1925 *Official City Plan* represents the agenda-setting type. It contained a list several pages long of needed capital investments. Updated in 1948 as the *Metropolitan Master Plan*, it continued the agenda-setting format with detailed maps of a full panoply of proposed capital investments ranging from transportation routes to schools, libraries, and health centers, all placed in their neighborhood locations. The plan's author, Ladislas Segoe, a consultant who had also helped with the 1925 version, noted that this plan was not merely "a package of plans," but the result of consultation with "the interested government agencies" and the "citizens of the area," who presumably helped determine the list of projects.

The *Extending the Legacy* plan for Washington DC (1996) explicitly outlines policy. The National Capital Planning Commission, which produced the document, describes it as

> a framework . . . not a blueprint that renders the future in precise and immutable detail. It is more like a map with a few dramatic highlights, corresponding to the best locations for museums, parks, bridges, transit

> stations and other public assets. It is both a guide to the big picture and a defense against the myopic quick fix.

The plan lists five general policies (and enumerates specific aspects of each): protecting the legacy, unifying the city and the monumental core, promoting economic development, renewing rivers and open space, and improving transportation.

Vision plans have been present from the turn of the twentieth century to today. Plans for Washington, DC (1901), and Chicago (1909) were early examples. Architect Daniel Burnham, who was involved in both, viewed this work in visionary terms. "Make no little plans," he proclaimed. "They have no magic to stir men's blood. Make big plans; aim high in hope and work, remembering that a noble, logical diagram, once recorded will never die." The documents are filled with glorious watercolor renditions of what the cities could look like – not precise designs, but blurred, romantic images of the possibilities.

Forty years later, Cleveland published a *General Plan* (1949) under the supervision of its planning director, John T. Howard. Howard presented a vision, one filled with threat, not glory. "Plan or perish," he exclaimed in the plan's opening sentence. He also offered a strategy to address the problems identified in the plan – poor housing in some areas was one issue. In a refined approach to residential land use, the plan ranked neighborhoods according to their physical characteristics. Categorizing them as "protection," "conservation," and "redevelopment," the plan anticipated urban renewal programs that were on the horizon by highlighting potential targets for improvement.

In 1996 the Regional Plan Association, a nongovernmental group, would also use the language of peril to bring attention to its vision. *Region at Risk: The Third Regional Plan for the New York–New Jersey–Connecticut Region* portrayed the New York region as a place that was "out of control" and in need of addressing problems related to globalization and unbalanced metropolitan growth: "Shifting employment prospects created by global competition for employment, industry restructuring, and immigration; growing disparities between poor central cities and inner suburbs and rich outer suburbs; environmental and transportation impacts of decentered automobile-based growth" were the threats to be met through a five-pronged campaign, encompassing strategies for the region's greenswards, centers, mobility, workforce, and governance.

The *Cleveland Policy Plan* (1975) was both a vision and policy document. Crafted under the direction of Norman Krumholz, director of planning, it offered a physical plan derived from specific social and economic policies. Krumholz's introduction to the plan explained its primary goal:

> In a context of limited resources, the Cleveland City Planning Commission will give priority attention to the task of promoting a wider range of choices for those individuals and groups that have few, if any choices. Given the disparities of income and power between the residents of the City of Cleveland and those of the surrounding region, this goal, in part, simply reflects our responsibility and commitment to serve the people of the City.

This plan had maps and charts like its predecessors, but their content was unprecedented, as they portrayed income, housing quality, and mobility, not the traditional land use and transportation data. The resulting recommendations were unprecedented as well.

The *Center City Philadelphia Plan* (1963) explicitly focused on design elements, as defined by this selection's author. Planning director Edmund Bacon and City Planning Commission chair G. Holmes Perkins oversaw the plan, which details the downtown part of the city's *Comprehensive Plan: the Physical Development of Philadelphia* (1960). Through language and analysis of center-city systems (transportation, activity centers, interrelated land uses – office, retail, open space, logistics, residential areas), it aims to build a well-oiled machine, the engine of the city and its surroundings. The plan's introduction is explicit about this approach: "The plan does not attempt to provide blueprints or cost estimates for detailed projects. Rather it presents a total set of relationships, a system of organization of part to part and part to whole, in which every element and every section of Center City plays a role."

Lewis Hopkins, the author of this selection, is Professor of Urban and Regional Planning emeritus, University of Illinois. In addition to his work on the idea of the plan, he has researched computerized decision support systems in planning. He edited *Engaging the Future: Forecasts, Scenarios, Plans, and Projects* (Cambridge, MA: Lincoln Institute of Land Policy, 2007) with Marisa Zapata. As a member of the Urbana Plan Commission, he participated in the commission's adoption of the city's *Comprehensive Plan of 2005*.

For additional reading on plans and planning, see William S. Saunders (ed.), *Urban Planning Today* (Minneapolis, MN: University of Minnesota Press, 2006), and Seymour J. Mandelbaum, "Plans," in *Open Moral Communities* (Cambridge, MA: MIT Press, 2000). Donald Krueckeberg's *The American Planner: Biographies and Recollections*, second edition (New Brunswick, NJ: Center for Urban Policy Research, 1994) has portraits of fifteen important planners offering their ideas for plans. For an international view of different plans, see Stephen V. Ward, *Planning the Twentieth Century City: The Advanced Capitalist World* (Chichester: Wiley, 2002).

How do plans work? Through what mechanisms or causal processes do plans affect actions? How can we explain why a particular plan is likely to have particular effects? [Plans work in] five different ways . . . : agendas, policies, visions, designs, and strategies. . . . Any one plan can work in one or several ways, which means that these are not categories for classification of plans but different mechanisms through which plans affect the world. . . .

An agenda is a list of things to do. An agenda works by recording a list to remind us what to do, or to share publicly a commitment to do these things. Agendas work when there are too many actions to remember or when there is benefit in gaining trust among people affected or legitimating actors as accountable. Publishing or publicly advocating an agenda serves both as a memory device and a commitment. . . .

[. . .]

The items in the agenda of a meeting have in common the timing of decisions at the same meeting and perhaps a common decision maker's authority, but the choice for one item need have no relation to the choice for another. Agendas are of interest to planners because they are a tool that focuses the attention of a constituency, whether an individual, a legislature, a group, an electorate, or the public at large. Setting agendas and pursuing agendas are thus ways of affecting the decisions that will be made. Agendas keep our attention focused on important actions or issues rather than merely on what "comes across our desk" at the initiative of others. An agenda is one way to focus the attention of decision makers on some decisions rather than others.

A policy is an if–then rule. A policy works by automating repeat decisions to save time or by ensuring that the same action is taken in the same circumstances, which yields fairness or predictability. Policies fit situations in which there are many repeat decisions and decisions are costly to make, consistency is viewed as fair, or predictability of repeat decisions is beneficial. For example, if the developer will pay for the cost of the sewer extension, then extend the sewer. This policy would save the costs of making this decision in each case, treat all developers alike, and make development actions predictable. Knowing the policies of other decision makers provides evidence for forecasting their decisions. Policies are distinct from regulations in that regulations change legally or administratively enforceable rights whereas policies identify standard responses for repeated instances of the same situation. . . . Policies work in three ways: saving decision costs, ensuring consistency (fairness), and increasing predictability.

A vision is an image of what should be. Visions compel actions. Visions work by changing beliefs about how the world works (beliefs about the relationships between actions and outcomes), beliefs about intersubjective norms (peer group attitudes about good behaviors), or beliefs about the likelihood of success (raising aspirations or motivating efforts). A vision could be interpreted as a normative forecast: a desired future that can work if people can be persuaded that it can and will come true. Visions, however, focus first on the outcome and then on the possibility of actions to attain this outcome. . . . Visions are useful in situations in which they can change beliefs and thereby change investment actions, regulations, or activity patterns of residents. Visions are distinct from target designs, which are focused on a feasible solution to a complex problem of interdependencies. Visions work their effect on beliefs, not by their feasibility of construction.

[. . .]

The Chicago Plan of 1909 is a familiar example of plan as vision. It included both graphic renderings of a physical vision and verbal descriptions of the characteristics of a great city. . . . The Portland 2040 Plan also uses images of the implied future to sell its less palatable actions. . . .

A design is a fully worked out outcome. Designs work by determining a fully worked out outcome from interdependent actions and providing this outcome as information before any action is taken. Designs fit situations in which there are highly interdependent actions, actions are easily inferred from information about the outcome, and there is little uncertainty about implementation of actions. We usually think of design as a process in which many ideas are tested and modified, but entirely in some simulated environment before any action is taken in the real world. . . . All the decisions involved in design of a single building are tested as hypotheses in combination through diagrams and calculations to see how they fit together before any action is taken to construct the building. Designs usually focus on patterns of capital facilities rather than on the human activity patterns that will occur given these facilities. Measures of success should, however, assess these human activity patterns.

Design works by figuring out a result for many interdependent actions before acting. . . . [Sometimes] the design concept breaks down over time in urban design but still results in somewhat coherent physical forms. A complete and coherent design for a section of a city is proposed. Some elements of the coherent design get implemented, but other elements do not because of citizen complaints, budget constraints, changes in government, or power relations. Then, situations change and new designs are proposed that in part relate to the elements of the previous design. Some of the elements of the new design are implemented. The realized urban form results from this sequence of dependent designs, none of which is implemented in its entirety.

As projects become more complex and more easily decomposed into actions that can be carried out separately (e.g., more than one building, phased buildings to be constructed with long periods between each phase), they take on the character of a sequence of design projects linked by strategies about related decisions. Although any architect designing buildings will point out that in many cases the design may be modified during construction and that the cost of design changes is not zero, these costs are small and these modifications are minor relative to the whole design. In larger urban development situations, actions taken at different times are each of similar magnitude, such as building an interceptor sewer now and an expressway later. Modifying the expressway capacity or service area before it is built to complement a sewer system designed to absorb twenty-five years of growth is a different level of relationship from modifying details as a building is constructed.

[. . .]

A strategy is a set of decisions that forms a contingent path through a decision tree. Strategies work by determining what action should be taken now contingent on related future actions. Strategies fit situations in which there are many interdependent actions under the authority of many actors and occurring over a long time in relation to an uncertain environment. In sequential decision making, at the time action is taken on a current decision, the future decisions have been thought through for each outcome from that current decision. Saying that we plan to do something means that we will take certain actions under certain conditions when the time comes. . . .

Strategy is arguably the most inclusive and thus fundamental notion of plans because it is the most explicit about the relationships among interdependent actions, their consequences, intentions, uncertainty, and outcomes. . . . In contrast, designs focus primarily on outcomes. Visions, agendas, and policies are often joint effects of plans that also work as strategies or designs. . . .

[. . .]

In contrast to plans, regulations set the rights of a decision maker by identifying what decisions are permitted and by setting the range of discretion of choices and criteria in making these decisions. Regulations are enforced by the state through its monopoly on the use of force. For example, zoning restricts the range of uses, building height, and land coverage that may be undertaken on a particular parcel. A subdivision ordinance restricts the patterns by which land can be divided into building lots. Regulations may be created by private groups under the force of contracts, which are in turn enforced by the state. Thus a homeowners' association may impose design regulations for its members. Regulations affect decisions by restricting the set of choices, whereas plans affect decisions by providing information.

In contrast to regulations, none of the ways in which plans work is inherently binding on actors. Plans that work as strategies set forth contingent decisions that affect choices made now, but there is no current or future change in the range of alternatives from which the decision maker is permitted to choose. The effect on current decisions is only through the decision maker's own assessment of related decisions. Regulations define the set of future alternatives from which a decision maker may choose, which can help to determine which decision is best for action now.

Plans also work as a focus of deliberation – discussion, argument, conflict, and resolution. Such work occurs both in the creation of plans and in their use to guide action.

INVESTMENTS AND REGULATIONS

Investments in physical infrastructure or facilities and regulations are widely recognized as the two major components of urban development plans. As in political interpretations, these different types of actions imply different tasks for plans. Investments, whether by public agencies or private firms, change the capital stock of infrastructure or buildings. Regulations change rights, the range of discretion in making decisions. . . . [The] pervasive focus of plans, whether made for governments or private firms or individuals, suggests that we should be able to explain why plans are made for investments in physical facilities and regulations rather than for other types of actions.

[. . .]

Investments in physical facilities mediate between geographic space and people's behaviors. Thus two kinds of decisions matter: the decisions to invest in infrastructure and the decisions to use the resulting infrastructure in particular ways. Indicators of quality of life depend on activities of populations, including their interactions with each other; the physical facilities in which they live and work, including the networks that connect these facilities; and the geographic locations in which these activities and facilities occur. Thus an indicator of vehicle miles traveled per person per day depends on where people who work downtown live and over what type of network they travel, which depends in turn on the geographic character of the site of the city. The important point is that investments occur in fixed locations and they create the physical context within which locational choice and daily behaviors occur. Whether investments are in buildings – housing, schools, treatment plants – or networks – roads, sewers, light-rail transit – they are fixed in place and cannot be moved without great cost. They are built with specific capacities, which cannot be changed without additional investment. . . .

People choose to live or work in facilities that exist at particular locations because someone invested in the facility at that location. People choose transportation mode and route over a network of streets and transit based on investments made to link locations by roads or transit routes. The outcome of the investment is realized only when the location choice and travel choice behaviors occur. We must therefore estimate these behaviors for given investments rather than trying to estimate the effects of investments directly.

This logic of plans for investments also applies to capital investments by the private sector. [T]he example of the creation of new nodes in a multinucleated city [is useful]. In many cases, no one developer has sufficient capital or land to build an entire new center alone. If several developers try to locate new subcenters when only one or two can be sustained, however, then some subcenters will fail. The capital invested will be lost and the underutilized land will displace other uses because of the high cost of conversion. Even the successfully established center will be slower in developing than necessary because some developments and tenants will have to move from other failed centers. The private developers have much to gain from figuring out ahead of time which new center will succeed and building there initially. Public infrastructure providers and house buyers would also be affected by the uncertainty of location of new centers.

Regulations have a structure similar to investments with two kinds of decisions: decisions to regulate and decisions to act given the regulations. A decision to zone a municipality by land use type and density is a decision to regulate. A decision to build a house in one of these zones is an action given the regulation. Usually the decision to regulate will be collective and the decision to act individual. In order to use regulations, decisions must be made about where to impose what regulations. . . . To implement a zoning regulation, we must consider a sufficiently large area to figure out a pattern of land uses that will reduce negative effects of adjacency of different uses and provide access to services. The area to be zoned must be considered in finite increments; it is indivisible. As with investments, regulations cannot, therefore, work by iterative adjustment. If a regulation is to reduce external effects of adjacent land uses, it will be effective only if it is imposed before the conflicting land uses invest in locating next to each other. If a regulation is to match density with infrastructure capacity, it can only be effective before investments are made.

[. . .]

DETERMINING WHETHER PLANS WORK

Do plans work? These explanations of how plans can work – as agendas, policies, visions, designs, and strategies – provide a means by which to assess whether plans do work. These explanations indicate what we can expect to observe if plans are working and how we can explain relationships among these observations. We can observe:

- Plan-making behaviors – the things planners and their collaborators do when they make plans.
- Plans – information available at particular times to particular people.
- People using plans while making decisions.
- Investments and regulations that may have been affected by plans, and
- Outcomes in terms of activity patterns resulting from these investments and regulations.

All of these observable phenomena provide opportunities for assessment. . . . There are four broad criteria for assessing whether plans work:

- *Effect.* Did the plan have any effect on decision making, actions, or outcomes? For example, if it was intended to work as an agenda, how many of the listed actions were taken?
- *Net benefit.* Was the plan worth making, and to whom? For example, if it was intended to work as strategy, were the gains in efficiency of infrastructure provision over time sufficient to compensate the costs of making the plan?
- *Internal validity (or quality).* Did the plan fulfill the logic of how it was intended to work? . . .
- *External validity (or quality).* Did the outcomes intended or implied in the plan meet external criteria, such as claims for a just society? For example, if it was intended to work as a vision, did the vision include equity? Ethical acceptability is a crucial component of external validity.

[. . .]

Did the plan have any effect on decision making, actions, or outcomes? Plans work by affecting actions, indirectly if not directly. Whether the actions taken yield intended outcomes is a distinct but important question. Good plans must not merely be more likely to affect actions. They must also be more likely to include actions that will yield intended outcomes. Does Chicago "look like" the Chicago Plan? Were the aspirations achieved based on some set of indicators? Does Cleveland "look like" the Cleveland Policy Report in the sense that indicators show an increase in choices for people who are least well off? Do the new towns of Reston, Virginia, or Milton Keynes in England look like the plans for them? To use this basis, we not only need to be able to measure outcomes pertinent to the plan, but also to provide an explanation of how the plan caused these outcomes. One difficulty is uncertainty in the relationship of actions to outcomes. Even if planned actions are taken, the intended outcomes may not occur. Even good choices in locating land uses relative to flooding or other natural hazards may yield larger losses over a given period than before the plan because of a particularly large flood or a cluster of hazard events. A plan that is based on the belief that people will use transit if they live in a transit-friendly environment may be used in decisions and affect actions but still not gain the outcomes that the plan sought because the belief about how the world works was wrong.

[. . .]

Each of the ways in which plans work implies an explanation of how a plan affects the world and thus an assessment based on that particular explanation. The measure of effectiveness for an agenda is whether the tasks were accomplished. We may also be able to observe whether actors or citizens, to sustain the implied commitment to the list, referred to the agenda as a reminder. Such observations would be evidence that the actions occurred because of the plan and because it served as an external memory device.

For policies, there are distinct measures of success for its distinct purposes. For decision efficiency, the measure of effect is whether decisions were made by reference to the policy rather than by considering the next decision situation from scratch. Reference by decision makers to the policy may be observable. Or, the policy may become habit and therefore not be directly observable, even though conformance with policy can still be observed. For decision fairness or consistency, the measure of effectiveness is whether the policy was applied accurately in similar situations. This can be determined by assessing a sample of situations in which the policy should have been applied.

Observing beliefs of the plan's target audience before and after the plan and asking whether beliefs changed can assess the vision mechanism. Beliefs might be elicited directly or inferred or revealed in actions. To determine whether these changes in beliefs also changed actions as intended would require observations of actions. Without observation of changes in beliefs or inference of such changes, however, we could not tell whether a plan was working as a vision.

For designs, the measure of success is whether the design is constructed or achieved. This measure of conformance has been used in several plan effectiveness assessments. It is generally not linked to a particular mechanism of how plans work, but rather a general notion of linking the plan directly to the outcome. Note that because the design mechanism is directly associated with the outcome, there is no intervening measure. The presumption is that we can recognize the outcome as resulting from the design because the design is sufficiently distinct that the outcome would not otherwise have occurred by chance. If design is not the mechanism by which a plan is expected to work, however, then conformance alone is not a sufficient measure of effect.

For strategies, the measure of success is whether the contingent strategy was pursued. Use of the strategy may or may not result in the most likely outcome being achieved. So for this explanation, the conformance measure is not directly pertinent.

[. . .]

We can determine whether a plan worked by linking three observable phenomena:

- Was the plan used? Or, a plan is good because persons use it in choosing actions.
- Were the actions taken? Or, a plan is good because the actions implied by the plan were taken.
- Were the outcomes achieved? Or, a plan is good because the outcomes sought by the plan were achieved.

The combination of these three types of observations can yield a persuasive argument that a plan affected decision making, actions, and outcomes in turn. They can test an explanation of how plans work and thus provide generalizable implications for other similar circumstances. Whether the outcomes were and still are valued and ethical is a question of external validity discussed below.

Was the plan worth making and to whom? Even if a plan is shown to have effects on decision making, actions, or outcomes, it may not be worth the cost of making the plan. There is so little empirical evidence of the effects of plans that it seems unnecessary to consider whether the effects compensate the costs. The question must be acknowledged, however, and effects should be identified in ways that might allow comparison to costs. Measuring the costs of making plans is conceptually straightforward. Measuring the benefits from the effects of plans, which might be negative, is a minefield of difficulties.

[. . .]

To value the effects of a plan requires considering all the ways in which the plan might work, distinguishing effects of the plan from what would have happened anyway, and estimating the value of these benefits. Uncertainty confounds these aspects further. . . . Clearly the value of the benefits is different across individuals and groups and raises all the problems of assessing changes in social welfare if we take a collective perspective. Even from the perspective of a particular institution, such as a sanitary district planning for sewers, the estimate of benefits is problematic. In practice, the most practical way to ask whether a plan was worth making is to estimate the costs of making it and then ask whether it is in rough terms plausible that the benefits could justify these costs. That is, it is unnecessary to estimate benefits any more precisely than whether they are greater or less than the costs. . . .

Was the plan internally consistent with the logic of how plans work? Internal validity depends on attributes of the plan itself. The internal validity of a plan can be determined by looking only at the plan. As with any decision in the face of uncertainty, the question is whether a good plan was made given the information available when it was made, not whether the outcomes that resulted were good outcomes. The typical approach is to ask whether a plan contains a certain set of components, such as transportation and land use, or has a particular set of attributes, such as being organized for reference by decision makers. A more careful interpretation would ask whether a plan fulfills at least one of the logics of how plans work. For the strategy aspect of a plan: Are the actions linked together in contingent strategies that meet the logic of decision analysis? Or for the design aspect: Are the elements combined into a designed target configuration that works, in which the interdependent elements should function as intended?

[. . .]

Did the plan seek outcomes that are ethically appropriate through means that are ethically appropriate? A plan that seeks to achieve equity for the least well off is a better plan than one that seeks to increase the efficiency of urban development in such a way that the efficiency gains accrue only to the most well off. Without elaborating ethical claims here, it is clear that a plan can affect decision making, actions, and outcomes, yield benefits sufficient to compensate its costs, be internally consistent in its logic, but still be a bad plan because of the goals it pursues or the means it employs. External validity calls a plan to the standards of ethics.

[. . .]

If we can explain situations in which plans are likely to be made and likely to work based on the first three broad criteria – effect, net benefit, and internal validity – then we can prescribe situations in which planners who measure success as plans affecting actions, and at costs that are compensated, should make plans. Such plans and planning are likely, however, to achieve what is easy and normal, not to accomplish unusual changes such as improvements in social equity. All four criteria – including external validity – are thus pertinent to evaluations of plans.

SUMMARY: PLANS WORK IN PARTICULAR SITUATIONS

Plans can work in more than one way. Given explanations of how plans work – explanations that link observable phenomena – it is possible to assess to what extent plans work in particular situations with respect to their effects, their net benefits, their internal validity, and their external validity. These explanations can also be used to predict that plans that meet these evaluation criteria will, in general, work in these ways in appropriate situations. They thus provide a basis for predicting what plans will be worth making. . . .

25

The Land Planning Arena

From *Urban Land Use Planning* (1995)

Edward J. Kaiser, David R. Godschalk and F. Stuart Chapin Jr

Editor's introduction

Urban planning is a highly politicized activity for the simple reason that it deals with land, and land has value – monetary and cultural. Different groups support one type of land use over another. A property owner has his view, an elected official hers – and those views may or may not be in agreement. Municipal government adjudicates the disposition of land use, following orderly procedures under local law. But politics often color the proceedings, and urban planners play multiple roles in the process. For discussion about their various roles, see Paul Davidoff's "Advocacy and Pluralism in Planning," Barbara Faga's "The Future of Public Participation," and William H. Lucy and David L. Phillips's "Sprawl and the Tyranny of Easy Development Decisions" in this *Reader.*

To understand the range of people who could be involved in a land use disposition question, consider a hypothetical case. Let's say a property owner proposes an apartment complex – four twenty-five-story buildings, a small park for tenants' use, a waterfront promenade, and a few thousand square feet of retail – for a five-acre former industrial site that he owns. Not only is the site a brownfield, it contains a vacant building eligible for landmarking, and has a mapped street running through it. One edge of the site is waterfront with two decaying piers. In addition, the site is adjacent to a low-income immigrant neighborhood, and the residents of this neighborhood fear the proposed project will trigger gentrification. They want the city to buy the land, turn it into a waterfront park, and restore the piers for fishing because there's no recreation space in the area. Conflict arises when the property owner seeks an amendment to the zoning law. This is when the fun begins.

Several urban planners are drawn into the situation. They represent the city, the neighborhood residents, and the property owner. Some work for the city's planning department, which oversees the approval processes. Serving as advisors to the city planning commission, the city council, and/or the mayor, these planners evaluate the project, judging its fit with the city's master plan or economic development goals. They also act as intermediaries for the various stakeholders, funneling residents' concerns to the decision makers, counseling the project developer about city requirements, and helping bridge communications with municipal departments that have their own interests in the case. The neighborhood planner not only articulates residents' opposition, but also develops an alternative plan that includes affordable housing and a public park. The planner working for the property owner helps with the site design and participates in all aspects of the approval process.

A number of key players have something to say about the proposal and planners come into contact with all of them. These participants include: elected officials; community representatives; environmental, preservation, and affordable-housing activists; federal and state officials, including the U.S. Army Corps of Engineers; the heads of city agencies such as Parks, Landmarks, Transportation, Fire, Police, and Environmental Protection. Each player has an interest in the project – for example, the Landmarks agency studies the abandoned building to see if it should be designated as a heritage site, while the Army Corps of Engineers oversees anything happening on the waterfront. Not surprisingly, the media (the local newspaper and television station) have become interested in the case. And public interest lawyers are monitoring the proceedings, ready to jump in if asked. Due to the high stakes of the proposal and the complications of the site, the situation has become quite politicized.

Sorting out their roles in a case such as this one challenges planners to be savvy and alert to the politics of the situation. At one time, planners considered themselves neutral, apolitical technicians dispensing advice to elected officials. But today they operate in a complex environment in which they have to identify and understand the motivations and behavior of the many people or stakeholders with whom they are interacting.

Political scientists, sociologists, and economists provide useful insights about the political arena that planners work in. Their theories on urban politics, the exercise of power, and the definition of public interest (or, in today's parlance, stakeholder identification) reflect different views of the nature and concerns of public decision making. Nonetheless all provide a helpful foundation that adds to today's planners' sophistication and understanding of the functioning of public decision making.

Pluralism reigns, according to political scientist Robert A. Dahl's *Who Governs? Democracy and Power in an American City* (New Haven, CT: Yale University Press, 1961), his classic study of power relationships in cities. Many people, not a small elite group, direct city policy, especially in urban redevelopment. Some make things happen, while others have only enough influence to resist, or cause modifications in decisions. This basic teaching still guides thinking about cities, but later scholars have detailed the dimensions of pluralism.

For example, in *The Contested City* (Princeton, NJ: Princeton University Press, 1983), political scientist John H. Mollenkopf identified a class of "political entrepreneurs" who align diverse, sometimes conflicting, groups to form powerful pro-growth coalitions who call all the shots in urban development and related matters. Sociologists John R. Logan and Harvey L. Molotch called the pro-growth coalitions "growth machines"; in their book *Urban Fortunes: The Political Economy of Place* (Berkeley, CA: University of California Press, 1987) they showed that these growth machines are composed of public and private sector leaders who exclude community-based interests from decision making. Government professor Clarence N. Stone went even further, alleging in *Regime Politics: Governing Atlanta, 1946–1988* (Lawrence, KS: University Press of Kansas, 1989) that the growth machine is actually a longlasting and tightly controlled regime, one that's anathema to plural democracy. Marxist theories that view urban politics (and related land use fights) as battles between capital and labor held sway in the 1980s and were articulated by sociologist Manuel Castells in *The City and the Grassroots: A Cross-cultural Theory of Urban Social Movements* (Berkeley, CA: University of California Press, 1983) and geographer David Harvey in *The Urbanization of Capital* (Baltimore, MD: Johns Hopkins University Press, 1985). Ten years later, when John Mollenkopf revisited the workings of urban power for *A Phoenix in the Ashes: The Rise and Fall of the Koch Coalition in New York City Politics* (Princeton, NJ: Princeton University Press, 1994), he concluded that the public and private sectors and citizen groups were not monolithic, but rather complicated and diverse. He observed that government agencies competed with each other, the private sector was variegated by size and type of industry, and racial and ethnic differences divided voters. This assessment stands today.

Through the next decade, researchers switched gears to explore how larger, sometimes global, themes play out locally, especially in public decision making revolving around planning. They pinpointed globalization (how competing in the worldwide economy shapes a city's choices to develop one kind of land use or another), immigration (how a city responds to the rival demands of an increasingly fractured population – the black/white dichotomy has become a "rainbow," increasing the number and complexity of views on land use questions), metropolitan governance (how cities and suburbs cooperate on land use and development issues), and sustainability (how cities can make land use decisions that not only satisfy the concerns listed above but also respond to climate change and the high cost of energy). (Dennis R. Judd and Todd Swanstrom, *City Politics: The Political Economy of Urban America*, sixth edition, New York: Pearson Longman, 2008, and Paul P. Kantor and Dennis R. Judd, *American Urban Politics in a Global Age: The Reader*, New York: Pearson Longman, 2008.)

Regardless of the complicated political nature of the arenas in which planners operate, they are still expected to serve the public interest. Their code of ethics demands it: "our primary obligation is to serve the public interest," and to "seek social justice . . . for all persons, [including] the disadvantaged" (American Institute of Certified Planners Code of Ethics and Professional Conduct adopted March 19, 2005). (See Carol Barrett, "Introduction," in this *Reader.*) While articulating this requirement, the code does not explicitly define the public interest, leaving planners wide discretion. But in land use decision making, discerning the public interest is difficult. Is it the sum of individual interests (as suggested in the classic study *Politics, Planning and the Public Interest: The Case of Public Housing in Chicago*, Glencoe, IL: Free Press, 1955, by Martin Meyerson and Edward C. Banfield)? Or does it equate with a city's deploying its land for the greatest gain (as Paul E. Peterson asserts in *City Limits*, Chicago: University of Chicago Press, 1981)? Is it planning a place that puts certain values like equity at the forefront (Norman Krumholz and Pierre Clavel, *Reinventing Cities: Equity Planners tell their Stories*, Philadelphia: Temple University Press, 1994, and Norman Krumholz and John Forester, *Making Equity Planning Work: Leadership in the Public Sector*, Philadelphia: Temple University Press, 1990)? To see how far apart definitions/views of the public interest can be, consider the U.S. Supreme Court case *Kelo v. City of New London*, 545 U.S. 469 (2005) that pit New London's economic development interests against property rights advocates – go to www.planning.org/amicusbriefs/pdf/stevens_kelo.pdf.

This selection – by Edward J. Kaiser, David R. Godschalk, and F. Stuart Chapin, Jr – comes from the fourth edition of the classic *Urban Land Use Planning* (Champaign-Urbana, IL: University of Illinois Press, 1995). It likens the decision-making process to a game, outlines the players, and describes the roles of public planners in it. Though sharply focused on land use planning, it applies to other planning endeavors that require public decision making. In reading it, think about the broader context of urban politics as outlined in this introduction. Try to identify the reasons why various people are involved in the process the authors describe.

Kaiser, Godschalk, and Chapin are professors emeriti, Department of City and Regional Planning, University of North Carolina, Chapel Hill. Over time, they have been involved, singly or jointly, as authors of the five editions of *Urban Land Use Planning.*

To catch up on the key works in contemporary political science, see Elizabeth Strom and John H. Mollenkopf (eds), *The Urban Politics Reader* (London and New York: Routledge, 2007). For more about urban planners and urban politics, read John Forester's pioneering *Planning in the Face of Power* (Berkeley, CA: University of California Press, 1988) and *The Deliberative Practitioner: Encouraging Participatory Planning Process* (Cambridge, MA: MIT Press, 1999), as well as: Michael P. Brooks, *Planning Theory for Practitioners* (Chicago: APA Planners Press, 2002) and Patsy Healey's second edition of *Collaborative Planning: Shaping Places in Fragmented Societies* (Basingstoke: Palgrave Macmillan, 2006). See also Lewis D. Hopkins and Marisa A. Zapata (eds), *Engaging the Future: Forecasts, Scenarios, Plans, and Projects* (Cambridge, MA: Lincoln Institute of Land Policy, 2007).

Local land use planning and decision making can be seen as a big-stakes game of serious multiparty competition over an area's future land use pattern. To win the game is to gain adoption and implementation of the future land use plan, development regulations, and development decisions that most benefit your role or organization. Land planners play the land use game in an arena with other players, each with resources and influence over decisions. However, planners also act as game managers, drafting and enforcing the rules of the game and advocating community cooperation to achieve multiparty benefits. Mindful of the inherent tensions between the player and manager demands of their roles, they must walk a tightrope between advocacy and neutrality. Effective land planners carefully watch and respond to the interests, actions, and alliances of the other players. Not to understand the game at every stage is to risk losing the planners' credibility and authority, and the broader public's stake in the community's future.

LAND PLANNING AS A SERIOUS GAME

The land-planning arena can be confusing and frustrating to the novice planner. Rather than an orderly and rational procedure of adopting land plans and policies based upon systematic technical studies aimed at the overall public interest, it often appears to be an ad hoc, political process based on influence and narrow interest group bias responding to the issue of the moment. The theories and statistical analyses taught in planning school may carry less weight with elected officials than the self-serving demands of a mob of angry speakers at a public hearing. The long-range projections of plans may fail as guides to decision making due to unforeseen changes in economic or social conditions. How can there be an art and science of land use planning in the face of such a politicized decision-making process and such an unpredictable future?

One premise . . . is that land use planning and decision making resemble a high-stakes *competition* over an area's future land use pattern. However, the players are locked together in a framework of interdependence in which they must gain agreement from other players in order to achieve their goals. Thus, the process's competition is tempered with the need for *cooperation* as well. Characterizing planning as a serious game helps to visualize the dynamics of the process and to see how to apply land use planning techniques to improve overall game outcomes. Understanding the nature of the land use game is the first step toward developing an art and science of land use planning and development management.

A second premise is that the land use plan is a key tool in coordinating community development activities. We disagree with those who hold that land use planning is *only* a process. We see it as a *process guided by a plan.* The plan fulfills many needs. It helps to turn competitors into collaborators through involvement in its preparation. It records a series of agreements among the players about ways to deal with their different objectives, serving as a community dispute resolution mechanism. It ensures that public interest goals are not overlooked in the rush to realize narrower aims, preventing the "tragedy of the commons" in which a valued community resource is destroyed by unbridled self-interest. It creates joint gains shared by various interest groups as well as the community as a whole. It acts as a platform for the application of analyses and technical studies. It lays out a vision of the area's future and a strategy for achieving it. As game conditions change, the plan is regularly revised in order to maintain currency and consensus. Just as the most successful athletic teams are those with effective game plans, so the most successful communities are those with effective land use plans.

In the land use game, planners are not only players, they are also game managers, drafting and enforcing the rules of the game and advocating cooperation among the players in order to achieve communitywide goals. Because of their responsibility for recommending and administering not only plans and regulations but also public participation processes for plan and policy review, planners have a unique position at the center of the land use game. They have inside information and privileged access to the other players. Land planners are expected to keep careful track of all stakeholders' interests, actions, and alliances. To lose track of the game status is to risk losing planners' credibility as experts, their authority as land use change managers, and their opportunities to facilitate cooperation among competing interests in building a better community.

In addition to planners, the major types of institutional stakeholders attempting to influence the direction of future urban growth and change are development market players, government officials, and advocates of various interests. Market-oriented players include private sector landowners, developers, builders, realtors, and others who seek to profit from land use change. Government players include public sector elected and appointed officials at the federal, state, regional, and local levels, who frame laws and make governmental land use decisions aimed at the overall public interest. Interest group players include representatives of special interests – neighborhood preservation, environmental conservation, economic development, farming, minority groups, and others – who view land use through the lens of their group's particular values. Land planners include those in government concerned with both current and future land use issues. These stakeholders compete over the content and procedures of land use regulations, plans, and development decisions. Sometimes they work together and sometimes they oppose each other in an ecology of games in which coalitions are framed by the issue at hand. The planner must understand the goals and interests of these institutional actors in order to be an effective player and manager of the land use game.

In theory, land use stakeholders are continuously in conflict, causing inherent tension for land planning. In practice, this conflict is moderated by the legal and governance systems – "the rules of the game." The rules turn conflict into regulated competition. Their constitutional provisions, laws, regulations, planning powers, and decision processes seek to protect overall public interests from the extremes of unregulated maximization of market, social, or environmental values. The planner must rely upon legal and governance systems to balance conflicts among values, to make hard choices about community priorities, and to ensure fairness in land use decisions. The planner is both a drafter and an enforcer of the game rules (in the form of development regulations and community objectives) but is not their final arbiter. That role is reserved for the elected officials of the community or the courts if the elected officials' decision is challenged. But the planner must understand the influences of legal and constitutional checks and balances upon the powers of land use plans to achieve community goals.

LAND PLANNING AND DEVELOPMENT MARKETS

The bottom line for land planning is its influence over the outcome of the land development process by which projects, neighborhoods, communities, and cities are built. A "good" land use plan with "good" implementation produces a "good" built environment. What is built, where it is built, and when and how it is built are all critical questions whose answers depend upon many actors, each with different definitions of what is good. The goal of the land planner is not simply to accommodate market demand for development, but to guide the market toward producing good communities.

In a democratic, capitalist society, business entrepreneurs assemble the land and capital to build the private sector portions of communities. They are the "engines" of community building, adding value to land through the improvements they place upon it. The most obvious measure of their winnings at the land use game is their profit from the sale of land and buildings. They gamble that they can anticipate market demand, producing developed urban space that the market will value at a higher level than its cost to them. They are risk-takers, willing to face financial losses and even bankruptcy in order to profit from their gambles. A more subtle measure of their success is their reputation as public-spirited "community builders" who create desirable living and working environments rather than simply market commodities.

Landowners, builders, and developers scrutinize land policies, regulations, and plans for their impacts on the monetary values of land. Landowners often include those whose land is presently in nonurban use, such as farms and forests, but who wish to preserve the option of conversion to urban use that would bring an increase in market value. Developers who make their livelihood from the conversion of land to more intense uses are also concerned with the effects of public actions on land prices. This group may be joined at times by other business interests affected by land policies, such as financial institutions, and by those who simply advocate the lessening of government intervention into the market as an ideological position.

Developers are constrained by both land planning and market demand. To succeed, their projects must pass both a government test and a market test. They must satisfy the intent of governmental plans and regulations adopted by the local elected representatives in order to get a development permit. They must satisfy the consumer's taste in order to sell and make a profit. They operate in a market of buyers and sellers that is influenced by public plans and service programs but not driven by them. For them, the driving forces are the growth of the population, the economy, and interest rates, which affect demand and capital availability.

[. . .]

Governments . . . take responsibility for seeing that the necessary community infrastructure – roads, water and sewer systems, schools, parks, and the like – gets produced. This infrastructure framework supports residential, commercial, and industrial development. And its influence over development's type, location, and timing is often more direct than that of land use plans or regulations alone. However, government infrastructure alone also does not drive development; that driving force comes from demand for the product of development – developed land in the form of building lots, houses, offices, factories, and stores.

Active land use planning affects the development market by identifying land that is available or planned for development; by limiting the type, location, timing, and density of development that can take place; by programming the infrastructure to support development and allocating its costs between the public and private sectors; and by specifying the standards under which development proposals will be reviewed. These actions define the supply of developable land and what can be built upon it. They have been described as "managing the market." Although that description is too extreme for most cases, it is clear that the active land planner is attempting to guide the process of structural change in accordance with community goals. In that sense, the land planner can be seen as both a "growth manager" and a "manager of change."

[. . .]

LAND PLANNING AND GOVERNMENTS

Government officials are major players in the politics of U.S. land use. Land planners must account not only for the policies of their individual governments, but also for the policies of governmental agencies at the federal and state levels and within multi-jurisdictional regions.

Federal government policy

The federal government sets national policy in response to perceptions of nationwide problems. This policy . . . is measured against the Constitution's broad framework and the specifics of legal principles and precedents. . . . Because of the complexity and competition within the federal establishment, federal policy affecting local land use is often fragmented and inconsistent.

The fragmentation of federal policy does not mean that it has nothing to say about land use decisions. . . . [F]ederal housing, community development, transportation and environmental programs have both direct and indirect land use policy implications [and] . . . provide the national perspective on land use. . . . Implementation of federal programs generally uses one or more of the following techniques:

1 Setting uniform national requirements . . . ;
2 Awarding grants and incentives . . . to meet federal objectives;
3 Withdrawing federal subsidies or assistance . . . ;
4 Regulating certain types of development through required permitting programs; or
5 Building certain types of public development projects.

Uniform national standards are typically used to implement environmental protection policies. For example, the Environmental Protection Agency (EPA) sets national standards for clean air, clean water, and safe disposal of solid and hazardous waste. Cities and states must meet these standards, often by means of land use regulations, or face penalties. . . .

Planning and management incentives are also used to implement environmental and land use policy. Federal grants are available for affordable housing subsidies, community development improvements, metropolitan transportation planning, and transit programs. Federal flood insurance is available to property owners in communities that implement floodplain regulations. . . .

Withdrawal of subsidies is another implementation method. For example, the Coastal Barrier Resources Act of 1982, administered by the Department of the Interior, identified coastal areas where further development should be discouraged. To reduce future federal losses from coastal storm damage on these hazardous coastal barriers, the program removes federal subsidies to development in the form of federal flood insurance and infrastructure funding.

Permitting of certain types of development involves federal agencies directly in land use decisions. For example, the Army Corps of Engineers issues permits for dredge and fill of waterways and for alteration of wetlands, requiring applicants to analyze and mitigate the impacts of their proposals upon these areas. . . .

Construction of public works also involves federal agencies in land use actions. The most prominent examples are the interstate highway system built by the U.S. Department of Transportation and the system of dams and flood control structures built by the Army Corps of Engineers. However, there are many other federal construction programs, ranging from military bases and installations to post offices and national parks.

STATE GOVERNMENT POLICY

State government policy-making processes rely primarily on state constitutions and state laws. Like the federal government, states also work within systems that balance the powers of the legislative, administrative, and judicial branches of government and often result in interagency competition. In terms of land policy, states fall roughly into two types: (1) growth-managing states, where fast growth and a high degree of development change have stressed local capabilities to the extent that new state growth management laws have been passed, and (2) growth-accommodating states, where growth and change have not been perceived as state problems.

Growth-managing states mandate actions by their local governments to manage development in accordance with state goals and objectives. During the 1960s and 1970s, five states enacted statewide growth management policies: Hawaii (1961), Vermont (1970), Florida (1972), Oregon (1973), and Colorado (1974). (Colorado did not continue to pursue growth management.) These were followed in the 1980s by five new state initiatives and one revised program: Florida (revisions 1984–86), New Jersey (1986), Maine (1988), Vermont (1988), Rhode Island (1988), and Georgia (1989). (Maine's program was cut back in 1992 for budgetary reasons.) The state of Washington joined this group with the enactment of growth management legislation in 1990 and 1991, along with Maryland in 1992. . . .

In addition to states with statewide growth management mandates, other states require local planning and growth management in specific regions or types of areas. Twenty-four have adopted coastal management programs under the Coastal Zone Management Act of 1972 in the form of state standards and procedures for local land use and environmental management and planning in designated coastal areas. Several have adopted mandates for protection of specific natural areas, including California and Nevada's Tahoe Regional Planning Agency (1970), New York's Adirondack Park Agency (1971), New Jersey's Pinelands Commission (1978), and Maryland's Chesapeake Bay Critical Area Commission (1984). Other states have adopted mandates for metropolitan area planning, including Minnesota's Metropolitan Council of the Twin Cities Area (1967), Georgia's Atlanta Regional Commission (1971), and Oregon's Portland Metropolitan Service District (1979).

[. . .]

Local planners in growth-managing states must plan by state rules, which in some cases put them in an adversarial position with the state planning agency. . . . [S]tate mandates may require local plans to:

1 Include certain elements – a content requirement;
2 Be consistent with state goals and policy – a consistency requirement;
3 Require infrastructure to be provided concurrently with development – a concurrency requirement;
4 Be coordinated with plans of neighboring governments – a coordination requirement;
5 Contain urban sprawl – a containment requirement;
6 Provide for affordable housing – an affordability requirement;
7 Broaden growth management to include promotion of economic development – a growth encouragement requirement; and
8 Protect natural resources – a conservation requirement.

The advantage of state guidance is that it can overrule local political decisions that do not meet statewide standards for sound land planning practice. . . . The disadvantage of state guidance is that it can reduce local initiative and creativity through the necessity of meeting uniform standards and timetables, resulting in local frustration with planning.

Growth-accommodating states follow the traditional role of providing a statutory base of planning and regulatory powers for those local governments that choose to exercise them. Enabling acts for planning, zoning, subdivision regulation, and health and building codes are enacted as permissive laws. The state government does not involve itself in local land use decisions, which are seen as the prerogative of the local governments. Either due to a lack of growth problems or to a political culture that stresses minimal government intervention into the market, the majority of states fall into the growth-accommodating camp.

Local planners in growth-accommodating states have considerable freedom to define planning content and processes. However, they lack the backbone of state-mandated principles and standards for acceptable plans, as well as the importance accorded to planning through state legislative attention. They also lack the potential coordinating power of required consistency between state, regional, and local plans, a particularly valuable power for planning infrastructure or environmental systems that cross jurisdictional boundaries.

Local government policy

Local governments are creatures of their states, deriving their power and authority from state legislatures. Standard state enabling Acts, adopted throughout the United States, make planning and development regulation legitimate functions of local governments. However, their ability to innovate beyond the standard powers is limited by the necessity to gain state legislative approval for major changes in their functions. Still, within the realm of the authority granted to them, they have considerable freedom to make decisions and considerable power to guide development. . . .

Growth-managing localities seek to influence the location, amount, type, timing, quality, and/or cost of development in accordance with public goals. These cities and counties place community and environmental goals on par with goals of the development market. They allocate land for development by means of a public calculus based not only on development demand but also on environmental protection, fiscal efficiency, neighborhood compatibility, limitation of sprawl, and aesthetic appeal. They carry out their land use plans with a wide range of tools – impact fees, urban limit lines, design guidelines, development agreements, capital improvements programs, and adequate public facility ordinances – as well as traditional zoning and subdivision controls.

Growth-accommodating localities seek to ensure a steady, unrestricted supply of land for development so that the market is not constrained. These areas limit the amount of regulation to that necessary for minimum health and safety protection. They allocate land for development by means of a marketability calculus based on ensuring a wide selection of sites sufficient to meet projected development demand with a generous margin for choice. Many carry out their land use plans with minimum zoning and subdivision regulations and also liberal terms for infrastructure provision.

Conflicts often occur between local governments whose growth policies are not compatible. These may be between adjacent local governments whose boundaries are coming together and who get into disputes over annexation or who differ on the growth management versus accommodation issue. They may be between cities and their parent county over differing urban and rural perspectives toward growth management. They may be between upstream and downstream areas over protecting water quality and maintaining stream flows in shared water bodies. Increasingly, local governments with overlapping jurisdictions, shared boundaries, or interests in area-wide resources are coordinating their development plans and policies through techniques ranging from informal staff consultation to multi-jurisdictional forums to formal intergovernmental agreements.

POLICY DYNAMICS

Elected and appointed policy makers at all three levels of government seek to influence land use in different ways, depending upon the problem of concern. For example, the federal government may pass a coastal zone management law offering grants to states to attack the problem of uncoordinated land use in coastal areas. Coastal states may respond by passing new state coastal management laws that establish procedures for local plans and for state review of certain land use proposals in critical coastal areas. Local governments in coastal areas must then incorporate the state and federal policies into their local land use plans. . . .

[. . .]

For the local planner, one of the most critical policy dynamics is the election of new local government representatives. Because of the power of local elected officials over the planning and development management processes, the outcomes of the elections that occur every two or four years can spell defeat or victory for a local planning effort. The knowledge, experience, and attitudes of elected officials about planning issues vary radically. Although the literature typically characterizes decision-making groups as homogeneous, in practice the members of elected bodies often disagree among themselves over planning proposals. Planners often find themselves acting as educators for elected officials who are unaware of the potential impacts of their planning decisions.

LAND PLANNING AND SPECIAL INTERESTS

In a democracy, land planning by law must afford opportunities for public participation in governmental decision making. Because land values – market, social, and environmental – are so pervasive, a variety of interests or publics have broad interest in land use decisions. As the impacts of both private and public development projects grow larger and more pervasive, more public constituencies are motivated to participate in the review process.

Market value advocates have already been discussed in terms of their advocacy for specific projects where they hold a financial interest and for community plans and regulations that favor market values. A broader type of public is also concerned with market values. This group may bring together businesses, chambers of commerce, economic development organizations, and citizens concerned with the monetary value of their land investments. Depending upon the issue, the group may include some strange bedfellows, as in the case of developers and low-income individuals joining together to oppose large lot exclusionary zoning.

Neighborhood residents typically advocate the preservation and enhancement of the social use values of communities. They scrutinize land policies and plans for impacts on their quality of life while also keeping an eye on impacts on the market value of their property. In the absence of an informed community consensus about future growth, these groups may mobilize to block or modify development. Neighborhood groups sometimes include those who seek to prevent any new development, or at least to prevent adjacent development at densities higher than theirs. . . .

Environmental groups range from those who seek to protect critical environmental areas to those who would stop growth altogether. Often these groups are local chapters of such national environmental advocacy groups as the Sierra Club. They view land policies and plans through an ecological lens, seeking the maximum preservation of existing natural environmental features such as wetlands, streams, and forests. They may form coalitions with neighborhood groups opposed to growth and even with business groups opposed to specific projects seen as antithetical to existing businesses.

[. . .]

Local land planners must work with many publics. To be acceptable and effective, land use plans must recognize and reconcile the pluralistic interests of these publics with those expressed by governments and markets. It is a major challenge to design a fair and orderly forum for expressing and reconciling the interests of those affected by land use decisions. Managing this complex public participation process makes land planning as much an art as a science, especially when planning concerns change so often.

[. . .]

26

Shaping Cities through Development Regulations

Redesigning Cities: Principles, Practice, Implementation (2003)

Jonathan Barnett

Editor's introduction

Urban planners, landscape architects, architects, historic preservationists, and even public artists intersect in an area called urban design when these professionals work with the form and structure of the physical environment. They operate on multiple levels, helping arrange blocks with buildings and their surrounding interstitial spaces, form the larger districts that make up a city, and forge regions shaped by open space, transportation and settlements, or centers. Other disciplines – including geology, economics, political science, sociology, and law – inform their work. But in the end, the planners, architects, and preservationists draw, map, and plan places, giving spatial identity to the ideas, principles, and laws that provide their foundation. The images they produce become the basis for discussion in citizen participation exercises and, ultimately, public and private development decisions.

These collaborative teams operate within identifiable constraints. While excellent aesthetics, however defined, is always the desired goal, many considerations have bearing on the outcomes, taking on varied importance in the judgment of the designers, their clients, and the law. Primary are the physical conditions of the sites in question, regardless of scale. The topography, the presence of water, wetlands, or seismic weakness, preexisting built conditions, and climate all are important drivers of design. Economic conditions also shape urban design. The costs of land and proposed improvements and the valorization of improvements are crucial. Culture and social conditions are also factors. Capturing the sense of place (or making a new one) in accord with the users' desires and needs is part of the design process. Finally, land use laws, especially development regulations, have an enormous effect on urban design solutions. For example, today's zoning ordinances, by designating the types and intensity of land uses, establish the character and appearance of places. Environmental regulations also affect location decisions, prohibiting the construction of buildings and roads in vulnerable places and shaping conservation lands. (Flood plains are a key variable here.) Subdivision regulations outline the layouts of new construction, while historic designations protect older built-up areas.

Of these laws, zoning is the most powerful. Ubiquitous and well enforced, it exists in all major U.S. cities except Houston. A zoning ordinance can be dynamic and malleable. Because zoning is a local law, a jurisdiction has the power to change or adjust it. Most municipalities make incremental changes – tweaking a provision or adding a new use – because overhauling an ordinance is a costly, time-consuming process. Nonetheless since the early 1990s, nineteen of the nation's fifty most populous cities (including Chicago, Denver, San José, and Boston) have modernized their zoning, and several others (including Philadelphia and Miami) are currently engaged in this process.

Zoning dates from the early twentieth century. (See Plate 1.6.) Its most widely adopted form, Euclidean zoning – named for the U.S. Supreme Court case that upheld it – focuses on the type and intensity of land uses and separates incompatible uses. (See Selection 11, Sullivan and Michel, "*Ramapo* Plus Thirty: The Changing Role of the Plan in Land Use Regulation.") A Euclidean zoning ordinance has two parts: a text that outlines the types of zones (land uses and their specifications) and a map that shows the location of the various types of zones (districts). Euclidean zoning governs each parcel in a district. Each designation has ten attributes, all of which can be adjusted or rewritten to meet urban design goals. They are: land use, lot size, setbacks, lot coverage, building size (floor area ratio), building height, density, parking, landscape, and, sometimes, signage. This set of attributes creates an envelope for each lot in which development can take place. For example, in a low-density residential zone, the ordinance would specify the use (e.g. single-family unit), setback dimensions (a house has to be a certain number of feet from the street and from the neighbors), the floor area ratio, or FAR (a calculation of the amount of allowable construction in relation to the lot size). It also outlines the amount of required parking and usually prohibits commercial signs. If a municipality wishes to encourage concentrated development in this area, all it would have to do is increase the number of allowable units, reduce the setbacks, and increase the FAR.

Over time, city planners and urban designers have created new forms of zoning: performance, incentive, and form-based. Performance zoning – invented in the 1970s by Bucks County, Pennsylvania, city planner Lane Kendig – organizes land uses according to their ability to meet clearly stated community goals, and ranks development proposals by their conformity to the goals. It aims to add flexibility to land use systems. Kendig outlined its principles in his book, *Performance Zoning* (Chicago: APA Planners Press, 1980). Since the book's publication, many cities have adopted performance zoning, often overlaying its provisions on pre-existing Euclidean codes for the purposes of natural resource protection. In Bucks County municipalities employ it in special development districts where they measure three criteria: open space, maximum density, and paved surface. For example, the Bucks County law categorizes maximum open space by natural resource type (e.g. floodplain has 100 percent open space, while a plot with a 25 percent slope calls for 80 percent open space). Laramie, Wyoming, uses a similar formula to protect its major water source, the Casper Aquifer District. The district has underlying five-acre zoning (that is, every development parcel must be at least five acres in size) supplemented with a performance-zoning overlay that encourages cluster development, open space, and paved surface. Fort Collins, Colorado, has developed a more elaborate system, the Land Development Guidance System (LDGS), also an overlay. Here a project is acceptable if it achieves a certain number of points based on meeting standards for neighborhood compatibility, public facilities and safety, resource protection, and specific design features.

Another type of performance zoning is the Planned Unit Development (PUD). This designation allows for the arrangement of a large, multi-acre site to meet standards as a whole even though individual lots may not be in compliance. A PUD provision might call for 30 percent open space but allow it to be met in a park, thus permitting clustered housing within the site. Master-planned and New Urbanist communities often rely on PUD designations.

Some cities have used incentive zoning to achieve urban design goals, offering developers bonuses, such as additional square footage, in exchange for the provision of a given feature. Invented in New York City in the 1960s, such bonuses are now widespread. The most dramatic examples are found in downtown skyscraper districts in hot real estate markets. Transit stations, public gardens, plazas, atriums or terraces, and retail space are typical features that a city might acquire through incentive zoning. Cities have also used incentive zoning to meet social goals such as affordable housing and childcare.

The results of such zoning are mixed, as shown by Harvard city planning professor Jerold Kayden in *Privately Owned Public Space: The New York City Experience* (Hoboken, NJ: Wiley, 2000). In a survey of 503 plazas and arcades comprising 3.2 million sq. ft of "bonused" public space, Kayden found designs of varying quality and major differentials in maintenance and the honoring of scheduled hours of public use. His study had been preceded by William H. Whyte's *The Social Life of Small Urban Spaces* (New York: Project for Public Spaces, 1980), which had resulted in major changes to the design specifications for New York's public plazas. Whyte also produced a documentary film of the same name, now a classic, that illustrates good and bad design brought about by this type of incentive zoning, highlighting elements that determine successful public space.

The newest zoning reform is the form-based code, which has emanated from New Urbanist thinking. (See Plate I.13.) Originated by Andrés Duany and Elizabeth Plater-Zyberk in the 1990s, it regulates the form, not the use, of development because its proponents view the original reasons for zoning, separation of noxious uses like housing and industry, an outdated idea. More important, they argue, is to use zoning to foster walking, neighborly visiting, and round-the-clock activity.

Form-based zoning codes usually have five components. They are: (1) the regulating plan that outlines the various districts and streets; (2) rules that outline the building types; (3) regulations for the dimensions and characteristics of streets and sidewalks; (4) landscape regulations that specify the species, size, and location of trees and other plantings; and (5) architectural regulations that designate design features, including walls, windows, fences, and roofs. Duany and Plater-Zyberk have developed a classification scheme, or taxonomy, that conceives of development along a six-sector continuum, or transect, ranging from the natural to the urban core, with each sector (numbered T1 to T6) having defined characteristics built into the form-based code. For example, a general urban zone (T4) has medium-sized blocks, is mixed-use but primarily residential with a wide range of building types. Although the code establishes common minimum standards per sector, its architectural and landscape guidelines allow local variation.

The adoption of form-based zoning is taking place slowly in the United States. Partial adoption has occurred in several towns in California (e.g. Petaluma, Sonoma, and Palo Alto), and also in Arlington County, Virginia, and Kendall, Florida. Miami, Florida, is in the midst of adopting a citywide form-based code as part of its "Miami 21" campaign, an overhaul of its land use and planning operations undertaken by Mayor Manny Diaz. Only California has passed special enabling legislation to authorize form-based codes, but Pennsylvania, Connecticut, and Wisconsin have some provisions in state law that accommodate their use.

Planners have packaged many of these newer forms of zoning, as well as other development regulations, into Smart Growth strategies. The goal: managing growth to harness sprawl. Many advocacy and professional organizations, along with federal, state, and local governments, are supporting and publicizing these tools in campaigns to conserve open space, transform contemporary development patterns, and promote sustainable development.

Jonathan Barnett, the author of this selection, is one of the few urban planners who appreciate the power of the regulatory process to shape the appearance and contents of urban places. When he first displayed his expertise in this area in *Urban Design as Public Policy* (New York: McGraw-Hill, 1974), he was fresh from serving as a member of New York City's trend-setting Urban Design Group, a special team created in the Department of City Planning during the mayoralty of John V. Lindsay (1966–74). Its charge was to institute innovations in the physical fabric of the city. Here he worked with twenty young professionals on several projects, including incentive zoning, creating special districts to preserve areas otherwise not attended to by existing zoning, developing design guidelines for streets and other areas of the public realm, and undertaking evaluative studies of crucial commercial corridors (Forty-second Street) or infrastructure investments (Second Avenue Subway).

Barnett went on to have an illustrious career in city planning and urban design, participating in the development of New Urbanism and authoring several books, many further exploring the power of development regulations as applied to contemporary growth. They include *The Fractured Metropolis: Improving the New City, Restoring the Old City, Reshaping the Region* (New York: Westview Press, 1996) and *Redesigning Cities: Principles, Practices and Implementation* (Chicago: APA Planners Press, 2003), from which this selection is drawn. Barnett is Professor of Practice, Department of City and Regional Planning, University of Pennsylvania, and affiliated with WRT, the Philadelphia-based planning and design firm. His latest book is *Smart Growth in a Changing World* (Chicago: APA Planners Press, 2007) with F. Kaid Benfield, Paul Farmer, Robert Yaro, Armando Carbonell, and Shelley Poticha.

For more on development regulations and their urban design implications, see American Planning Association, *Planning and Urban Design Standards* (Hoboken, NJ: Wiley, 2006); Jon Lang, *Urban Design: A Typology of Procedures and Products* (Oxford: Elsevier Architectural Press, 2005); Eran Ben-Joseph, *The Code of the City: Standards and the Hidden Language of Place-making* (Cambridge, MA: MIT Press, 2005);

and Donald Watson, *Time-saver Standards for Urban Design* (New York: McGraw-Hill, 2003). For a basic introduction to zoning, see Dwight Merriam, *The Complete Guide to Zoning* (New York: McGraw-Hill, 2005). For a discussion of form-based zoning, see Robert J. Sitkowski and Brian Ohm, "Form-based Land Development Regulations," *Urban Lawyer*, 28: 1 (2006), pp. 163–72, and Charles C. Bohl and Elizabeth Plater-Zyberk, "Building Community across the Rural-to-urban Transect," *Places Magazine*, 18: 1 (2006), pp. 4–17. The literature on Smart Growth is extensive; to read about the regulatory tools promoting its goals, see Stuart Meck, *Growing Smart Legislative Guidebook 3* (Chicago: APA Planners Press, 2002). To read actual codes, see the U.S. Department of Energy's Smart Growth Network site, land use section, at www.smartcommunities.ncat.org/landuse/lucodtoc.shtml.

Development regulations are among the most powerful forces shaping the built environment. Almost every city, county, and town in the United States has a code that separates manufacturing, commercial, and residential zones. These zoning codes also contain height and setback requirements that have a decisive effect on the placement and appearance of individual buildings. Subdivision codes govern the layout of streets and lots on greenfield sites. Many other laws affect development in specific locations. Landmark buildings and historic districts are protected. The environmental impact of large federally funded projects must be assessed before they go forward, and some states have comparable requirements. Other federal laws specify what can be built in floodplains and how to build there, and what develops in a coastal zone is subject to review. More indirect, but still decisive requirements come from federal laws regulating air and water pollution, and from state growth management legislation.

What is being built today is very much the product of the limits and instructions written into codes. If we don't like what is happening, we need to adjust the regulations.

ZONING: SEPARATING USES, PROTECTING LIGHT AND AIR

Zoning originated in Germany and the Netherlands at the end of the nineteenth century as a way of keeping heavy industries that were intruding into cities away from historic or residential districts. Separating factories from upper-income neighborhoods was one purpose of the 1916 New York City zoning, an early and influential ordinance in the United States; but its framers also were interested in stopping tall buildings from blocking too much light and air from streets and neighboring properties. They turned for precedents to the laws governing the height and roof setbacks of buildings in Paris, which had been regulated since the eighteenth century. The Parisian system related the height of a building to the width of the street, with the roof sloping back above the height limit at a predetermined angle.

In New York City, the same system was applied, but at the larger scale of elevator buildings, with the first setback occurring on wide streets at about the eleventh floor, and then, instead of an attic, a series of additional setbacks under an imaginary "sky-exposure plane" set, like the Parisian roofline, at a predetermined angle from the street. To keep every building from turning into a pyramid, towers could go straight up once setbacks had reduced their area to no more than a quarter of the building lot. The design of the Empire State Building, a relatively slim tower on a big base, is closely determined by this ordinance.

Similar height and bulk controls were enacted in many other U.S. cities. Architects became interested in the designs that derived from applying these regulations. . . . Much of what we now think of as the Art Deco style of skyscraper design was influenced by the streetwall height limits and setbacks of zoning codes.

ZONING CODES AND MODERN DESIGN

The modern movement in architecture rejected these Art Deco shapes in favor of straight towers surrounded by open plazas. While it was possible to build sheer towers under the old zoning codes, the setback requirements limited the tower size. . . .

The owners gave up potential space on the lower floors in order to achieve the desired shape and still meet the code. A revision to the 1916 zoning ordinance . . . in 1961 in New York City [changed the approach]. Instead of relying on height and setback regulations to control building size, the newer code uses a floor area ratio. The ratio is perhaps better described as a multiplier. If the area of the building lot is 10,000 sq. ft, and the floor area ratio is 10, the permissible floor area is ten times the lot area, or 100,000 sq. ft.

Floor area ratios are intended to limit the number of people accommodated on a site, not the building's external shape. Unoccupied areas like mechanical equipment floors, basements, stair towers, and elevator shafts are generally exempted from floor area calculations. Parking garages often are exempted also, even if they greatly increase a building's size.

[. . .]

New York City's . . . ordinance [also] removed the 25 percent restriction on tower sizes as related to the lot, going first to 40 percent and then, by amendments for various circumstances, to almost full coverage. To encourage the appropriate modernist setting for the new sheer towers, the New York City code gave as much as a 20 percent bonus in floor area to owners who provided a ground-level plaza. . . .

THE EFFECT OF MODERNIST ZONING CODES ON CITIES

Most other cities have, like New York, switched from relying only on height and setback controls to setting absolute limits on building size with floor area ratios. Some emulated New York City in encouraging plazas. The tower and plaza combination was fashionable and was used in many situations that were not influenced by zoning requirements.

Enabling architects to go from decorating a basic shape, predetermined by zoning, to controlling the form and design of the whole office tower, hotel, or apartment house permitted individual creative decisions. Some of the resulting buildings were very good, others were not. In most cases the image of the individual building took precedence over the image of the city. The predictable, if often dull, relationships of urban buildings were lost. The skylines became more spectacular, but the street frontages could become fragmented, confusing and discouraging to pedestrians. In residential neighborhoods, modernist apartment towers, enabled by the new codes, intruded into established relationships of houses and older, lower-scale apartments.

. . . Modern zoning codes have been much more explicit about separating different types of industry, and segregating industry from other uses. They also designate different zones for different kinds of commercial activities, and create a large number of residential districts based on lot size and the number of housing units permitted.

[. . .]

MAKING CORRECTIONS TO MODERNIST ZONING

Only a few years after New York City's modernist zoning came into effect . . . , weaknesses in the code had appeared: Its expectations about land use were simplistic, and the new separate towers encouraged by the ordinance did not fit well into the existing city. By the late 1960s, New York was already enacting amendments to preserve theaters in the theater district and shopping on Fifth Avenue. Over time, the plaza regulations were modified to require not just square feet of open space, but open space with a favorable orientation equipped with seating, lighting, and landscaping.

The importance of continuity in street frontages, especially where there are ground-floor shops, became evident as soon as a few plazas had interrupted them, and New York City eventually came up with requirements to maintain the "street wall," that is, build up to the front property line for at least a large percentage of the building, on streets with a strong retail presence. Other zoning measures have limited the heights of new buildings in relation to the surrounding context.

[. . .]

SUBURBAN ZONING AND SUBDIVISION AS A CAUSE OF SPRAWL

Development regulation in the suburbs is just as influential as it is in cities. Much of the recipe for urban sprawl can be found in local zoning and subdivision regulations. . . . [T]he endless ribbons of commercial development along highways all follow zoning and so do the big tracts of suburban houses, each the same size on the same-sized lots. The subdivision ordinance, by setting a maximum grade for streets, is usually the reason for the drastic stripping and bulldozing of the landscape that so often happens when a suburban area is developed.

These problems come from two basic flaws in current suburban zoning and subdivision laws: They treat land as a commodity and not an ecosystem, and they protect the neighbors from negative influences rather than creating a positive template for the whole community.

[. . .]

[For example,] the subdivision ordinances that set the rules for laying out streets and dividing tracts of land into individual lots do recognize that land has contours, by setting a maximum gradient for streets. If all streets in a subdivision may not slope at a grade of more than 5 percent, and the site is divided up by a conventional street system, the developer is likely to bulldoze the high points of the landscape down into the low-lying areas so that the lots have the same general gradient as the streets, which requires removing all the trees and other vegetation. It also means stripping the topsoil and putting any streams that cross the site into culverts.

The alternatives in local zoning ordinances that permit cluster development are useful, but they have not, by themselves, been the expected cure for suburban development problems. The units of planning follow property lines, which seldom relate to natural boundaries like watersheds, so a site plan that appears to have some advantages in preserving the environment may not make sense in a larger context.

[. . .]

The biggest problem with typical regrading practice is the cumulative effect on the regional ecosystem. Trees and shrubs retain rainwater, when they are removed the flow of water across the landscape is accelerated. Land contours have reached an equilibrium over time; when they are disturbed and then subjected to powerful flows of water, erosion can take place with great rapidity.

An individual subdivision or planned unit development may have an internally consistent grading system, but what about its relationship with surrounding properties? Have they also been regraded? If not, there is likely to be an escarpment or a ditch at the property line, both unstable, erosion-prone conditions. If the neighboring property also has been regraded, the problems of accelerated runoff are likely to be multiplied. Many localities are discovering that floods of a magnitude that was once expected to occur every 100 or even 500 years are now a frequent occurrence.

PLANNED UNIT DEVELOPMENT

The problems created by applying conventional zoning and subdivision to any kind of terrain more complicated than a flat open field were recognized back in the 1960s. The cure was supposed to be an alternative procedure called Planned Unit Development, which most local communities have since incorporated into their zoning codes. As the name suggests, streets, lots, and buildings may be planned as a unit in ways that are otherwise not permitted; the site plan of streets and lots becomes the zoning and subdivision for the property. The procedure is sometimes called cluster zoning, as it is often used to cluster the permitted number of houses . . . into smaller lots on the most buildable portions of a site, leaving the most environmentally sensitive parts of the property undisturbed.

ENVIRONMENTAL ZONING AND SUBDIVISION

Why shouldn't there be a reduction in the permitted development, if the carrying capacity of the land doesn't support it? . . . If a 100 acre parcel has only seventy acres of buildable land, why shouldn't the calculation of permitted development be based on the seventy acres? Why create development rights on unbuildable land, and then transfer them to the land that can be developed, causing it to be built on at a higher density than is specified in the zoning?

These are good questions. In most zoning ordinances, the area of land under development is the basis for calculating how much building is permitted. [One proposal is] a simple amendment to local zoning that discounts the land area for calculation purposes, based on the land's sensitivity to environmental damage. Land under water would be discounted 100 percent, hillsides above a certain steepness 85 percent, lesser slopes a lesser percentage, and so on. . . .

This environmental zoning procedure clearly protects public safety, health, and welfare by reducing erosion and flooding. It is elegantly simple. It is based on objective considerations. It applies uniformly to everyone. In other words it should meet the constitutional tests for zoning. . . .

[Similarly], the subdivision ordinance can specify that there should be no major changes to natural drainageways and steep hillsides, and require the developer to show how buildings will be kept away from sinkholes, sites of previous landslides, or any floodplains and wetlands. There also can be restrictions for places with easily erodible soils, or land formations susceptible to erosion.

In addition, there can be water retention requirements that say that water should leave the property no faster after development than it did before. If this goal cannot be attained by conserving the natural landscape, it is possible to channel runoff water into detention ponds.

[. . .]

GRADING AND TREE-CUTTING ORDINANCES

What prevents a developer from stripping and bulldozing a property before applying for zoning and subdivision approval? Nothing, unless a community also has laws that require permits for grading and for cutting down trees larger than a specified size, usually a 6 in. or 8 in. caliper. The key phrase in such regulations is to require that grading or tree cutting be done in accordance with an approved development plan. Good draftsmanship can provide exceptions for working farms and for individual homeowners who want to do a small amount of tree cutting and clearing.

While it makes sense for every community to adopt environmental protection provisions, such codes, by themselves, may only make sprawl worse by spreading development more thinly. They need to be accompanied by other code provisions that encourage neighborhoods and compact business centers.

TRADITIONAL NEIGHBORHOOD DEVELOPMENT AND NEIGHBORHOOD ZONES

[. . .]

. . . [A] more generic . . . code combines elements of both zoning and subdivision, calling it a Traditional Neighborhood Development ordinance, or TND. . . . Like a subdivision ordinance, the TND has instructions about street layout and width, block sizes, requirements for open space. Like zoning it specifies the location for different building sizes and different mixes of activities. TNDs have now been adopted by a number of jurisdictions. They are generally available as an alternative to conventional zoning, and are thus an option like a planned unit development, but with a lot of the rules for approval spelled out in advance. . . .

A Neighborhood Zone would be a mixed lot-size residential district that applies as of right to all properties within the given area. The community can still limit the overall density of an N district. For example, a district designated N-4 can have an average density of four housing units to the acre. However, within the district, properties could have a variety of sizes, not just the quarter-acre lots of conventional R-4 zoning. For the variety to be meaningful, the overall size of a district would have to cover a whole neighborhood. . . . [I]f a neighborhood's size is determined by walking distances it would be about 160 acres. The neighborhood could include row houses, apartments, small houses, big houses, and even big estates. In order to be fair to all property owners, the zoning district would have to be accompanied by a plan that covered the whole area and assured all property owners the right to build at a lower baseline level. Once the combination of built units plus permitted baseline units had reached a total of 640, the maximum permitted in N-4, the neighborhood plan would be complete.

[. . .]

STREET PLANS

The ability to map streets is one of the oldest and most unquestioned powers available to local government. However, since World War II, it has become customary to wait to map local streets until the owners of the properties apply for subdivision approval. The subdivision ordinance regulates the way larger pieces of land are subdivided into individual streets and lots to meet the zoning code. It contains performance standards and dimensions for streets that will be acceptable to the local government, but the actual layout is left to the planners and engineers employed by the applicant, subject to review and approval.

. . . [L]eaving the streets to individual owners inevitably disconnects each development from the one next door. As only the major streets are mapped by the local authorities, and all the subdivisions are connected to these main streets, most travel from one subdivision to another takes place on these major streets. . . .

[. . .]

If development regulation is to be a positive template for development, communities need to resume responsibility for the design of more of the street network. To do this effectively requires an understanding of the terrain and it also means making assumptions about future land use and densities.

SPECIFIC PLANS OR SPECIAL DISTRICTS

[Another] strategy . . . for redesigning commercial corridors as a series of mixed-use zones located at key intersections, will usually require that special provisions be added to local development regulations. In California there is a well-established procedure called the Specific Plan, which is intended for just such a purpose. The specific plan is comparable to a planned unit development, but it applies when an area is divided among different owners. . . . A local government can initiate a specific plan, including streets, site plan, location and size of buildings; if it is adopted, it becomes the zoning for the area and is binding on all the property owners. Oregon and Arizona have enabling legislation that also permits specific plans.

In other states it is possible to get to the same result, but it requires more steps. A community can adopt a master plan for a specific area. The master plan can include a street plan and a zoning plan. As zoning is always supposed to be the implementation mechanism for a master plan, the existing zoning districts within the area covered by the plan can be changed to reflect the plan's objectives. If the right kinds of zoning districts do not exist in the current code, new zoning districts can be created.

In addition, a local government can enact a special zoning district, which is an overlay that applies only to the area in the plan. The special zoning district can include provisions like build-to and setback lines to govern building placement, height limits, required pedestrian connections and other similar provisions. Because they are part of zoning, these overlay districts are necessarily more abstract than the drawings in a planned unit development or a specific plan, but, if appropriately crafted, they can achieve similar objectives.

The major obstacle to the approval of such a plan is, of course, the assent of the property owners. The owners can be overruled by the specific plan or by a master plan and zoning overlay, but the decisions become more difficult politically the more objections are voiced by property owners. It helps if the specific plan or new zoning increases development over what was permissible before the plan was drawn. It also may be necessary to add incentives to make it worth while for property owners to agree to the plan. The local government can provide assistance in building streets and utilities, and in dealing with parking and storm water management. As the plan is likely to increase future tax revenues, these revenues can be set aside to pay back the costs for elements – like streets and utilities – needed to make the project happen. This technique is known as tax-increment financing. The tools for such proactive local planning are generally available, if the community has the political will to use them.

A special zoning district is more general and abstract than a specific plan, but can be used to promote a special design character for a particular area. A good example is the special district enacted for Times Square in New York City. Historically, Times Square had been the center of a district of big electric advertising signs: "The Great White Way." As new corporate office buildings were constructed in the area, the Municipal Art Society and other citizen groups became concerned that Times Square's distinctive character would be sanitized out of existence. In response an ordinance was enacted that specified minimum areas for electric signs on the lower floors of building facades. Initial resistance from developers vanished when they discovered how much money could be made from leasing the signs. Ingenious methods were devised to permit occupants on lower floors to look out their windows through the signs. The character of the area has been reestablished by the zoning. There are more bright lights on Broadway and Times Square than there ever were before.

[. . .]

HISTORIC DISTRICTS AND DESIGN REVIEW

Historic districts are among the earliest forms of development regulation to have a specific design component. The Old and Historic District in Charleston, South Carolina, was first enacted in 1931, the Vieux Carré in New Orleans became a historic district in 1937, the historic district in Alexandria, Virginia, dates from 1946, the preservation of Beacon Hill in Boston from 1955. In all these places the existing building fabric was to be maintained and new construction reviewed to ensure that the design would be in harmony with its historic surroundings.

The Federal Historic Preservation Act of 1966 included guidelines for the establishment of local historic districts and the preservation of individual buildings. This act also established the National Register of Historic Places. A listing on the National Register affects eligibility for tax subsidies and can protect a building or district from federally funded actions, but the real task of preservation and review is left to landmark and district designations established by local governments.

In a historic district there are two basic design questions. The first has to do with the appropriate way to treat an existing building: What paint colors are correct, how to make replacements of structurally deteriorated elements, what to do about the accretions from different periods? The second question concerns filling in gaps within the district. Should a new building counterfeit the appearance of an old building? If not, what constitutes an appropriate architectural expression?

[. . .]

DESIGN REVIEW AND DESIGN GUIDELINES

The courts have upheld historic districts and historic district review procedures over the years because the criteria are based on well understood general principles that apply to all properties within a district. Design review outside of a historic district has a somewhat different basis.

1 Design review as a condition of ownership

In an urban renewal district, when a public entity assembles property and sells it to a private investor, design conditions can be made part of the property transaction. The deed can require the developer to fulfill conditions; the public agency can review plans for a proposed development to make sure it meets the conditions in the purchase agreement. The design guidelines for Battery Park City in New York are based on transaction requirements. When you buy the property, you buy the guidelines. . . . The design codes that are part of many planned communities have a similar basis as part of a purely private transaction. When you buy a lot in Celebration or at Seaside, you buy into their requirements. . . .

[. . .]

2 Design review as a condition of public action

Cincinnati's Urban Design Review Board advises the city manager on any important downtown project that requires a discretionary approval or any kind of city action or subsidy to support the project. It becomes more immersed in the actual design process than most review boards. Recent design review legislation in Seattle gives the Review Board some of the powers of a zoning board of appeal. The design review process can lead to exceptions from the zoning, granted for appropriate design, rather than for hardship, which is the basis for exceptions from a board of appeal.

[. . .]

3 Design review as a general requirement

The public interest in a more general kind of design review comes from the "police powers" of the state: The state delegates its "police power" to the local government, enabling it to insure public safety, public health, and general welfare. The more objective and generally applicable the review criteria can be, the more likely the review process can be sustained if there is a challenge to it in court.

Norfolk, Virginia's, zoning code makes all development downtown subject to a discretionary Downtown Development Certificate. The code lists the criteria by which a proposal will be judged, and the design review committee and the planning commission adopts specific guidelines that apply to each project at the beginning of the review process. The intent is to make sure that applicants understand the criteria on which they will be reviewed before design and budget commitments have been fully established.

[. . .]

Many other communities have a design review commission or design review board that reviews proposals for compliance with special area plans, compliance with requirements of a particular ordinance, or even for consonance with the established character of the community. Ordinarily the statute setting up the board sets qualifications so that some percentage of the members are design professionals. Almost all such boards have both preliminary and final reviews, so that the applicant has a chance to take the board's guidance into account before becoming too committed to a particular design. Often the board is advisory to the planning commission: Projects are referred to the board for comment before the planning commission takes action. The advantage of a design review process is that it permits development requirements to be administered in a flexible way that relates to the specifics of a particular site and building program.

[. . .]

Representative plans, techniques and results

Growth of a typical Northeastern city: 1890. (Drawing by Andrew Whittemore)

Plate 17 *Society Hill, Philadelphia, 1950s and 2007.* In the 1960s Society Hill was the site of one of the nation's first urban renewal programs to combine historic preservation, clearance, new construction and a carefully designed open space network to create a strong neighborhood that has had important spillover effects in the surrounding area. Today this neighborhood has contributed to the Philadelphia's high number of downtown residents. (Photo Edward Hille, courtesy Eugénie L. Birch)

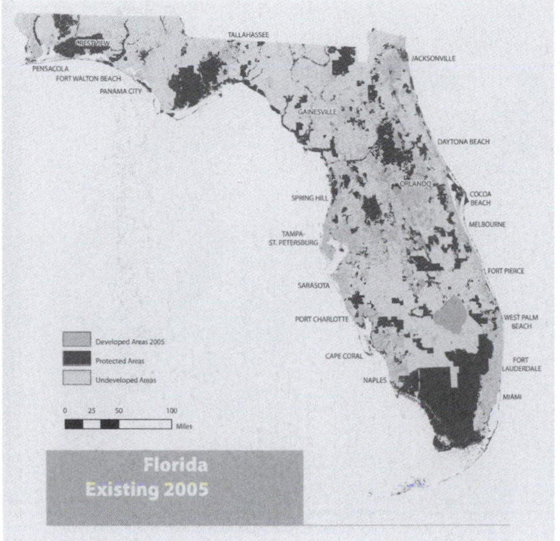

(a)

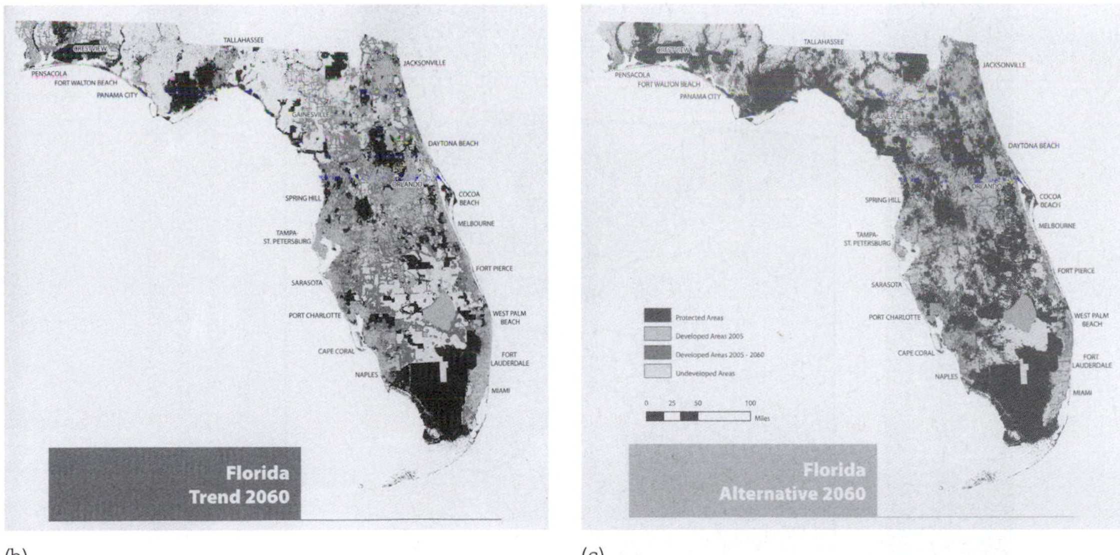

(b) (c)

Plate 18 *Growth Scenarios in Florida.* These growth scenarios of Florida, developed for 10,000 Friends of Florida in 2007, illustrate development choices for the state and its local jurisdictions. (a) The state's current land use. (b) If these stakeholders follow current trends for the next fifty years, urban land consumption will cover much of the state. (c) If they make strategic investments in mass transit and open space and enact regulatory amendments to encourage more dense development around transit stations or within existing cities, urban land consumption would decline dramatically. (Courtesy Department of City and Regional Planning, University of Pennsylvania)

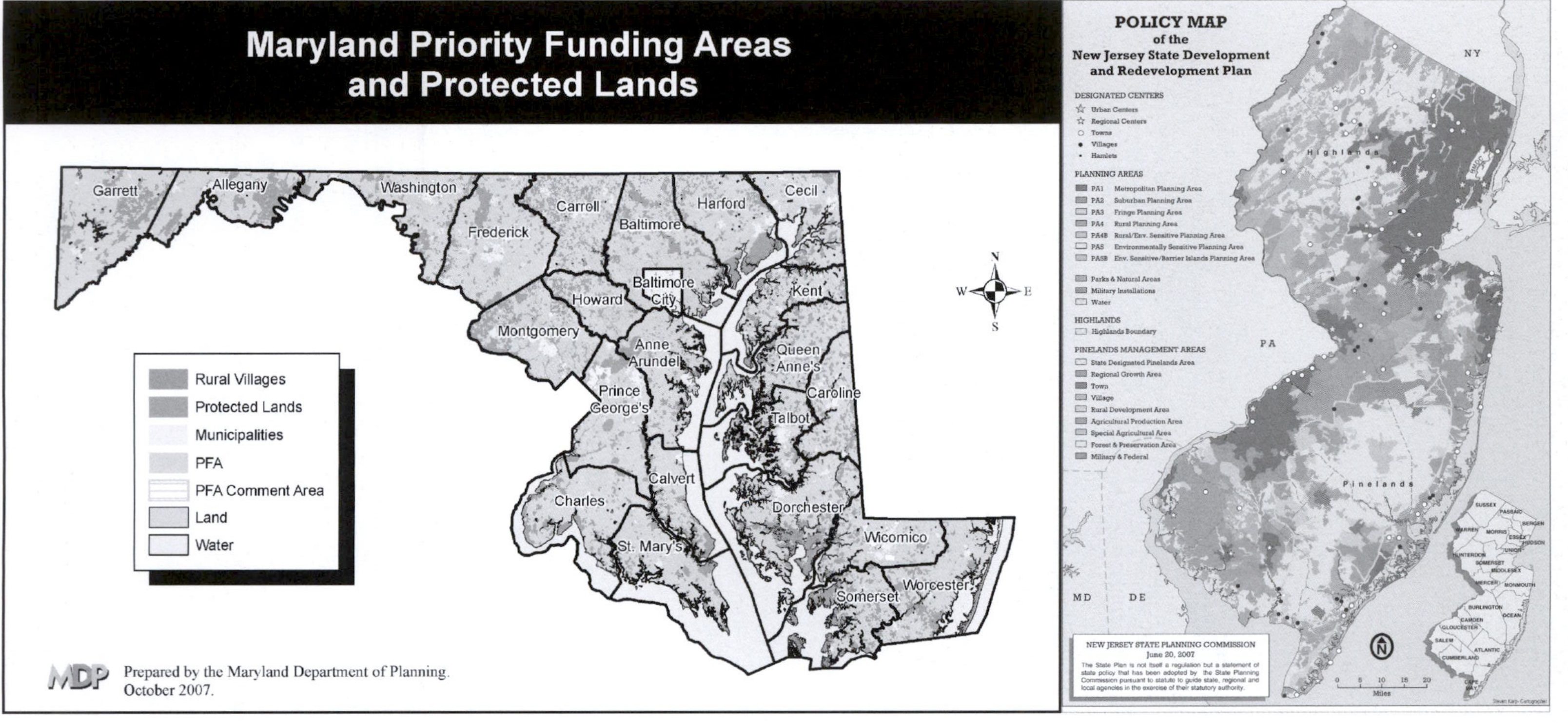

Plate 19 *State Plans for Maryland and New Jersey*. Under its Smart Growth Priority Funding Areas Act (1998) Maryland is attempting to target developed areas for growth while protecting sensitive areas by directing funding to specific areas, as illustrated on the map (*left*). Authorized in 1985 under the State Planning Act, the seventeen-member State Planning Commission approved the first state plan for New Jersey in 1992 and amended it in 2001 to differentiate among five areas (metropolitan, suburban, fringe, rural and environmentally sensitive), as illustrated on the plan's policy map (*right*). (Courtesy Maryland Department of Planning and New Jersey Office of Community Affairs)

Plate 20 *Listening to the City.* In July 2002, 4,000 "Listening to the City" participants assembled in New York City's Javits Convention Center to express their views on six preliminary reconstruction proposals for the World Trade Center, destroyed by terrorist action on September 11, 2001. Employing computer technology for rapid communication, this citizen engagement exercise displayed the group's responses to a number of planning issues instantly displayed on the large screens pictured. In a dramatic conclusion to the day, they expressed such strong opposition to the proposed schemes that government officials reopened the planning process, creating a widely publicized international competition for the site plan, ultimately won by Daniel Liebeskind. (Courtesy Jacquie Hermmerdinger)

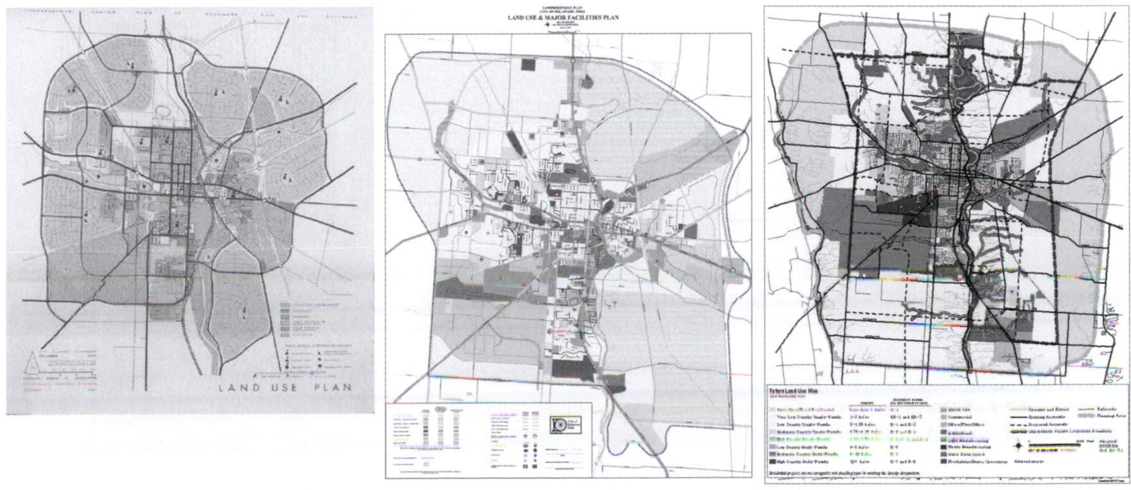

Plate 21 *Delaware, Ohio, Comprehensive Plans.* These three plans for Delaware, Ohio, dating from (*left to right*) 1964, 1996, and 2004, illustrate how this small city thirty miles north of Columbus, the state capital, is coping with growth. Since the 1960s it has experienced a 131 percent increase in population. In the 1960s, the plan envisioned growth in low-density neighborhood units evenly dispersed through the city (*left*) and by 2004 it calls for a more varied, mixed-use pattern. (Courtesy City of Delaware, Department of Planning and Community Development)

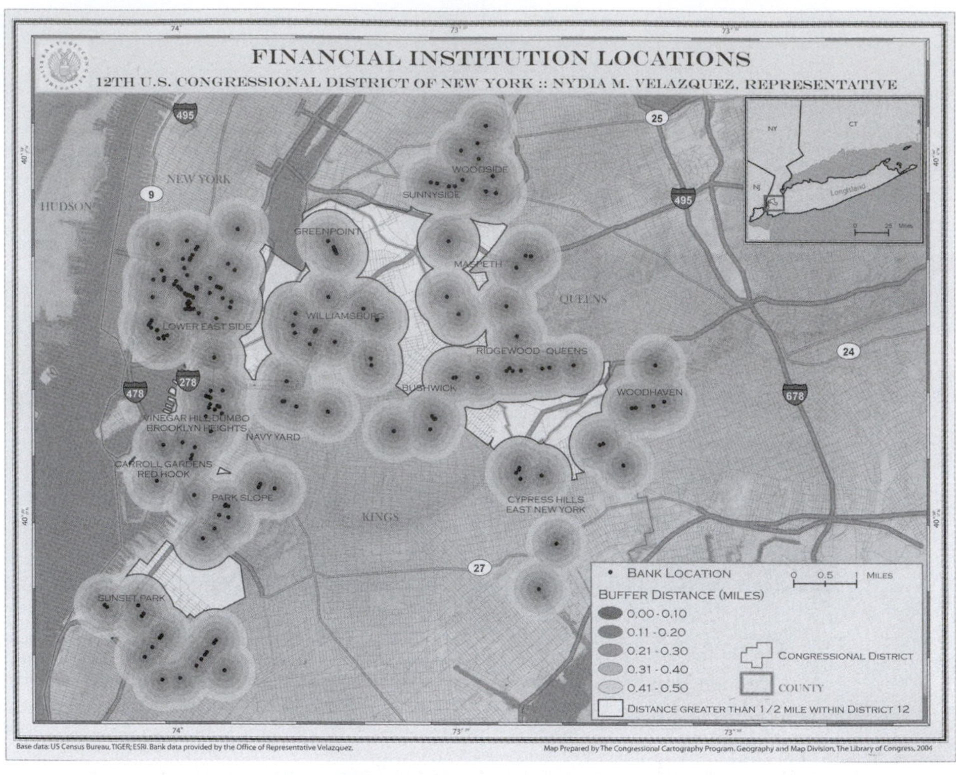

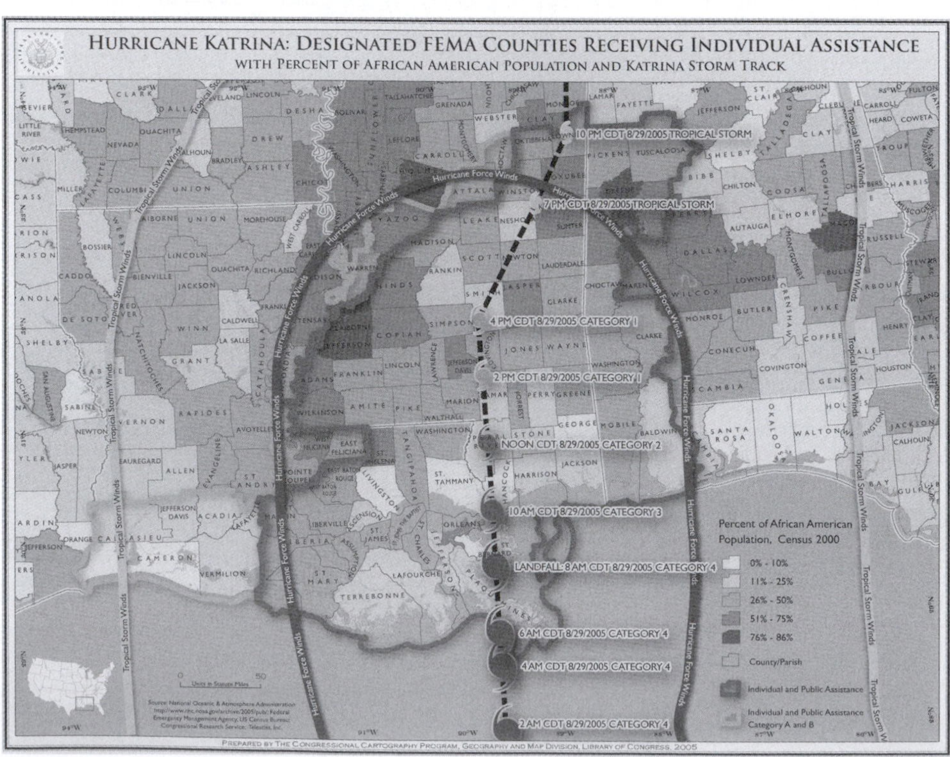

Plate 22 *GIS and Visualization.* In displaying many kinds (or layers) of data in a single map, as do these, geographic information systems technologies provide spatial information that assists decision makers. The map of bank locations (*left*) demonstrates underserved areas, an important issue in low-income communities. The map of the path of Hurricane Katrina (*right*) not only tracks the storm but illustrates the racial and income characteristics of the residents in its path. (Courtesy ESRI)

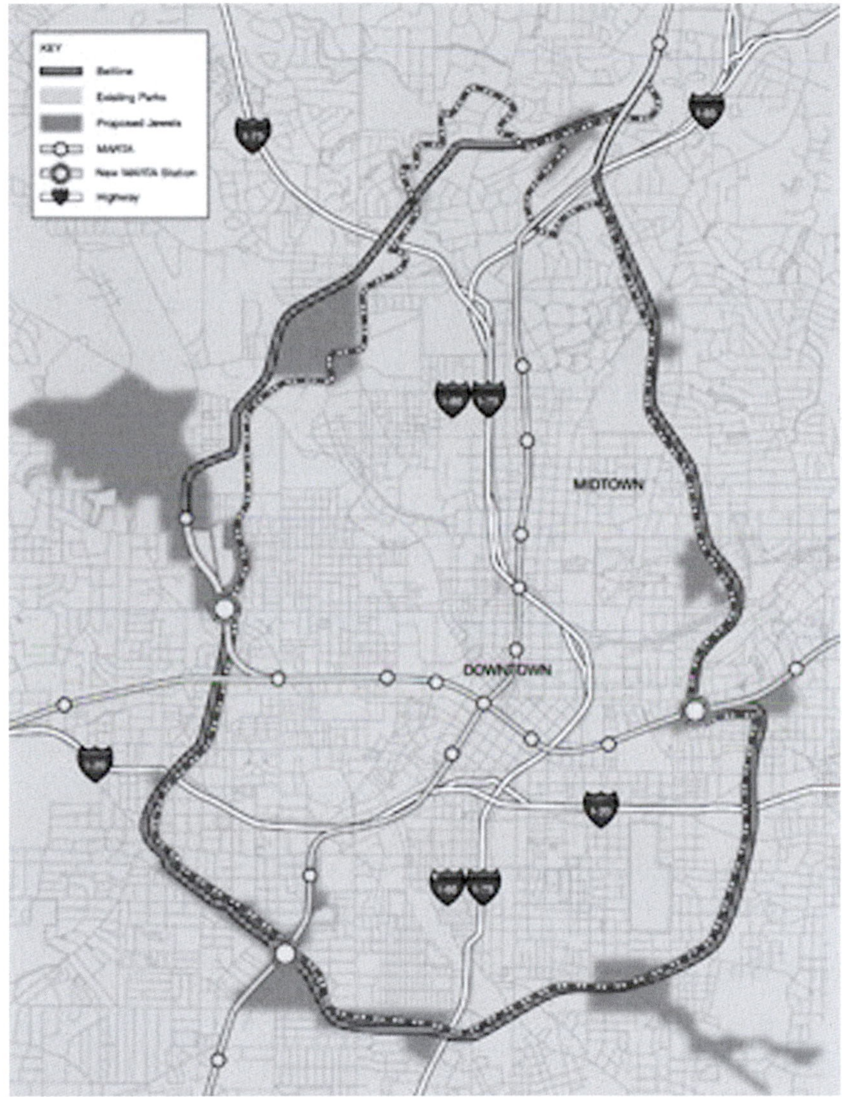

Plate 23 *Atlanta's Emerald Necklace Beltline.* Inspired by an urban planning graduate student's thesis that outlined the city's unused rail lines, and subject of a public realm analysis by Alexander Garvin & Associates, the Atlanta Emerald Necklace BeltLine will transform the city with twenty-two miles of mass transit, trails and bikeways connecting forty-five neighborhoods. To finance the investment, voters approved a 6,500-acre tax allocation district, a bonding scheme whose revenue streams will cover 60 percent of the $1.7 billion project costs. When it is complete the city will have gained 1,200 acres of open space, $240 million of investment in workforce housing and other transit-oriented development. (Courtesy Alex Garvin & Associates)

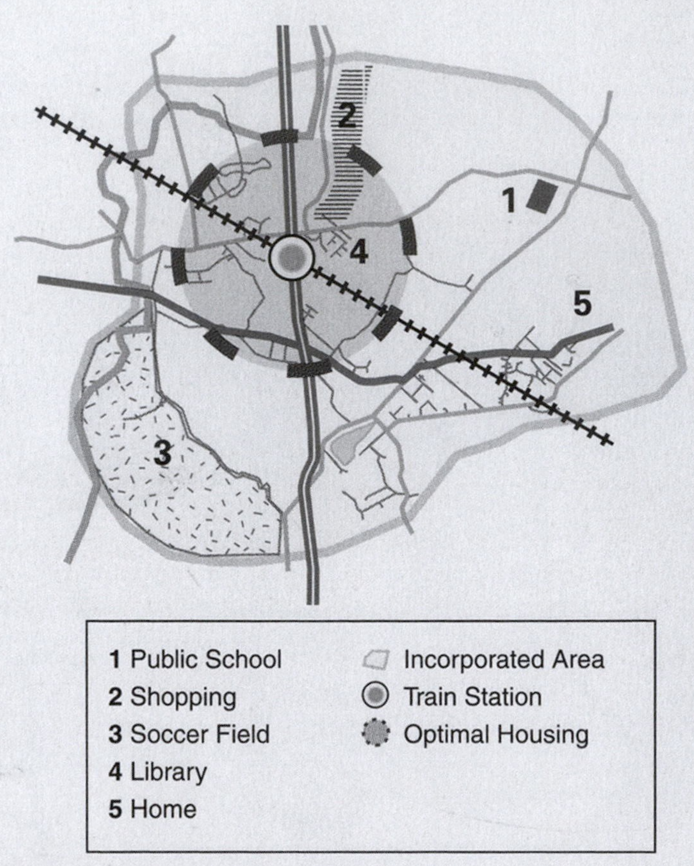

Compact Development

If housing and other local destinations are concentrated at the outer edges of incorporated areas and away from transit centers, trips to typical destinations will increase traffic congestion on arterials and expressways.

Plate 24 *Compact Development.* Drawn from an award-winning regional plan, this illustration outlines how to retrofit an auto-dependent community by building housing and associated retail adjacent to a transit station in order to minimize vehicular traffic. Planners label this "compact" or "transit-oriented" development (TOD). (Courtesy Northeastern Illinois Planning Commission/Chicago Metropolitan Planning Agency)

Plate 25 *Chattanooga Waterfront.* Named the nation's most polluted city in 1969, Chattanooga's leadership undertook a major waterfront reconstruction effort framed by the river. The River Park Master Plan (1985) was completed with the inauguration of the last segment of a twenty-two-mile long River Walk trail in 2005. Pictured is its centerpiece, the wildly successful Tennessee Aquarium, opened in 1992 and host to its ten millionth visitor eight years later. The 129-acre redevelopment contains other venues, including a museum, carousel, and Imax theater. Public realm improvements include plazas, terraces and a rehabilitated pedestrian bridge connecting parks on both sides of the Tennessee River. Adjacent to these facilities are retail, hospitality, and residential uses. (Courtesy Chattanooga Pulse/Culture Systems)

Plate 26 *Rincon Hill and Transbay Transit Center, San Francisco.* To accommodate new population growth within its limited forty-six square miles, the City and County of San Francisco Planning Department worked with citizens, the redevelopment authority, and other government agencies to develop two important plans, adopted in 2005. The Rincon Hill plan, the first neighborhood plan approved in more than ten years, covers a fourteen-block area south of downtown. It promotes a new mixed-use development residential neighborhood, allowing housing for 20,000 residents, providing a funding mechanism for major open space and streetscape improvements and a community center. The San Francisco Redevelopment Agency's plan for the Transbay Redevelopment Project area, that abuts and includes part of the Rincon Hill area, creates a tax increment district to support public investment in a major multimodel transit station to serve municipal and regional rail, in supportive amenities for the area's mixed use, including streetscaping and other open space attractions. (Courtesy San Francisco Redevelopment Agency)

Plate 27 *Herr Island, Pittsburgh, before and after.* Guided by a master plan and employing performance zoning and strategic public investment, Pittsburgh's forty-two-acre Washington's Landing, two miles from downtown's Golden Triangle, was once the city's meat-packing district (*left*). (Courtesy Urban Redevelopment Authority of Pittsburgh.) Public investment of $27 million in brownfield remediation, streets, and open space leveraged $44 million in private investment in housing, retail, light industry, and recreational facilities (*right*). (Courtesy Rubinoff Company)

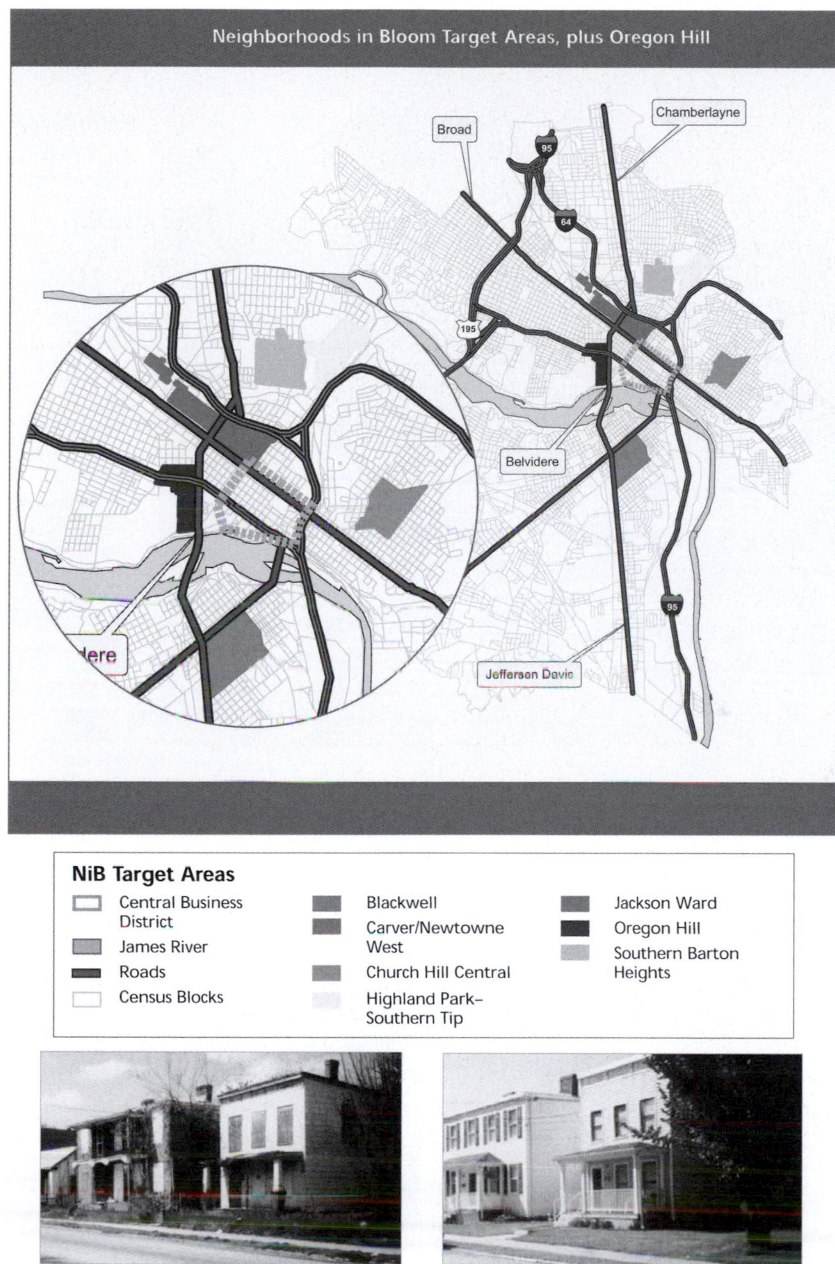

Plate 28 *Richmond: Neighborhoods in Bloom.* Noting the failure of dispersing its limited housing funds equally among neighborhoods, the City of Richmond, Virginia, launched its "Neighborhoods in Bloom" program in 2000. Following a three-year inclusive planning process, it targeted 80 percent of its community development expenditures in six to twelve block areas within six neighborhoods. It supplemented the funding with increased crime prevention and code enforcement, provided fifteen-year tax abatements, low-cost mortgages, and homeownership counseling. Within six years it saw dramatic improvements, measured in property values, in these areas. (Courtesy Federal Reserve Rank of Richmond)

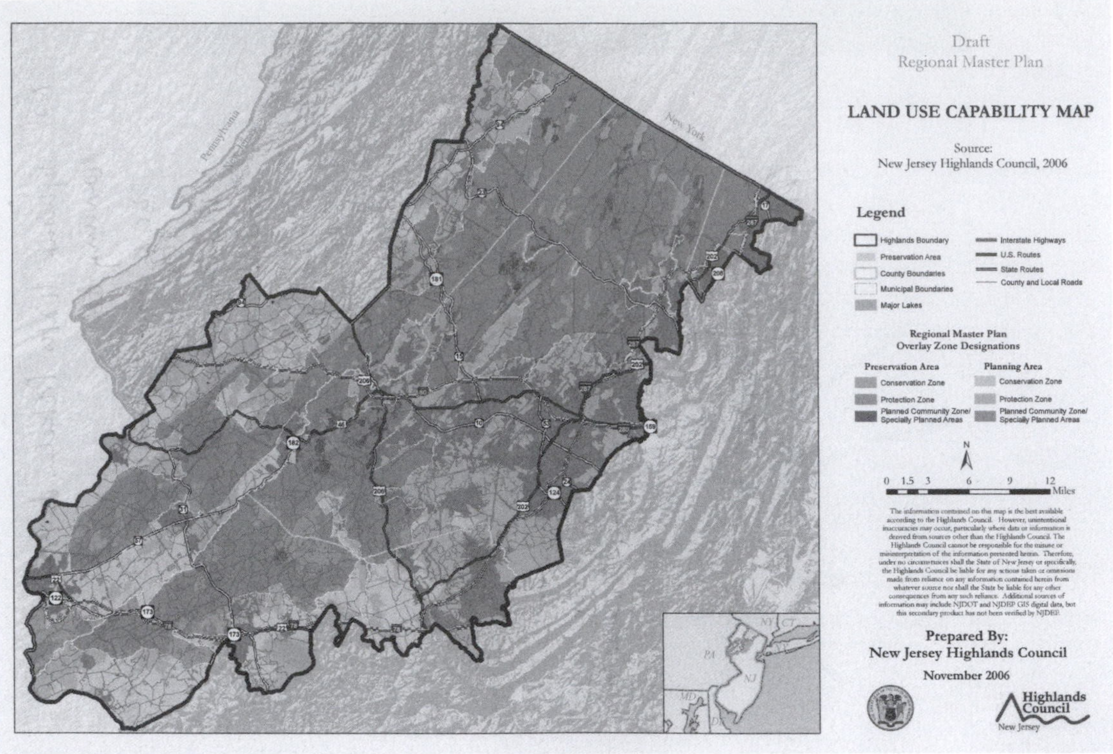

Plate 29 *New Jersey Highlands Plan.* In 2004, the state of New Jersey enacted the Highlands Water Protection and Planning Act to protect the Highlands region, a 1,250 square-mile region that supplies water to 65 percent of the state's residents. The Highlands region contains eighty-eight municipalities in seven counties and is threatened with overdevelopment: between 1995 and 2000 the Highlands lost 25,000 acres of forest and farmland. The Highlands Act created a fifteen-member council, appointed by the Governor, whose members are charged with developing and implementing the regional master plan. The Highlands Regional Master Plan, Final Draft, November 2007, includes the Land Use Capability Map (*above*), which outlines three primary overlay zones: Protection Zone (natural resources), Conservation Zone (agriculture interspersed with woodland) and Existing Community Zone (concentrated development) and three subzones (Conservation Zone – Environmentally Constrained, Existing Community – Environmentally Constrained, and Lake Communities). (Courtesy New Jersey Highlands Water Planning and Protection Council)

Plate 30 *Anchor Institution, University of Pennsylvania.* Anchor institutions – universities, medical complexes, performing arts centers, museums, stadiums or other sports facilities – are increasingly dominant employers in their cities and large property owners. Many have led revitalization efforts in Philadelphia. The University of Pennsylvania's West Philadelphia Initiative (2004 to the present) and Penn Connects (2007 to the present) exemplify two strategies. One focus is local community development (with programs targeting housing, education, crime, sanitation, labor force enhancement) and the other is city economic development (land purchase/mixed-use adaptive reuse, and new construction with public – private partnerships, brownfield remediation, and establishing physical downtown links). (Courtesy the University of Pennsylvania)

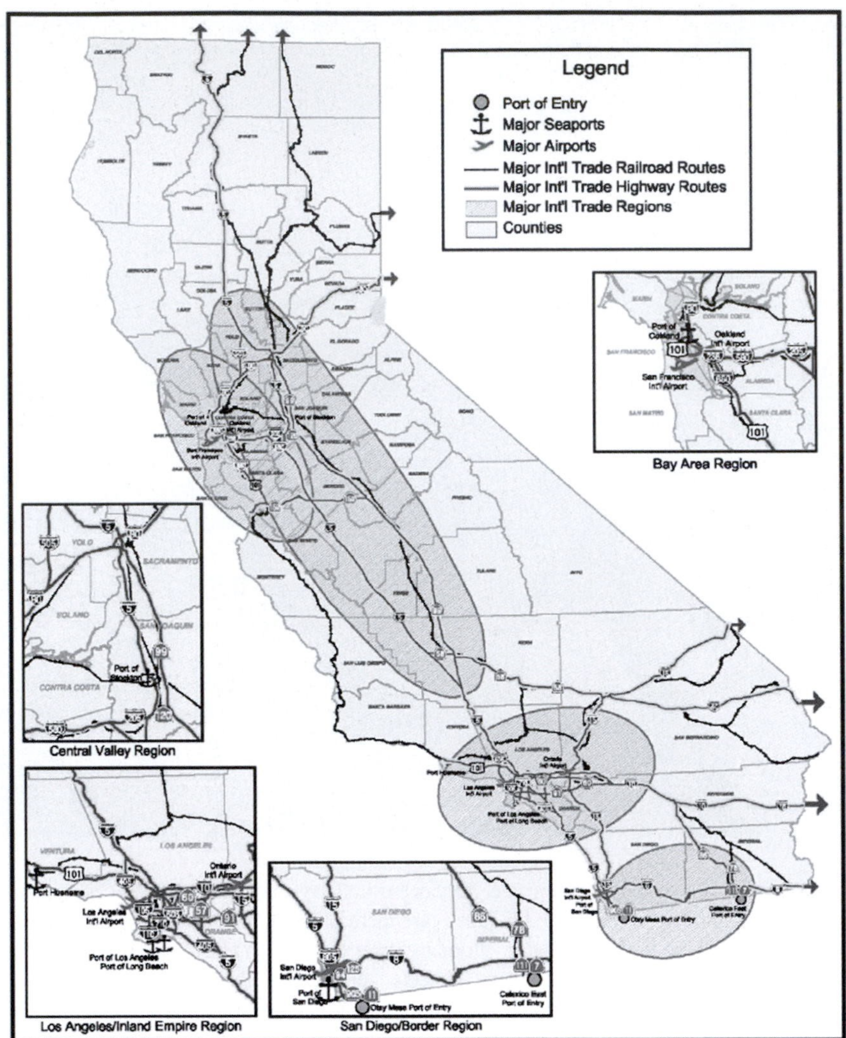

Plate 31 *California State Transportation Plan, 2007.* Growing foreign and domestic trade in California has put enormous pressure on its transportation routes. In an unusual move, the state responded with an integrated transportation plan for four critical areas – Bay Area, the Central Valley and the San Diego/Tiajuana and Long Beach/Los Angeles regions. The plan is notable because it takes into account transportation *systems,* incorporating airports, seaports, highways, and rail, and prioritizes funding for the coming decades. Formerly, transportation planning did not consider the entire network but looked at individual modes independently. (Courtesy State of California, Business, Transportation and Housing Agency)

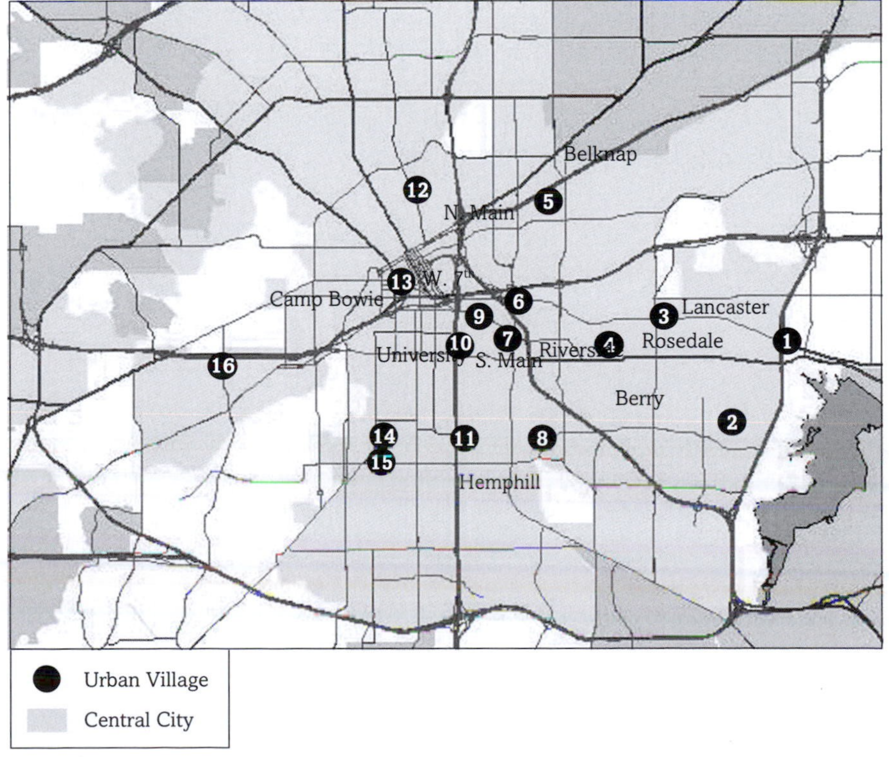

Plate 32 *Designated Urban Villages, 2008.* Fort Worth, Texas, has selected sixteen places, designated as urban villages, as targets for denser, mixed-use zoning and capital investment in order to counteract sprawling development at its periphery. The upper photo illustrates the hoped-for changes, the lower one is of existing conditions. The map shows the location of the urban villages. (Courtesy City of Fort Worth Planning and Development Department)

PART FIVE

Planning Practice and Methods

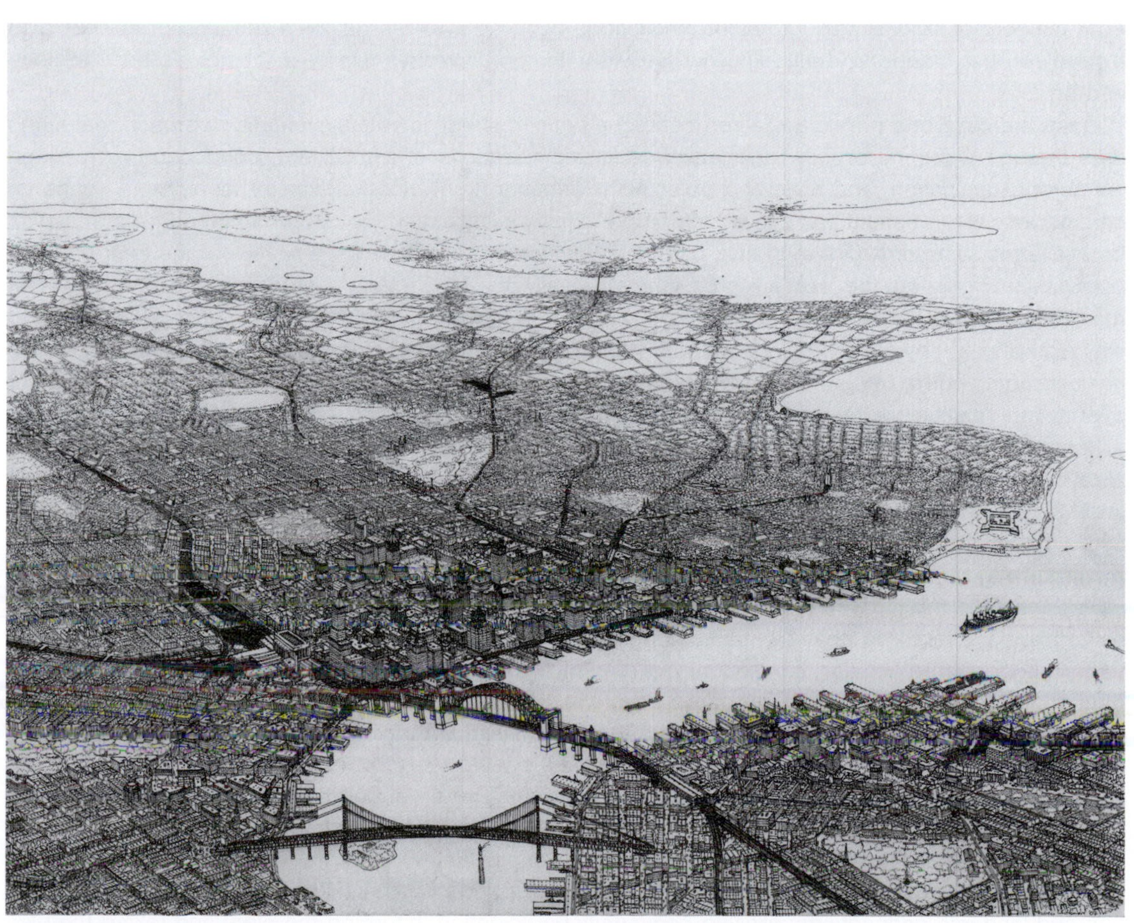

Growth of a typical Northeastern city: 1925. (Drawing by Andrew Whittemore)

INTRODUCTION TO PART FIVE

What does an urban or regional planner have to know to become a practitioner? So far, the *Reader* has covered key aspects of the field – its current context, its history and theory, and its major product (the plan). Now it's time to focus on planning as a profession. To begin, let's define "profession": it is a line of work whose practitioners possess expertise or specialized knowledge, develop a standard means of transmitting their knowledge to future practitioners, and adhere to licensing and ethical codes.

Each element of this simple definition deserves further discussion. Consider the notion of expertise. For some, it is built on the scientific method, a systematic means of developing theories or generalized principles to be used as a basis for solving a field's problems. For others, expertise derives not only from mastery of scientifically grounded knowledge but also involves art, or a process of attending to a problem that blends scientific knowledge with intuition. In reality, expertise combines both science and art.

Take the case of a physician. When treating a patient, she follows a systematic approach, gathering data through a routine process that includes asking questions of the patient; conducting a thorough hands-on examination; and reviewing objective information from blood-chemistry tests, bone scans, or cardiograms. If the patient has a fever, after a preliminary examination the doctor might administer aspirin to relieve the symptoms (knowing that, barring any unusual circumstances, it will lower the temperature), but works to figure out the fever's source. She eventually develops a treatment plan based on her assessment of the data. However, while this physician, like all professionals, has the specialized training to solve problems, she does not operate with complete independence; external factors will also shape her decisions. For example, the patient's medical insurance provider may constrain her ability to gather information (limiting lab tests) or her choice of treatments (refusing to pay for experimental protocols); a hospital oversight committee will insist on monitoring her activities; and peers from related specializations will have their opinions. So expertise in this case combines technical knowledge with judgment and is responsive to exogenous influences.

An urban planner's practice is not unlike a physician's. Theory informs his expertise on technical matters. Judgment and sensitivity to external factors are also key. To test your understanding of these professional realities, tease out the planner's expertise in the following scenario. A mayor and city council ask a planner director to deal with traffic congestion in a particular neighborhood. The planner examines the situation, searching for the sources of the problem. He conducts a field survey. He analyzes data on the origins and destinations of existing traffic. He estimates future traffic patterns based on forecasts of population and economic growth and anticipated land use changes, often using computer-based modeling programs to simulate various alternatives. To relieve the immediate symptoms, he suggests a traffic management plan that adjusts the timing of traffic lights – the equivalent of the aspirin that the physician dispenses for her patient's fever. But he knows that a long-term solution is required.

To solve – not just alleviate – the traffic problem, the planning director initiates a process whose end product will be a plan. The process involves tapping people with technical knowledge about transportation and land use connections; eliciting community participation in forging the plan; and taking the plan through the various approval processes in the community. So what exactly does he do? He hires outside consultants to work with his own staff and related city agencies to develop technical solutions. He engages neighborhood residents in a lengthy visioning exercise to educate them about the issues and coalesce support around various proposals. He discusses the proposals with elected officials and city administrators who will be involved in implementation. Once he has a proposed plan in hand, he will run it through an environmental review, and, after participating in required public hearings, bring it before the city planning commission and city council for approval.

How did this planning director learn the varied skills to undertake this project? More important, how did he become the city's planning director? In all likelihood, he qualified for this job through education and training. Most urban planners graduate from one of the seventy accredited university-based planning degree programs. These programs are monitored by a neutral board, the Planning Accreditation Board (PAB), which is empowered by the field's three professional organizations: the American Planning Association (APA), the Association of Collegiate Schools of Planning (ACSPO), and the American Institute of Certified Planners (AICP). The PAB verifies that each program delivers the specified knowledge and basic skills that make up the field's expertise. It expects each graduate to be knowledgeable in six areas: (1) the structure and functions of human settlements, (2) the history and theory of planning processes and practices, (3) the administrative, legal, and political aspects of plan making and policy implementation, (4) collaborative problem solving, plan making, and program design, (5) how to apply this knowledge in real-life situations, and (6) ethics. In addition, it requires graduates to have basic skills in (1) problem formulation, (2) research methods and data gathering, and (3) written, oral, and graphic communication.

If an individual does not get a degree from one of the PAB-approved programs, he can qualify as a planner by demonstrating competence in certifying examinations; having a university degree in a related profession like architecture, engineering, or law; and having on-the-job training. For example, the AICP extends membership to those who have passed a rigorous written examination and have professional work experience. It allows a member to place the AICP initials after his name to signify his being certified as a professional planner. It also requires members to maintain their certification with continuing education.

Finally, like other professions, planners have a code of ethics that governs their conduct. Dating from 1978, the code has been developed and is administered by the AICP. In addition to advising practitioners on their proper roles, the *AICP Code of Ethics and Professional Conduct* (2005) informs public expectations about planners' performance.

The seven selections describe professional practice and its methods. In "Introduction: Planning for People and Places," Charles J. Hoch, Linda C. Dalton, and Frank S. So profile urban planners and their work. Although this essay, drawn from the most recent handbook of urban planning published by the International City/County Management Association, *The Practice of Local Government Planning* (Washington, DC: 2000), deals with public sector planning alone, that sector accounts for 75 percent of planners working today. Frank S. So and Judith Getzels's essay "Planning Environments," which describes the planning process and outlines different types of plans, comes from an earlier version of the ICMA handbook. But its message about the importance of developing plans systematically is crystal-clear.

In "A Realistic Approach to City and Suburban Planning . . . Ingredients of Success," Alexander Garvin urges planners to focus their expertise. He circumscribes urban planning as "public action that generates a sustained and widespread private market reaction which improves the quality of life of the affected community, thereby making it more attractive, convenient, and environmentally healthy." In light of that definition, he lists six items that planners must take into account when intervening in the physical environment: the market, location, design, financing, entrepreneurship, and timing.

Dowell Myers has another take on what planners need to know. He drills down into population analysis in fresh ways. In "Demographic Futures as a Guide to Planning: California Latinos and the Compact City," he challenges planners to continue to make the best projections possible but to lend great care in their interpretations, since past assumptions about behavior do not necessarily apply today.

The next two essays elaborate other skills demanded of today's planners. Ann-Margaret Esnard, Nancy Sappington, and Milton R. Ospina in "Geographic Information Systems" outline the capabilities of computer mapping in planning analysis. Barbara Faga in "The Future of Public Participation" sets current progress in civic engagement against trends in globalization, growing cultural and ethnic diversity in communities, and the increasing use of technology.

Carol D. Barrett's "Introduction" brings the discussion home. In this deceptively simple but profound essay, Barrett reminds readers of the importance of ethics for responsible urban planning. After all, making great communities depends on moral values, which is why they are at the heart of everything a planner does.

"Introduction: Planning for People and Places"

From *The Practice of Local Government Planning* (2000)

Charles J. Hoch, Linda C. Dalton, and Frank S. So

Editor's introduction

Urban and regional planning has evolved over time. Modern planning began as an endeavor to meet the needs of urbanized, industrial societies, a goal that remains today. Its nineteenth-century beginnings, described in Part Two of this *Reader*, were rooted in Europe and the United States, which experienced rapid metropolitan growth through the economic transformation known as the industrial revolution. A population explosion placed pressure on municipalities to deliver basic services (housing, sewerage, potable water, open space), while the economic demands tested the logistical limits (transportation, land use) of newly industrialized places. Responding to these problems, urban and regional planning emerged to control and guide development.

In the United States the urban planning function is decentralized in accordance with the governmental structure outlined in the U.S. Constitution. The Constitution, on which all U.S. law is based, reserves certain powers, such as the right to declare war and govern interstate commerce, for the federal government, and gives all other powers to the state governments. State governments then allocate power to local units (e.g. counties, municipalities, and special districts) either through a city charter, "home rule" laws, or special legislation. In their legal form, cities are municipal corporations, distinct from private corporations and required to maintain a nonprofit status in all operations.

Currently, 87,525 local governments exist in the United States. Of these, 44 percent are general-purpose governments, with 92 percent (35,933) of these municipal governments, and the remainder county governments. More than 174 million people (or 62 percent of the U.S. population) live in municipal government jurisdictions. Of the remaining 56 percent of local government units that are not general-purpose, 72 percent (or 35,052) are independent school districts and the remainder are special districts, primarily oriented to undertake a single function such as protecting a natural resource like an aquifer or providing a service like fire protection or sewage.

In the United States, cities are the "creatures of the state." Under a late nineteenth-century judgment known as "Dillon's Rule," upheld by the Supreme Court in the early twentieth century, cities' powers are limited to those expressly given by the state, those implied in the expressed powers, and those indispensable for carrying out their functions. The resolution of any dispute about the powers between a state and a municipality favors the state. And if a city wishes to change or add powers, it must obtain state permission. In the early twentieth century, cities interested in instituting urban planning and zoning as municipal functions lobbied their state governments to pass enabling legislation for authorization. (See selections "*Ramapo* Plus Thirty: The Changing Role of the Plan in Land Use Regulation" and "The Master Plan: An Impermanent Constitution" for more on state enabling legislation and the legal origins of planning and zoning.) Although no authoritative count exists, the majority of the more than 38,000 units of county and municipal governments have state permission to perform planning functions and legislate land use.

Three forms of municipal government are dominant in the United States: the mayor/council, the commission, and the council/manager. Under the mayor/council form, which is what the majority of municipalities have, citizens elect a mayor at large as well as members of the city council, usually through district representation (though sometimes there are a few at-large seats). The mayor exercises executive powers (appoints department heads, establishes the budget, has a veto), and the city council is the legislative branch and initiates laws and programs, approves appointments and budgets, and, in some cases, may override mayoral vetoes. Some places give more power to the mayor in executing these tasks, and others less. In the commission form of municipal government – first employed in Galveston, Texas, in 1901, and popular in the early twentieth century – citizens elect three to seven (depending on the law) commissioners, each of whom is responsible for one or more municipal departments. The chair of the commission may serve as a mayor who has no additional power but functions as a ceremonial head. Most commission governments have evolved into council–city manager arrangements. The only large city to retain the commission form is Portland, Oregon, governed by a mayor and four commissioners who constitute the city council. The council–manager form of government consists of the city council and a hired city manager who administers city departments and is accountable to the council. About 3,000 U.S. municipalities (16 percent) have this type of government.

This selection provides background on the planners and their duties in the nation's local general-purpose governments. Along with the selection that follows ("Planning Environments"), it focuses on the profession's membership, functions, theaters of operation, and value systems. It comes from the basic handbook of urban planning, *The Practice of Local Government Planning*, that the American Planning Association and the International City/County Management Association have jointly published for more than sixty years under a variety of titles. Known as the "Green Book," for the color on its cover ever since it was first published in 1938, it represents practitioners' consensus about the field's range of concerns. (See the selection "Practitioners and the Art of Planning" for more on the Green Book.) The next edition of the Green Book, *Local Planning*, edited by Gary Hack (senior editor), Eugenie L. Birch, Paul Sedway and Mitchell Silver, was published by ICMA/APA in 2008.

Jointly authored, this selection reflects the views of the three practitioners who edited the eighth edition of the Green Book in 2000: Charles J. Hoch, Professor, Urban Planning and Policy Program, College of Urban Planning and Public Affairs, University of Illinois, Chicago; Linda C. Dalton, AICP, Vice-president of Planning and Enrollment Management, California State University, East Bay, and former Chair, Seattle Planning Commission; and Frank S. So, FAICP, former Executive Director, American Planning Association. Hoch is author of *What Planners Do: Power, Politics and Persuasion* (Chicago: University of Chicago Press, 1994), Dalton has written extensively on planning education and planning theory, and So has edited three previous editions of the Green Book.

Other works describing U.S. planning include Barry Cullingworth and Roger W. Caves, *Planning in the USA: Policies, Issues and Processes* (London and New York: Routledge 2003), and John M. Levy's *Contemporary Urban Planning*, sixth edition (Upper Saddle River, NJ: Prentice-Hall, 2002). Both volumes survey the field and its components. For information about other countries' approaches, see Bishwapriya Sanyal (ed.), *Comparative Planning Cultures* (London and New York: Routledge, 2006), which presents case studies of selected countries in Europe, Asia, Middle East, Oceania, and the Americas. Sir Peter Hall's *Urban and Regional Planning*, fourth edition (London and New York: Routledge, 2002), provides an overview of the history and development of the Western planning tradition, with chapters contrasting U.S., British, and French experience. For more about local government, see the authoritative law text by Gerald E. Frug, Richard Ford, and David Barron, *Cases and Materials on Local Government Law* (Egan, MN: West Publishing, 2006). Frug's *City Making: Building Communities Without Building Walls* (Princeton, NJ: Princeton University Press, 2001) explores how to overcome the limits of the U.S. municipal legal arrangements. Finally, for more data on units of local government, see Census Bureau, *2002 Census of Governments*, Vol. 1, No. 1 (Washington, DC: U.S. Department of Commerce, 2002).

Modern urban and regional planning emerged more than a century ago in response to urban problems. The growth of planning practice, planning policies, and planning institutions has helped to address many of these problems, but others remain. The early, widespread belief in the power of planning to engineer permanent solutions has evolved into a more realistic and pragmatic approach. Cities [regions, suburbs and other places that engage in planning] pose a number of challenges for planners. . . . First, their complexity and variability defy understanding across scales. . . . Second, the longevity of existing urban features gives rise to inertia . . . , so instead of starting from scratch, plans must fit the new to the old. Third, cities are the points of convergence for a multitude of differences – among people, institutions, and landscapes – that intersect, overlap, and change over time. The complexity, longevity, and diversity of cities resist the implementation of carefully prescribed plans.

[. . .]

Planners cannot predict the future, but they do use rational analysis and practical judgment to anticipate and envision it. Planners estimate population trends, model urban development patterns, compare competing development goals, and create diverse remedies for the problems that arise from the interplay of people and place. But planning includes more than analysis: successful planning involves the people and organizations that are willing and able to put plans into action. Planners advise, negotiate, and otherwise communicate among property owners, developers, citizens' groups, technical specialists, and elected and appointed officials to translate planning promises into reality.

Planners also assist municipalities, counties, and special districts in exercising their taxing and regulatory authority wisely: planners analyze fiscal resources, regulate property development, create and evaluate long-range growth alternatives, assess environmental risks, help educate the public, and much more. When planners from overlapping or adjacent jurisdictions share information and jointly develop creative alternatives to current policies, potential conflicts can be transformed into opportunities for cooperation, and everyone can benefit: planning processes that rely on the logic of watersheds, air basins, labor markets, and commuter patterns to assess urban problems – rather than on narrowly defined notions of "turf" – can provide a useful antidote to unnecessary and wasteful disputes. Although the results of regionwide rational planning analysis may not be pleasing to local officials seeking competitive advantage at the expense of their neighbors, planning proposals that equitably distribute the burdens of environmental protection, the advantages of economic development, and the responsibility for affordable housing across an entire metropolitan region accurately reflect the realities of ecological balance and economic interdependence.

But because future risks may pale in comparison to the benefits of current development practices, some communities are reluctant to embrace the inclusive and future-oriented vision of planning until they are faced with the ravages of a natural disaster, environmental pollution, or economic depression. Other communities build planning into the routines of local management and policy development to achieve their aspirations for an improved quality of life.

WHO PLANNERS ARE AND WHAT THEY DO

According to the occupational analysis of the U.S. census, planning grew rapidly between 1960 and 1980, when more than 13,000 men and women identified themselves as planners. Growth then slowed during the 1980s, although professional membership in the American Planning Association (APA) climbed to over 20,000 in 1995.

Most urban planners enter the field after receiving a professional education in planning, but planning jobs – unlike those of some other professions – are often open to people who have no planning education. Planning organizations hire not only planners but also other specialists, such as landscape architects, economists, hydrologists, and demographers. Similarly, planning expertise has value outside planning occupations: many people who were trained as planners do not work in planning occupations but find their training in that field of great value in other fields.

Most planners (65 percent) have a master's degree, but 21 percent have their degrees in fields other than planning. Trained planners generally join APA, which is open as well to students, citizens, and officials who have an interest in planning. . . . Professional members of APA may apply to test for entry into the American Institute of Certified Planners (AICP), the professional certification arm of APA.

The typical professional member of APA is a white man with ten or more years' experience. The proportion of women planners increased steadily between 1981 and 1995; in 1995, more than 25 percent of planners were women, although only about one in six planning directors were women. Over that same fourteen-year period, however, the proportion of minority planners remained steady at about 7 percent. Because studies of planners' values show that background characteristics (as well as political orientation) make a difference regarding the issues that planning should address, statistics on current trends among ethnic and racial minorities in the planning profession are cause for concern: both participation and leadership among minority members have been essentially stagnant.

More than 70 percent of professional members of APA work for government agencies. In 1995, about a third worked for municipalities and another sixth worked for counties. In addition, planners work for regional planning agencies, special districts, and special-purpose authorities. Special-purpose authorities focus on specific functional activities, such as housing, redevelopment, schools, parks and recreation, water, or transportation. Public planning agencies vary significantly in size and character: a small town may have a consultant on retainer, while a big-city planning agency may have hundreds of employees. In 1991, planning agencies in communities whose populations ranged from 50,000 to 250,000 employed between one and thirty-four planners, with an average of 7.7 and a mode of 4.

[. . .] In the private sector, planners may work for real estate development or utility companies as well as for private consulting firms, including environmental or financial analysts. Few APA members work in universities (just over 3 percent) or nonprofit organizations (under 3 percent).

[. . .]

In a 1992 survey, most professional members of APA reported devoting most of their time to government regulation and plan making, a pattern that seems especially prevalent in growing communities. Detailed analysis of the 1992 survey reveals planners who were confident in their ability to analyze issues, develop goals, identify alternatives, assess project impacts prepare reports, and make public presentations. These tasks require planners to be able to write clearly, synthesize information, and interact effectively with a variety of people. Many respondents also emphasized the growing importance of negotiation, mediation, financial analysis, and computer literacy. These areas draw on a combination of communication, visualization, and rhetorical skills required to speak to the special needs of clients, citizens, and public officials.

THE FRAMEWORK OF LOCAL GOVERNMENT PLANNING

The framework within which local government planning occurs is determined by larger political institutions. Although the federal government does not have a unified planning department or agency, many national programs, policies, and regulations include planning requirements. For example, the Environmental Protection Agency, the U.S. Army Corps of Engineers, and the Departments of Agriculture, Housing and Urban Development, and the Interior all require, fund, and evaluate plans. If local governments hope to tap federal power and funds for a wide assortment of local improvements – including wildlife protection, historic preservation, neighborhood redevelopment affordable housing, freeway expansion, and flood control – they must adapt to federal guidelines.

Federal and state judicial decisions are another influence on local planning actions: the 1926 U.S. Supreme Court decision to uphold the legality of local zoning powers is an example. Courts review issues of procedural and substantive fairness. For example, have all affected parties had a chance to review a development proposal? Does a planning regulation unfairly reduce the discretion of a private property owner? The courts may also help settle disputes among stakeholders pursuing different objectives.

State governments authorize the formation of local governments and, in so doing, usually provide authorization for planning as well. Although municipalities and counties do most local planning, state agencies generally get involved when localities undertake large projects, such as arterial roadways, rapid transit systems, stadiums, and regional parks.

Moreover, since urban populations predominate in most states, state officials tend to give significant attention to the needs of urban jurisdictions. And when local problems give rise to regional disputes, state governments may intervene. Some states, including Hawaii, North Carolina, Oregon, and Washington, have undertaken state-level planning to manage growth. Other states, such as California and Florida, have mandated specific features and requirements for local planning processes. Nevertheless, most state officials are reluctant to preempt local planning powers: the nature and quality of municipal planning depend on local support.

Because most planning originates at the local level, problems that do not fit neatly within jurisdictional boundaries are difficult to address. For example, regional and metropolitan planning agencies created to deal with air pollution, traffic congestion, or flooding typically lack statutory authority to enforce their plans; instead, they must rely on the voluntary cooperation of their constituent local governments, which may be reluctant to support regional policies that conflict with the interests of their own communities.

Planning disagreements originate in a variety of ways: local governments may resist state or federal requirements; neighboring local governments may disagree regarding a highway route; organized community groups may use established rules (such as environmental review requirements) to challenge the actions of developers or the government. When planning disputes reach the courts, attorneys and judges may become unwitting "planners"; the decisions that result would, in most cases, have been better handled through careful planning intervention. During the 1980s and 1990s, planners realized that leaving disputes to legal adjudication often leads to poor decisions or to costly settlements that leave citizens and governments at odds and fail to embody good planning principles. Planners then began to work with the legal community to find more constructive means of resolving disputes.

PLANNING IN DIFFERENT PLACES

Despite the homogeneity that interstate highways and national retail chains have imposed on much of the American landscape, significant differences – among communities, metropolitan areas, and regions – still remain. These differences affect the nature and scope of planning activities.

Older central cities, for example, tend to have similar sets of interrelated problems: aging infrastructure, abandoned housing, declining central business districts, and obsolete factories. In the face of widespread poverty and shrinking fiscal resources, planners generally focus on economic revitalization, which may include a range of programs – from housing rehabilitation to small business loans to employment programs.

Although large Sunbelt cities continue to attract downtown convention centers and regional corporate headquarters, these cities are prone to real estate boom-and-bust cycles, freeway gridlock, and a shortage of public transit. Moreover, the prosperity that these cities enjoy may not extend to all their inhabitants: racial and ethnic minorities often suffer the same fate as their counterparts in northern industrial cities – for example, poor schools, inadequate social services, and a lack of affordable housing. Planners in such communities may find themselves facing runaway growth one year and a recession the next.

Suburbs are another major planning area. In many metropolitan areas, the populations of maturing suburbs have become more diverse: once the exclusive enclave of the middle class, suburbs are now home to lower-income, blue-collar, and minority residents as well. In addition to harboring more diverse populations, suburbs have become home to more diverse uses: in the years immediately following World War II most suburbs were predominantly residential, but today's suburbs are home to regional shopping malls, multistory office buildings, and sprawling industrial parks. Some of these larger, rapidly growing suburbs on the periphery of major metropolitan areas have come to be known as "edge cities" because of their location and diversity of land uses. As retail establishments and other economic activities move out of older suburban downtowns and into the outer suburbs, many older suburbs are experiencing the patterns of disinvestment that once characterized only inner cities.

Small rural communities pose particular challenges for planners. Those that are located near metropolitan areas may become suburbanized unexpectedly, as rapid growth envelops them. Those that are isolated may suffer from economic stagnation and outmigration of young adults. Rural communities of both types rarely have an established planning function and may need to turn to planning consulting firms for help.

Resort communities – communities near seashores, ski slopes, or other recreational amenities – present special problems for planners because of wide seasonal fluctuations in population. Typical issues that arise in these communities include seasonal versus year-round employment, seasonal traffic problems, and housing costs that are often beyond the means of middle- and lower-income year-round residents.

PLANNING FOR PLACES WITHIN PLACES

Planners and planning agencies plan not only for different types of communities but also for different parts of communities, such as districts and neighborhoods. When detailed plans are prepared for part of a community, such as a central business district, or when a development proposal has significant impacts on existing development, large- and small-scale planning issues become intertwined.

In many metropolitan areas, decades of suburbanization have left urban downtowns – once centers for retail sales, business, government, and entertainment – struggling to maintain an economic base. Planning efforts in many central business districts focus on attracting or retaining specialized retail and office functions, promoting tourism, seeking convention business, or encouraging the development of new entertainment centers, including sports stadiums. Some plans for downtown revitalization include initiatives to increase housing as well. Residential development downtown adds evening and weekend activity, increasing safety and encouraging the use of new downtown amenities, but needs to be supported by grocery stores and other neighborhood services.

Historically, areas surrounding central business districts were used for wholesaling and light industry. As these businesses moved further out, vacant buildings were left behind. Some communities are facilitating the transition of such areas to other uses, including office space, housing and gallery space for artists, and loft residences.

Moving outward from the center of the typical city reveals older residential neighborhoods that may exhibit various states of health. In some cities, for example, mansions and other historic structures have been successfully converted to condominium residences or to office space for nonprofit organizations; in others, older housing has deteriorated as more affluent populations have moved away. Declining middle- and working-class neighborhoods are often the focus of community development efforts designed to improve both the quality of housing and the health of local businesses. . . .

PLANNING WITH DIFFERENT PEOPLE

A community's physical aspects – its size, climate, location, and geography – shape many of its planning issues. Other, less immediately visible, factors – such as a community's history and social and demographic makeup – also contribute to its uniqueness. The populations that planners work with are diverse by any measure: age, race, ethnic origin, religious affiliation, and political belief. Thus, planning necessarily engages people with varied backgrounds, values, expectations, and needs.

Nevertheless, planning practice often fails to take account of such variations. Indeed, many of the principles underlying dominant elements of planning, such as land use and zoning regulation, housing planning, and transportation planning, have reinforced societal stereotypes for women. For example, although most mothers now work outside their homes and urgently need access to day care, many communities limit childcare facilities to commercial zones. Such regulations discourage family day care homes, which are more geographically accessible and may be more affordable to low-income parents. When land use controls and building regulations, such as those governing day care, are left unexamined in the face of demographic change, the result may be inadvertent discrimination against people whose needs were not envisioned when the regulations were first developed.

Research shows that how we experience the physical environment varies with age and background. For example, a study of Los Angeles public parks found that Hispanic families use public parks intensively for sizable social gatherings, whereas African Americans are more likely to visit parks with same-sex peers. Non-Hispanic whites tend to use parks for solo recreation and for outings with their children. In contrast, elderly Chinese have little interest in using "American" parks because their ideal park is an aesthetic garden rather than a recreational facility.

[. . .]

Social and demographic changes affect every aspect of planning, and part of a planner's task is to recognize and address differences in how various segments of the population may be affected by plans. . . . Planners must learn to plan for a population that is diverse *now* and that will continue to change over time: the proportion of residents in one age group or another may shift; resident ethnic or racial makeup may change; and old-time residents who move away may be replaced by newcomers with new values. Planners need to be alert to the characteristics and needs of the populations they serve, and, where inequities exist, planners need to help ensure that they are addressed.

28

Planning Environments

From *The Practice of Local Government Planning* (1988)

Frank S. So and Judith Getzels

Editor's introduction

The basic definition of the planning process, types of plans, and context of planning in this selection comes from the sixth edition of *The Practice of Local Government Planning*, known as the Green Book. (See "Practitioners and the Art of Planning" for descriptions of the various editions of Green Book and their contents as they evolved over time.) The publishers of the Green Book, the International City/County Management Association is a ninety-four-year-old professional organization, founded as the International City Managers' Association. It is an excellent source for information on public administration best practices and it sponsors executive training, publications, and advocacy programs for good government. It publishes the Green Book in collaboration with the American Planning Association.

The planning process described here is often labeled the "rational decision-making model," based on work of Nobel Prize winner Herbert A. Simon (1916–2001), Carnegie Mellon University Professor of Computer Science and Psychology, published in *Administrative Behavior*, third edition (New York: Free Press, 1976) and adapted for planning. The model serves as a normative guide for developing plans and, as the authors note, its various steps may be abbreviated or skipped.

Over time, the rational model has had its critics. Charles Lindblom, former Sterling Professor of Economics and Political Science at Yale University, attacked it in an imaginatively titled essay, "The Science of Muddling Through" (1959), positing that the model cannot be relied on because a decision maker can never master all aspects of the issues and alternatives at hand. Lindblom suggested that "disjointed incrementalism," or making decisions based on limited knowledge, was a more accurate description of what actually took place. Despite these debates (which are covered in Part Two of this *Reader*), the use of the rational decision-making model has prevailed in planning practice, with practitioners assuming that the range of alternatives and choices is limited. Planners apply the model in crafting plans with a variety of time horizons, scales, scopes, and foci.

The authors of this selection describe the five major types of plans. While many plans, especially the comprehensive, area, and system types, emanate from municipal planning departments, other government agencies and even nonprofit and private groups may produce any of the kinds of plans outlined here. Plans generated by governmental agencies are often adopted by municipal, county, regional, or state governing bodies. While they serve many purposes, they are primarily advisory documents.

At the municipal level, state legislation usually embeds the plan-making function in two bodies: a staff agency – a department of city planning, whose director is an executive (mayoral or city manager) appointee – and a planning commission headed by the planning department director and made up of citizens appointed by the executive and approved by the legislative (city council) branches. The commission commonly has a mandate to produce a city plan while the department supplies staff assistance and research to support the plan-making role. Other city agencies often prepare plans for their areas of expertise. For example, a transportation department may develop a plan for mass transit, a housing agency may create a plan for affordable housing.

The authors edited the sixth edition of the Green Book that included this selection. Frank S. So started his thirty-four-year administrative career as a researcher with the American Society of Planning Officials (ASPO) and continued with its successor organization, the American Planning Association (APA), serving as its Deputy Executive Director and, later, Executive Director, retiring in 2001. In addition to this Green Book, So edited its companion, *The Practice of State and Regional Planning* (1986), and its successor, *The Practice of Local Government Planning* (2000). Getzels also joined ASPO and continued with APA, holding the position of Director of Research of the American Planning Association. Together So and Getzels oversaw the Planners' Advisory Service, the professional organization's technical arm, founded in 1949, which has published more than 500 reports on specialized planning issues, exemplified by Getzel's *Zoning Bonuses in Central Cities* (1988) and *Home Occupation Ordinances* (1985).

For additional readings on the role of planning in municipal governments, see Herbert H. Smith, *The Citizen's Guide to Planning* (Chicago: APA Planners Press, 1993). Review Melville Branch's classic *Comprehensive City Planning* (Chicago: APA Planners Press 1985) for an explanation of the general outlines of the field.

Anyone considering becoming a planner should see Warren W. Jones and Natalie Macris, *A Career Worth Planning* (Chicago: APA Planners Press, 2000), which describes the professional development of a planner. The Association of Collegiate Schools of Planning publishes a *Guide to Graduate and Undergraduate Education in Urban and Regional Planning*, twelfth edition (2006) that can be viewed at www.acsp.org/Guide/guide_index.htm. This guide explains the multiple roles that planners can fill and describes all accredited planning programs in the United States.

To sample exemplary plans, take a look at the National Awards program of the American Planning Association at planning.org/awards/2006winners.htm. The website has links to outstanding plans ranging from the Northeastern Illinois Planning Commission's *2040 Regional Framework Plan*, an effort covering growth in six counties in three states, to New York City's West Chelsea/High Line Plan, focused on the redevelopment of a mixed use neighborhood around the transformation of an abandoned one-and-a-half-mile elevated rail line into a linear park.

PLANNING AND PLANS

Planning and plans are common to life, business, and government. To most of us, planning conveys the idea of preparing for the future or getting from here to there. This seems simple enough. However, within the context of local government, and to the various practitioners who work in it, planning and plans can have a variety of meanings depending on the situation. The discussion that follows describes what planners mean by "the planning process" and what various levels of "plans" are.

Planning as a process

At the most general level of planning and management, the planning process is divided into five major steps:

1 *Basic goals.* For local planning, determining basic goals may mean asking questions such as the following: Do we want to grow? Do we want to arrest decline? Do we want to be a center for high-tech industry? What balance do we want between investment in highways and in mass transit?
2 *Study and analysis.* Among other things, planners study land use, population trends, the economic base of the community, and physiographic features.
3 *Plan or policy preparation.* A plan or policy is prepared for the community as a whole or for a segment of it. It is a basic statement of how the community will develop, in what direction, and perhaps at what pace.
4 *Implementation and effectuation.* To carry out plans, planners use tools such as zoning ordinances, land subdivision regulations, capital

improvements programs, and general guidelines for private development and public investment.

5 *Monitoring and feedback*. The last step determines, for example, how well the plans and policies are being carried out, whether the goals were realistic, and whether the study and analysis foresaw new occurrences. Feedback may become the basis for a redesign of the plans and even of the planning system.

These five steps describe the planning process in the abstract. In reality, most local planning consists of three steps: (1) examining inventories and trends in land use, population, employment, and traffic; (2) forecasting the "demand" (in some ways this is a "free market" approach); and (3) planning facilities and services of sufficient capacity to accommodate future demand. This traditional method of planning is an inexact process, depending as it does on forecasting. No matter how sophisticated the forecasting methods, trends are difficult to forecast – especially when a freewheeling entrepreneur makes a location decision that will have major consequences for a city. Nevertheless, planners must become adept at analyzing a number of dependent interrelated systems.

[. . .]

Plans

. . . At the broadest and most general level is the comprehensive plan, at one time called the master plan and also known as the general plan. More recently, products of this kind have been called strategic plans. The comprehensive plan is a document (in multiple volumes for very large jurisdictions) that is the result of lengthy and intensive study and analysis. The geographic scope is the entire community and its regional environment. The time scale is long-range or indefinite. Such a plan is comprehensive in that it tries to link long-range objectives to a number of interdependent elements, including population growth, economic development, land use, transportation, and community facilities.

The comprehensive plan discusses the principal issues and problems of growth or decline facing the community. It will also indicate the main trends that seem inevitable and that need the attention of public or public–private programs. Plans usually contain a mixture of suggestions, proposals for which the means of implementation have not yet been identified, and actual commitments of local governments. For at least a decade plans have tended not to be presented in map form because of the difficulty of identifying precise sites or dimensions; instead, most plans are a series of policy statements.

Depending on the size of the community, a comprehensive plan may also contain chapters on public utility and facility systems if general policies have been established for these systems. However, the degree of detail is nowhere near that in system plans, shown in the second level of Figure 28.1. System plans are more detailed than the comprehensive or general plans yet do not necessarily approach engineering specifications. A system plan summarizes the plans, policies, and programs for a specific network of communitywide facilities – the entire sewer system, the entire fire protection system, or the park and recreation system. System plans deal with specific facilities in specific places.

At the third level are area plans, which include details about certain geographic parts of the community. The most common type is a plan for the central business district. Other area plans are for industrial districts, civic and cultural centers, individual neighborhoods, the community or district encompassing several neighborhoods, and waterfronts. Details in area plans may go down to the block level. The area plan identifies the issues and problems facing a geographic area, develops plans for the future, and provides guidance to local decision makers reviewing private development proposals.

The style of planning can vary with the area. For example, a plan for a central business district may emphasize urban design components and the involvement of the private sector, whereas neighborhood-level planning involves citizen groups and neighborhood organizations.

The fourth type of plan is a detailed engineering plan for components of subsystems, such as trunk sewer lines, major water mains, and street extensions. These plans are based on systemwide

<table>
<tr><td>1 Comprehensive plans
Plans containing basic policies for land use—residential, commercial, industrial, public—and general policies for public systems</td><td>3 Area plans
Central business district
Industrial districts
Civic and cultural centers
Community districts
Waterfronts
Neighborhoods</td></tr>
<tr><td>2 System plans
Land use
Sewers
Water
Storm drainage
Transportation: streets and mass transit
Solid waste
Parks and recreation
Libraries
Government administration
Police
Fire
Health
Public works buildings and facilities
Cultural
Institutional</td><td>4 Plans for subsystem components
Engineering plans for construction of trunk sewer line, major water main, or a street extension

5 Site plans
Library
Fire station
Neighborhood park</td></tr>
</table>

Figure 28.1 Types and levels of plans

plans and typically are prepared under the direction of the public works department.

At the fifth level are site plans for facilities such as libraries, fire stations, and neighborhood parks. Such facilities require land purchase, architectural and engineering drawings, scheduling in the capital improvements program, possible public hearings for site selection, meetings with neighborhood or community groups, and coordination with neighborhood development plans. System plans are considered separately from site plans, because linear systems provide the basic network for planned development. If the systems are not in, development may not occur. After they are constructed, the availability of utilities may accelerate development. In other words, these linear systems shape urban development.

THE CONTEXT OF PLANNING

Public planning is never as simple and direct as a textbook makes it sound. A planner must learn to be a jack-of-all-trades. Today's planner works in a complex intergovernmental web of plans, policies, and regulations. Planners must have the patience and the understanding to work in the changing world of politics. Planners must comply with a large body of land use law and must be ready to contribute to the making of that law. To make reasonable forecasts, planners need a working knowledge of practical economics. Planners must understand how a community's fiscal situation will limit or support the implementation of a plan. Planners must understand how to protect the natural environment and how to encourage development that most people will find aesthetically attractive. . . .

A Realistic Approach to City and Suburban Planning . . . Ingredients of Success

From *The American City: What Works, What Doesn't* (2002)

Alexander Garvin

Editor's introduction

Change is the *modus operandi* of planning. Urban and regional planners are engaged in it through their analytical studies, plans, and implementation programs. Consider the concluding sentence of the mission statement of the American Planning Association, which pledges to "contribute to the public well-being by developing communities and environments that meet the needs of people and society more effectively." The use of the phrase "more effectively" implies that the *status quo* is unacceptable and calls for change.

Garvin demystifies them, forging a theory of urban and regional planning practice for the United States and other places with market economies. First, it chooses physical development as the suitable arena for planning. Second, it translates an abstraction, "change," into a measurable concept (sustained private sector investments in a city or suburb stimulated by public sector action aimed at producing healthy and safe places that promote the general welfare of their inhabitants). Third, it sees planning as figuring out what, where, when, and how to take public action. Fourth, it identifies six factors that shape public action: the market, location, design, financing, entrepreneurship, and time. Fifth, it argues that the test of the success of planning is a positive answer to the question: "Did it stimulate the desired private investment?" This approach is the foundation of Garvin's planning practice, notable for its plans for Atlanta, Georgia (2004), New York City (2006), and elsewhere.

The Beltline Emerald Necklace: Atlanta's New Public Realm (see Plate II.7) is illustrative. Commissioned by the Trust for Public Land as a conceptual scheme, it has received major municipal support. In this plan, Garvin concentrates on open space and transportation, including parks, highways and streets, transit systems, and other public capital facilities to stimulate the physical reorganization of the city and associated private investment. Here, he is emulating the early proponents of city planning, the authors of the *Standard City Planning Enabling Act* (1928) and their followers, who clearly identified what Garvin calls "the public realm" as the key element in the master plan. Garvin, in emphasizing its importance for the twenty-first century, is reinvigorating this tradition.

Atlanta is emblematic of twenty-first-century urbanism. Infamous for its traffic congestion and air pollution – in 1999 the federal government suspended several funding programs until the region complied with the Clean Air Act – it anticipates that its population of 5 million will rise to 7.5 million in the next three decades. It is spread out, auto-dependent, and favors single-family housing on large lots. Its in-town neighborhoods have not been able to attract new residents at the same rate as the suburbs. While there are many reasons for this growth pattern, one clear indicator of the city's lack of competitiveness is a quality-of-life measure: open space. Atlanta ranks in the bottom 15 percent of the nation by this measure. Its 7.8 acres per resident is well below the national average (16.2 acres). And only five of the nation's large cities (Mesa, San José, Fresno, Nashville, and Tucson) have less open space as a percentage of total land area.

Atlanta is at a crossroads. It can go on growing as it has (though its urban problems will increase), or it can reinvent itself through planning and investment, as suggested in *The Beltline Emerald Necklace: Atlanta's New Public Realm.* The plan presents a strategy to enhance the city's existing features and advance its development, seizing underused assets, such as a twenty-mile rail corridor encircling downtown and 600 acres of disparate parks. With strategic public investments, 3,000 acres (or about 4 percent of the city's land area) could be transformed into an interrelated system of parks, transit lines, and transit-oriented development sites. The additional 1,400 acres in new parkland, a circular transit line, and three new rapid transit stations will not only link the city's forty-five neighborhoods but also enable the city to reorient itself "from a city framed by highways to a city framed by a magnificent public realm."

While Garvin frankly admits his debt to Frederick Law Olmsted and Boston's Emerald Necklace for the vision of his plan, its strong implementation section is strictly modern. In late 2005, responding to a positive vote by the citizens, the City, Fulton County, and the Atlanta Public Schools created a Beltline Tax Allocation District (TAD) to fund the project, estimated to cost approximately $2.5 billion over the next twenty-five years. The city quickly geared up: the mayor appointed an advocacy body, the Beltline Partnership, a public–private partnership to oversee ongoing planning and project development, and the Atlanta Development Authority, authorized to administer the TAD funds, created Atlanta Beltline Inc. as the project's implementing arm. By fall 2006, Atlanta Beltline Inc. had a draft work plan, one involving extensive community input, and technical studies for the Beltline's first stages, aiming to finalize the work plan after a public comment period. To follow the progress of this important effort, refer to the Partnership website: www.beltlinepartnership.org.

Alexander Garvin, Yale professor and consultant, has developed his view of planning by drawing on his experiences as a practicing planner, public servant, university professor, and student of cities. He served as vice-president for planning and development for New York's Lower Manhattan Development Corporation (LMDC), charged with overseeing the selection of a plan to reconstruct the World Trade Center site. Under his guidance, the LMDC ran the competition that resulted in the choice of Daniel Liebeskind's scheme. As chief designer of New York 2012, he conceived the design for the New York City Olympic bid (though New York's scheme was not chosen). Distilling years of observation, Garvin's book, *The American City: What Works, What Doesn't* (New York: McGraw-Hill, 2002), from which this selection is drawn, evaluates past models of planning. Notably, he has visited each place about which he writes, exploring its layout, observing every element – topography, design devices, scale, integration into the surrounding fabric, and use by people.

To delve into the history of thinking about the public realm, see Charles E. Beveridge and Carolyn F. Hoffman (eds), *The Papers of Frederick Law Olmsted*, Vol. I, *Writings on Public Parks, Parkways, and Park Systems* (Baltimore, MD: Johns Hopkins University Press, 1997). References on particular Olmsted works include Cynthia Zaitzevsky, *Frederick Law Olmsted and the Boston Park System* (Cambridge, MA 1982), and Roy Rosenzweig and Elizabeth Blackmar, *The Park and the People: A History of Central Park* (Ithaca, NY: Cornell University Press, 1992). For a discussion of modern greening systems, see Blaine Bonham, Jr, Gerri Spilka, and Darl Rastorer, *Old Cities/Green Cities: Communities Transform Unmanaged Land* (Chicago: APA Planners Press, 2002), and Alexander Garvin, *Parks, Recreation and Open Space* (Chicago: APA Planners Press, 2001). For an excellent collection of cutting-edge open-space projects around the world, see Peter Reed, *Groundswell: Constructing the Contemporary Landscape* (New York: Museum of Modern Art, 2005). Public Broadcasting System's "Edens Lost and Found" has an accompanying monograph, *Edens Lost and Found: How Ordinary Citizens are Restoring our Great American Cities* (White River Junction, VT: Chelsea Green Publishing, 2006) that profiles open-space efforts in Chicago, Philadelphia, Los Angeles, and Seattle. To obtain a copy of *The Beltline Emerald Necklace: Atlanta's New Public Realm*, go to the Trust for Public Realm website: www.tpl.org.

There is agreement neither on what to do to improve our cities and suburbs nor on how to get the job done. Some believe the answers are a matter of money; others believe they involve politics, or racial and ethnic conflict, or some other factor. One thing most people share, though, is disillusionment with planning as a way of fixing the American city.

This disillusionment with planning is far from justified. There are dozens of projects that are triumphs of American planning:

- Chicago would not have twenty-three miles of continuous parkland along Lake Michigan if this land had not been included in the city's comprehensive plan of 1909.
- The glorious antebellum sections of Charleston, South Carolina, would not have survived if the city had not adopted zoning in 1931.
- Pittsburgh would not rank sixth in the nation as a major corporate headquarters center if it had not virtually rebuilt its downtown during the 1940s and 1950s.
- Portland, Oregon, would not be a lively retail and employment center if during the 1970s and 1980s it had not enriched its pedestrian environment, built a light rail system, and reclaimed its riverfront.

Such triumphs are easy to overlook. Once a problem is solved it disappears and is forgotten. Even local excitement over a successful project rarely spills over into national publications other than those with a narrow group of readers (preservationists, environmentalists, realtors, lawyers, architects, bankers, or some other group that is intimately involved with specific sets of city problems).

Many people are disillusioned with planning because so many of its promises are not kept. Usually these promises are made in good faith by planners who believe that their job is to establish municipal goals and provide blueprints for a better city. Too often these planners' efforts end before they consider how they will obtain political support for their proposals, who will execute them, or where the money to finance them will come from. Disillusionment with planning also develops when physical improvements fail to solve deep-seated social problems. This is not the fault of planning. After all, fixing cities does not fix people. The disillusionment in this case is a product of misplaced expectations. Crime, delinquency, and poverty are afflictions of city residents, not of the cities themselves. Such problems can be found in suburban and rural areas as well.

We need more realistic expectations of what planning can accomplish. While it cannot change human nature and is therefore not a panacea for all urban ills, it surely can improve a city's physical plant and consequently affect the safety, utility, attractiveness, and character of city life. When Chicago began creating its waterfront parks, for example, large sections of the shoreline of Lake Michigan were being used as rail yards and garbage dumps. Simply removing these uses reduced hazards and made neighboring property more attractive.

We also need a better understanding of how effective planning is translated into a better quality of life. Planners operating in a vacuum do not accomplish it. By themselves, planners cannot accomplish very much. Improving cities requires the active participation of property owners, bankers, developers, architects, lawyers, contractors, and all sorts of people involved with real estate. It also requires the sanction of community groups, civic organizations, elected and appointed public officials, and municipal employees. Together they provide the financial and political means of bringing plans to fruition. Without them even the best plans will remain irrelevant dreams.

Finally, the planning profession itself needs to improve its understanding of the way physical changes to a city can achieve a more smoothly functioning environment, a healthier economy, and a better quality of life. For example, the restoration of Charleston's historic district generated substantial tourist spending and the reconstruction of the bridges and highways leading into downtown Pittsburgh reduced the cost of doing business and initiated an era of major corporate investment. These and other successful planning strategies are too frequently ignored in the search for more innovative prescriptions.

At its best, planning alters the very character of city life. During the 1970s and 1980s, Portland completely reorganized vehicular and pedestrian circulation. The business district was encircled by a ring road that greatly improved motor vehicle accessibility. A light rail system provided transit service from the suburbs. Pedestrian precincts were established by transforming the old downtown highway into riverfront park, by eliminating private motor vehicles from two downtown streets and repaving them as transit ways, and by acquiring several downtown blocks and converting them into new public parks. As a result, Portland became a safer, more convenient, more beautiful city. It also became a more attractive destination for the city's rapidly growing metropolitan region, drawing tens of thousands of additional weekday shoppers and weekend visitors.

Despite many remarkable successes, American planning has been plagued with continuing mistakes. These mistakes were and are avoidable. More than four decades have passed since Jane Jacobs in her pioneering book, *The Death and Life of Great American Cities*, observed that we had spent billions of dollars for:

> Housing projects that are truly marvels of dullness and regimentation. . . . Civic centers that are avoided by everyone but bums, who have fewer choices of loitering places than others. Commercial centers that are lack-luster imitations of standardized suburban chain store shopping. Promenades that go from no place to nowhere and have no promenaders. Expressways that eviscerate great cities. This is not the rebuilding of cities. This is the sacking of cities.

Four decades and hundreds of billions of dollars later, her criticisms still ring true. Most cities continue to lack housing, civic and commercial centers, places to congregate and promenade, traffic arteries. In too many cases, the attempt to remedy the situation constituted further "sacking of cities." These attempts may have been financially and politically feasible. However, they failed because they were conceived without proper consideration as to whether they would benefit the surrounding city.

DEFINING THE PLANNING PROCESS

Much of the nation's unsuccessful planning arises from the enormous belief that project success equals planning success. Highways that are filled with automobiles, housing projects that are fully rented, and civic centers with plenty of busy bureaucrats maybe successful on their own terms. The cities around them, however, may be completely unaffected. Worse, they may be in even greater trouble than they were prior to these projects.

Only when a project also has beneficial impact on the surrounding community can it be considered successful planning. Thus, planning should be defined as *public action that generates a sustained and widespread private market reaction* which improves the quality of life of the affected community, thereby making it more attractive, convenient, and environmentally healthy. That is precisely what has occurred wherever urban planning has been successful.

- When Chicago transformed its lake shore into a continuous park and drive, the real estate industry responded by spending billions to make it a setting for tens of thousands of new apartments. Hundreds of thousands of people responded by coming to its lovely lakeshore parks.
- When Charleston preserved its old and historic district, it retained an extraordinary physical asset that, decades later, would attract a growing population, draw millions of tourists and provide the basis of a thriving economy.
- When Pittsburgh cleared its downtown of the clutter of railyards, and warehouses; reduced air and water pollution; and built new highways, bridges, and downtown garages, businesses responded by rebuilding half the central business district.
- When Portland invested in a riverfront park, a light-rail system, and pedestrianized streets, the private sector responded by erecting office buildings, retail stores, hotels, and apartment houses. Downtown Portland was transformed from a place that had been deserted from Friday evening to Monday morning into an area that attracted people seven days a week.

[. . .]

. . . This broad definition highlights the fact that planning is about *change*: preventing undesirable change and encouraging desirable change. It may involve a tax incentive, a zoning regulation, or some other technical prescription, but only as a mechanism for instigating change. The important element is change itself. Planners obtain changes in safety, utility, and attractiveness of city life through strategic public investment, regulation, and incentives for private action.

[. . .]

A NEW APPROACH TO PLANNING

We need a new approach to planning that explicitly deals with both *public action* and the *probable private market reaction*. Such change-oriented planning requires general acceptance of the idea that while planners are in the change business, others will actually effect the changes: civic leaders, interest groups, community organizations, property owners, developers, bankers, lawyers, architects, engineers, elected and appointed public officials – the list is endless.

Being entirely dependent on these other players, planners must concentrate on increasing the chances that everybody else's agenda will be successful. They may choose to do so by targeting public investment in infrastructure and community facilities, or by shaping the regulatory system, or by introducing incentives that will encourage market activity. But whatever they select, their role must be to initiate and shepherd often-controversial expenditures and legislation. More important, the public will be able to hold them accountable by evaluating the cost effectiveness of the private market reaction to their programs.

Only when this approach to planning takes hold will we get beyond the technical studies, needs analyses, and visions of the good city that currently masquerade as planning and get on with the business of fixing the American city.

There is no formula that guarantees a desirable private market reaction in response to public action. However, there are six ingredients that must be intelligently dealt with for any project to succeed. They are: market, location, design, financing, entrepreneurship, and time.

The need to consider these ingredients may seem obvious. Unfortunately, the proliferation of stillborn projects reveals how little they are understood. Otherwise, why would there be housing for which there is no *market*, commercial centers that are in the wrong *location*, civic centers for which *financing* is not available, places whose *design* makes them unpleasant and unsafe areas in which to congregate, economic development projects whose completion is beyond the *entrepreneurship* of the responsible public agency, and public works whose *time* has passed but are still under way.

If any of these six ingredients is absent or if they are not combined in a mutually reinforcing fashion, the project will fail. Even when all the ingredients are properly combined, they may be insufficient to guarantee project success because city planners, unlike chefs, cannot keep unexpected ingredients from getting into the pot. Nevertheless, an intelligent mix of market, location, design, financing, entrepreneurship, and time is the key to success. Thus, an understanding of how these elements operate and interact will increase the likelihood of favorable results.

Market

The existence of a market for any urban planning prescription is primary, for without it there is no reason even to consider action. The word "market" is not synonymous with population. It means a specific population's desire for something and its ability and willingness to pay for it in the face of available alternatives. Nor is market synonymous with "need". Too often what one person calls a need is really a preference for what other people ought to have.

To be successful, an urban planning prescription must reflect both market demand and supply. The demand side requires a user population with enough money to purchase what it desires and the willingness to spend it. It also requires enough spending to support both the capital cost and operating expenses of a project. Without this critical mass of customers, if private action is involved, in addition to user charges there will have to be subsidies; if it is a public project, the electorate will have to be willing to pay the necessary taxes.

[. . .]

Location

Location consists of two elements: a site's inherent characteristics and its proximity to other locations. Site characteristics alone may be sufficient to make it attractive. A spectacular view is an example. Another is an architecturally distinctive housing stock, such as the one that made renovation particularly inviting in the historic districts of Charleston.

Site conditions can also ruin an otherwise desirable location. During the first half of the twentieth century, air pollution in downtown Pittsburgh was so serious that streetlights often remained on twenty-four hours a day. Raw sewage polluted both riverfronts. Daytime traffic congestion seriously restricted both circulation and business activity. In order to alter these inhibiting conditions, the city obtained state legislation that allowed it to regulate air and water pollution, rebuild its highways and bridges, create more than 5,000 parking spaces and clear away the tangle of rail yards, dilapidated warehouses and obsolete manufacturing lofts. Once these site conditions were eliminated, property owners invested hundreds of millions in redevelopment. Within a couple of decades, more than half of the business district had been rebuilt.

Proximity involves both time and space. The temporal dimension is shaped by technology and can be understood in terms of available means of conveyance. During the eighteenth century, when people were concerned with walking distances, cities had to be compact and densely built up. By the end of the twentieth century, when distance is measured in driving time, the resultant landscape is "spread city."

The spatial dimension of proximity involves interdependence with neighboring areas. An obvious example is the relationship between movie theaters, parking facilities, and eating places. On a larger scale, nineteenth-century warehouse and manufacturing districts often developed in close proximity to waterfront areas through which they received and shipped goods and materials.

[. . .]

Design

The most misunderstood of the six ingredients of success is design. Too often, it is thought of as decoration that can be applied after the important decisions have been made. In fact, design is the physical manifestation of any prescription and, therefore, is integral to its success or failure from the time of inception.

Design is not just a matter of architectural style. Styles go in and out of fashion; successful planning has to survive for decades. Other more enduring aspects of design are more important. They include the arrangement of project components, the relative size of those components, and their character. Each element affects a project's utility, cost, and attractiveness. When they are organized in a mutually supportive manner, the result is an identifiable destination that provides an auspicious place for the activities occurring there. When arranged to fit the right combination of market, location, financing, entrepreneurship, and time, the result is a successful project.

[. . .]

Dimensions are as important as the arrangement of components. They have to be correct from the beginning. For example, traffic engineers suggest a width of 12 ft for every lane of traffic. That may not be enough on busy streets where trucks keep stopping to unload merchandise. Similarly, building codes mandate a minimum height for every habitable room (usually 8 ft). How much higher should one build? The answer should vary with the type and floor area of each room. But, whether the product is a traffic artery or a residence, the quality of the results will depend on dimensional appropriateness.

[. . .]

Financing

Every prescription for fixing cities requires financing. When this involves governmental action, as is the case with parks, the financing comes from taxes. . . .

[. . .]

Financing is also essential to private sector activity. Privately financed projects need *capital* to cover startup costs, a short-term *development loan* to pay expenses until it is operational, and a *permanent mortgage* to replace the other two when the project is complete and tenanted. The obvious place to obtain financing is a bank. Banks lend their depositors' money to developers whose projects pay a large enough return to keep depositors happy and contribute towards covering the costs of bank operations. In other words, developers pay banks for the use of their money. The price will depend on its assessment of the risk involved. If the deal looks too risky, the bank will not lend a penny.

Most banks will not lend enough to cover project cost. The rest of the money, the *equity* investment, usually comes from the developer and from investors who have confidence in the venture. Investors know that the bank has not lent enough money to complete the project. They know that if the venture fails, the bank may recoup its investment but they may not. They also know how much the bank is getting for its money. Consequently, developers have to pay investors a higher price for equity funds than they are paying for bank money. Developers will put up their own money if bank mortgages and investor equity do not cover all development costs. Typically, if something goes wrong and the investment has to be liquidated, the bank mortgage will be repaid first, then the equity investors, and finally the developer. Since the developer is taking the greatest risk, he or she will not go into the venture unless the return is better than that of the bank and the equity investors.

In other words, money is obtained at different prices, depending on risk and availability. Mortgage money is usually the least expensive. Equity money is more expensive. The developer's money is the most expensive. The greater the proportion of project costs that comes from other sources (the greater the *leverage* of the developer's cash investment), the more attractive the venture will appear to be to the developer. Government can increase the likelihood of project success by creating an investment climate in which bank financing is readily available and developers maximize leverage.

[. . .]

Entrepreneurship

No prescription is self-implementing. Each requires talented public and private entrepreneurs. Without them a perfectly appropriate prescription will not get off the ground. Entrepreneurs conceive projects, often when others are unaware that there are any opportunities available. They assemble and coordinate the various players who will execute whatever needs to be done. Without the extra drive that entrepreneurs supply, these other players would be overwhelmed by the uncertainties of the marketplace.

Entrepreneurs do not appear automatically whenever there is unfulfilled demand for something. They have to believe that the risk of failure is minimal and the rewards that come with success are generous. Unless such favorable conditions are prevalent, entrepreneurs will exploit other, more attractive opportunities.

Public projects often fail because public officials ignore the role of entrepreneurship. They mistakenly believe that once a project has been assigned to a government agency, its role is purely administrative. In fact, public entrepreneurs are needed to assemble, coordinate, and inspire all the participants in the development process. . . .

While it is easier to understand the role of an entrepreneur in the private sector, it is essentially the same . . . performed by public officials. . . . The role includes coordinating a plethora of participants, dealing with uncertainty, recognizing available opportunities that have not yet been exploited, and frequently accomplishing things in ways that have never been tried before. The difference between private and public entrepreneurial activity is only in the form of payment. The private entrepreneur is paid in hard currency; the public entrepreneur, in power. The sort of people capable of getting things done, however, will have to be extremely well paid in their respective coin. In many cases private and public entrepreneurs work side by side. . . .

[. . .]

Time

Three time sequences affect success. The first is relatively brief: the period during which a person passes through an area. The second takes into account what will occur twenty-four hours a day, seven days a week. The third may take decades, during which political and financial climates will certainly change many times.

Developers of retail shopping facilities are perhaps the most skilled in predicting a person's activity pattern within an area. They have to be skilled in dealing with this brief time period because their tenants' profits are dependent on transient customer activity and their own profits are dependent on tenant success.

[. . .]

The movements of a single individual, on the other hand, are irrelevant in planning for a twenty-four-hour day and a seven-day week. Such planning requires providing a suitable environment for a wide variety of users on a continuing basis. Thus, the crucial questions revolve around the people who are likely to be in an area over a seven-day period, what will they want to do, how many people are needed to support those activities, and in what ways should the environment be organized to accommodate satisfactorily those people and activities.

[. . .]

MANIPULATING THE INGREDIENTS OF SUCCESS TO OBTAIN DESIRABLE PRIVATE-MARKET REACTION

. . . City and suburban planning . . . must be evaluated in terms of the cost-effectiveness of the induced private-market reaction and civic gain (whose value is harder to assess). The same ingredients that determine the community impact of profit-motivated projects determine that reaction.

While private developers rarely seek to generate and sustain a widespread private-market reaction, some of their projects make profound changes to surrounding communities. . . .

The only way to ensure that market demand will spill over into the surrounding area is to plan *not* to satisfy that market within the project. Then there will be a reason for people to go elsewhere. . . .

[. . .]

For a project to generate a sustained market reaction in surrounding areas, it must exploit linkages to those areas.

[. . .]

Private real estate ventures . . . do not provide financing or entrepreneurs for other projects. At best, they demonstrate the potential of further real estate activity, perhaps attracting other developers and reducing the wariness of previously skeptical lending institutions.

Government programs, on the other hand, can manipulate financing and entrepreneurship in a manner that affects market activity. . . .

The only period of time during which a project can affect surrounding market activity is the period during which it is in operation. Its impact, however, is particularly important when it supplies neighboring businesses with additional customers during slack periods. The increased consumer spending may support neighboring businesses whose market would not otherwise be large enough. . . .

THE ROLE OF GOVERNMENT

Public action is increasingly difficult to justify if it achieves only a single purpose without providing additional benefits to the surrounding community. Too many cities want a plentiful supply of electricity but are unwilling to permit power plants within their boundaries. Too many neighborhoods want clean streets but bitterly resist sanitation garages. . . . Citizens regularly defeat such single-function public projects. . . .

Three questions need to be answered positively to justify public action. Can it be implemented? Will it generate a private market reaction? Is that private market reaction worth it? . . . When all six ingredients of success are properly combined, a project can be successfully implemented and can generate a private market reaction. As a result of Philadelphia's urban renewal strategy for Society Hill, for example, between 1960 and 1990 the number of housing units in the neighborhood more than doubled and their value increased twenty-five times at a time when values citywide increased only six times. . . . The revitalization of Society Hill . . . [was] purchased at bargain-basement prices: the city's share of urban renewal subsidies was approximately $9.5 million.

Some cities have enacted zoning ordinances that allow parking requirements to be satisfied at offsite locations. This increases pedestrian traffic between those parking facilities and the consumer's ultimate destination. Other cities offer a bonus of additional rentable space to developers who provide suitably designed open space, thereby increasing pedestrian traffic to and from more congested nearby locations.

The preceding examples involve the use of *investment* (housing subsidies), *regulation* (parking requirements), or *incentives* (a zoning bonus) to alter four of the ingredients of success (market, location, design, time of operation). Success in generating further market activity may also require the other two ingredients: financing and entrepreneurship. . . .

We can overcome citizen opposition and ensure project feasibility if we stop thinking solely in terms of individual projects. Instead, we must make decisions that are *also* based on the probability of a desired private-market reaction. Then the public dialogue will shift from consideration of the project itself to the ways in which its market, location, design, financing, entrepreneurs, and times of operation will benefit the surrounding community. More important, we will increase financial and political feasibility while simultaneously increasing the likelihood of the desirable, sustained, and widespread market reaction that is characteristic of good city and suburban planning.

30

Demographic Futures as a Guide to Planning California: Latinos and the Compact City

From *Journal of the American Planning Association* (2001)

Dowell Myers

Editor's introduction

Demographic analysis is a fundamental tool for urban and regional planners. They incorporate data regarding an area's current and future population and economic base in preliminary planning studies, recommendations, and plans.

Generally, these data come from the *Decennial Censuses of Population and Housing*, the surveys taken by the U.S. Census Bureau. The Bureau also contributes many other useful databases, including the *American Community Survey* (a sample of selected population characteristics covering populous places) and the Economic Census (a survey of business establishments taken every five years for the nation, states, metropolitan areas, counties, places, and zip codes). Other agencies and private groups – including regional councils of governments, metropolitan planning organizations (MPOs), utility companies, and private firms such as Woods & Poole Economics, Moody's Economy.com, and Claritas – also collect and distribute data.

Mandated by Article 1, Section 2, of the Constitution of the United States, the federal government has carried out a national census every ten years since 1790 in order to apportion seats in the House of Representatives among the states. Thomas Jefferson, as the Secretary of State, supervised the first census when a handful of horse-riding U.S. marshals counted 3.9 million inhabitants. Today, the Census Bureau, lodged in the U.S. Department of Commerce since its creation in 1902, supervises the count. For the 2000 census, the Bureau hired 860,000 temporary workers to help enumerate the nation's nearly 300 million people.

When undertaking the decennial census, the Bureau has mailed either the Short Form (sent to 83 percent of all households in 2000) or the Long Form (sent to 17 percent in 2000). The Short Form asks the names, ages, genders, races, and Hispanic ethnicities of the members of the household; relationships among household members; and tenure of the housing unit. The Long Form includes the Short Form questions and other questions on the social, economic, and housing characteristics of individuals and of the household. In tabulating the data, the Bureau aggregates figures for a variety of geographies, ranging from the nation to the block. (It discloses any geographic changes from period to period in the various categories, so analysts can account for variation.) The Bureau disseminates the information in hard copy through 1,400 federal depository libraries and 1,800 government offices, and, since 2001, digitally through its portal American Factfinder (factfinder.census.gov/). American Factfinder has data from the 1990 and 2000 censuses, the *American Community Survey*, the *Economic Census*, and other reports.

To reduce the occurrence of counting errors – especially in immigrant communities where citizens are not accustomed to answering questionnaires or do not understand the importance of the survey – the Bureau does widespread outreach to identify and instruct non-English-speaking residents. In addition, many cities undertake extensive preparations, including address verification, to ensure a high response rate.

Over the years, the government has added and subtracted questions, making the census a dynamic instrument, reflective of the national interests of a given time. For example, in 1810 it inventoried manufacturing, in 1850 it added questions on crime, in 1930 it asked about radio ownership, and in the 1950s and 1960s it assessed housing conditions. By the 1980s, as the nation was more attentive to population diversity, it expanded questions relating to race and ethnicity, adding categories including Asian and Hispanic. In 2000 it allowed respondents to indicate whether they were mixed-race or mixed-ethnicity.

The Census Bureau first published data on urban population in 1930, indicating those living in incorporated places having 2,500 or more inhabitants and unincorporated places having 10,000 or more with a density of 1,000 per square mile. It calculated urban trends back to 1880. And it offered a national portrait of urbanites with data on race, gender by race, age distribution, marital status, and number of dwellings and tenure. It also outlined the level of urban population by state.

Clearly, the Census Bureau delivers a massive amount of information. For the Short Form alone, it produces 286 tables and many reference and thematic maps. The Bureau also has an estimate program, issuing revised numbers annually for the larger geographies (the nation, metropolitan and micropolitan areas, states, counties, cities, and towns). In 2010, in an effort to provide more timely information, the Bureau will eliminate the Long Form, replacing it with the *American Community Survey* (ACS). At present, the ACS, a 3 percent sample, provides socioeconomic data for geographic areas with populations of 65,000 and up. In the future, it will issue reports on such smaller geographies as the census tract and block group after aggregating data over three- and five-year intervals.

Demographers, economists, urban and regional planners, and others use all this data to make estimates, projections, and forecasts about population. An estimate is a calculation that refers to the present and past numbers, undertaken as a substitute for a direct count. A projection is a precise measurement of a future number based on a set of assumptions. A forecast is also a measure of the future, but includes an assessment of the likelihood of its assumptions proving true. (All forecasts are projections, but not all projections are forecasts.) All three assessments employ statistical techniques such as trend analysis, simulation, or modeling and assign ratios, or shares, of a larger population.

Urban and regional planners have employed demographic analysis since the early days of their profession, writing instructional books to outline their methods. A quick review of these texts – starting with Ladislas Segoe's *Local Planning Administration* (Chicago: International City Managers' Association, 1941) and moving to Krueckeberg and Silver's *Urban Planning Analysis: Methods and Models* (New York: Wiley, 1973), Richard Klosterman's *Community Analysis and Planning Techniques* (Savage, Md: Rowman & Littlefield, 1990), and Charles Hoch, Linda Dalton, and Frank S. So's *The Practice of Local Government Planning* (Washington, DC: ICMA Press, 2000) – shows a continuing concern with population change (its size and composition) and spatial distribution.

As data analysis and planning have become more sophisticated, planners' attention has moved from assessing gross trends pertaining to the total population to reviewing more fine-grained tendencies. Here is a hypothetical example: Recent research shows that between 1990 and 2000 central business districts in forty-five cities experienced an 8 percent increase in households. The question is: Will this movement continue? Seeking a response, an urban planner ascertains that this behavior applies to certain segments of the population (people aged twenty-five to thirty-four and forty-five to sixty-four). He or she then takes a look at the population projections for these groups in the future and makes a judgment about the proportion of the groups that can be expected to make the same choice. So if 10 percent of the fifty-five to sixty-four-year-olds in a metro area moved downtown between 1990 and 2000, the planner might conclude that 10 percent of the forty-five-to-fifty-four-year-olds, the next generation to become empty nesters, would act similarly between 2000 and 2010.

Dowell Myers, the author of this selection, is professor of Urban Planning and Demography, School of Policy, Planning and Development, University of Southern California. He has written extensively on the use of data in planning, focusing on immigration, housing demand, urban growth, and sprawl – issues he explores in *Analysis with Local Census Data: Portraits of Change* (San Diego, CA: Academic Press, 1992). In 2001, Myers edited "Putting the Future Back in Planning," a special section of the *Journal of the American Planning Association* (67: 4). And in 2007 he published the prize-winning *Immigrants and Boomers: Forging a New Social Contract for the Future of America* (New York: Russell Sage Foundation). In this selection, he reviews the advantages of the census data (continuity, comprehensiveness, reliability, and accuracy) and their limits (issues of timeliness and the disadvantages inherent in any numbers-based assessment not tempered by qualitative data).

For a masterly demonstration of the use of census data to track the transformation of American life over time, see Michael B. Katz and Mark J. Stern, *One Nation Divisible: What America was and What it is Becoming* (New York: Russell Sage Foundation, 2006). For additional reading on the use of demographic analysis in urban and regional planning, see Arthur C. Nelson, *Planner's Estimating Guide* (Chicago: APA Planners Press, 2004). Philip R. Berke, David R. Godschalk, and Edward J. Kaiser, with Daniel A. Rodriguez, *Urban Land Use Planning*, fifth edition (Urbana, IL, and Chicago: University of Illinois Press, 2006), and Jerry Weitz's *Jobs–Housing Balance* (Chicago: APA Planners Press, 2003) apply census data to specific fields. For discussions about projection and estimating techniques, see Andrew Isserman's thoughtful articles, including "Projection, Forecast and Plan: On the Future of Population Forecasting," *Journal of the American Planning Association*, 50: 2, (1984); "The Right People, the Right Rates: Making Population Estimates and Forecasts with an Interregional Cohort-component Model," *Journal of the American Planning Association*, 59: 1 (1993); and "Economic Base Studies for Urban and Regional Planning," in Lloyd Rodwin and Bishwapriya Sanyal (eds), *The Profession of City Planning: Changes, Images, and Challenges, 1950–2000* (New Brunswick, NJ: Center for Urban Policy Research, 2000).

The overarching purpose of planning is to meet the needs of residents in communities more effectively. As the nature of planners' clientele changes over time, so must the plans themselves. Effective planning and policy making require us to think prospectively about changing demographics. At a minimum our "current" data are always out of date. By the time the 2000 census data are released in detail, it will already be 2002 or 2003. Unfortunately, relying on such *uncurrent* data continually places our understanding behind reality.

Even an accurate picture of the present is hardly enough. So often we forget that projects or plans designed today will not yield their intended benefits until a future date. Who will be the users at that time? In effect, we should aim our plans to satisfy the needs of the future population, not the people we know from the present or the past.

Demographic futures – which include population projections, detailed descriptions of changing characteristics (population analysis), and normative interpretations – have special advantages for planners. They direct our attention to the future residents of the community and keep our attention focused on people as the object of planning. Not only does population analysis keep planners focused on the changing needs of residents, but it also provides much needed insights about likely future *impacts* for which we should prepare. Planning analysts regard population statistics as integral to virtually all aspects of planning. Indeed, it has been argued that the changing population profile of a community should provide as central a context for planning as does the land use map. Just as land use underlies most planning activities, so do changes in the number and types of residents. There are direct consequences for all planning functions: housing and neighborhood planning, travel patterns, commercial development, employment patterns, parks and recreation, schools, health care, and virtually all activities that occupy space or impact land use patterns.

Despite the obvious importance of demographic change for planning and policy making, substantial disagreements surround the *interpretation* of ongoing changes. In fact, it is the very importance of those changes that leads to such disagreements.

Since the changes are so multifaceted, they lend themselves to many different stories of the future, each emphasizing a different aspect of change, each interpreted from a different value set, and each emphasizing different policy responses. Thus, demographic futures constitute more than a simple prediction of the changing population; they amount also to a construction of the changing identity of planners' clientele and an assessment of the urgent priorities for policy attention. Even with a single scenario of population change, given the diversity of viewpoints among various stakeholders, multiple demographic futures will always exist.

How well can we know these demographic futures, and how should they be judged? Political commentators, scholarly analysts, and outspoken citizens alike all thrust competing interpretations in planners' paths, making it difficult for us to proceed with a clear sense of direction. . . . An important skill planners can bring to the table is expertise in assembling, arranging, and interpreting demographic projections. General projections of growth and change are a useful starting point and become the basis for several different stories about the future. With increasingly detailed information, the extravagant range of competing stories may be narrowed considerably, but there remain clear dilemmas about which planners must be versed. Through their keen insights on demographic change, planners can participate directly or indirectly in the authoring of alternative stories of demographic futures.

CALIFORNIA: POPULATION GROWTH AND A CHANGING BASE OF RESIDENTS

These points are well illustrated by the urban planning issues that surround the changing population of California. The state's rapid growth and the predominance of Latinos in that growth provide a clear example of the value of demographic futures. . . . Recently prepared official state projections foresee an increase of 15.5 million Californians between 1990 and 2020, rising from a base of 29.9 million in 1990. . . .

[. . .]

The prospect of adding 15 million additional residents between 1990 and 2020 – a 50 percent increase in population – has planners in California transfixed. "How are we to accommodate so many new residents?" they seem to say. If the newcomers will have the same needs as each of the 1990 residents that would imply a 50 percent expansion in all services and facilities. Viewed another way, within a span of just thirty years California's borders must house an additional population the size of Florida's, which in 2000 was the fourth largest in the nation. One-third of the way through this growth period, California's track record of accommodating this growth is dismal, as will be described.

WHO WILL THE RESIDENTS BE?

The first question planners need answered is: Who will be the clientele for their plans of the future? The simple answer that most seem to grasp is that in California the number of people will be *more*. . . .

Yet that simple view of growth ignores dramatic changes taking place *within* the population. The clientele of planners is rapidly changing, undermining presumed patterns of needs and shifting traditional expectations of impacts from growth. Thus, there is a second set of answers to who the residents will be in California. More than just being more numerous, they also will be different in their characteristics. Among other factors, differences in race/ethnicity, nativity, and age composition. . . .

Race/ethnicity

. . . Both White and Black populations will grow slowly, but their numerical increases will be dwarfed by that of Latinos, whose share of all growth from 2000 to 2020 is projected to be 65.7 percent. (Asians follow with 22.9 percent of all growth.)

Nativity

Rapid immigration to California has expanded the foreign-born population. The new arrivals each decade are added to a growing stock of previously settled immigrants. . . . Between 1990 and 2020, the proportion of the state's population that is foreign-born will have expanded only slightly, from 22 percent to 24 percent of the total. At the same time, the proportion of all the foreign-born who are new immigrants (fewer than ten years in the United States) will have declined from 50 percent to 22 percent. As shown later, differences between new immigrants and settled immigrants have great significance for urban development patterns. In 2020, the great majority of these foreign-born residents (51.7 percent) will be Latino and another 33.4 percent will be Asian.

Age

A different perspective on the changing composition of the population focuses on the racial/ethnic makeup of different age groups. . . . Although the non-Latino White population greatly dominated the older age groups, under age forty the number of Latinos surged upward, and under age twenty it equaled the number of Whites.

[. . .]

Overall

Summing up, we could say that the future of California is the future of the Latino population. At present, many would say that this future is one of economic gloom. As a group, Latinos are characterized by high poverty, low education, and low-skilled occupations. Whether those disadvantages will persist among future generations of Latinos in California is a critical question. Also important to planners is how the urban planning impacts of Latinos might change in the future. For these and other reasons, the status of Latinos is important not only to Latinos but also to their fellow residents with whom they will share resources.

BUILDING CITIES FOR FUTURE RESIDENTS

Currently the prospects for the future look muddled in California. An alarming shortage of housing and a mismatch of housing with population needs have generated a widespread perception of crisis. At the same time, environmental and quality-of-life concerns are creating a surge of support for new urban forms that emphasize compact cities and sustainability over the traditional pattern of sprawl. Surely the state is in transition, both in its population base and in its planning goals and desires. From the current disarray, a consensus may yet emerge. A key element of this resolution could be the role of the growing Latino population in leading the way to more compact cities.

Lagging housing construction

One-third of the way into the 1990–2020 growth period, California has been very slow to expand housing, transportation, or other services to accommodate the needs of its growing population. Most alarming is the low level of housing construction during the decade of the 1990s. For the decade, building permits averaged around 100,000 per year, far below the expected level. To meet projected housing needs would require construction of 220,000 units per year from 1997 to 2020. Such a level of residential construction has rarely been achieved in any single year since 1970 and has never been sustained for more than two consecutive years, let alone twenty-three! More than a problem of production capacity, such an acceleration of construction would constitute a sharp break with the slow development patterns to which citizens and elected officials have grown accustomed in recent years, and many would appear reluctant to embrace such a sharp increase. Especially trouble-some is the decline in the construction of multifamily housing. Whereas multifamily housing maintained a 45–8 percent share of all construction in each of the preceding three decades, during the 1990s its share of an already diminished production fell to 24 percent. This decline runs counter to the population changes under way in the state, and the shortage forces the burden of finding affordable housing on to renters. The continuation of a relatively high level of single-family housing construction sustains homeownership opportunities for better-off residents, but it also encourages lower-density development. Although smaller lot sizes in single-family developments certainly can help to increase densities, only a shift toward more apartment construction can substantially increase development densities and lead to more compact cities.

Compact cities or sprawl?

Given the enormity of the projected population growth, planners have been searching for the most efficient means of coping. Sentiment in California has been growing against continued reliance on low-density development patterns characterized as sprawl. The solution favored by many citizens and planners is to encourage more compact development in housing. This solution goes hand-in-hand with increased emphasis on public transit in urban areas across the state. Although there is no shortage of land for construction in most counties, a high-density urban pattern is more easily serviced, consumes fewer resources of many different kinds, and is presumed to be more sustainable.

Others have resisted this prescription, arguing instead that the great success of low-density residential development in the state is the pillar of the American Dream and should be celebrated, not denigrated. . . . Meanwhile, lower-density development continues apace for political as well as market reasons. Despite the merits of apartment construction, citizens and local officials who govern land use on specific sites remain reluctant to approve developments at higher densities.

Latinos and the compact city: key indicators

Even without the creation of high-density housing patterns, California's population growth is increasingly supportive of compact cities. The Latino population, which comprises the bulk of the growth ahead, has a propensity for lifestyles that are compatible with compact cities. This is evident from three key indicators: mean persons per household, multifamily housing, and compact commuting. . . . Data on these indicators were collected from the 1990 census Public Use Microdata Sample file for the State of California. . . .

Comparison of Latinos to non-Latinos

Household sizes were substantially larger for Latinos than nonLatinos, implying that more than one-third fewer housing units are required to house the same number of Latino residents than nonLatinos. The same difference held at both higher and lower income levels, suggesting that the household size differential is not a simple function of poverty. Instead, it is likely a matter of custom among Latinos to live in larger family groups. Similarly, more Latinos than nonLatinos lived in multifamily housing. 37.2 percent of Latino households lived in apartments, versus 29.1 percent for nonLatinos, a ratio proportionally larger by more than one-quarter. Taken together, these two indicators show that Latinos tend to live in residences that are substantially more compact than those of nonLatinos.

The proportion of Latinos in multifamily housing differs sharply by income level. Whereas 37.2 percent of all Latinos lived in apartments, that was true of 45.5 percent of lower-income households, but only 20.3 percent of higher-income households. Moreover, the higher prevalence of Latinos on this indicator seems to dissipate when income is controlled. This suggests that apartment living may be less a preference and more a necessity for Latinos. Nevertheless, as long as Latinos have lower than average incomes, their residences will be concentrated in apartments.

Turning to the commuting indicator, Latinos were nearly twice as likely as nonLatinos to commute to work by public transit, bicycle, or walking (13.1 percent versus 7.2 percent). At higher incomes this compact commuting behavior decreased but was still half again as likely for Latinos as for nonLatinos (8.6 percent versus 5.3 percent). Thus Latinos' commuting behavior reinforces their residential behavior in supporting compact-city lifestyles.

Comparison of native-born to foreign-born Latinos

We now compare the compact-city lifestyle indicators of native-born and foreign-born Latinos. A large share of Latinos in California are recent immigrants whose characteristics shape the averages for all Latinos. Yet, over time, as new immigrants become longer settled, their lifestyles may change accordingly. Incomes also tend to rise as immigrants settle in, and that may contribute additional changes that planners should expect.

. . . [F]oreign-born Latinos had much larger household sizes than native-born Latinos, and the more recent immigrants had the largest sizes of all. These differences did not decrease when income was controlled. If anything, they increased. In part this may reflect stronger cultural preferences for large families among the newcomers. It also may reflect the younger age of newcomers and their likelihood of having children still living at home.

The foreign-born were also much more likely to live in apartments than the native-born. In fact, the native-born Latinos' rate of apartment residence was below even that of nonLatinos, with or without income controls. The newest immigrants showed the highest rate of apartment residence by far (63.2 percent overall, 66.3 percent at lower incomes, and 45.2 percent at higher incomes). In contrast, the longest settled immigrants had apartment residence rates well below those of nonLatinos. Thus we conclude that Latinos' predominance in apartment housing is driven entirely by immigrants of fewer than twenty years residence in the United States. The data suggest that most of the longer-settled immigrants' departure from apartments is associated with increasing duration of residence at the same income level, while a small additional contribution is associated with rising income, if in fact they are able to move to higher income.

Finally, with regard to commuting behavior, foreign-born Latinos were more than twice as likely as the native-born to practice compact commuting. As with apartment residence, native-born Latinos' rate of compact commuting was below even that of nonLatinos, with or without income controls. The highest rate of compact commuting was again that of the newest immigrants (24.2 percent overall, 26.4 percent at lower incomes, and 18.3 percent at higher incomes). Among the longest-settled immigrants, compact commuting declined sharply but remained above that of nonLatinos. Again we must conclude that Latinos' predominance in compact commuting is driven largely by immigrants of less than twenty years' residence in the United States. Most of the longer-settled immigrants' departure from compact commuting is associated with increasing duration of residence at the same income level, while a small additional contribution comes from rising income, should they be able to advance to a higher income.

COMPETING STORIES OF THE FUTURE

The preceding analysis shows that Latinos dominate California's population growth and that Latinos lead lifestyles more compatible with compact cities than do nonLatinos. They occupy fewer housing units relative to their population size and live more often in apartments, requiring less land for their housing. Latinos are also more likely to be compact commuters. Accordingly, not only the sheer magnitude of California's population growth but also its makeup argues for more compact development in the future.

Competing stories, however, have been proposed about California's future and the role of Latinos. These stories describe alternative futures by relying upon sets of facts that overlap one another but selectively draw attention to particular facts and valuations. The role of Latinos in developing compact cities is necessarily cast within these broader interpretations. Each has substantial adherents, and without agreement on a single story of the future to guide planning, it is difficult to gain consensus on key decisions that will alter the status quo.

. . . [A]lternative versions of California's demographic future are outlined here and can be labeled for conciseness as:

1 Economic polarization.
2 Growing assimilation or splintering multicultural norms.
3 Upward mobility and arrival in the middle class. . . .

All of these stories of the future share some elements, but they emphasize very different plot lines and interpretations, making the stories powerful for channeling the debates and focusing areas of discussion. Compact cities and the desirable role for Latinos are viewed differently in the context of each of these interpretations of the evolving future.

Economic polarization

More than racial and ethnic composition is changing in the population. A deepening polarization based on unequal income distribution and poverty is mounting in California. . . . This deep economic division is founded on highly unequal rates of educational attainment, largely associated with differences between immigrants and others. At the same time that California boasts higher than average college completion rates – 26.4 percent of adults age twenty-five and older with a Bachelor's degree, versus 24.4 percent in the nation as a whole – it also claims one of the lowest high-school completion rates – 19.9 percent without a high-school degree versus 17.2 percent in the nation. Thus, the state is building a polarized labor force concentrated at both the high and low ends of the skill ladder. This polarization is attributed to economic restructuring, globalization of the economy, and employers' readiness to exploit a low-wage labor force drawn from neighboring Mexico.

Overall, growth appears more rapid in the bottom than the top segment, and the shift in California has been quite rapid. Throughout the mid 1980s, California's poverty rate was lower than the nation's (12.7 percent versus 13.3 percent), but then it surged. By 1996–98, even after recovery from the recession, the poverty rate stood at 16.3 percent while the nation's was 13.2 percent. The growth in impoverished people creates a context that planners cannot avoid.

The scenario of a shrinking middle class and a growing poor class intersects and modifies the compact-city scenario. Construction has continued apace for single-family homes, enabling homeowners to maintain their established lifestyles, but it has plummeted for multifamily residences, as discussed above. This pronounced shortage of multifamily housing undermines the compact-city vision of high-density development. Instead, the shortage requires renters with lower incomes to double up in even higher-density households. Thus California's housing development is polarized, with one population segment, largely Latino, following the compact-city vision via high household densities, while others continue with traditional residential patterns.

Growing assimilation or splintering multicultural norms?

This polarization confuses planners about what are appropriate responses, making it difficult to respond to California's demographic future.... The debate about California's future is about the appropriate norms to use when evaluating the desirability of current and future development patterns....

[...]

Crowded housing highlights the dilemma for planners. If the situation is not viewed as a problem by the affected cultural group, should planners still treat it as a problem? In a multicultural planning context we may have multiple norms that create complex situations of intergroup relations. More than the enhancement of individual well-being, planning is focused on the management of relations between neighbors, and the "overcrowding problem" must be judged by these external implications as well. When neighbors hold different norms about crowding, noise, and other elements of lifestyle, planners' jobs are made more difficult.

[...]

Upward mobility and arrival in the middle class

A third story of California's future emphasizes the strong upward mobility that carries immigrants to reduced levels of poverty and increased rates of homeownership as they live longer in the state....

Of course, the upward-mobility story of the future downplays the fears cited in the competing story of polarization. The more likely future contains a mix of both, with upward mobility softening initial class divisions, but with a long-term persistence of relative disadvantages. Similarly, the story of arrival in the middle class emphasizes assimilation over multicultural differences. Again, the more likely future contains elements of both, with new arrivals to the middle class retaining many of their ethnic distinctions.

The upward-mobility story conflicts with visions of compact-city lifestyles as well.... [I]mmigrants tend to abandon their compact-city lifestyles as they live longer in California, regardless of changes in income. At present, compact-city living is highly dependent on the large share of new immigrants in the population.... [T]he evidence shows that Latino immigrants do begin to travel and live in apartments like native-borns after a decade or two. Because in the future a smaller share of the population will likely be newly arrived immigrants, upward mobility will erode the current client base for compact cities. If planners seek to build more compact cities in the future, they will need to incorporate the middle class and the native born in that vision.

[...]

WHAT METHODOLOGY FOR GAUGING THE FUTURE?

If the preceding arguments about the future are good "stories," but hardly constitute the proof needed to plan the future, what real methodology can be employed? . . . These stories provide a narrative that ties together many different facts and values, making sense of the complexity for the benefit of decision makers. They embed facts drawn from focused analytical studies prepared through careful use of both quantitative and qualitative analyses.

Census data and demographic projections

Planning and policy making rely largely on quantitative arguments about "how many" and "how much." At the core of planners' investigations of the future lie census data and demographic projections. Census data leave out much detail, such as information on the attitudes and values of residents, but they provide an indispensable backdrop for assessing the future. For this use their most important quality is consistency of measurement at different points in time, thereby allowing measurement and assessment of trends. Census data are also valued for their comprehensive coverage; there is no other source of data that offers so complete a record of the population, its housing, its employment, and its use of transportation.

Closely coordinated with the census data on population size and characteristics are population projections. This dimension of the future is better known than any other, in part because of the great predictability of the aging process, and because of other predictable life cycle-related population behavior. In fact, planners may have taken population projections too seriously, by assuming that mechanical projections are both accurate forecasts and descriptive of desirable futures. . . . [P]lanners should . . . take the projections simply as tools for evaluating possible futures. Within feasible limits, different futures can be made to come true, but . . . we must beware the abuse of projections that are quietly adjusted to achieve political ends. Planners need to become much more sophisticated in the use of projections than they have in the past if they are to reclaim their leadership of the future.

[...]

CONCLUSION

Demographic futures are vital for focusing attention on the future clients of plans that are currently being debated. Emphasis on the demographic characteristics of planners' clientele is useful for quantifying how many and what kind of residents will be affected by these plans. The demographic approach has the farther advantage of exploiting the unique temporal focus of demography (year, age, duration, cohorts, and periods), something ideally suited to looking toward the future. This stands in contrast to the strengths of other fields, such as the spatial emphasis of geography or the emphasis on income of economics. To date, the demographic approach has been much less utilized by planners, but then again, planners have also underemphasized the future.

[...]

Mastery of demographic futures provides an important vehicle for planners to exercise leadership in building the future. Planners have long emphasized the importance of equity and meeting diverse needs. Effective and legitimate plans for the future require knowledge of the changing population mix and the behavior patterns of many groups. This information is a useful contribution to the collaborative planning process and is not likely to be gained from other sources. Planners also require familiarity with the narratives, debates, and stories that are being spun with regard to changes under way in their region. There is no need to choose the one right story; rather, what is required is to be conversant with the many stories within which planning issues are embedded. The future is being forged out of the competition among many conflicting voices. With planners' expert participation, we can help keep the process more informed, balanced, and directed.

31

Geographic Information Systems

From *Planning and Urban Design Standards* (2006)

Ann-Margaret Esnard, Nancy Sappington, and Milton R. Ospina

Editor's introduction

Geographic information systems (GIS), or digital mapping, are widely used in urban and regional planning. Local and county governments employ this technique for such routine planning functions as inventorying land use, mapping demographic information, displaying zoning regulations, identifying environmentally sensitive land, plotting streets and sewers, tracking capital projects, and outlining service needs related to crime, disease, or poverty. (See Plate II, 6.) They draw on federal, state, and local agencies for the data required for the mapping. For example, since 1989, the U.S. Census Bureau has provided TIGER (Topologically Integrated Geographic Encoding and Referencing) files that allow GIS mapping of census data. The Department of Housing and Urban Development has mapping capabilities ranging from showing target areas for low-income housing tax credits (the nation's principal affordable-housing program) to the location and extent of subprime mortgages. In 2005, in an effort to enhance e-government services, the Department of the Interior launched the Geospatial One-stop portal that consolidates federal GIS data in a single site (gos2.geodata.gov/wps/portal/gos).

Urban planners employ GIS, often linking with other tools like Google Earth, for non-routine matters, including visioning and modeling exercises, and at variety of scales from the neighborhood to the megaregion. For example, they might use GIS in modeling manmade and natural disasters. The terrorist attacks on the World Trade Center and Pentagon led to GIS mapping of sites vulnerable to assaults. Hurricane Katrina stimulated several studies looking at potential hurricane paths and damage. Finally, today's global warming discussions have yielded maps of anticipated environmental changes, including rising sea level in urban places. An excellent example of the latter is Jonathan Barnett and John Beckham's "Coastal Cities See Rising Seas," in *Planning Magazine*, August–September 2007.

Although GIS is less than fifty years old, it has transformed urban planning practice. With its ability to display layers of spatial data easily and quickly – showing, for instance, the place-based implications of current socioeconomic characteristics, or presenting an inventory of an asset like open space – it has given professionals and the public more information with which to make decisions.

GIS has roots in large-scale transportation and landscape studies and the rise of computers and the digital revolution that occurred in North America after World War II. For example, in the 1950s the U.S. Federal Highway Administration underwrote several ambitious metropolitan transportation studies designed to gauge the level and location of the demand for highways, measuring the effects of economic growth on land use and transportation. Researchers employed mathematical formulas to show and predict transportation and land use patterns, modeling growth in major American cities. The resulting studies – the Detroit Metropolitan Area Transportation Study (DMATS, 1953), the Chicago Area Transportation Study (CATS, 1955), and the Philadelphia/Camden Study (Penn-Jersey, 1959) – are exemplary. Their innovation was to divide the metro areas into small units (transportation analysis zones), and collect and map data on many behaviors, including trip origin and destination, choice of mode of transportation, and increases in economic activities. At first researchers used hand calculations to record the data, but soon turned to the computer. Although the lumbering machines were primitive and slow – researchers used punch cards to feed in data – they demonstrated their usefulness, thus beginning a new tradition of integrating large databases into planning analysis.

But the real foundation of modern GIS was laid in Canada, where in the 1960s the national government mandated an inventory and suitability analysis of rural land. The Department of Forestry and Rural Development undertook a three-decade-long study that surveyed 30 percent of the country's undeveloped territory. Analysts created a land classification system to inventory potential uses (agriculture, forest, recreation, and wildlife protection) and used computers to handle the masses of information. By 1971, they had invented the Canada Geographic Information System (CGIS), now called the Canada Land Data System (CLDS), the world's first general-purpose GIS.

Simultaneously, Howard Fisher, a Harvard-trained architect and part-time lecturer at the Technological Institute, Northwestern University, secured a Ford Foundation grant to take a computer-mapping project called SYMAP that he had developed in Chicago to Harvard's Graduate School of Design (GSD). With this funding Fisher founded Harvard's Laboratory for Computer Graphics and Spatial Analysis. There, he and his associates refined SYMAP and several other programs that became today's GIS. For example, in conjunction with student city planning and landscape architecture studios, they studied the Delmarva (Delaware, Maryland, Virginia) peninsula, applying SYMAP to environmental planning issues, layering soil suitability, hydrology, and other factors. Landscape architect Ian McHarg, GSD graduate and University of Pennsylvania professor, worked at times with the Harvard group, demonstrating the application of layering techniques in his best-selling *Design with Nature* (New York: Wiley, 1969). (See Selection 19.)

In 1969, Jack Dangermond, also associated with the Harvard lab, founded the Environmental Systems Research Institute (ESRI) to focus on the commercial application of GIS to land use analysis projects. After several years of research and development, ESRI released ArcINFO (1982), soon adapted to desktop computer use as ArcGIS, thus enabling today's widespread utilization of GIS. Located in Redlands, California, about sixty miles from Los Angeles, ESRI now employs more than 4,000 staff and actively supports the development and dissemination of GIS through annual conferences and publications.

While GIS began as the domain of experts lodged in university research settings and government agencies, another branch, public participation GIS (PPGIS), emerged in the 1990s. Expert users and researchers, understanding the power of GIS in explaining contemporary conditions, began to explore ways to provide for its wider dissemination to nonexpert members of the public, or to grassroots organizations, for use in discussions about public policy and planning. Over time, various PPGIS suppliers emerged. They include community groups, university–community partnerships, publicly accessible GIS facilities in libraries, universities, the internet, and neighborhood GIS centers. PPGIS fosters two-way communications, disseminating information to the community, and, at the same time, incorporating information the community supplies.

This selection, taken from an encyclopedic handbook, *Planning and Urban Design Standards*, provides a basic outline of GIS. Author Ann-Margaret Esnard is Associate Professor and Director, Visual Technology Planning Lab, Florida Atlantic University. Nancy Sappington and Milton R. Ospina both work at ESRI. Sappington has written *The ESRI Map Book*, Vol. 21 (Redlands, CA: ESRI Press, 2006), and *Mapping for Congress: Supporting Public Policy with GIS* (Redlands, CA: ESRI Press, 2007), with Milton Ospina, Christopher Thomas, Brent Roderick, Ray Carnes, Michael Law, and Polly Barbee, from the Library of Congress Cartography Program.

To read more about GIS, see George Korta's *The GIS Book*, fifth edition (Clifton Park, NY: OnWord Press, 2000). The ESRI website has a comprehensive bibliography – see the bookstore section of the site. Of particular interest to planners are: Juliana Maantay and John Ziegler, *GIS for the Urban Environment* (Redlands, CA: ESRI Press, 2006), Ayse Pamuk, *Mapping Global Cities* (Redlands, CA: ESRI Press, 2006), and Richard

T. LeGates, *Think Globally, Act Regionally* (Redlands, CA: ESRI Press, 2005). For a broad discussion of PPGIS, see William J. Craig, Trevor M. Harris, and Daniel Weiner (eds), *Community Participation and Geographic Information Systems* (London and New York: Taylor & Francis, 2002). Finally, for a wonderful narrative of the GIS pioneers, read the firsthand account of Nick Chrisman, one of the original researchers at the Harvard Laboratory for Computer Graphics, *Charting the Unknown: How Computer Mapping at Harvard Became GIS* (Redlands, CA: ESRI Press, 2006).

A geographic information system (GIS) is a tool that connects databases to maps. It combines layers of information about where things are located with descriptive data about those things and their surroundings. Information such as where a point is located on a map, the length of a road, the size of a parcel, and the number of square miles a community occupies can all be stored in digital format in layers, also called themes, of the GIS. By combining a range of spatially referenced data and analytic tools, GIS technology enables people to prioritize issues, understand them, consider alternatives, and reach viable conclusions.

The capability of a GIS to link data sets together by common location information facilitates the sharing of information, such as interdepartmentally within an organization or via the internet with the public. Participatory GIS efforts involve partnerships among GIS technical professionals, planners, and community members to ensure adequate neutrality, confidentiality, and objectiveness of information, as well as adequate grounding in community perspectives to explore all possible scenarios that may apply during any planning process.

GIS COMPONENTS

The main components of a GIS are hardware, software, data sources, including metadata, and data structure types.

Hardware

The hardware consists of a computer that meets the software system requirements and other equipment, such as printers, scanners, or digitizers. Check with the particular software vendor to make sure that the equipment being purchased has the specifications that the software needs. Unlike in the past, when there were cost issues regarding the most appropriate hardware, today many systems have increased technological capacity at decreased cost.

Software

GIS software provides the functions and tools necessary for storing, analyzing, and displaying spatial information. These include a graphical user interface, a database management system, tools for entering and manipulating geographic information, and query functions.

Data sources

To determine the type of data needed, one needs to first determine the types of products a GIS will produce. Data can come from existing sources, such as spreadsheets, relational databases, images, or computer-assisted drawings (CAD) files; paper or mylar maps that are scanned and digitized; field-collected data using global positioning systems (GPS); and remotely sensed information, such as satellite imagery, aerial photography, digital elevation models (DEMs), and orthophotos.

Metadata

Metadata are perhaps the most critical part of a GIS, and often the most neglected. Metadata provide specific information about a data set, such as its creation date, scale, projection, resolution, and accuracy. This information is necessary to make sure the data can be accurately used for the analysis. The Federal Geographic Data Committee (FGDC) metadata standard provides guidance on required content, particularly for posting data sets on internet-based data clearinghouses. In addition, each state has metadata rules that should be consulted.

Data types

There are two types of data: vector data and raster data. A GIS will integrate both types of data.

Vector data

Vector data contain information about points, lines, or polygons, which are stored as *x* and *y* coordinates. Below are examples of the types of data that would be represented as points, lines or polygons:

- *Points:* trees, benches, street lights, fire hydrant locations.
- *Lines:* streets, water and sewer lines, electrical networks.
- *Polygons:* census tracts, parcels, building footprints, building layouts, and landscape planted areas.

Raster data

Raster data are values assigned to evenly spaced cells in an image. They contain information about the Earth's geographic features that is stored in cells within a grid or matrix. Each cell stores a single numeric value representing data, such as land use, vegetation, slope, elevation, or aspect.

GIS FUNCTIONALITY

Some of the various GIS software functions commonly used by planners and designers are described here.

Query by location

A query by location can be a search on a point within a polygon, or a radius search from a specific point. For example, a query to identify all parcels less than 0.5 acres in size.

Query by attribute

A query by attribute searches on data stored in the table; for example, a query to identify all parcels owned by people with the last name Smith.

Boolean queries

Boolean queries are a combination of location queries and attribute queries. They specify relationships between fields and values, using common operators such as "and" (both expressions are true) and "or" (at least one expression is true). For example, a Boolean query could involve a search for all parcels that are less than 0.5 acres in size and are owned by people with the last name Smith.

Buffers

A buffer is a region around a geographic feature or phenomenon. Buffers can be one ring or multiple rings. Also called proximity analysis, buffer analysis can be used to depict spheres of influence, areas that are more significantly impacted by a given phenomenon than those on the outside, and zones of protection, areas that are protected from impacts from a given phenomenon.

Address matching

This process matches the location of an event recorded as a street address, latitude and longitude position, or milepost locations along a route in a table, to a street centerline, zip code, or other administrative zone. The common result is a GIS layer with points corresponding to these events. A common use for this is taking name and address information collected at a public meeting and correlating this with the GIS database.

Address matches are often imperfect, because of inconsistencies between the address table and the digital street map, and errors such as misspellings or other data entry errors. Most software packages can do rematching when partial or unmatched records result. Depending on the sensitivity of the planning analysis, there might be a need to balance specificity with confidentiality. Offsetting points, matching to more generalized boundaries such as a zip code or raster-based density mapping, might be necessary.

Measuring distance

Distance can be measured as Euclidean distance, which is the distance of straight or curved paths, to indicate how far apart things are; and cost distance, which is the distance measured that involves the least effort in moving across a surface.

Overlays

The overlay operation is central to many GIS applications. It is one of the original motivations for creating GIS software technology. An overlay uses a geometric process to associate all the attributes at each location and then uses a rule to combine these attributes.

Suitability analyses

Suitability analysis is commonly used for finding optimum locations for a project, based on a combination of map layers. For example, suitability analysis can be used to locate the most suitable location for a new park or school. This technique offers the opportunity to use GIS in conjunction with public participation, bringing together stakeholders to solve a community problem.

Often, some objectives and their related data layers have more importance to the overall suitability model. To account for this, one can rank and weight the data sets. Ranking and weighting are subjective processes, which differ based on the analyst's background, expertise, and preferences.

[. . .]

32

The Future of Public Participation

From *Designing Public Consensus: The Civic Theater of Community Participation for Architects, Landscape Architects, Planners and Urban Designers* (2006)

Barbara Faga

Editor's introduction

Modern citizen participation has come a long way since it broke into public consciousness in the mid twentieth century. Many interrelated phenomena stimulated its growth. They include reactions to high-handed decision making emblematic of early urban renewal and highway construction projects, increasing democratization demanded by labor and civil rights activists and informed by modern political theory, environmental activism, widely publicized disclosures of abusive corporate and government experiments on humans, civil disobedience and antiwar demonstrations, and urban riots. The confluence of these disparate elements, especially in the 1960s, conspired to challenge traditional authority. In urban planning, these challenges came from writers (such as Jane Jacobs, the author of *The Death and Life of Great American Cities*, New York: Random House, 1961, which railed against the status quo), sociologists (such as Herbert Gans, who wrote *The Urban Villagers: Group and Class in the Life of Italian-Americans*, New York: Free Press, 1963, which took Boston planners to task), as well as practitioners like lawyer Paul Davidoff (see Selection 18, "Advocacy and Pluralism in Planning").

But citizen participation is rooted not just in trends, but also in law. It dates from the 1954 Housing Act that mandated maximum feasible participation as a prerequisite to receipt of federal urban renewal funds. Other federal and, later, state and local legislation continued to require inclusive public decision making in the city-building, public works, and environmental arenas. The laws secured either indirect or direct public participation.

Indirect citizen participation brings systematic disclosure to public decision making through analytical processes that outline the impacts of a proposed project and alternatives. The primary example is the environmental impact study (EIS), most frequently employed with public projects as designated by the National Environmental Protection Act (1970) and subsequent state and local legislation. (Some places undertake fiscal impact studies, but no federal legislation requires them.) An EIS assesses effects of a project and its alternatives related to the environment, economy, social, cultural and historic resources, and other elements specified by the law. In undertaking an EIS, a sponsoring agency follows three steps: scoping (deciding the geographic area, the alternatives, and elements of the project to be considered), presenting the draft EIS, and finalizing the report to be submitted to the decision makers.

The direct approach endeavors to insert citizen voices in the conceptions and choices involved in a given project. While a worthy goal, it is difficult to achieve, as history has borne out. Those involved in advancing direct citizen participation have struggled to address many definitional and procedural issues. They include timing (identifying and locating appropriate stakeholders), balance (mediating among and between local or neighborhood interests and citywide and regional needs), process (choice of types and place of citizen output), power (locus of decisions), and the role of professionalism (what or whose expertise reigns). Due to these complex concerns, citizen participation can take a long time, is "messy," and often nonlinear. In addressing these concerns, city planners are important players, serving, for example, as facilitators, expert advisors, advocates, and/or strategists.

Early direct citizen participation approaches focused on designating citizen committees that would meet with municipal officials to offer advice or be consulted on various aspects of a given project. In the mid 1960s the drafters of the Demonstration Cities Act (Model Cities Act) attempted to redefine citizen participation by empowering local citizen committees to submit model cities applications directly to the federal government. This experiment failed for many reasons, not the least being widespread objections of elected city officials. When President Richard M. Nixon abruptly terminated urban renewal and other city-focused categorical grant programs in 1973, direct federal intervention in local matters related to citizen participation disappeared.

Localities, however, continued to seek ways to strengthen but channel citizen participation. For example, in 1975 New York City amended its charter to institutionalize citizen participation. First, it established fifty-nine community districts, each having a city-appointed district manager and community board with members selected by the area's elected officials. Second, it gave advisory powers to the community board on a range of issues (capital budget, zoning, and other planning actions) prior to city council and mayoral approval and created the Uniform Land Use Review Process (ULURP), a time-limited approval process for large-scale projects in which the community board has a vote. Third, it allowed community boards to produce district plans for City Planning Commission approval. Although advisory, these plans become part of the public record when any project is under review. Subsequent charter revisions strengthened these powers and have resulted in a structured, highly participatory planning process that actively engages citizens at the local level. Other municipalities – such as Minneapolis, Atlanta, Portland, and Los Angeles – have adapted some variation of these neighborhood planning devices.

Over time, city planners and others have come up with many ways to enhance the quality of citizen participation. They range from sponsoring charrettes (intensive design exercises) to engaging in various consensus-building decision-making activities (group exercises using gaming techniques – with names like Delphi, swap-sort, force field, positive–negative, PINC filter – that allow participants to select among options) to interactive visualization activities.

Advances in technology have had an enormous effect on the development of these methods, especially in the area of visualization and consensus building. Computer programs like CommunityViz and ArcGIS 3D Analyst help the nonprofessional see what a proposed project or land use arrangement might look like. These programs can model various effects, including shadows, sunshine, and wind, and they can also insert improvements like trees. Beyond simulation, Google Earth has also become an important tool for situating projects or identifying prized resources, especially open space and water bodies, to assist in ongoing citizen dialogue about planning.

One of the most important technology-based breakthroughs has been "e-democracy" exercises. A dramatic and highly publicized example occurred in July 2002 as part of the World Trade Center reconstruction. Due to the emotional intensity and political sensitivity of the post-terrorist attack planning process, government officials sought to maximize citizen participation. They supported an exercise, "Listening to the City," conceived by the Civic Alliance, a coalition of civic and professional groups seeking to use technology to solicit widespread citizen participation. Having witnessed an experimental version several months earlier, government officials commissioned the Civic Alliance to convene a mass meeting (4,000 attendees) in the city's convention center to review proposals for the site. (See Plate II.4.) Assisted by America Speaks, a Washington-based firm, the sponsors arrayed the participants in tables of ten equipped with a facilitator and a laptop wired to a central control system. As the tables responded to an array of questions (ranging from demographic information about attendees to substantive questions about the future needs and role of the WTC site) the facilitators transmitted the answers to the central control. Within minutes the results splashed across huge screens viewed by everyone. The drama grew as the questions moved toward the big issue: voting for the best plan. (See Plate II.4.) When that time came, the citizenry's opposition to the alternatives was unanimous. The WTC planners went back to the drawing board and crafted a new design competition that later received worldwide publicity. After participating in "Listening to the City," attendees recorded their satisfaction with the process and its tangible results. They felt that, despite their large number, their individual voices had been heard.

This type of citizen engagement is powerful, but not predictable. It is also expensive, so it is best reserved for extraordinary events rather than routine matters. In fact, the WTC planners did not employ it in later efforts to solicit opinion about the submissions in the design competition. Instead, they produced an exhibit that ran for an extended period and asked viewers to submit paper or internet votes. But a version of "Listening to the City" did take place in Louisiana after Hurricane Katrina.

The design charrette is another citizen participation method that has benefited from the digital revolution. It expands on the usual process of assembling citizen teams, assisted by designers and planners, to explore alternative ideas for a building or urban design. The general format is to have an intense experience (a day or two) at which the teams sketch their schemes, making them ready for exhibit and discussion. The new dimension is to e-mail the sketches to a drafting firm in a different time zone whose employees transform the sketches to renderings, giving more substance to the ideas.

Bringing planning and design directly into the community on a regular basis is another means to improve citizen participation. In several cities, professional associations or universities have sponsored storefront studios, open to the public and the locus of exhibitions, charrettes, lectures, or citizen workshops revolving around timely issues. One of the original ones, the Planning and Design Studio, Chattanooga, Tennessee, played an important role in that city's downtown redevelopment. After several years as an extension activity of the University of Tennessee, it became a unit of municipal government. A more recent example of a studio is Arizona State University's Phoenix Urban Research Laboratory, located in a high-rise in the center of the city and charged to serve as a thinktank, project center, and classroom as ASU opens up a downtown campus. Other citizen participation projects are the HUD-sponsored Community Outreach Partnership Centers Program (COPC), funded in the 1990s, that partners universities with community groups and has, in some instances, involved planning exercises that enhance citizen participation.

Finally, in many places, citizen participation efforts have become larger citizen engagement projects. This change in terminology represents a shift in thinking. Experts now view participation as part of an extended program designed to inform, educate, and solicit opinion through a variety of devices. Exemplary is the work of PennPraxis, the practice arm of the School of Design, University of Pennsylvania, that has developed a well-thought-out process involving partnerships with the local press, expert-based informational programs, site visits, field trips to other places, internet messaging and communications, political outreach, design charrettes, exhibits, and other means of soliciting and expressing public opinion (see www.planphilly.com/about). This type of program is especially useful in places where institutionalized citizen participation is weak or the concerned population is dispersed or seemingly disinterested.

Citizen participation experiences have stimulated a good deal of theoretical and practical literature. Examples of the theoretical output are Patsy Healey's *Collaborative Planning: Shaping Places in Fragmented Societies*, second edition (London: Palgrave Macmillan, 2006), and Maarten A. Hajer and Hendrik Wagenaar (eds), *Deliberative Policy Analysis: Understanding Governance in the Network Society* (Cambridge: Cambridge University Press, 2003). The authoritative work on the practical side is Barbara Faga's *Designing Public Consensus: The Civic Theater of Community Participation for Architects, Landscape Architects, Planners, and Urban Designers* (Hoboken, NJ: Wiley, 2006).

Faga, author of this selection, is executive vice-president of EDAW, the international planning and design firm. She has had extensive experience in designing and facilitating citizen participation efforts for the firm's projects, which include the Wharf District Park, Rose Kennedy Greenway, Boston; the Atlanta Emerald Necklace Beltway; the Diagonal Mar Parc (Barcelona); and the Mutirao 50, a "defavelization" or slum upgrading project (Fortaleza, Brazil). In this selection Faga highlights three important trends now shaping citizen participation: globalization of planning, growing diversity, and increasing technology.

To read more about citizen participation, see Lewis D. Hopkins and Marisa A. Zapata, *Engaging the Future: Forecasts, Scenarios, Plans and Projects* (Cambridge, MA: Lincoln Institute of Land Policy, 2007). For more in-depth assessment of various techniques like mediation and consensus building, see Lawrence Susskind and Jeffrey Cruikshank, *Breaking Robert's Rules: The New Way to Run your Meeting, Build Consensus, and Get Results* (New York: Oxford University Press, 2006). To design a charrette, see Bill Lennertz and Aarin Lutzenhiser, *The Charrette Handbook: The Essential Guide for Accelerated, Collaborative Community Planning* (Chicago: APA Planners Press, 2006). To follow progress in this area, see the websites of the National Charrette Institute (www.charretteinstitute.org) and the Consensus Building Institute (www.cbuilding.org). Finally, the Lincoln Institute of Land Policy has inaugurated a website dedicated to visualization – see www.lincolninst.edu/subcenters/VTP.

Public process is a moving target, altering constantly to meet local demands and playing out differently everywhere. It would take a clairvoyant to tell us where it is going, but I am convinced of one thing: There will be more public process in the future, and it will be more complicated. We can identify a number of current developments that are already affecting our work. Three of the most notable are:

- Globalization of planning and design projects.
- Growing diversity and cultural and social variations within our communities.
- Emergence of new technologies to communicate with the world.

The cases that follow illustrate how some of these developments are transforming the public process.

The trend toward globalization is one of the hallmarks of our time, and the design and planning professions are already feeling the effects of this movement. Many more professionals are involved in international projects, and these demands will only increase as the resources grow to develop public infrastructure and spaces in such places as China and Brazil. In developing countries, we may find ourselves working to meet pressing housing, transportation, economic development, and environmental needs. Some of these projects will combine familiar public participation challenges with new issues such as intercultural sensitivity, deeply held traditional values, and illiteracy. In some areas, working with residents can become another way to draw them into public life and teach them about democratic process. We'll need to think creatively to adapt our tried-and-true public process techniques in an appropriate way to projects in diverse communities around the globe.

We're already facing language and cross-cultural issues at home, with increasing immigrant populations in many of our urban and even suburban areas, coupled with a growing wave of interest in urban revitalization and redevelopment. . . . [W]e may be faced with language barriers, unfamiliar patterns of social interaction, and a very basic need for inclusion in decision making. We will need to take extra time and care to understand the needs and viewpoints of these communities.

Another harbinger of the future is the rapid advance in communications technology. Innovations that include live TV audience polling, Web-based information services, multifunction cellphones, satellites, global positioning systems (GPS), instant messaging, streaming video, and other innovations have raised public expectations of instant input and instant access to information. Web-based public processes . . . demand a higher level of sophistication regarding communications and interactive capabilities. A growing segment of the public – and certainly younger generations – now presume that participation will be possible worldwide, 24/7, and faster than the speed of light. A rapid response from us in the design and planning professions is now required.

[. . .]

OPEN PROCESS IN VISITACION VALLEY

In the wake of the high-tech boom-and-bust cycles that fueled controversial building conversions, demolitions, and displacements in San Francisco's eastern neighborhoods, in 2001, the city found it necessary to impose interim zoning controls and initiate a community planning process. Residents, business and property owners, and other stakeholders in five districts have since taken part in community workshops to ensure that future growth and development actually meet neighborhood needs. In 2002, EDAW led a team of planning consultants that included Nelson\Nygaard out of San Francisco and Strategic Economics out of Berkeley, to produce a strategic concept plan to redevelop a former industrial site in Visitacion Valley, a remarkably diverse working-class neighborhood. One community newspaper described Visitacion Valley as "a community of new immigrants and people of color – including large populations of Chinese, Filipinos, Samoans, and Latinos – [that] already represents the true diversity that America is expected to see in the next fifty years." The scope of the project included the industrial site of the Schlage Lock Factory, which closed in 1999, and neighboring parcels totaling about twenty acres.

When Home Depot proposed to build on the site, neighborhood opposition turned the megastore away. As part of the study of the site and its environments and planning for the future, EDAW's team produced a week-long intensive community design workshop to develop an inclusive vision for the site. . . .

We had no fixed plan going in; instead, the process was an open, dynamic one, which allowed the community to create its own vision. A storefront studio was set up a few hundred feet from the site on Visitacion Valley's main commercial street. Throughout the week, all meetings, consultant discussions, and sketching were conducted here. Drawings, goals, and maps were displayed in the windows for passers-by to see, and the doors were always open. . . .

A few community events during the week were held in other locations throughout the neighborhood. These larger public meetings included a kickoff event with formal presentations by city staff, consultants, and technical staff. Smaller groups of residents brainstormed their ideas for the site. A Saturday workshop was held at the Schlage Lock site to introduce a framework plan based on input from the first meeting. In the afternoon, participants broke into small groups and worked with facilitators to create their own plans. These were later displayed on the walls of the storefront studio.

Through this open process, residents were able to educate the planning team about their priorities, values, and the special features of their neighborhoods. In turn, residents learned about the complexities, constraints, and trade-offs of creating a plan. Rather than resistance to plans that had come down from developers or city hall, the residents were able to participate in a transparent and aboveboard process. Because this was a very diverse community, community volunteers were on hand at every meeting to translate the presentations and Q&A exchanges into the various languages represented. . . .

A final meeting was held to present "community consensus points" regarding uses for the site, as well as drawings illustrating important design considerations. The community's vision was to reinvent the site as the heart of the community, with an emphasis on cleanup of soil and groundwater at the site, housing, neighborhood retail, community services such as day care and a library, and open space.

The storefront design studio, open meetings, ongoing informal contacts with residents, and translations of materials and discussion into multiple languages all contributed to an open and transparent process. Another key element was the genuine openness on the part of the professionals to create a plan that reflected the vision of the community. The diverse and involved public of Visitacion Valley responded in turn, with support for the strategic plan to guide the city in rezoning the site. . . .

VIRTUAL PARTICIPATION IN NEW YORK

Imagine New York, a project of the Municipal Arts Society, was conceived to "give voice to the people's visions" for the future of Lower Manhattan. . . . [T]he Imagine New York process cast a very large net to gather the ideas and impressions of the public in the aftermath of 9/11. In 2002 and 2003, Imagine New York hosted an interactive website and hundreds of public workshops across the tristate region, to produce an "innovative and replicable alternative model to traditional public hearings."

. . . [O]ne of the main objectives was to ensure that the public's ideas would be incorporated into decision making about the World Trade Center site. The process provided multiple opportunities for members of the local and global public to share their ideas and proposals for memorializing the tragedy at the World Trade Center and rebuilding downtown. During the first phase, in spring 2002, the website and meetings collected more than 19,000 comments, proposals, and drawings. These were incorporated into an online gallery of ideas, developed by Neighborhood America as a centerpiece of Imagine New York. . . .

The web-based system was also designed to manage information gleaned in public meetings and hearings. It collected and analyzed input from various sources throughout the project's life cycle, including online submissions of ideas, images, and graphics. The system supported management of a public involvement website, where members of the public could view documents, browse through a huge collection of ideas and images, send a letter, and submit their comments. . . .

Far more than an online repository of information, the Imagine New York public involvement site is highly interactive. It captures data from meetings and online submissions and organizes all public input by themes. This helps to maintain unity and consistency between the ongoing discussions in meetings and online. The site also serves to integrate project management with the public process and provides comprehensive documentation and institutional memory of the process. The database is searchable by text or image, keyword, issue, workshop, or name of submitter. The extensive data-organizing capabilities of the system enabled two or three full-time administrators to manage a public process that supported hundreds of local public meetings and tens of thousands of comments and images from all over the globe. . . .

As an online site for virtual participation, Imagine New York demonstrated the flexibility and reach of this new technology. It served the public by affording a structure to participate at any time in the public discussion, from next door or from half a world away. Those with something valuable to say who could not attend meetings or speak comfortably in public could find their voice in the virtual forum. Online, everyone has an equal chance to speak, without constraint and with ample time for reflection and forethought. . . .

The use of online technology also provides a more dynamic visual and auditory experience, thereby enhancing understanding of the project or plan. PowerPoint, streaming video, narratives by planners explaining the design, and textual narratives enrich the presentation of the design or conceptual plan. . . .

An interactive website for public involvement also creates its own public space. The site can shape communities of interest and common understanding, bringing disparate and far-flung people together with a common thread of information. In the case of the World Trade Center, this community helped the healing process by letting people express themselves, share their thoughts, and interact with each other on this topic of deep concern to them all. Imagine New York also stands as a record for the future. It takes only a few clicks to discover the thoughts and dreams of tens of thousands of people, drawn together by a common purpose. . . .

A NEW ERA

. . . Our televisions, computers, and cellphones have become interactive tools that enable virtual conversations, meetings, and exchanges of information on any and every topic. What do these developments portend for the public design process? Soon no one will need to spend four or five hours attending a public meeting when he or she can sit in a favorite chair and respond directly through cable or broadband. What this means for public participation remains open for discussion. Some say only zealots show up at public meetings. Does this mean, then, that more zealots will participate once the public can respond virtually, or that more of the general public will get involved? Will virtual participation take the place of meetings, or will it continue to supplement personal encounters? Can we do without the informal contacts and conversations that occur when mingling in person? . . .

The internet offers new opportunities to collect and store information, provide easy access 24/7, and draw in more diverse participation. It can also save on costs. Downloading documents electronically in large planning projects can provide tremendous cost savings, as compared to more traditional hard-copy publications. Kim Patrick

Kobza of Neighborhood America cites a large regional planning project in central Florida, www.myregion.org, which involved more than 100 public agencies and many of the most visible private employers. Over a period of two years, the site supported 300,000 document downloads. Think of the money saved. . . .

But the internet cannot solve all our problems of reaching out to the public; and it continually creates new demands for transparency, speed, and visual sophistication. We now deal with public audiences that are extremely well versed in video technology and computer capabilities. Expectations are high, and younger generations will put even greater demands on our ability to convey information dynamically. Fortunately, new technologies for visualizing what plans and designs will look like in the real world (such as computer-generated visual simulations, animation, streaming video, and fly-throughs) help to meet these expectations and enable us to better communicate with the public overall. . . . But we, the designers and planners of the future, will have to move faster and work smarter in order to keep up.

At the same time, we also have a number of important reasons to slow down, think situations through, and take time to listen. As new projects increasingly involve diverse, complex, and multicultural public audiences, our participation programs will have to incorporate an important layer of communication and understanding to enable us to effectively cross cultures, ages, and socioeconomic divides. And as public participation broadens, to include the young and the elder, immigrants, and low-income groups, it will require of us greater flexibility and adaptability; at the same time it raises the significance of our capability to listen and empathize while we do our jobs. Furthermore, we will have to approach each project and locality with an acute awareness of differences in context, values, and priorities – all of which we will have to explore and comprehend in order to produce a satisfactory result. These are not trifling issues.

Even in these very different and diverse environments, the lessons for us professionals have a very familiar ring. We have to build the public's confidence and trust, learn new ways of doing things, gain acceptance from the community, and listen to what people have to say. We have to be sensitive to what many may mistakenly deem as minor issues: addressing people correctly, dressing appropriately, and understanding the best places and times to schedule meetings. These are not so different from the adjustments we have always been required to make to work in any community; but these differences will be wider and harder to gauge in other countries and cultures and may require a more subtle response from us. Just as we have to accept that we may be wrong sometimes, and members of the community may be right, we will have to recognize at times that the way we do things in our own milieu won't work in the favelas or the Bahamas or the big city or the Old South. It takes courage and awareness to step off that familiar path.

Finally, we can safely assume that the public is now in charge. Our talents and experience as planners and designers are already being tested every day by the need to create buildings and spaces that truly respond to the public's needs and desires. With a few notable exceptions, the days of "voila" design and centralized control of planning are behind us. In order to accomplish anything, our work now requires skillful navigation of the political and social rapids. In the future, these challenges will only intensify. And the future is now.

33

"Introduction"

From *Everyday Ethics for Practicing Planners* (2001)

Carol D. Barrett

Editor's introduction

From ancient days to the present, philosophers have struggled with questions of moral values and ethics, seeking to understand how humans govern their behavior as individuals and in society. (Note the origin of the word *moral*, which is rooted in the Latin *moralis*, related to doing one's duty, and *mores*, referring to laws or customs; *ethics* comes from the Greek *ethikos*, meaning showing moral character.) Their studies deal with distinguishing right and wrong and determining standards for judging right and wrong. In the Western tradition, a direct line of thought about ethics flows from Aristotle to medieval scholars like Thomas Aquinas to the Renaissance and the Enlightenment through modern political philosophers like Thomas Hobbes and Immanuel Kant. There are similar strains in the Eastern tradition. Contemporary philosophers have divided the field into three areas: metaethics (study of the origin of ethical concepts), normative ethics (study of standard principles that guide behavior), and applied ethics (study of specific cases, often those related to the professions).

Stung by major ethical lapses among the nation's public, corporate, and professional leaders, public intellectuals and university scholars have recently opened up a new area of inquiry: practical ethics. This branch merges political philosophy and applied ethics to address complex questions of modern society. The goal is to push professionals in business, law, design, and other fields to extend their ethical boundaries to consider how their work affects broader social questions like income inequality or the environment. Bioethics is a particularly vibrant example.

In the past few decades, more than thirty university centers were founded to pursue ethical questions, and many have taken up practical ethics. Among the most active centers in this area are the Olsson Center for Applied Ethics (founded in 1966), University of Virginia; the Center for the Study of Ethics in the Professions (1976), Illinois Institute of Technology; the Edmond J. Safra Foundation Center for Ethics (1987), Harvard University; the University Center for Human Values (1990), Princeton University; and the Joan and David Lincoln Center for Applied Ethics (2000), Arizona State University. The Association for Practical and Professional Ethics (founded in 1991) at Indiana University facilitates communications and joint ventures among these centers and other related organizations.

Practical and applied ethics are of direct interest to urban and regional planners, who often face difficult dilemmas in their work. When dealing with land use and other planning issues, planners make or influence decisions that affect the economic, social, and political conditions of people. Their choices are not value-neutral – siting a halfway house for drug rehabilitation might entail balancing the desire for homogeneous family households by neighborhood residents with the social workers' desire to separate the recovering addicts from their problematic environments. Most importantly, planners' work evokes ethical questions involving basic rights in a democratic society – those of property owners, third parties affected by a decision, the disadvantaged, and future generations.

In addressing today's complex problems, a planner relies on his technical skills and substantive knowledge and turns to his ethical code for a moral compass. This code has roots in the early twentieth-century explosion of professional societies in many areas (law, engineering, architecture and urban planning). In coping with the complicated issues that had sprung up in the United States, which had rapidly urbanized and industrialized, these groups emerged to advance their knowledge (sponsoring publications and meetings to share innovations and best practices), monitor training of new entrants (stimulating university-based educational programs), and guarantee the competence and integrity of their members to the public and to each other (adopting codes of ethics).

The American Medical Association led the way in the area of ethics. Issuing the nation's first ethical code in 1847, the very year the AMA was founded, the group established a code that was widely replicated. This model had four parts: an explanation of the association's governing principles, followed by statements about physicians' duties to their patients, to each other, and to the public (*Code of Medical Ethics of the American Medical Association: Originally Adopted at the Adjourned Meeting of the National Medical Convention 1847*, Chicago: American Medical Association, 1847).

Other professional groups eventually adopted similar codes, including the American Bar Association and American Institute of Architects in 1908, and accountants and consulting, electrical, civil and industrial engineers between 1910 and 1919. Certain trades, such as life insurance underwriters, joined in. By the end of the 1920s more than twenty professions and seventy-eight trades had codes of ethics (Andrew Abbott, "Professional Ethics," *American Journal of Sociology*, 88: 5, 1983, p. 876.) These twentieth-century codes not only regulated professional behavior but also provided procedures for issuing sanctions for misconduct, with the most egregious violations resulting in expulsion from practice. In some professions, such as law, medicine, and architecture, the codes supplemented licensing requirements.

Urban planners were rather late to the game. In public affairs, only the International City/County Management Association, the organization of city managers, had an ethical code early on (1924). The American Society for Public Administration did not have one until 1985. In 1962 the American Society of Planning Officials (ASPO) became the first professional planning group to adopt one. And the American Institute of Planners (AIP) acted even later, in 1971. In part, these groups were reluctant to take this step toward professionalism because they were not even sure they qualified as professionals. Notably, they were not only late to adopt ethical codes, they also did not foster accreditation of their educational programs until 1984 (Eugénie L. Birch, "Advancing the Art and Science of Planning: Planners and their Organizations, 1909–1980," *Journal of the American Planning Association*, 46: 1, 1989).

However, when ASPO and AIP merged in 1978 to form the American Planning Association (APA) and its professional arm, the American Institute of Certified Planners (AICP), the AICP immediately took action. The AICP issued the *American Institute of Certified Planners Code of Ethics and Professional Conduct*, and has since revised the code two times, in 1981 and 2005. The code has three parts: (1) principles covering responsibility to the public, to clients and employers, and to the profession and colleagues; (2) twenty-five rules of conduct; and (3) administrative procedures. Furthermore, in 1987 the APA circulated *Ethical Principles in Planning* (amended in 1992) to guide all involved in planning – citizens, government administrators, city planning commissioners. Thus, the major professional planning organization in the United States has two ethical statements, one an advisory guide without sanctions, and the other a code with enforcement powers. View the APA and AICP documents at www.planning.org/ethics/.

Today as in the past, serving the public interest is at the forefront of all professional codes of ethics. Professions acknowledge this duty in exchange for the authority, status, and quasi-monopoly rights accorded to them by society. For planners, serving the public interest encompasses supporting fair, open, and inclusive planning processes; promoting excellence in design as well as protecting the natural and heritage elements of the environment; and expanding choice and opportunity for the disadvantaged.

Although the *AICP Code of Ethics and Professional Conduct* clearly articulates these values, how individuals observe them in practice is another matter. In the early 1990s, planning ethicist Elizabeth Howe surveyed about 100 planners from several states, asking them to explain how they acted in the public interest. The respondents fell into three groups: (1) those who cited loyalty to the employer (defined as an elected official who represented the citizens' will), (2) those who pointed to treating every member of the community equally, and (3) those who indicated helping the disenfranchised or disadvantaged secure more community resources (*Acting on Ethics in City Planning*, New Brunswick, NJ: Center for Urban Policy Research, Rutgers University, 1994). If someone were to do another survey today, these same groupings would appear, but planners might add "pursuing sustainability" or "environmental justice" to the last category. The selection that follows illustrates how the varied interpretations of the public interest will color different planners' answers to ethical questions.

Carol Barrett, FAICP, author of this selection, is Planning Manager, City of San Gabriel, California. She is extremely active in the APA and AICP locally and nationally, having served as president, National Capital, and the Texas Chapter of the American Planning Association and on national committees for professional education and ethics. She is author of *Everyday Ethics for Practicing Planners* (Chicago: APA Planners Press, 2001).

To place planning ethics within the larger stream of ethical thinking, read Lawrence C. Becker and Charlotte B. Becker (eds), *A History of Western Ethics* (New York: Routledge, 2003). Martin Wachs (ed.), *Ethics in Planning* (New Brunswick, NJ: Center for Urban Policy Research, Rutgers University, 1985) and Susan Hendler (ed.), *Planning Ethics: A Reader in Planning Theory, Practice and Education* (New Brunswick, NJ: Center for Urban Policy Research, Rutgers University, 1995), offer classic readings on the subject. For an excellent, up-to-date overview of planning ethics, see Carlos Nunes Silva, "Urban Planning and Ethics," in Jack Rabin (ed.), *Encyclopedia of Public Administration and Public Policy* (New York: Taylor & Francis, 2005).

For reading on the general importance of ethics and moral values, see Derek C. Bok's classic essay "Can Ethics be Taught?" (*Change*, October 1976). Pioneering planning ethicist Jerome Kaufman offers a planning-focused discussion about this topic in "Reflections on Teaching Three Versions of a Planning Ethics Course," *Journal of Planning Education and Research*, 12: 2 (1993), and urges educators to recognize their special obligation to instruct students and new entrants to the field about ethics and moral values. From time to time, bloggers on *Planetizen* (www.planetizen.com) take up the issue. And, for fun, go to www.studystack.com/studytable-14636, a clever web page with games – hangman, word scramble, flash cards, a crossword puzzle, and Scrabble – based on the *AICP Code of Ethics and Professional Conduct.*

For reading about moral values and civic responsibility in a liberal democratic society, see John Rawls's *A Theory of Justice*, revised edition (Cambridge, MA: Harvard University Press, 2003). Follow with the report of the American Political Science Association's Standing Committee on Civic Education and Engagement, ed. Stephen Macedo *et al.*, *Democracy at Risk: How Political Choices Undermine Citizen Participation, and What we Can Do about It* (Washington, DC: Brookings Institution Press, 2005), especially the chapter on the American metropolis.

Finally, the following books by practical ethicists are instructive for planners: Arthur Isak Applbaum, *Ethics for Adversaries: The Morality of Roles in Public and Professional Life* (Princeton, NJ: Princeton University Press, 1999); Archon Fung, Mary Graham, and David Weil, *Full Disclosure: The Perils and Promise of Transparency* (New York: Cambridge University Press, 2007); and Kwame Anthony Appiah, *Cosmopolitanism: Ethics in a World of Strangers* (New York: Norton, 2006).

Planners are not expected to be the most perfect of society's members. As authority and influence accrue to a profession, public expectations rise. This public trust imposes increased responsibility on planners to offer an example of impeccable conduct. . . .

Most of those in the planning profession accept the concept of ethical conduct for planners as important. Most planners also believe that what they do is ethical simply because they deliberately chose a specific course of action. At the same time, in any office, one will observe a variety of different behaviors provoked by the same circumstance. For example, Sandy rigidly insists that each issue can be properly analyzed and that any course of action can be determined to be either right or wrong. Gary is more elastic in his views: life consists of variables, not constants. Mike is skeptical that a workable ethical code can be formulated for the entire planning community. Paula snickers at the lofty aspirational tone of the AICP Code. Caroline thinks that no planner, threatened with job loss, can afford the luxury of ethics. And Bernie rejects the entire concept of ethics. He says that life, planning included, is a zero-sum game. In his view, bluffing, deception, and corner cutting are acceptable to achieve planning objectives. The attorney, meanwhile, advises that it is sufficient to stay within the letter of the law.

Some benchmarks are needed to sort through this plethora of perspectives. Trying to ignore ethical problems is akin to getting rid of a boomerang by throwing it away. Ethical dilemmas just keep intruding. . . .

As an ethical resource, this . . . discussion . . . does not contain the definitive answer to every ethical question or problem. There is no such thing. . . . [It] can help professionals clarify their ethical responses to the ever-present challenges of practicing planning. For example, you are the planning director and it is your job to make a recommendation on a conditional use permit for a manufacturing plant that emits a chemical not regulated by the Environmental Protection Agency. Project opponents present information showing that the emission is a health hazard.

- What are the facts? Is there a health hazard?
- Is this an ethical issue for the planner?
- Are there genuine moral issues at stake?

Suppose the compound is a virulent carcinogen and is emitted in large quantities. Most people would probably agree it should be stopped. The general prohibition against knowingly harming other people would necessitate this conclusion. But, suppose the emission produces only mild respiratory problems in a small percentage of the population. Further, suppose that removing the compound is so expensive that the plant could not be built at all. Additionally, imagine there is a severe unemployment problem in your community. Will there be serious disagreement among the planning staff about the appropriate recommendation?

Planners rightly believe that they have an obligation to protect the physical health of the community. But doesn't the staff also have an obligation to be concerned with the economic health of the community as well?

In this instance, there are conflicts among competing obligations, both of which appear to be valid. The conflicts that give us trouble are not those between good and bad, but between comparing goods, both of which cannot be fully realized.

When faced with two competing obligations, each of which appears to be justified, one approach is to try to find a way to satisfy them both. While it often is not possible to satisfy all moral requirements in a pure form, it sometimes is possible to satisfy them in a modified form.

Most general moral rules have an "all other things being equal" qualifier implicitly attached to them. In real life, all things are often not equal. For example, the threat to the safety, health, and welfare of the public may be both uncertain and minimal. Whether or not the emissions are in fact harmful may be a matter of controversy, and there may be, as of yet, no evidence that the emissions produce fatal diseases, only significant respiratory problems in a certain small percentage of the population. Furthermore, the obligation to the community may include an obligation to its economic well-being. So the question is, what in this case is the planner's obligation to the public?

This issue involves morals, ethics, and the law. The term "morals" refers to generally accepted standards of right and wrong in a society. Ethics refers to more abstract principles that appear in a religious, societal, or professional code. Moral and ethical statements are distinguishable from laws. The fact that an action is legally permissible does not establish that it is morally and ethically acceptable. The fact that it may be legal for the plant to emit the chemical doesn't mean that it is morally permissible to do so.

People often view ethical reasoning as fuzzy and imprecise, and it certainly is true that the qualitative thinking involved in ethics is not susceptible to the same kind of precision that can be achieved in arithmetic. For example, factual disagreements may be a component of an ethical dilemma. Answers are based on what can be known and documented. In evaluating moral disagreements, appeals are made to broader and more basic moral principles that are organized into theories. The three most common moral theories are utilitarianism, the Golden Rule, and deontological principles or:

- End-based decisions.
- Care-based decisions.
- Rule-based decisions.

In end-based decisions (utilitarianism), actions are right when they produce the greatest total amount of human well being. A utilitarian analysis of a moral problem consists of three steps. The utilitarian must determine:

- The audience of the action or policy in question – those who will be affected for good or ill.
- The positive and negative effects of the alternative actions or policies.
- The course of action that will produce the greatest overall utility.

There are two drawbacks to the utilitarian perspective on morality. First, implementation of the utilitarian perspective requires extensive knowledge of facts, and sometimes this knowledge is not available. If we do not know the long-term positive and negative consequences of an action or policy, we do not know how to evaluate it from a utilitarian perspective. Sometimes utilitarians are reduced to a best-guess approach. This is not very satisfactory. Utilitarianism also can lead to injustice for some individuals. Mining operations that cause black lung disease in some of the miners may produce more utility than harm from an overall standpoint, but would be unjust to the miners.

In making care-based decisions (following the Golden Rule), actions are right when they equally respect each human person as a moral agent. To consider the effects, you must place yourself in the position of those who could be affected by your actions. Philosophers have proposed a hierarchy of rights that should be protected through the application of the Golden Rule. The first is the most basic: life, physical and mental health. The second involves maintaining self-fulfillment through honest and truthful interactions. The third involves rights necessary to increase one's self-fulfillment, such as the right to property and to equal treatment.

Application of the Golden Rule morality involves these steps:

1 Determine the audience for the action. (Similar to the audience whose rights are under the utilitarian analysis.)
2 Evaluate the seriousness of the rights infringements each action will impose.
3 Choose the course of action that produces the least serious rights infringements.

There are two principal difficulties with Golden Rule morality. First, it is sometimes hard to apply the rule in a way that leads to a clear conclusion. This is especially true if the rights violations are merely potential rather than actual, or if the action is only a slight infringement on a right. Also, Golden Rule moral philosophies can produce implausible results. To protect lives, cars can be made so safe that no one would be able to afford one.

Following rule-based analysis, or deontological principles, our actions must be translated into universal principles of action without regard for the consequences. For example, it is right for me to toss my biodegradable lunch leftovers out of the car window only if I believe it would be correct for everyone driving on this road every single day to do the same. Critics of this method of decision making argue that it can result in a mindless bureaucracy.

For most choices to be made by planners, the utilitarian model, with its emphasis on information, is the most useful. Let's return to the problem facing our planning director and briefly analyze the dilemma in terms of this theory.

END-BASED DECISIONS/ UTILITARIANISM

Who is the audience for the conditional use approval? (Who will benefit and who may be harmed?)

Beneficiaries:

- Contractors who will build the plant.
- Business owners from secondary and tertiary economic spending.
- Residential property taxpayers as a result of an improved tax base that may reduce their burden.
- Other companies that may now be able to locate in the community given that a precedent has been established.
- Schools, as a result of an increase in the tax base.
- The unemployed and their families.
- Purchasers of the product made by the plant.
- The manufacturing company.

Losers:

- Those with existing health problems that may be exacerbated or those who will have new health problems – primarily the elderly and children.
- Adjacent property owners, who may see the value of their property decline because of their proximity to a perceived health hazard.
- Taxpayers, who may have to cover the cost of increased health and social services to the ill, and the families of those who become sick.
- Those who will pay higher taxes and utility rates to cover new infrastructure costs to support plant and related economic/residential development.

Consider the positive and negative effects of the alternative actions. [See Table 33.1.] Decide which course of action produces the greatest overall utility. In this case, are there ways that the negative outcomes or costs can be shifted to the project and away from the community? If so, the end-based decision may be even easier, particularly if the plant were to offer to sign a legally binding agreement with the community stipulating that it would:

- Continue to look for new technologies that might, in the future, eliminate emissions of potential hazardous materials.
- Fund annual physicals for those affected by the plant and be responsible for long-term health care for those individuals whose health is harmed by the plant.
- Pay development impact fees to cover infrastructure improvements needed to support the plant.
- Agree to operating stipulations that limit emissions during active daytime hours and permit them only during the night when children and the elderly are less likely to be outdoors.

With such an agreement in place, and considering the need for jobs in your community, it would be much easier for you as the planning director to recommend the approval of the conditional use permit for the manufacturing plant.

[. . .]

A SUMMARY OF THE AICP CODE OF ETHICAL CONDUCT

For planners who have joined AICP, the adopted "Code of Conduct" establishes a set of community norms for guiding behavior. Key elements of the code appear below. . . . The code has been in force since 1978, when first adopted by AICP. Since then, the code has been refined and enhanced through the adoption of amendments and advisory rulings. Adherence to the code is one requirement of membership in AICP. Procedures exist to provide counseling, to judge alleged misconduct, and to discipline members in the case of misconduct.

[. . .]

Table 33.1

	Positive	**Negative**
Approve	Jobs Tax base Economic development	Loss of reputation as clean Some air pollution Probable health effects for some percentage of the population
Disapprove	Protected health for the most vulnerable members of the community – elderly and children	Loss of jobs
	Avoid increased public health care costs	Discourage future economic development of similar jobs
	Preserve and build upon reputation as clean community	Loss of investment and spin off opportunities

THE APA STATEMENT OF ETHICAL PRINCIPLES AND ITS RELATIONSHIP TO THE CODE

The APA Board of Directors first adopted a "Statement of Ethical Principles" in 1987 as a guide to all APA members. It complemented the AICP Code, but did not include all of the same elements. The statement was more specific in identifying ethical behavior for public planning officials. Over the next several years, the members of AICP and APA worked to craft a single statement of ethical principles. The American Planning Association adopted a "Statement of Ethical Principles in Planning" in May 1992. . . . The statement, as distinct from the code, serves as a guide to ethical conduct for everyone who participates in the process of planning as advisors, advocates, and decision makers. The introduction sets forth its intent by presenting a set of principles to be held in common by certified planners, other practicing planners, appointed and elected officials, and others who participate in the process of planning.

The code and principles contain many common elements. It is important to keep in mind that the code, which is a component of the ethical principles document, is formally subscribed to by each certified planner and includes enforcement procedures. The APA principles are advisory and members of the organization are expected to adhere to the standards although there is no enforcement mechanism. Nonetheless, there is significant value in having a unified statement. The aspirational contents of the planning professional's code are made clear to all and the standards that should inform the behavior of all participants in the planning process are set forth. . . .

[. . .]

TOOLS FOR LEARNING ABOUT ETHICS

There are many ways of learning planning ethics – reading, attending lectures, and listening to the experiences of others. Ever since the first fires were built in caves, storytelling has been a powerful teaching tool. . . .

Planners are challenged to consistently and systematically analyze situations to develop a basis for informed ethical judgment. Rarely will planners be confronted with a dilemma such as the one described above whereby a plant that may emit carcinogens wants to locate in town. Much of the ethical conflict will be of a more basic variety that is guided by the code. In these situations, planners must first recognize what the ethical issues are. Next, they must examine and rank their values with respect to a particular decision. Planners perform their work in an environment that allows for a broad range of administrative discretion. As a consequence, planners need some benchmarks for relating the various, and sometimes competing, claims of different values that can enter into official actions. There are steps that lead to practical moral reasoning by which planners can evaluate the appropriateness of a particular course of action. These steps will help the planner develop a keener moral imagination. The steps allow a planner to integrate internal sources of values with external demands while maintaining a sense of integrity.

1 Define the problem. . . .
2 Clarify your primary goal. . . .
3 Examine all the facts in light of this primary goal. . . .
4 Brainstorm alternative courses of action that you might pursue. . . .
5 Evaluate the alternatives and the consequences that may result from your course of action. Compare these alternatives to the code. . . .
6 Select the preferred course of action and give it another test: publicity. How would you defend this decision in public if asked to do so?
7 Implement your plan of action. . . .

These steps are similar to those undertaken in many types of planning analyses. They are recommended here because they work to help identify and analyze choices. The preferred course of action is the one most consistent with the requirements of the code, and that has the greatest potential for a personally satisfying decision. In the section that follows, the technique is described as it might be used on the job.

SAMPLE SCENARIO: WHAT ARE YOU WILLING TO DO TO KEEP YOUR JOB?

[. . .]

You're a newly hired planner and the boss drops by your office to chat one Friday afternoon. He mentions that a development review assignment will be on your desk early the following week. He says, "Mr Jones has a small subdivision proposal. He was the treasurer of the mayor's last reelection campaign. The mayor just called to remind me of his support for good new development and his belief that Mr Jones's subdivision would be an asset to the city. I told the mayor that I did not foresee any problems." As the director gets up to leave, he turns and, smiling, says, "Have a nice weekend. Relax and come back to work ready to do your job. See you first thing Monday morning."

Although new to the agency, you are not new to the profession. You believe that the director has ordered you to make a favorable recommendation on Mr Jones's subdivision. You are extremely uncomfortable. You wonder what you should do next.

1 Define the problem. The mayor's relationship with the applicant seeking development review is subverting the normal process of analysis and recommendation. Your boss has alerted you to the relationship and encouraged you to offer a favorable review.
2 Clarify your primary goal. You would like to keep your job and to behave ethically. Given this goal, you will seek resolution without confrontation.
3 Examine all the facts. In this case, you are missing several pieces of information. For example, the project might actually conform to all of the development requirements and the mayor is correct in his assessment regarding the project's desirability. Also, what is the director's reputation? Is he usually a "good guy"? Or is there a history of staff coercion? This absence

Table 33.2

Alternative courses of action	Possible outcomes
Hear no evil. See no evil.	Keep job/lose integrity.
Do full analysis to see what the subdivision proposed is really like.	You have more specific info about whether or not the project is in compliance. By supporting your argument with facts, the likelihood of a favorable outcome increases.
Clarify what the planning director really wants.	You may find out that you misunderstood or overreacted.
Resign.	Unemployed with a loss of income.
Prepare a memo to the file that outlines the assignment you have been given.	Establishes a paper trail in the event of future problems.
Confront the planning director.	Unclear. You will clarify your personal stance but may damage future working relations.
Consult with others.	Find out if the inference is standard operating procedure or whether you have misunderstood the conversation. You may find moral support as well.
Delegate upwards. Ask your boss to take the lead on this. Explain that, because of your new status, you are unsure of your ability to do the assignment.	The director might refuse and your days could be numbered. The director might accept and you would be off the hook. The director might choose to reassign the task to someone else.
File grievance.	You are likely to lose such a grievance. You may retain your integrity, but you will have to resign because the work environment will be extremely uncomfortable after a grievance procedure.
Check personnel policies for protection.	Understand your rights and responsibilities, and those of your employer.

of information dictates your immediate course of action – collecting more data.

4 Brainstorm alternative courses of action. You consult with trusted colleagues. You eventually come to the conclusion that the director is asking you to make a finding unsupported by the facts of the case. You weigh the alternatives in terms of the requirements of the code.

[. . .]

5 Evaluate the consequences.

6 Select the preferred course of action and give it another test: publicity. Based on all of the factors set forth, you decide that you will ask for a meeting with the director to explain your position and to ask for a response. If the director remains unmoved, you will decline the assignment. Adding the step of making absolutely sure there is no misunderstanding makes you feel comfortable with how you would explain or defend your actions if asked to do so.

7 Implement your plan of action. You call the director's office and make an appointment. You also begin reworking your résumé, understanding that you have embarked on a course of action that may well cost you this job.

PART SIX

Key Topics in Urban and Regional Planning

Growth of a typical Northeastern city: 1950. (Drawing by Andrew Whittemore)

INTRODUCTION TO PART SIX

When Bruce Katz, Director of the Brookings Institution's Metropolitan Policy Program, unveiled an important position paper entitled "Blueprint for American Prosperity," in fall 2007, he asserted: "The very assets needed to prosper today are rooted in our metropolitan areas – the nation's interconnected web of cities, suburbs, and counties now home to more than eight in ten Americans and their jobs." Although Katz was referring to the country's 362 metropolitan areas, he quickly turned a spotlight on the top 100 metros, noting that though they have two-thirds of the U.S. population, they contribute *three-quarters* of the gross domestic product. Further, he observed, these 100 metros have a disproportionate share of the qualities needed to sustain the United States' technological, knowledge-based economy – 75 percent of those holding graduate degrees, 78 percent of the patents, and 94 percent of venture capital (Katz 2007).

Katz was not sanguine about these metros' continued economic strength, however. He described them as weakened by human capital shortfalls (low educational and income attainment among minority groups – "you earn what you learn"), inadequate infrastructure (aging highway systems, congested airports, underutilized and obsolete railways), and questionable quality of place (environmentally irresponsible metropolitan development), and he called for an alignment of federal, state, and local policies to attack these enduring problems. Leadership and capital to strengthen education, support infrastructure investment, and finance large open spaces and other public goods should come from the national government and flow to the localities. Public–private partnerships should provide additional funding and knowhow. Most important is the focus and consensus to move ahead strategically.

The current situation may be even more urgent than Katz's analysis indicates. The U.S. Census Bureau anticipates that the country's population will rise to more than 400 million by 2050. And about 70 percent of that growth will probably occur in ten megaregions, the very places afflicted by the already identified problems (*America 2050*, 2004). (See Plate 1.11.)

Urban planners are well positioned to address two of the problem areas: infrastructure and quality of life. In terms of shaping physical development, they are on the front lines, helping formulate, interpret, and implement policies and programs in transportation, land use, urban design, and environmental planning on the federal, state, regional, and local levels. Planners' role in the delivery of social services is more limited. While a planner might site a neighborhood school, he would have no say in pedagogical matters. A planner may manage a community development corporation, whose activities range from building affordable housing and promoting local retail to job training, but she would not be involved in programs related to employment readiness or tutoring.

Where urban planners have special expertise is enhancing the movement of people and goods, revitalizing downtowns and neighborhoods, protecting and improving the environment, and designing urban spaces. Over time, planners have deepened their knowledge and skills in these areas. For example, in transportation, they have developed many ways of dealing with vehicular traffic congestion, ranging from traffic-light timing to broad schemes that disperse travel on different modes (buses, mass transit); they also know how to adjust land use patterns through zoning or capital investment to address the sources of transportation problems. In urban revitalization, they have overseen projects ranging from mixed-use and mixed-income neighborhoods to waterfront reclamation.

The selections in this section explore major subfields of planning – transportation, environment, land use, and urban design. Many present advanced thinking in their areas. Collectively, they show the varying spatial applications – cities, suburbs, and regions – of the specializations. Vukan R. Vuchic's "Implementing the Solutions: Measures for Achieving Intermodal Balance" focuses on reducing automobile dependence in metropolitan areas, especially in cities. He argues for planning that incorporates many modes – car, transit, bikes, and walking. He outlines administrative and design approaches that will lead to a more balanced transportation network in urban places.

In "Environmental Planning" Thomas L. Daniels and Katherine Daniels outline the national environmental policies that shape metropolitan areas. They cover ecosystem issues ranging from air and water quality to open-space protection and brownfields reclamation.

Eugénie L. Birch zeroes in on cities in "Hopeful Signs: Urban Revitalization in the Twenty-first Century," arguing that cities have to focus on making improvements to their major, highly prized asset, land, to create local tax revenues to support the rising costs of municipal services. Employing a broad definition of revitalization, she categorizes municipal efforts as: catalytic (or transformative), downtown, neighborhood, and project-focused.

Moving to the suburbs, William H. Lucy and David L. Phillips, in "Sprawl and the Tyranny of Easy Development Decisions," assert that today's development patterns are the product of many unilateral choices made by local governments without consideration of neighboring jurisdictions. The authors describe this process as an intricate dance among three categories of actors (government officials, developers, and consumers) who respond to trends in housing markets, job location/commuter sheds, and local political attitudes.

Andrés Duany, Elizabeth Plater-Zyberk, and Jeff Speck's "How to Make a Town" looks at development on new sites, prescribing important qualities – pedestrian orientation, higher density, and mixed land uses – that taken together are labeled "New Urbanism." The authors call for plans that make the neighborhood the basic building block and incorporate transit-oriented development whenever possible.

Finally, Gary Hack's "Planning Metropolitan Regions" challenges contemporary thinking on regionalism. It asserts that older monocentric models have given way to polycentric arrangements. Today's metropolitan regions are not organized around a dominant single city but incorporate many cities of varying size and function. They also [feature] the dispersion of middle- and upper-income populations to the suburbs, concentrations of poverty in central cities, spread-out employment, jobs/housing mismatch, and ineffective transportation routes. These conditions call for a new brand of urban and regional planning.

Thus these selections contribute ideas to respond to the challenges that Bruce Katz has put forward. They show how to improve transportation, environmental quality, urban centers, new suburban development and regions. They also offer insights into how land use decisions are made, essential knowledge for urban planners.

34

Implementing the Solutions: Measures for Achieving Intermodal Balance

From *Transportation for Livable Cities* (1999)

Vukan R. Vuchic

Editor's introduction

The United States has a massive, well developed, but unbalanced transportation network that is naturally of great import to urban and regional planners. It is composed of roads and highways; rail lines; navigation channels and ports; air routes and airports; mass transit lines and pipelines. Planners are involved in the maintenance, improvement, and expansion of various parts of this network, and also in seeking ways to integrate its different modes and deal with their impact on land use and urban development. To understand the conditions that planners work with, consider the size and composition of the transportation network/system/systems in this country.

In infrastructure terms, highways dominate. Four million miles of roads connect all parts of the country. Of these, 160,000 miles are part of the National Highway System (federally designated in 1995 as arterials important to the U.S. economy, defense, and mobility), including the 47,000-mile Eisenhower Interstate Highway system. The Interstate, the nation's largest public works project, built between 1956 and 1972, was purposely designed to link the nation's cities of 50,000 people or more. Traveling along this road system are approximately 247 million motorized vehicles (almost one for every American), including 137 million cars, 95 million vans or SUVs, 8 million trucks, 6 million motorcycles, and just under a million buses. Measured in passenger miles traveled, the road system accounts for 88 percent of the total U.S. total travel.

Air travel comes in a distant second, accounting for 11 percent of the nation's passenger miles traveled. There are more than 5,000 public-use airports in America – 1,200 are for commercial carriers, divided between hubs (130 hubs handle 97 percent of all passengers) and nonhubs.

Supplementing the roads and airports are about 400,000 miles of rail, water, and mass-transit routes and about 1.5 million miles of oil and gas pipelines. Only 23 percent of the nation's 162,000 miles of rail lines are for passengers; the remainder is used for freight. Along the United States' 26,000 miles of navigable channels are 9,000 commercial port facilities (wharves, piers, docks) and, of these, the top twenty handle the lion's share of the nation's freight (94 percent of all container tonnage). Finally, mass transit has 176,000 miles of routes, of which 96 percent are for buses, 2 percent commuter rails, and 1 percent subways or light rail.

These figures, which come from the U.S. Department of Transportation's *Pocket Guide to Transportation* (Washington, DC: Bureau of Transportation Statistics, 2008), certainly prove that the United States is auto-dominated. On average, households spend about 18 percent of their income on transportation – and 94 percent of that figure on vehicles. Furthermore, transportation is a huge part of the U.S. economy – generating

11 percent ($1.4 trillion) of the national gross domestic product (GDP); only housing, healthcare, and food are larger. This sector includes everything from vehicle purchases to freight and passenger travel.

The car's hegemony, which has shaped today's metropolitan settlement pattern, resulted from a combination of factors. First and foremost, in America's strong economy, industrialists produced inexpensive cars, the nation's political leaders responded to the public clamor for better roads, and transportation finance experts crafted a dedicated funding stream, the Highway Trust Fund, from user taxes (on gasoline, tires, the weight of trucks and buses) to pay for new highways. In this environment, nimble actions by the housing industry, banks, and government mortgage insurance providers and the availability of accessible undeveloped land also contributed to population and employment dispersion. In wake of this decentralization, cities tried to remain competitive with suburbs in the office, retail, and tourism markets. Municipal leaders successfully lobbied for Interstate ring roads and radial extensions for easy access to their cities, especially their downtowns. They cleared blocks of old offices to make room for parking, and devoted public and private funds to build new office complexes, convention centers, stadiums, and festival malls like Boston's Faneuil Hall, creating today's well known urban landscape.

U.S. transportation policy has been an important driver of today's settlement patterns. Between the mid 1950s and the 1970s, the national government spent $425 billion on the Interstate, providing 90 percent of the construction costs to states so long as they conformed to minimum standards – such as lane widths and signage – and engaged in long-term planning. (The federal government had no comparable program for mass transit.)

After a twenty-year hiatus, the federal government took on major new transportation initiatives – to the tune of $759 billion – through three pieces of landmark legislation: the Intermodal Surface Transportation Efficiency Act (ISTEA, 1991), Transportation Equity Act for the twenty-first century (TEA-21, 1998) and the Safe, Accountable, Flexible, Efficient Transportation Equity Act: A Legacy for Users (SAFETEA-LU, 2005). Unlike the 1956 Federal-Aid Highway Act and its successors, these laws allowed the Highway Trust Fund to spend more money on mass transit – by the end of the 1990s, 37 large cities embarked on light-rail projects. They supported programs to relieve congestion, protect the environment, and increase research on transportation issues. For example, they created the Bureau of Transportation Statistics, provided funding for up to 60 university research centers, and authorized the National Surface Transportation Policy and Revenue Study Commission to advise on major policy matters.

Planners have been deeply involved in local, regional, state, and national transportation planning since the 1960s when the federal government mandated long-term plans as a pre-requisite for federal funding. The U.S. Department of Transportation (DOT) – established in 1966 and today an agency with a $65 billion annual budget, 60 percent of which goes to the Federal Highway Administration – monitors these planning activities. Over the years DOT has pushed states to undertake regional planning. In 1975, it began requiring them to form Metropolitan Planning Organizations (MPOs), special regional planning agencies covering urbanized areas, and ordered each MPO to develop two planning documents: the Regional Transportation Plan (RTP), with a twenty-year time horizon (revised every three years) and the four-year Transportation Improvement Program (TIP) of priority projects. (For examples, go to www.dvrpc.org for the Philadelphia region or www.sandag.org/index.asp?fuseaction=about.home for the San Diego area.) DOT has also made the MPOs responsible for seeing that their regions conform to several other federal mandates, including clean air.

The selection focuses on cities and the reduction of car use, an increasingly important aspect of transportation planning. This selection, concerning the reduction of auto-dependence in cities, is timely because of current interest in finding ways to become sustainable and reduce the carbon footprint of urban areas. Its author, Vukan R. Vuchic, is an international expert in urban transportation systems and UPS Foundation Professor of Transportation Engineering, School of Engineering and Applied Science, University of Pennsylvania. His most recent books are *Transportation for Livable Cities* (New Brunswick, NJ: Center for Urban Policy Research, 1999), *Urban Transit: Operations, Planning, and Economics* (Hoboken, NJ: Wiley, 2005), and *Urban Transit: Systems and Technology* (Hoboken, Wiley, 2007).

For a concise account of federal transportation policy from the 1930s to the 1990s, see Edward Weiner's *Urban Transportation Planning in the United States: An Historical Overview* (Westport, CT: Praeger, 1999).

In "New Mobility: The Next Generation of Sustainable Urban Transportation," *Frontiers of Engineering: Reports on Leading-edge Engineering from the 2006 Symposium* (Washington, DC: National Academies Press, 2007), Sharon Zielinski pinpoints three key challenges for twenty-first-century transportation policy: perfecting the computer-based modeling on which it does projections; figuring out how transportation demand can be changed with land use (e.g. more density) and advances in telecommunications (e.g. reductions in the need for travel); and developing new business models to pay for needed improvements (e.g. public–private partnerships).

In *The Geography of Urban Transportation*, third edition (New York: Guilford Press, 2004), editors Susan Hanson and Genevieve Guiliano air a related policy question: should government promote accessibility (equalizing the ability of all to get to places) or mobility (enhancing the ability of most people to get from place to place)? Three recent books discuss transportation policy in cities: Sigurd Grava's *Urban Transportation Systems: Choices for Communities* (New York: McGraw-Hill, 2002), Donald Shoup's *The High Cost of Free Parking* (Chicago: APA Planners Press, 2005), and J. H. Crawford's *Carfree Cities* (Utrecht: International Books: 2002). Anthony Downs addresses suburban congestion in *Still Stuck in Traffic: Coping with Peak-hour Traffic Congestion* (Washington, DC: Brookings Institution Press, 2004). Peter Newman and Jeffrey Kenworthy's *Sustainability and Cities: Overcoming Automobile Dependence* (Washington, DC: Island Press, 1999) is a pioneering study on sustainability issues. The same authors have written a chapter, "Greening Urban Transportation," in the Worldwatch Institute's *State of the World 2007: Our Urban Future* (New York: Norton, 2007). Myer Kutz (ed.) covers biofuel and hybrid cars and new congestion management techniques in *Environmentally Conscious Transportation* (Hoboken, NJ: Wiley, 2008). For detailed technical information on the relationship between global warming and transportation, see *Potential Impacts of Climate Change on U.S. Transportation*, Special Report 290 (Washington, DC: Committee on Climate Change and U.S. Transportation, 2008), an important report by the Transportation Research Board, one of the six divisions of the National Research Council and the division on earth and life studies.

For popular writings on auto dependence and its effect on the American landscape, see Jane Holtz Kay's classic *Asphalt Nation: How the Automobile took over America and How we can Take it Back* (Berkeley, CA: University of California Press, 1998) and James Howard Kunstler's *The Geography of Nowhere: The Rise and Decline of America's Man-made Landscape* (New York: Simon & Schuster, 1993).

TRANSPORTATION IN URBAN PLANNING

In designing new developments or improvements to existing urban areas, planners must evaluate the role and characteristics of the transportation system in light of the goals adopted for the character of the city and metropolitan area. Lifestyle and functions required for individuals – as well as social, commercial, recreational, and other activities – lead to formulation of the transportation services needed and to the transportation modes that can deliver them.

To achieve an intermodal system that can utilize the benefits of diversity provided by a combination of modes, each mode must be planned separately and then integrated with other modes through an iterative planning and design process. It is important, however, that the street/freeway system, which in the context of the total transportation system actually is a subsystem, initially be planned as a system and network. Similarly, a transit subsystem must be planned as a system, then integrated with the street/freeway subsystem. This is especially true when the transit system has a special infrastructure: separate rights of way (rail), stations, even special lanes on streets and freeways (buses). If transit consists only of street modes (buses, streetcars, or tramways), coordination of the car/street and transit subsystems will be mostly at the operational level, rather than in planning infrastructure – that is, street and freeway networks.

Transportation systems and the composition of modes vary not only among cities, but also for different sectors within metropolitan areas. In central cities, the street network is ubiquitous because it must provide access to all lots; but in addition to the access function, it must serve the traffic-carrying function. In revising an existing street network, an important goal is usually to convert a street network with numerous streets of equal significance into a network of local streets and arterials. That way, through traffic is discouraged or eliminated from local streets, while the arterials are upgraded through design and traffic engineering measures to handle large traffic volumes and, secondarily, to provide access to the abutting land and buildings.

Construction of freeways (motorways) into central cities is a critical decision for a city's character and livability. During the freeway construction era, many large North American cities, such as Houston, Boston, and Philadelphia, designed three freeway rings, or "beltways," as parts of a saturation-type freeway network for the entire metropolitan area. This was based on the belief that all travel in the future would be by private car, supplemented by buses during peak hours and for "transit-captive" riders only at other times. Unlimited travel by car was considered a basic need of metropolitan areas, more important than historic buildings, amenities, environment, or quality of urban life. . . .

[. . .]

These extensive networks bring and distribute large volumes of vehicular traffic into the heart of the city or throughout the region. Although convenient for drivers, these designs, especially radial ones, have three harmful consequences. First, because they require the destruction of buildings, parks, and residential areas, they have a harmful physical and environmental impact on the central city as well as other highly concentrated areas through which they pass. Second, they create physical barriers between areas, often becoming economic and social dividers in the process. Third, they overload street networks with heavy traffic volume and require high parking capacities, further disrupting the urban texture and destroying the city's human character.

[. . .]

Different decisions about the role of street, arterial, and freeway networks can and do have highly different impacts on cities. Construction of several radial freeways converging on an inner freeway ring results in a car-dominated city with extensive parking garages and lots. Conversely, central cities without freeways are more compact, and concentrated volumes of travel must be handled by high-performance transit, usually one or several rail transit modes.

[. . .]

The extent and the form of freeway networks depend on the demand for travel by car, as well as on the physical limitations of the area and its individual corridors. Generally, freeways should be placed around major activity centers and corridors. Because of the large area they consume, freeways should not go through centers of activity. In addition, the impact of large traffic volumes can be ameliorated if freeway traffic is distributed from interchanges through arterials and streets, rather than brought directly into high-density activity areas. It is important that each network – street, arterial, and freeway – be designed to meet its set of requirements; however, all three must be considered together. They should be complementary, and their contact points, intersections and interchanges, should have balanced capacities.

Transit must also be planned with respect to the two basic functions it serves. Local, mostly short-distance, travel usually is served by street transit, which utilizes streets and arterials and requires little separate infrastructure. A high-performance or high-speed network serves trips of all lengths, but predominantly medium to long ones. It has "high performance" because it operates on independent rights of way. . . . These lines must be incorporated with urban land uses; unlike freeways, they optimally should go through the cores of high-density activity areas, since rail lines rely on pedestrian access and directly support activities in station vicinities. In addition, these lines should be coordinated with local transit and have joint stations where passenger transfers take place.

In planning new developments, towns, and cities, it is important to develop not only arterial street networks but also transit rights of way. . . . This ensures that an independent, congestion-free transit system will be provided. At the planning stage, these rights of way involve low investment costs. If they are not provided, however, transit will indirectly be given a secondary role, because it cannot compete with car travel. Provision of an independent transit system may be necessary later, but then it will involve much higher costs than if the land was reserved for it at the beginning.

[. . .]

Neglecting to consider transit access is a very common problem, and its consequences are greatly underestimated. The widespread construction of large suburban shopping, office, residential, and other developments leads to transportation problems that cannot be resolved physically or financially. Sprawl in medium- and high-density developments is even more serious than the "traditional" sprawl of single-family housing, for several reasons. In addition to extensive land consumption, high-energy consumption, and heavy environmental load, low-occupancy cars become the only physically feasible mode of access. The result is high generation of VKT [vehicle kilometers traveled], a need for extensive highway construction (which the population needs for travel, but which is rejected for environmental reasons), inability to provide either transit or paratransit without extremely high subsidies, extensive chauffeuring of passengers (businesspersons, children, the elderly), and not surprisingly, minimal use of walking and bicycling.

The last aspect, use of walking and bicycling, is an extremely important, albeit overlooked, component of transportation planning. Whereas inadequate land use planning is caused by a lack of adequate controls on land use as well as by political and commercial pressures, failure to provide for convenient walking and bicycling is usually a result of professional lapses and misguided thinking (for instance, that walking "happens" and needs no major attention in planning and design). Sound planning, therefore, must include not only land use/transportation at the macro scale but also at the design level of individual buildings, developments, streets, and other transportation facilities. Requirements and standards for pedestrian and bicycle transportation must be included in every plan for the design or redesign of facilities and urban developments.

[. . .]

FINDING THE OPTIMAL ROLE FOR THE CAR

With its unique ability to provide superb personal mobility, the car is a fundamental element of our civilization. Its availability and use are a major asset and element of our living standard. Its utility to users depends on the value of mobility in absolute terms, as well as relative to other expenditures such as housing, education, and recreation. . . .

[. . .]

. . . The reduction of car trips and VKT is . . . the main objective in efforts to achieve a desirable balance between the private car and other modes and to reduce the negative impacts of excessive car use. Two sets of measures – *car use disincentives* and *incentives to use alternative modes* – are complemented by long-term planning for reduction of the demand for car travel. The measures directly concerning the use of the car, particularly in its single-occupancy version, are reviewed here, grouped in three major categories . . .

- Car use pricing, or charges for driving and parking . . .
- Reduction of VKT through organizational measures . . .
- Reduction of demand for car use through planning and design . . .

CAR USE PRICING

With the realization that urban transportation problems cannot be solved merely by building more roads, it has become clear that conditions could be vastly improved if travelers would to some extent face cost elements reflecting market conditions. . . .

[. . .]

Various methods for pricing the use of cars have been used. Some, such as road tolls and parking charges, are conventional, while others have been used successfully in a limited number of cities. These include area licensing schemes, "ecotaxes," and other programs. Finally, a number of pricing methods have been proposed but not yet implemented. In recent years, the need for increased user charges in highway and urban transportation has grown due to two trends – the need for additional funding of transportation systems, and the need to increase the efficiency of system operations by raising the direct costs of driving and compensating for currently unpaid social and environmental costs. These trends are likely to lead to the introduction of innovative pricing methods. The different types of charges for car use can be classified by their objectives, rationales, and methods in four main categories:

1 Restructuring of car user costs – increasing the share of direct costs.

2 Conversion of subsidies into user charges.
3 Impact charges – payments for presently unpaid costs; and
4 User charges to influence travel behavior.

[. . .]

Reduced parking subsidy and supply. One of the most effective intermodal balance tools in cities . . . is the reduction of parking subsidies and supply. In most U.S. cities, however, parking is oversupplied and underpriced and thus represents one of the largest subsidy items in urban transportation. . . .

Oversupplied and underpriced parking causes widespread economic inefficiencies, social inequalities, negative environmental impacts, and urban transportation problems. For example, . . . about three-quarters of "suburban economic center parking" is in surface facilities. In many suburban areas, all parking is on the surface. As a result, a typical suburban shopping center dedicates 55 to 70 percent of its total area to parking and driveways, and an office park 50 to 60 percent. In addition to this land consumption, the cost of providing a parking space varies from $2,000 in suburban areas to $20,000 in urban areas – not counting maintenance, supervision, and so on. When the cost of these facilities is absorbed by the building owners and businesses, every customer or employee who comes by transit, bicycle, or walking actually subsidizes the parking of those who arrive by car. Not only is this inequitable, it stimulates the modal split toward the car rather than its alternatives. . . .

[. . .]

. . . [P]arking measures can be organized and introduced fairly easily. The main step to be taken is to adopt a policy toward the supply and pricing of parking. Several measures can be used to correct present inefficiencies and integrate parking into a coordinated transportation system.

The federal tax bias toward car use. The federal tax bias in favor of the car should be eliminated. Until 1991, all costs of employee parking were exempt from federal business taxes. Employers' contributions to transit passes were limited to a mere $15 per month. An employer contributing more than this amount would lose even this small exemption. The Intermodal Surface Transportation Efficiency Act (ISTEA) limited the parking exemption to $155 per month while increasing the transit and employee-provided van pool exemption to $60. Although this is an improvement over the previous unlimited federal subsidy of employee parking, it nevertheless represents a subsidy for the car that is two and a half times that of its alternative – an illogical measure from a transportation policy point of view. The first step in correcting this counterproductive federal government subsidy would be to eliminate the pro-car bias by making exemptions equal for all modes. . . .

Parking rates. Parking rates can have a major impact on modal split in a particular area. Two elements are important: (1) parking rate level and (2) structure. The rate level influences the relative attractiveness of car use in relation to other modes, as well as whether or not a trip will be made. Structure, on the other hand, can also influence trip and modal split fluctuations in time. For example, a flat parking rate (independent of parking duration) encourages car commuting while discouraging short-duration parking, which usually serves trips for shopping or business meetings.

. . . Graduated parking rates, sometimes even progressive ones (hourly charge increases with duration) can therefore be used effectively to reduce peak-hour congestion and all-day parking, which is highly undesirable, especially in urban centers. At the same time, it means easy and economical parking for many trips, such as visits, shopping, and meetings.

Parking validation schemes. These plans usually offer partial or full payment for shoppers or customers. Many merchants consider this an effective marketing tool. . . . [A] correction would be that every store that validates parking must also compensate passenger fares for customers who come by transit.

[. . .]

Parking "cashout." Parking cashout is an arrangement whereby employers who provide their employees with either free or greatly discounted parking convert this benefit into a "travel allowance" – an amount equal to the market value of a parking space – which each employee can then choose to use for parking or for transit fare, or can take as cash, then ride a bicycle or walk. . . .

[. . .]

Reduced fee or free parking for HOVs [high occupancy vehicles]. This measure, applied successfully in many cities, corrects the bias toward SOVs [single occupancy vehicles] by favoring HOVs, which require much less space per passenger and thus increase efficiency. . . .

Fuel taxes. Today, gasoline and diesel fuel taxes represent the largest source of revenue paid by car and truck drivers for the public portions of the highway transportation system. . . . [T]he fuel tax is a constant charge for travel regardless of location, facility, and time. Revenue collected is used, however, for a certain category of highway investments at any location in the country. . . . [F]uel costs do not reflect differences in the impacts of driving on congestion (a social cost); nor do they reflect environmental externalities, except for a small difference in increased fuel consumption in driving on highways under congested conditions. . . . [T]heir amount is so low that the revenues collected cover only a portion of total costs of highway infrastructure and operations. . . . [S]everal studies have shown that the shortfall in revenues is large even if the extensive externalities are not included. . . . [T]he direct cost of car use, even with fuel taxes, is so low that it involves a high consumer surplus. This leads to . . . excessive driving – the main component of urban transportation problems. Finally, with oil imports accounting for about 50 percent of oil consumption in the United States and continuing to increase, this country's dependence on foreign oil represents a major economic and political liability.

[. . .]

In recent years, sensational media, populist dogma against all taxes, and lobbies interested in perpetuating total dependence on cheap car travel have generated strong negative publicity about any increase in gasoline taxes. This has been carried to such an extreme that, in many areas, highways and streets are poorly maintained because of insufficient funding, while increases of gasoline taxes, which would be moderate by any standards, are rejected as a "hardship for motorists." The hardships caused by poor roads, inadequate pedestrian facilities, and the absence of alternative means of travel tend to be overlooked.

[. . .]

The visibility of a gas tax has both positive and negative features. Its advantage is that it is precisely what is needed to produce an effect on the reduction of VKT. Its disadvantage is that its opponents skillfully use the tax's high profile to point out its impact on individual drivers – without explaining its positive side in achieving the desirable goal of VKT reduction. Introduction of taxes will therefore take political leadership and a responsible, professional explanation of facts in order to overcome the populist, dogmatic opposition to all taxes.

[. . .]

Area licensing scheme. . . . (ALS) is a special form of road – or, more precisely, area – pricing aimed at discouraging the use of cars in a potentially congested area and ensuring efficient, congestion-free streets. . . .

Road pricing . . . is a form of user charge for driving that is more specific than a fuel tax because it can be related to the type of facility used. . . . In its most common form – tolls for highways, bridges, tunnels, and similar facilities – pricing is used for its initial purpose: to pay for the construction and maintenance of the facility. . . . The next most common objective of road pricing may be to reflect the indirect costs and externalities of car use. The purpose of these charges is twofold: to recover the imposed costs (and sometimes to finance their alleviation or compensation); and to discourage excessive travel, thus preventing generation of externalities.

[. . .]

If road pricing is so important and potentially so effective, why does it remain mostly in theoretical studies and academic papers (where there is a considerable consensus about its desirability across the ideological spectrum, from liberals to conservatives)? Traditionally, there have been only two obstacles (albeit serious ones) to its implementation:

1 *The physical difficulty of collecting charges* that could apply to entire areas and change with locations or times. Stopping all traffic for toll collection in a dense urban area has often been considered physically and operationally infeasible.
2 *Opposition by the driving public,* which has an aversion to out-of-pocket payments. This opposition creates a disadvantage for the areas served by tolled facilities versus those with free services.

The first obstacle may appear more difficult than it actually is. Singapore has shown that a large central area can be cordoned off by a ring of "gates," which actually consist only of overhead signs. Drivers make payments prior to approaching these locations, so that entry does not require them to stop. Moreover, the invention of "smart cards" has made feasible efficient collection of charges in any amount, variable by location, time, and vehicle. Implementation of the system for charge collection and control does, of course, require careful planning and the installation of extensive electronic or other control systems; but the entire concept can be applied on most streets in urban and suburban areas.

The second obstacle, public opposition, remains serious, although some inroads have been made. Similar to the problem of introducing a fuel tax, the introduction of new charges is strongly opposed because "free driving" is considered by many to be a basic feature of our society and living standard. . . .

[. . .]

Peak-period road pricing. Increasing prices during peak traffic periods . . . [aims] to reduce peak-period congestion by influencing the time distribution of travel (a "flattening of peaks") to achieve more efficient traffic flow. . . .

[. . .]

Congestion pricing. Congestion pricing has the same basic objective as peak-period pricing. The main difference between the two is that congestion pricing is applied not for fixed time periods but as a direct function of traffic volume on a highway or in an area.

[. . .]

Ecotaxes. These are surtaxes imposed in some countries with the specific goal of reducing car use for environmental protection. . . . They are representative of the taxes imposed to reflect and reduce the negative impacts . . . of car use . . . [on] the man-made and natural environment.

Registration fees. Fees contingent on car kilometers (miles) driven would make the present fixed registration cost more proportional to the extent of car use and thus shift a portion of fixed costs into variable user costs. . . .

[. . .]

REDUCTION OF VKT THROUGH ORGANIZATIONAL, PLANNING, AND DESIGN MEASURES

The measures directly or indirectly aimed at reducing VKT and increasing the livability of cities include a range – from administrative, mostly palliative prohibitions or limitations on car use to long-term area or transportation system designs that generate fewer and shorter car trips.

Even/odd license numbers. This driving restriction is a purely regulatory, usually temporary measure to enforce reduction of VKT. The concept is simple: a decree is made that on alternate days only cars with, respectively, even or odd license numbers may be driven. Thus, half the cars are simply prohibited from driving in the city. . . .

Driving restriction by license number has been used effectively on occasion in a number of metropolitan areas that faced emergency conditions and where reduction of VKT was urgently needed. It was first introduced during the energy crisis of the 1970s as a fuel conservation measure. Seoul, Korea, used it as a traffic reduction measure during the 1988 Olympic Games, and some large employers have continued to apply this restriction to reduce the enormous pressures of street congestion in that city. Mexico City and Paris have applied this measure to reduce the volume of exhaust during periods of severe air pollution.

Parking supply standards. These standards represent a powerful tool in controlling the use of cars in cities and suburbs, and thus in controlling overall modal split. . . . Parking policies and standards, usually implemented by ordinances, have undergone an interesting evolution. When increased use of cars caused congestion on urban streets, cities mandated that new construction provide a certain minimum number of offstreet parking spaces to prevent further congestion. . . . Many cities built so many parking lots and garages that even their CBDs became dominated by parking facilities – such was the case in Dayton, Houston, and Syracuse, for example. In addition to the physical damage to an area's human character, such ample parking makes use of any other mode very inefficient.

Recognizing the negative impacts of excessive parking in CBDs and other major activity centers, cities began to reverse the regulations. Instead of requiring a minimum number of parking spaces for each new development, they introduced a maximum number of spaces that could be provided. . . . Many cities, including Hamburg, London, Boston, and Portland (Oregon), have a maximum number of parking spaces allowed in their central areas.

The parking problem in suburban areas is somewhat different. Town ordinances often prescribe not only the minimum number of spaces required for any new development, but even the dimensions of parking stalls. Typically, the required number is quite high and the dimensions, based on the standard large car of the 1960s, are larger than necessary for today's car sizes. . . .

[. . .]

Traffic calming. . . . Also known as "traffic taming," traffic calming is a comprehensive set of techniques for control of vehicular traffic. It is usually applied in residential streets and areas to enhance their safety and livability by protecting pedestrian activities and children's play. . . . A "calmed" street has a low speed limit, and drivers expect to find various speed control devices. These range from signs and pavement markings to street realignment – curves in the roadway, narrow locations, and physical "bottlenecks" (islands in the middle of the roads, speed bumps, large flower or bush containers, widened sidewalks with benches, and frequent crosswalks).

[. . .]

Auto-restricted zones (*ARZ*). ARZ is a broad set of traffic calming measures – including regulatory, design, and sometimes pricing measures. . . . Usually, this is applied to urban areas that have a residential or local activity character, though sometimes it is also used for shopping and business zones. . . .

It is important that the movement of transit vehicles, usually buses, through ARZs be planned so that they are not made inefficient by traffic diversions and limitations. On the contrary, by exempting buses or rail vehicles from traffic restrictions, transit becomes relatively more attractive than car travel within ARZs. This can be done by providing surface rail transit tracks through diversion barriers, or by "traps" – wide-wheel path surfaces with a sand ditch in the middle that buses can negotiate but cars cannot (used, for example, in suburban areas of Calgary, Canada).

[. . .]

Traffic cells. [T]raffic cells [prohibit] traffic except transit vehicles among the cells . . . [and] usually are applied to the urban core zone, often the central business district. Circumferential arterials around this central zone are improved, and parking facilities are provided along its periphery. The zone is then divided into several "cells" – areas that vehicles can enter from the circumferential arterial, but from which they cannot travel to other cells. Transit vehicles follow the dividing lines between the cells, so that they are not intersected by vehicular traffic. . . .

[. . .]

Traditional neighborhood development (*TND*). Traditional neighborhood development (TND) and transit-oriented development (TOD) . . . provide mixed land uses and nearby location of residential buildings with stores, neighborhood schools, and various services. Many trips that residents make are thus within walking distance, children can use bicycles, and transit stops are placed at the core of the developments. . . .

[. . .]

INCREASING THE ROLE OF TRANSIT

The important (and stable) role that transit should play in increasing the livability of all medium-sized and large cities can be achieved through a number of measures. Appropriate measures range from intermodal urban transportation policies and their implementation via regulatory measures to the construction of high-performance systems. These incentives, or measures for increasing the use of transit, should be based on a rational, comprehensive transportation policy that encompasses a clear definition of the role of transit in the city and its relationship to other modes. The policy should include numerous measures and actions, which can be classified by duration as short-, medium-, and long-term transit incentives. The major categories are listed below and briefly described.

[. . .]

SHORT-TERM TRANSIT INCENTIVES

Priority for transit vehicles. . . . The popular name for giving transit priority treatment is "Transit First," a set of primarily low-investment cost measures that allow transit vehicles to travel faster and to experience fewer delays along their routes. Transit vehicles are separated at intersections or along streets by transit, or "diamond," lanes or by curb-protected rights of way. At signals, they either get a green light before other traffic or have a "full override" (an immediate green light) as they approach an intersection. Transit stops are located conveniently for pedestrians, transit vehicles can make turns or travel through certain blocks where other vehicles are prohibited, and so on.

[. . .]

Intermodal integration. Intermodal integration is paramount in increasing the quality of transit services and attracting ridership. Competing with the ubiquitous travel capability of the car, transit must also provide service that is "seamless," involving a minimum of interruption. To achieve this, transit must consist of an integrated network of one or several modes provided by one or more public agencies or private companies. When a passenger wants to travel by transit, he or she should be able to view transit as a unified system rather than several different systems or modes that require special knowledge about which lines to use, how to transfer, and how to pay fares. . . .

Organizational integration. Transit services must be provided either by one agency or by an umbrella organization that unifies all functions affecting passengers. This is achieved by having agencies and private companies make special arrangements for joint terminals, fares, and so on. . . .

Operational integration. In operationally integrated service, different lines and modes are coordinated by network layout and schedules designed for the most efficient and convenient transfers at stations or terminals. Fares are paid only once for each trip, regardless of the modes used. Information about services is also fully integrated.

Physical integration. In physically integrated systems, the location and design of stations and the control of vehicle movements are planned and designed so that passenger transfers are safe, fast, and convenient.

Integrated transit systems are much more capable of attracting passengers and competing with other transport modes than are separate modes. It is therefore important to provide integration even if there are several transit service providers. Deregulation usually destroys service integration. . . . Privatization may, but does not necessarily, lead to problems of disintegration. If a central coordinator is retained, it is entirely possible to achieve cost savings while retaining integrated services, even when services are provided by several public agencies and private companies. . . .

Fare policy. The level and structure of fares, as well as innovative collection methods, can have a significant impact on the attraction of travel by transit. The level and structure of transit fares should represent an optimal compromise between the need to maximize revenues and the dominant goal of transit services – to attract as many trips as possible. . . .

To enable transit to compete with low direct car user costs, there has been a trend toward increased use of transit passes – weekly, monthly, and, in some cities, even annual. With a pass, the urban traveler selecting a mode of travel faces a decision between more comparable expenditures: transit travel, similar to driving with many subsidies, does not involve a direct payment for each trip. To compensate for the indirect subsidies of car travel (such as parking or use of a company car), employers in many cities have begun to contribute to the cost of transit passes. This contribution may be in the form of "parking cashout" schemes, or supplying transit passes through monthly payroll at a discounted rate.

MEDIUM-TERM TRANSIT INCENTIVES

[. . .]

Introduction of "smart cards." Smart cards are credit or charge cards that can be used for transit or, even more effectively, for all transportation charges, including gasoline, tolls, and parking. . . . It will make feasible various innovations in charge differentiation, such as peak-hour pricing and commuter discounts, in addition to making it easier to implement and change these charges as required.

Intelligent Transportation System (ITS). . . . ITS can provide significant improvements in the precise tracking of transit vehicles. This facilitates the operation of control centers that follow bus travel and intervene when delays occur, detours are needed, or other unusual events must be handled.

[. . .]

LONG-TERM IMPROVEMENTS TO UPGRADE TRANSIT

Adequate and stable financing. Availability of needed funds is a basic condition for implementing the permanent provision of attractive services that can respond to increasing demands for high-quality, high-volume public transportation. As with most transportation modes, methods of financing transit systems differ for capital expenditures and operating expenditures.

Capital expenditures include purchases of vehicles and construction of facilities and infrastructure for high-performance systems, such as rights of way and stations for rail systems and exclusive busways. These expenditures usually exceed the financial capacity of local governments; thus, state and federal governments must participate with significant, often dominant contributions. . . .

Operating assistance for transit should be a result of intermodal policies that dictate how favorable the policy is toward attracting travelers to transit and what level of cost recovery through fares is required. While efficiency of transit system operation is a major factor in the coverage of operating costs through fares, the degree of subsidy or charges for car use also has a major impact on the required level of cost recovery.

[. . .]

High-performance transit. . . . The basic feature of high-performance transit – its physical separation from general traffic . . . provides service that is reliable, high-speed, and independent of congestion, and that has a strong image, so that it becomes competitive with car travel. Transit modes in this category are usually rail systems – LRT, metro, or regional rail – but buses on busways, AGT systems, and monorails may also be included.

Because of separate rights of way, all these systems require substantial investment; but their construction represents a fundamental change in the character of the transit system, as well as an upgrading of its role in the city. This has been demonstrated in dozens of cities that have constructed new rail and busway systems in recent years – from Ottawa and Curitiba with busways to San Diego and Calgary with LRT, to Munich, Washington (DC), and San Francisco with their metro and regional rail systems.

[. . .]

The task of constructing a high-performance transit network appears daunting for cities that have only buses on streets; recent developments, however, show that a consistent policy and program for implementation of a balanced system is achievable. For example, some fifteen to twenty-five years ago, Atlanta, Washington, San Diego, and Calgary started by building single rail lines. Today, they have sizable high-performance rail transit networks and are far ahead of their peers who stayed with buses only, such as Seattle and Houston, in offering the diversified mobility of a balanced system.

From the considerable variety of modes available to cities wishing to upgrade their transit systems, the most common selections are among the following.

Buses on busways. Busways, including guided buses (O-Bahn), are effective in networks with numerous branches. Some branches may be either busways or regular streets with exclusive bus or general-purpose lanes. Exemplified by the systems of Ottawa or Curitiba, these systems can serve a network of diametrical lines covering the city and suburbs, rather than only commuter runs into and out of the CBD.

Although buses on busways may require investments similar to those for LRT with similar ROWs [right-of-ways] on trunk lines, . . . buses on busways can provide a more extensive network for a given investment than any other mode with a separate ROW. Moreover, the introduction of busways is simpler because no new technology is needed, with the possible exception of somewhat differently designed buses to better fit the operating conditions of the upgraded network. . . .

Light rail transit. LRT is the most diversified rail transit mode, due to its ability to utilize different alignments and types of ROW and to provide a broad range of capacity, performance, and types of service. It is the fastest-growing high-performance mode in North America and many other countries because it fits the need for upgrading transit in many cities served only by buses. These cities need transit service superior to that provided by buses, but with a larger network and much lower investment than metro systems require. . . . LRT is superior to buses in riding comfort, vehicle performance, system image, and other qualities that the electric traction of LRT offers: quiet operation, absence of air pollution, the ability to operate in tunnels. . . .

[. . .]

Metro or rapid transit system. The highest-performance, highest-investment transit mode is a metro or rapid transit system. It is the optimal choice in corridors with high passenger volumes, with potential for strong coordination of transit with intensive land uses, as well as major intermodal transfer points in suburban areas. . . .

In U.S. cities, the role of metros has increased with the growth of suburbs, which has created a need for transit systems "between" metro and regional rail. San Francisco's BART, Washington's Metro, and Atlanta's MARTA are examples of this new breed of systems that serve networks resembling regional rail, but that have controlled stations and operate with high frequency. . . . In planning such systems, joint planning of metro lines and stations with developments in CBDs and major activity centers throughout the region is important.

Regional rail. Regional rail lines are playing an increasing role in many metropolitan areas. Typically, radial railroad lines on which these services are operated serve areas with high car ownership. However, regional rail can attract riders . . . with its high level of service: a separate ROW allows high speed and reliability, and the large cars are very comfortable.

The traditional commuter rail service, consisting of long trains operating mostly during commuting periods at long headways (intervals), is not the only type of operation that should be considered. With the growth of regions and the increasing diversity of travel directions, there is a need for all-day service at twenty-to-thirty-minute headways. . . . With self-service fare collection and self-powered cars, it is increasingly becoming possible to provide all-day, transit-type service that attracts not only traditional center city commuters but also "reverse commuters" working in suburban areas, as well as off-peak travelers.

[. . .]

Automated Guided Transit (AGT). AGT is a viable option for lines that have short station spacings, go through areas with space for aerial ROWs, and that have considerable volumes of travel. In such cases, AGT, typically utilizing medium-sized rubber-tired or rail vehicles, can offer more frequent service at lower cost than conventional metro. . . .

Promoting transit use in the city. . . . The great variety of policy, planning, and design measures needed to ensure the creation of urbanized areas that can rely on an efficient, balanced intermodal system can be classified into two major categories: (1) macro-scale planning, involving urban form and its relationship to the networks of different modes; and (2) micro-scale design, which makes housing, commercial, business, and other developments pedestrian-friendly and which physically provides for easy and convenient use of transit. . . .

Macro-scale planning. Planning land use and transportation networks must be based on the principle that adequate, efficient accessibility is a basic requirement for every development. This is a requirement similar to that of providing water, electricity, and other utilities, fire protection, and other amenities needed for buildings and complexes to function. In urban areas, accessibility should consist of car, transit, and other modes of service, their relative roles varying with the type of development and local conditions.

To provide highly efficient access while allowing diverse development with respect to functions, densities, and environmental conditions, a transit network must be designed to serve all major activity centers. These include city and suburban town centers, commercial and business buildings, shopping malls, high- and medium-density residential areas, and so on. These developments should be built only at locations that have adequate access not only by street and road networks, but also by high-performance transit. Thus, land use must be coordinated with a present or planned transit network.

At the same time, the operational requirements of the network must be considered, particularly if it is a rail system. These include the requirement that lines be reasonably direct along major passenger-demand corridors. Network density and direction of lines should be planned so that they maximize area coverage and provide convenient transfer points among lines. The operational feasibility of line interconnections and the number of branches to suburban areas must also be considered.

Major activity centers or town centers in suburban areas can often be planned as the locations of intermodal centers – transfer points between rail transit and bus, park-and-ride, and other feeders. Moreover, a large bus stop operated as a timed-transfer facility among many routes can be integrated as a "focal point" in a major retail, business, or residential development.

[. . .]

The present total car dependency in many suburban and urban areas, which results in a continuous increase in VKT on one side, and lack of mobility for some groups on the other, can be stabilized or reduced in the long run only by improved coordination of land use planning with transit services. Based on the definition of access as a basic requirement for the functioning of diverse activities, every plan for a new development should be evaluated with respect to its accessibility. This must include not only car access but also intermodal access, which serves all potential users and is compatible with the present and future transportation system. The evaluation must therefore be based on an analysis of trip generation, modal split, and assignment, and must include street capacity and the quality of transit services.

Micro-scale planning and design. This planning and design includes access roadways, parking areas, transit stops or stations, pedestrian paths, and other traffic facilities for immediate access and local circulation within major activity centers or residential areas. These elements often have a major impact on modal split because of access trips, as well as on trips within activity centers.

[. . .]

To reduce car dependency by encouraging the use of alternative modes, the design of a development must not only protect pedestrians but make walking, transit, and bicycle use convenient and attractive. This is achieved by developing layouts with buildings that are clustered or connected by attractive walkways and with bus or rail stations in the "center of gravity" of the development – that is, with easy access to all trip-generating buildings and areas. . . .

ENCOURAGING THE USE OF OTHER MODES

While car and transit modes dominate urban passenger travel, other modes may have superior performance for certain categories of travel. The efficiency of commuting traffic can be increased by car and van pooling. Several types of specialized services are often provided most effectively by paratransit; bicycles offer the most economical type of vehicular travel and are especially convenient for trips in neighborhoods, small towns, or high-density areas. Finally, walking represents the basic, unique mode, one that complements all other modes and which performs many short trips more efficiently and attractively than vehicles.

[. . .]

35

Environmental Planning

From *The Environmental Planning Handbook for Sustainable Communities and Regions* (2003)

Thomas L. Daniels and Katherine Daniels

Editor's introduction

In the short time since 1984, when the United Nations named the Prime Minister of Norway, Gro Harlem Brundlandt, to head its new Commission on the Environment and Development, and 1987, when the Brundlandt Commission, as it came to be known, issued its report, *Our Common Future*, a series of catastrophes occurred: Africa suffered a drought in which a million people perished; a leak in an Indian pesticide factory killed 2,000 and injured another 200,000; some 60 million people, primarily children, died of diseases related to unsafe drinking water and malnutrition. These tragedies only added urgency to the Commission's plea for nations to support sustainable development, defined as "meeting the needs and aspirations of the present without compromising the ability to meet those of the future" (*Our Common Future*, Oxford: Oxford University Press, 1987).

Since the Brundtland report appeared, achieving sustainable development, especially in cities, has became an important goal of urban and regional planners. Recent discussions about global warming and peak oil have intensified their interest in the topic. Consider these facts to see why planners are concerned with urban sustainability. Today, more than half of the world's population lives in cities – and 80 percent of the U.S. population is urban. Fifty-three percent of Americans live in coastal areas and their annual per capita gas consumption is 1,635 gallons (versus 174 for the world). (See "The State of the World's Cities" and "The Promise of Urban Growth" in this *Reader*.)

By their very nature, cities strain natural resources: In the United States, each person uses 1,500 gallons of water and generates 4.3 lb of garbage daily – which means a city of a million people needs to deliver 1,500 million gallons of water and dispose of more than 2,000 tons of garbage every day. And this is just the beginning of a long list of environmental issues related to urban living: cars and industry pollute air and water; dense settlements require costly sanitary sewerage treatment. Urbanization threatens natural habitats, including coastal lands, wetlands, farms, and forest. It increases impervious surface, paved-over areas that contribute to non-point pollution (toxic runoff streaming into a city's water bodies) and slow aquifer recharging because they do not absorb the rain and snow melt.

So what do urban and regional planners do about these problems? Plenty. They conduct land suitability studies to locate development in places that are not flood-prone, too steep, or endowed with special natural features that should be preserved. They promote compact development by changing zoning to permit dense, mixed land uses. They reduce auto dependence, traffic congestion, and air pollution by targeting transportation investments in sidewalks, bike trails, car-share pods, and mass transit. They engage in cleanup, water conservation, and recycling programs. They support adaptive reuse of obsolete buildings and brownfields remediation. They design open-space systems that use nature to absorb rainwater and air pollutants – planning parks, employing porous asphalt on playgrounds, and encouraging green roofs, revegetation, riparian buffers, and urban farms.

Among U.S. cities, Portland, Oregon, has been a leader in sustainable development and environmental planning. In addition to its widely heralded growth management program dating from 1971, it has addressed climate change since 1993, when it adopted its own plan to respond to global warming. (See Oliver Gillham, "Regionalism," in this *Reader.*) Since then, Portland has created an Office of Sustainable Development (2000) to step up its activities in recycling, energy conservation, technical support for developers of green buildings; convinced its county (Multnomah) to join in adopting the *Local Action Plan for Global Warming* (2001), that set a goal for GHG reduction of 10 percent below its 1990 levels; and established a Special Peak Oil Task Force whose report "Descending the Oil Peak: Navigating the Transition from Oil and Natural Gas. Report of the City of Portland Peak Oil Task Force" (March 2007) set clear directions for city policy, including urban planning initiatives. The city monitors its progress towards its sustainability goals and anticipates its meeting the 10 percent GHG reduction objective by 2010.

Of course, innovative sustainability work is not only taking place in U.S. cities. Planners around the world are directly involved in projects ranging from dramatic ecocities – Dongtan, outside of Shanghai, designed by Arup; Masdar, Abu Dhabi, by Norman Foster; and Celebration, Florida, by Cooper Robertson and Robert A. M. Stern – to important retrofitting programs in Curitiba, Brazil, and Bogotá, Colombia, which have replaced cars with buses in their center cities, encouraged coordinated dense development, built new open spaces, and planted thousands of trees.

Sobering announcements from the International Panel on Climate Change (IPCC) in 2007 have intensified urban planners' engagement in sustainable development. The IPCC has verified the higher temperatures, rising sea level and more extreme weather events (record heat and cold waves, more violent precipitation and hurricanes) in the past five years caused by a dramatic increase of heat-trapping greenhouse gases (GHG). (GHG emission production rose twice as fast between 1995 and 2004 than between 1970 and 1994.) (*Climate Change 2007 Synthesis Report*, Geneva: IPPC, 2008).

The GHG story is quite clear. Carbon dioxide (CO_2) constitutes three-quarters of all greenhouse gases. It comes from burning fossil fuels and rises when forests that would ordinarily absorb the pollutant are cut down. Methane (CH_4) and nitrous oxide (N_2O) by-products of agriculture produce 22 percent of greenhouse gases. Leading the world in GHG emissions are the United States, Europe, China, Russia, and Australia. The United States produces 25 percent of the total greenhouse gases, due to its high use of petroleum and coal that respectively supply 40 percent and 23 percent of its energy. Americans use these fuels to heat, cool, or light buildings (40 percent of the total, divided between commercial [18 percent] and residential [21 percent]); industry (32 percent) and transportation (28 percent).

With experts predicting that by 2030 carbon dioxide will increase by 55 percent if nothing is done to stop it, global warming threatens future generations – a clear violation of sustainable development principles (International Energy Agency, *World Energy Outlook*, London, 2007). But tragedy can be averted if the world holds its GHG production constant, even with its projected population growth. And urban planners are poised to help achieve this goal.

Princeton University's Robert H. Socolow and Stephen W. Pacala structure an approach to solving global warming problems that has a key role for urban planners. They lay out a portfolio of fifteen proven technologies that individually cannot solve the problem but taken altogether or in significant parts can. Each of fifteen elements constitutes what they call a wedge and represents the reduction of 25 billion tons of carbon dioxide in fifty years. (This "wedge" analogy has caught on among scholars and practitioners engaged in climate change discussions, see John R. Nolon's piece in this *Reader.*) Prominent among the wedges is reduced use of vehicles: 2 billion cars traveling 5,000 miles (not the current 10,000 mile average) per year. Planners can help enact transportation and land use reforms to achieve this goal. Another wedge specifies adding 2 million windmills (fifty times current capacity) as a substitute for coal power. Again, planners are important players here – siting windmills has major land use implications, which will call on planners' abilities to find locations that are environmentally safe and to negotiate the NIMBY (not in my backyard) issues that are likely to erupt ("Stabilization Wedges: Solving the Climate Problem for the Next Fifty Years with Current Technologies," *Science* 305 (August 2004).

In this selection, authors Tom Daniels and Katherine Daniels describe the federal, state, and local laws that undergird environmental planning in the United States. They also present a straightforward account of the tools and methods that environmental planning offers to promote sustainable development and address the problems associated with climate change and peak oil.

Thomas L. Daniels is Professor of City and Regional Planning, University of Pennsylvania. He has written extensively on land conservation, small-town planning, and environmental planning. Among his books are *Small Town Planning Handbook*, third edition (Chicago: APA Planners Press, 2007) with John Keller, Mark Lapping, Katherine Daniels and James Segedy, *When City and Country Collide: Managing Growth in the Metropolitan Fringe* (Washington, DC: Island Press, 1998), and *Holding our Ground: Protecting America's Farms and Farmland* (Washington, DC: Island Press, 1997). Katherine Daniels is Senior Planner, New York Planning Federation, and adjunct professor of environmental planning, SUNY/Albany. Along with Tom Daniels, she is author of *The Environmental Planning Handbook for Sustainable Cities and Regions* (Chicago: APA Planners Press, 2003).

For additional reading on environmental planning, see John Randolph, *Environmental Land Use Planning and Management* (Washington, DC: Island Press, 2003). To find out how European countries approach the field, see Donald Miller and Gert De Roo, *Integrating City Planning and Environmental Improvement: Practicable Strategies for Sustainable Urban Development*, second edition (Aldershot: Ashgate, 2004). Lisa Benton Short and John Rennie Short's *Cities and Nature* (New York: Routledge, 2008) compares environmental issues in cities of the global North and South.

With regard to monitoring progress towards addressing sustainability issues, the *2007 State of the World: Our Urban Future: A Worldwatch Institute Report on Progress Toward a Sustainable Society* (New York: Norton, 2007) provides best practices and a useful timeline for the year.

A number of books on reconciling design and the environment have appeared since Ian McHarg's classic *Design with Nature* (New York: Doubleday, 1971), including Douglas Farr, *Sustainable Urbanism: Urban Design with Nature* (Hoboken, NJ: Wiley, 2008), and James A. La Gro, Jr, *Site Analysis: A Contextual Approach to Sustainable Land Planning and Site Design* (Hoboken, NJ: Wiley, 2008).

For a basic discussion of global warming, see Albert Gore's *An Inconvenient Truth: The Planetary Emergency of Global Warming and What we Can Do about It* (New York: Rodale Press, 2006) and its accompanying video. For serious scientific coverage, go to the website of the Intergovernmental Panel on Climate Change (www.ipcc.ch/). Gore and the Intergovernmental Panel on Climate Change shared the 2007 Nobel Peace Prize for this work.

Finally, a number of popular books on climate change have appeared. Examples are: Environmental Defense Fund president Fred Krupp and journalist Miriam Horn's *Earth: The Sequel. The Race to Reinvent Energy and Stop Global Warming* (New York: Norton, 2008); Jeffrey D. Sachs, *The Common Wealth: Economics for a Crowded Planet* (New York: Penguin Press, 2008); and John Kerry and Theresa Heinz Kerry's *This Moment on Earth: Today's New Environmentalists and their Vision for the Future* (New York: Perseus Books, 2007).

Environmental planning involves deciding how to use natural resources, financial capital, technology, and human resources to achieve and maintain healthy communities and a high quality of life. The environment is made up of three broad land use types.

- *Natural areas* that include wildlife habitats, wetlands, water supplies, most coastal and riparian regions, national and state parks, and wilderness areas; natural areas also include floodplains, steep slopes, and landslide areas.
- *Working landscapes* that include farms, rangeland, forests, mines, and recreation areas.
- *Built environments* of cities, suburbs, and towns that encompass buildings, transportation systems, sewer and water facilities, and public spaces and parkland.

How these three broad land use types interact with one another affect a community's appearance, operating efficiency, and environmental quality. Deciding how, when, and where these land use types should or should not change is the primary challenge of environmental planning.

Environmental planning looks to shape a community or region by protecting and improving air and water quality, conserving farming, forestry, wildlife, water, and energy resources, reducing exposure to natural hazards, and maintaining the natural features and built environment. Environmental protection and natural resource management are necessary for long-term public health, safety, and economic success.

Environmental planning relies on information from a variety of sciences, including biology, botany, zoology, chemistry, agronomy, meteorology, geology, epidemiology, hydrology, forestry, engineering, and ecology. There is much about the environment that is not known with certainty. Scientists may disagree about the sources of environmental problems, possible solutions, and what is an acceptable level of risk to humans or wildlife from polluting activities. Environmentalists often support the "precautionary principle" which holds that the absence of complete scientific certainty should not be used as an excuse to avoid prudent action.

RECENT HISTORY OF ENVIRONMENTAL PLANNING IN AMERICA

The twentieth century marked America's growing recognition of the need to protect the environment and conserve natural resources. The rise of the automobile, extensive highway networks, and steady population growth fueled rapid changes in both cities and suburbs, and assaulted the nation's air, water, working landscapes, and wildlife. During the twentieth century, the nation's population increased nearly fourfold to 281 million people. Today, America boasts 200 million cars and trucks, more than 43,000 miles of interstate highways, and more than one million miles of other roads. Many cities in the Northeast and Midwest have lost population since 1950 while their suburbs have continued to spread outward. Cities in the West and South, built around the automobile, have sprawled into the countryside at relatively low densities. By 1990, America had become a suburban nation with more people living in suburbs than in cities. Motor vehicle-dependent suburban sprawl has increased air pollution, threatened water supplies, converted working farm and forestlands, and destroyed wildlife habitat.

It was not until the 1960s that serious degradation of the nation's air and water and loss of open space was fully recognized as a serious threat to public health and the well-being of future generations. In response, Congress enacted a series of laws to clean up the environment, set standards for environmental quality, and conserve valuable landscapes. The National Environmental Policy Act of 1970 (NEPA) established a process to review proposed federal actions that could significantly affect environmental quality and result in the irreversible loss of natural resources. The heart of NEPA is the Environment Impact Statement (EIS) process, which screens all proposed federal projects, policies, and funding for potential environmental impacts, and requires an evaluation of alternatives and a mitigation of impacts.

The U.S. Environmental Protection Agency (EPA) was created through an Executive Order by President Nixon in December of 1970, partly in response to the first Earth Day on April 22, 1970. The agency has broad regulatory powers that affect nearly every industry and government in the nation. The EPA has the authority to implement and enforce a wide variety of environmental laws, such as the Clean Water Act, Safe Drinking Water Act, Clean Air Act, Resource Conservation and Recovery Act (RCRA), Comprehensive Environmental Response, Compensation and Liability Act (CERCLA), among others.

THE CLEAN AIR ACT

The Clean Air Act of 1970 authorized the EPA to establish emissions standards for motor vehicles and stationary sources, such as factories and power plants, as well as set national ambient air quality standards (NAAQS) for six criteria pollutants: nitrogen oxides, sulfur dioxide, lead, carbon monoxide, particulates, and ozone. Under the Clean Air Act, the EPA has also set national fuel efficiency standards for cars and trucks. The EPA has not yet set standards for carbon dioxide emissions, which is the main greenhouse gas that contributes to global warming.

THE CLEAN WATER ACT

In the late 1960s, about 60 percent of America's water bodies were not fit for either drinking or swimming. The Clean Water Act of 1972 authorized billions of dollars to state and local governments for the construction of sewage treatment plants to improve water quality. Section 402 of the Clean Water Act prohibits the discharge of any pollutants into navigable waters from a point source, such as a factory or municipal sewage treatment plant, unless the discharge has been authorized in a National Pollutant Discharge Elimination System (NPDES) permit, issued by the EPA or a state environmental agency. Beginning in 1977, the EPA has turned over most of the NPDES permitting, monitoring and enforcement to the states. Under the Clean Water Act and the Clean Air Act, businesses have been compelled to invest billions of dollars to reduce air and water pollution from factories and power plants.

RCRA AND CERCLA

The Resource Conservation and Recovery Act of 1976 required safeguards in disposing of municipal solid waste and handling toxic materials. RCRA established standards for municipal landfills that handle solid waste, and a cradle-to-grave tracking system for the production and disposal of hazardous waste. The CERCLA Act of 1980 authorized the creation of a national priorities list of heavily contaminated hazardous waste sites, and created a "Superfund" to help pay for the clean up of these sites. CERCLA imposed liability for cleanup on polluters and owners of polluted sites, but the EPA has also spent several billions of dollars to clean up Superfund sites.

STATE LAND USE PLANNING

The 1960s and 1970s also marked the start of state-level land use planning efforts, with an emphasis on land conservation and environmental quality. Hawaii (1961), Vermont (1970), Florida (1972), and Oregon (1973) adopted pioneering programs, which provide broad land protection and development standards for local governments and private development. Several other states later adopted statewide planning programs. Twenty-two states have adopted State Environmental Policy Acts (SEPAs), which are similar to NEPA in requiring a review of state projects and actions and sometimes local government and private projects and actions that could adversely impact the environment.

A CHANGING FEDERAL ROLE

The 1990s saw a shift in environmental responsibility from the federal government to state, regional, and local governments. States gained primary control of air and water quality programs. Metropolitan regions drafted transportation plans that were required to meet air quality goals. Local governments began to implement growth management and "smart growth" programs. In 1998 through 2002, voters in more than thirty states approved over 500 ballot measures involving more than $20 billion for land conservation and "smart growth" projects.

SPRAWL AND POPULATION GROWTH

As the twenty-first century begins, America faces several environmental challenges, most of which are related to the outward spread of suburban development, better known as sprawl. In 2000, the U.S. Bureau of the Census estimated that the nation's population will grow to just over 400 million by 2050, an increase of 120 million people – the equivalent of adding the population of nearly *four Californias.* Where will these people live, work, and play? How will new developments affect air and water quality, food and water supplies, and plant and animal habitats? Will the environment be sacrificed to accommodate this surge in growth? Or will growth have to adapt to the finite carrying capacity of the natural environment?

The answers to these questions will have to be worked out among the federal, state, and local governments, private landowner, businesses, and the general public. But because most land use planning is controlled at the local level, local governments will be asked to shoulder much of the burden for accommodating the expected growth.

URBAN ENVIRONMENTAL PLANNING

There are three main environmental issues in cities and adjacent older suburbs: (1) air quality; (2) water quality and supplies; and (3) brownfields redevelopment.

Air quality

Air quality in most American cities has improved since the 1972 passage of the Clean Air Act. Yet, three major problems remain: nitrogen oxides, particulates, and ozone. Nitrogen oxides (NOx), which come mainly from the exhaust of cars, trucks, and buses, and factory smokestacks, are the main ingredients in smog and acid rain. Factories, coal-fired power plants, and the growth in the number of motor vehicles and vehicle miles traveled were the main reasons why the 2000 level of nitrogen oxide pollution was 20 percent higher than the 1970 level. Over those thirty years, the nation's population grew by more than 50 million people, but the number of cars nearly doubled, and vehicle miles traveled soared by 143 percent.

Particulates are microscopic particles of dust, soot, smoke, and minerals that are released from fireplaces, smokestacks, processing plants, farm fields, cars, buses, and diesel trucks. They are the leading air pollution health threat in America, causing nose and throat irritation, respiratory ailments such as asthma, and premature death.

Ozone (O_3), a poisonous form of oxygen created by nitrogen oxide and hydrogen compounds, reacts with sunlight to produce photochemical smog. Ozone sources include internal combustion engines, factories, paints, solvents, glues, fire-places, and wood stoves. Ozone can cause cancer and lung ailments, as well as damage to crops, trees, and other plants. In 2001, the EPA reported that 62 million U.S. residents were living in counties that did not meet national ambient air quality standards for at least one of the six principal pollutants for compliance with the standards for the major criteria pollutants.

The 1990 Clean Air Act Amendments allow the EPA to withhold federal funds for new highway and industrial construction in areas with air pollution levels that exceed the federal standards. In July of 1998, the EPA withheld federal highway funds from thirteen Georgia counties for noncompliance with federal standards. This forced the state legislature to create the Georgia Regional Transportation Authority (GRTA), with the power to block highway projects and major construction and to push for mass transit alternatives. Based on GRTA and a state plan to improve air quality, the EPA lifted the ban on federal highway funds in July of 2000.

Water quality

When the 1972 Clean Water Act Amendments were passed, point sources – such as power plants and factories–posed greater threats to water quality than nonpoint sources, such as stormwater runoff. Today, the opposite is true, in part because the NPDES program and the construction of wastewater treatment plants have significantly reduced point source pollution. But fertilizers and pesticides from suburban lawns and farm fields, oil and salts from streets and parking lots, and soil erosion and sedimentation from construction sites have remained problems. The 1987 amendments to the Clean Water Act added Section 319 to require states to develop and implement plans and programs to control nonpoint source pollution and to provide federal loans and grants for the control of nonpoint sources. Since 1990, the EPA has required an NPDES stormwater permit on construction sites that involve clearing, grading, and excavating five or more acres of land. In 1999, EPA extended the rule to apply to most construction sites that disturb between one and five acres of land. Also since 1990, NPDES permits have been required for storm sewer systems that serve 100,000 or more people. Combined storm and sanitary sewers are vulnerable to overflow and the release of untreated sewage into waterways when there is a heavy rainstorm. The EPA requires communities with combined sewer systems to develop long-term overflow control plans. Starting in early 2003, permits are required for systems that serve fewer than 100,000 people.

The Clean Water Act has resulted in a substantial improvement in the nation's water quality. Yet, in 2000, about 40 percent of America's waterways were still not fit for drinking or swimming; and about four out of every five U.S. residents were living within ten miles of a polluted lake, river, stream, or coastal area.

The primary goal of public water system planning has been to protect the water quality of the systems, rather than maintaining or increasing water supplies. Congress passed the Safe Drinking Water Act (SDWA) of 1974 with amendments in 1986, 1988, and 1996 to reduce contaminants in public drinking water supplies. The SDWA has enabled the EPA to: (1) set national drinking water quality standards and maximum levels for ninety different contaminants; (2) require water quality monitoring, water treatment, and the public reporting of contaminants in drinking water systems; (3) fund source water protection programs to protect watersheds, aquifers, and wellheads from potential contamination; (4) limit the underground injection of hazardous wastes; and (5) regulate some 170,000 public water systems that provide 90 percent of the nation's drinking water. Where contaminants are found above the maximum acceptable levels, the water must be treated to achieve the applicable standards prior to distribution to customers.

Most major cities rely on surface water supplies. In 1996, EPA adopted the Enhanced Surface Water Treatment Rule under the SDWA to require such communities to filter their water before it is distributed. The EPA may allow a waiver if a public water system has good water quality and the community in which it is located has an active watershed protection program and demonstrated ability to control potential contamination. For example, New York City is attempting to avoid building a $6 billion filtration plant by protecting surface water supplies in the Catskill and Delaware watersheds.

Brownfields

Brownfields are former industrial sites that show a moderate amount of contamination from hazardous waste, but usually not high enough levels to be placed on the EPA's priority list for the Superfund program. Still, brownfields may pose a threat to public health. Nationwide, there are an estimated 500,000 brownfields, with most located in cities and older suburbs. If brownfields could be successfully rehabilitated for new businesses and housing, then there would also be less demand for sprawling development in the outer suburbs and countryside.

Brownfields redevelopment occurs as a partnership between public regulatory and funding agencies on the one hand, and private investors, developers, and neighborhood groups on the other. Successful redevelopment depends on a reliable assessment of the contamination, risk-based cleanup standards, limits to current and future cleanup liability, and financial incentives. Since 1993, the EPA has provided grants of up to $200,000 to states, cities, and towns for the inventory and assessment of brownfields sites as a first step toward redeveloping those properties. The City of Baltimore used an EPA grant to set up a GIS system to identify available vacant and underutilized properties to help developers find, assess, and redevelop properties.

In 1996, Congress limited the liability of lenders who make loans for brownfields redevelopment or foreclose on mortgages on brownfields sites. In 2002, Congress passed the Small Business Liability Relief and Brownfields Revitalization Act, which removed the unlimited liability feature of CERCLA for owners of brownfields in order to encourage redevelopment of the sites. In addition, as of 2002, forty states had initiated voluntary cleanup programs; as long as the landowner cleans up the hazardous waste no financial penalty or liability is assessed.

The EPA has established a revolving loan fund to lend money to cities and towns, which in turn lend these funds to developers interested in redeveloping brownfields. In 2002, Congress authorized $1.25 billion over five years in grants to state and local governments for brownfields cleanup and redevelopment.

REGIONAL ENVIRONMENTAL PLANNING

Watersheds, airsheds, and wildlife habitats are examples of important regional environmental systems. Regional transportation networks strongly influence land use patterns and can have major environmental impacts. As regions grow, they face the loss of open space, and working farm and forest landscapes. Regional thinking also means a concern with social justice through equal access for everyone to a quality environment.

AIR QUALITY AND TRANSPORTATION

In many metropolitan regions, cars and trucks generate most of the air pollution. The [U.S. Department of Transportation requires] all metropolitan areas to establish a Metropolitan Planning Organization (MPO) to qualify for federal transportation funding. The MPO has the responsibility to plan for transportation projects that will maintain the region's air quality or move the region toward attainment of federal air quality standards. The nation's 341 MPOs draft . . . transportation plans: (1) a twenty-year regional transportation plan; (2) a three-year Transportation Improvement Plan (TIP); and (3) a scheme showing that individual transportation projects, such as new roads, and bus or rail lines, are consistent with the state transportation plans and air quality improvement plans. The TIP must show that any increase in vehicle miles traveled and vehicle trips will not jeopardize the improvement of air quality. The TIP also includes transportation projects recommended for federal funding and projects to be paid for entirely with state, local, or private money.

SOLID WASTE

America's "throwaway" culture has caused a growing shortage of landfill capacity. Although recycling efforts are improving to where almost 30 percent of municipal waste is recycled, many cities and metro areas are shipping out their solid waste to regional landfill sites more than 100 miles away. Solid waste disposal is typically the third largest municipal expense after education and transportation. But solid waste disposal is a regional and even multistate problem. Some states are expected to run out of landfill capacity within the next ten years.

WATER SUPPLIES

Adequate supplies of clean water are becoming increasingly important as America's metropolitan population continues to grow. Residents of the eastern United States often take water supplies for granted because of normally adequate rainfall and an abundance of rivers and lakes. But in the west, access to water can mean the difference between valuable real estate and desert. The lack of adequate long-term water supplies limits growth. Identifying the location, amount, and quality of water supplies and protecting them over time are essential actions in planning for the future of a metropolitan region. Water supply planning is especially important in places experiencing rapid population growth and development. In the 1990s, Las Vegas, Nevada, was the fastest-growing metro area, with an 86 percent jump in population. Several other Sunbelt metro regions, such as Atlanta and Austin, Texas, also experienced substantial growth. But Sunbelt regions from Florida to California are facing water supplies that may not be able to keep pace with the population growth. A number of cities have tapped distant water supplies, such as the use of Colorado River water in Phoenix and greater Los Angeles. But landowners and governments with existing water rights are often reluctant to yield them or even sell them to others. In the near future, many major cities may face difficulties in trying to expand their water supplies. Water shortages and rationing may become common.

OPEN SPACE

Protection of open space has emerged as a major planning concern in metro areas throughout the United States. Metro regions produce about one-quarter of America's food, and most of its fruits and vegetables. Nationwide, farmland losses to urban uses exceed one million acres a year. Several metro areas have lost a significant amount of tree cover over the past thirty years. Tree cover not only provides wildlife habitat and a natural air filter, but also shade that keeps cities and suburbs from becoming "heat islands" that require additional air conditioning. The loss of natural areas and wildlife habitat to human developments is the leading threat to biodiversity and rare and endangered species. To many ecologists, "biodiversity" – the diversity of plant and animal species – defines the health of ecosystems. The loss of certain species or the disruption of ecological processes can reduce the ability of an ecosystem to recycle waste, to cycle vital nutrients, and to support a variety of life forms. Humans have a very real interest in maintaining healthy, functioning ecosystems that are biologically diverse. The Endangered Species Act (ESA) of 1973 was passed to avert the extinction of plant and animal species on public and private land. Since the early 1990s, over 400 Habitat Conservation Plans have been forged between developers and local and state governments and the U.S. Fish and Wildlife Service to identify areas where development can occur with minimum impact and where essential habitats are off-limits to development.

ENVIRONMENTAL PLANNING IN PRACTICE

Many planners associate environmental planning with site planning: organizing the development of a tract of land to minimize stormwater runoff, avoid steep slopes and wetlands, and blend in with the natural surroundings. However, environmental planning also determines how the development of one site fits in with neighboring properties, available public services, and the whole community or region. In addition, it encompasses a regional environmental perspective because few ecological systems are contained solely within a single political jurisdiction. Moreover, large developments, such as shopping malls and major highways, have impacts on air and water quality and development patterns that are felt in more than one township or county.

Local governments are becoming more aware that they must recognize their dependence on one another if they are to achieve effective regional environmental planning. For example, in New York State, three towns, Suffolk County, and the state government jointly formed the Long Island Pine Barrens Commission to strictly limit development in a 100,000 acre area with a major aquifer drinking water source and a high concentration of rare and endangered plant and animal species.

TECHNIQUES OF ENVIRONMENTAL PLANNING

Since local governments have primary control over land use decisions in America through the comprehensive plan and such implementation tools as zoning, subdivision regulations and review processes, they are in the front line of environmental planning. One of the most important techniques used to incorporate environmental considerations in the comprehensive plan is the Natural Resources Inventory. This identifies the location, quantity, and quality of air, water, soils, geologic formations, farmland, forests, minerals, wetlands, and plant and animal species. It catalogs their vulnerability to development or overuse. It also identifies natural hazards and development constraints such as steep slopes and floodplains. Often such an inventory takes the form a composite map of natural resource layers, as pioneered by Ian McHarg in *Design with Nature*, and now generated by a GIS system.

The Natural Resources Inventory yields a *land and water suitability analysis*, a primary building block of the comprehensive plan, which identifies those areas of the community that are appropriate for development, places that have moderate limitations to support buildings, and areas that should be protected in their natural state because of productive land capabilities, important environmental functions, or severe environmental constraints. Together with a community's future population projections, the land and water suitability analysis provides important information on the *carrying capacity* of the community; that is, how many people and how much development the community can support before serious negative impacts on the natural environment occur.

A *vision statement* for the community or region describes the desired future quality of the natural, working, and built environments of the community or region. The land and water suitability analyses help to formulate the environmental goals and objectives of the comprehensive plan that expand upon the vision statement. They typically advocate four outcomes: compliance with state and federal environmental standards, a healthy, sustainable environment, a sustainable economy, and a good quality of life for all citizens.

The environmental goals and objectives reflect community desires and priorities, and provide direction for elected officials on public spending, taxation, and land use regulation. They usually encompass the full range of environmental issues facing the community or region by building on strengths (such as a good water supply) and addressing weaknesses (lack of parkland), opportunities (wildlife and ecotourism), and threats (groundwater pollution). Pursuing the goals and objectives involves employing the Natural Resources Inventory and analysis to draft a future land use map to serve as the basis for zoning regulations and other measures to implement the comprehensive plan.

Implementing the plan involves several tools, including zoning provisions that protect sensitive environmental areas. The usual approach is to create overlay zones that typically apply to one or more environmental features. The overlay functions as a double zone with separate requirements for the "base" zone (such as single family residential) and others for the environmentally sensitive zone (such as a floodplain).

Other implementing tools include subdivision and land development regulations that incorporate environmental impact reviews of proposed developments to assure that potential adverse impacts are mitigated and inappropriate sites avoided; and public facilities ordinances that require concurrency of infrastructure before new development is approved, thus avoiding premature development and leapfrog development patterns.

Innovative techniques for capital improvement programs include urban and village growth boundaries, public acquisition of land and development rights, transfer of development rights, impact fees, exactions, and property tax incentives, among others.

ENVIRONMENTAL PLANNING TRENDS AND THE FUTURE

Today, important new trends in environmental planning are emerging. They build upon national legislation, state and local planning efforts, innovations by companies, and greater involvement by nonprofit organizations and concerned citizens. Perhaps the single most encouraging trend is that communities and regions are enacting their own programs to protect the environment, especially through brownfields cleanup, greenways to buffer waterways, and the preservation of open space.

Improvements in analytical tools, especially GIS systems, remote sensing, and a host of internet sites, are assisting planners, local officials, developers, and the public in developing environmentally sensitive comprehensive plans, zoning and subdivision regulations, and capital improvements programs. These tools are also helpful in tracking development, setting benchmarks for environmental preservation and improvement, and monitoring progress toward those benchmarks.

Hopeful Signs: U.S. Urban Revitalization in the Twenty-first Century

From *Land Policies and their Outcomes*, ed. Gregory K. Ingram and Yu-hung Hong (2007)

Eugénie L. Birch

Editor's introduction

When columnist Stewart Alsop wrote "The Cities are Finished" in the April 5, 1971, issue of *Newsweek*, he seemed to be voicing a widely felt sentiment. Political scientist York Y. Willbern had asserted that cities were no longer delivering services – suburbs and special districts had replaced them (*The Withering Away of the City*, Birmingham, AL: University of Alabama Press, 1964). Urbanists from the Harvard–MIT Joint Center for Urban Studies, writing in the aftermath of the race riots of the 1960s, had warned that concentrated poverty, violence, and crime were destroying cities (James Q. Wilson, ed., *The Metropolitan Enigma: Inquiries into the Nature and Dimensions of America's "Urban Crisis,"* Cambridge, MA: Harvard University Press, 1968). Sociologist Alvin Toffler had predicted that telecommunications would make cities obsolete because people could work anywhere (*Future Shock*, New York: Random House, 1970).

Demoralized mayors, overwhelmed by tribulations, were dubious about the fate of their cities. Plummeting populations, disappearing jobs, shrinking tax bases, and escalating demand for expensive municipal services were some of their problems. Rampant property abandonment was vexing. In 1971, when the mayor of Seattle viewed a particularly devastated area in New York City as part of a U.S. Conference of Mayors' meeting, he gasped, "God, it looks like Dresden"[1] (Reeves, "Eleven Mayors Warn Here of Collapse of U.S. Cities," *New York Times*, April 22, 1971, p. 1).

Even the federal government seemed to have given up hope. A year after the *Newsweek* column appeared, U.S. Department of Housing and Urban Development secretary George Romney threw up his hands, declaring that restoring prosperity to American cites would take $3 trillion – a sum more than double the nation's gross domestic product (GDP) at the time. (U.S. Department of Housing and Urban Development, *Newsletter*, June 12, 1972.) Within a few months, under orders from President Richard M. Nixon, he closed down all federal urban renewal and housing assistance programs. For cities, this was a doomsday scenario.

Fast forward to 2008. Cities in the United States are alive and well. They have grown in number and population. (See "Megaforces Shaping the Future of the Nation's Cities" in this *Reader.*) And they have changed in other ways. Today they are integral parts of metro areas, not the solo behemoths of earlier days. Although metro areas may be referred to by the names of their central cities, what's known as "Philadelphia" is in reality Philadelphia–Camden–Wilmington, while "Seattle" is Seattle–Tacoma–Bellevue. More important, cities are at the core of the metropolitan areas that drive the national economy. America's 100 largest metros generate 75 percent of the GDP, and all 360 metros contribute 87 percent of the GDP – which in 2006 amounted to $11 trillion.

Metros contain cities of varying sizes and offerings. Larger centrally located cities are home to museums, performing arts centers, and specialty hospitals, drawing their audiences or clients from a large radius, while smaller, scattered cities have grocery stores, dry cleaners, and gas stations, with customers coming from shorter distances. In the 1930s, geographer Walter Christaller first noted the existence of "central places" and "systems" of cities (*Die zentralen Orte Süddeutschland* (translated *Central Places in Southern Germany*, Jena: Gustav Fischer, 1933), and Sir Peter Hall and Kathy Pain (eds) updated this concept with an important European Union-sponsored study, *The Polycentric Metropolis: Learning from Mega-city Regions in Europe* (London: Earthscan, 2006). In it, they point to strong symbiotic relationships among cities in a region, noting that within an extended geography, large cities have business linkages with smaller ones that enlarge the economic capacity of each city and the region as a whole. London's strong connections in the Southeast with Reading, Cambridge, Southampton, and six other places is an example of this kind of symbiosis.

While the Hall and Pain analysis helps account for the existence of "superstar," or global, cities – like London, New York, and Tokyo – that are at the center of extended networks, it does not explain why lesser cities exist. Urban economists, however, offer insights. Agglomeration, or clustering, of economic activities, opportunities for knowledge exchange and serendipitous face-to-face contacts are a few reasons why people and jobs concentrate in cities (Edward L. Glaeser, "Are Cities Dying?" *Journal of Economic Perspectives*, 12: 2, 1998). But these same economists posit that agglomeration can lead to overdevelopment that can then be a drag on city economies. High costs of housing and of doing business (costly wages, expensive land and taxes) can drive individuals and firms out of cities. In weighing whether to stay or leave or where to move, they consider bottom-line issues like operating costs and sunk capital investments and indirect items like amenities that may affect their living arrangements or business success (Edward L. Glaeser and Joshua D. Gottlieb, "Urban Resurgence and the Consumer City," *Urban Studies*, 43: 8, 2006). Cities, large and small, try to lure people and jobs with favorable tax policies, business retention programs, excellent public services (especially schools and police), and quality-of-life features, including parks, entertainment, and cultural facilities.

But the question remains: Did the doomsday scenario of the 1970s have a happy ending for cities or for metro areas in the 2000s? The story is mixed. When middle-class populations and jobs migrated from cities in the 1970s, the suburbs supplied the inexpensive land for housing and businesses that enabled the economic growth of the entire region. To survive, many cities reinvented themselves, retaining some old businesses and developing new ones. They made room for immigrants fired up with entrepreneurial energy. They encouraged anchor institutions – universities, hospitals, museums, libraries, performing arts centers, stadiums – to become their new basic industries. The latter elements have become quite important in attracting the so-called Creative Class, highly educated professionals contribute mightily to a city's (and the nation's) economy (Richard Florida, *The Rise of the Creative Class: And How it's Transforming Work, Leisure, Community, and Everyday Life*, New York: Basic Books, 2002). They invested in themselves by undertaking a variety of projects in downtowns and neighborhoods.

By no means have all cities overcome the problems that surfaced thirty years ago. And they have new problems to cope with as well. They try to balance downtown and neighborhood revitalization, attract new populations without displacing the old ones, support the needs of immigrants, address issues of institutional expansion, gentrification, and provision of workforce housing. As this selection argues, cities have only one thing to work with: their land, in pursuit of their redevelopment objectives. They can change land use, but the choices can be difficult, especially if different groups dispute the proper use of the same piece of land, as often happens. Here is an example: Some may think that a site should be used for a fancy office or expensive restaurant to attract the spending power of their neighbors in the metro by providing the needed environments for agglomeration and face-to-face contact or by attracting dollars from suburban consumers. Others think it should be used for affordable housing, daycare and an AIDS treatment center to improve housing conditions and stimulate local retail and other economic development projects.

Urban planners are at the heart of these revitalization activities in the public, private, and nonprofit sectors. They are involved in all the areas mentioned above: community or economic development, urban design, historic preservation, brownfields restoration, land use, and zoning. They work on large-scale projects involving office space, arts and culture, sports, and downtown living. They engage in neighborhood redevelopment involving commercial corridors, affordable housing, and open-space planning.

Eugénie L. Birch, author of this selection, is the Lawrence C. Nussdorf Professor of Urban Research and Education, University of Pennsylvania. She is also editor of this *Reader.* Her most recent books are *Rebuilding Urban Places after Disaster: Lessons from Hurricane Katrina* (Philadelphia: University of Pennsylvania Press, 2006) ed. with Susan M. Wachter and *Growing Greener Cities: Urban Sustainability in the Twenty-first Century* (Philadelphia: University of Pennsylvania Press, 2008) ed. with Susan M. Wachter.

For more on metro areas and their role in the U.S. economy, see Alan Berube, *MetroNation: How U.S. Metropolitan Areas Fuel American Prosperity* (Washington, DC: Brookings Institution Press, 2007). This study is part of the Brookings Institution's Metropolitan Policy Program's "Blueprint for American Prosperity: Unleashing the Potential of a Metropolitan Nation," overseen by the program's director, Bruce Katz. For more information, go to www.brookings.edu/projects/blueprint. Also take a look at the U.S. Conference of Mayors' study: Global Insight, *U.S. Metro Economies: GMP–The Engines of America's Growth* (Washington, DC: U.S. Conference of Mayors, 2007) at usmayors.org/metroeconomies/.

To explore the underpinnings of urban economics, see Masahia Fujita and Jacques-François Thisse, *Economics of Agglomeration: Cities, Industrial Location, and Regional Growth* (Cambridge: Cambridge University Press, 2002). And for a demographic overview of metropolitan areas, see Bruce Katz and Robert E. Lang (eds), *Redefining Urban and Suburban America: Evidence from Census 2000* (Washington, DC: Brookings Institution Press, 2006). Contrast today's conditions with the portrait of the late 1960s in *Report of the National Advisory Commission on Civil Disorders* (New York: Dutton, 1968); the commission was chaired by Illinois governor Otto Kerner, Jr, and many refer to its report as the Kerner Report.

Many works focus on revitalizing older industrial cities. The American Assembly, Columbia University, has published several position papers on this subject: the most recent is *Retooling for Growth: Building a Twenty-first Century Economy in America's older Industrial Areas* (2007, americanassembly.org) and an associated volume of the same title edited by Richard M. McCahey and Jennifer S. Vey published by the Brookings Institution Press.

Neighborhood-focused urban revitalization is the topic in *Comeback Cities: A Blueprint for Urban Neighborhood Revival* (Boulder, CO: Westview Press, 2000) by Paul S. Grogan and Tony Proscio, and *House by House, Block by Block: The Rebirth of America's Urban Neighborhoods* (New York: Oxford University Press, 2003) by Alexander von Hoffman. For downtown-oriented efforts, see *New Downtowns: The Future of Urban Centers* (Princeton, NJ: Policy Research Institute for the Region, 2006), ed. Jonathan Oakman, and *Cities Back from the Edge: New Life for Downtown*, new edition (Hoboken, NJ: Wiley, 2000) by Roberta Brandes Gratz with Norman Mintz. For coverage of current progress in middle- and upper-income housing, see Ann Breen and Dick Rigby in *Intown Living: A Different American Dream* (Westport, CT: Praeger, 2004) and Mark Hinshaw, *True Urbanism: Living in and near the Center* (Chicago: University of Chicago Press, 2007).

For case studies on the rise of new players in city revival, see Elizabeth Currid's *The Warhol Economy: How Fashion, Art, and Music Drive New York City* (Princeton, NJ: Princeton University Press, 2007), Judith Rodin's *The University and Urban Revival: Out of the Ivory Tower and into the Streets* (Philadelphia: University of Pennsylvania Press, 2007), and Shannon Christine Mattern's *The New Downtown Library: Designing with Communities* (Minneapolis, MN: University of Minnesota Press, 2007).

A useful website for monitoring a brand of urban redevelopment known as equitable development is PolicyLink (policylink.org). For an example of the type of work they sponsor see Lorlene Hoyt and André Leroux, *Voices from Forgotten Cities: Innovative Revitalization Coalitions in America's Older Small Cities*, New York: PolicyLink, 2007. And for housing and community development news: knowledgeplex.org. The Urban Land Institute website, www.uli.org//AM/Template.cfm?Section=Home, sells books related to urban revitalization.

Note

On February 13–15, 1945, the Allies firebombed the German city of Dresden, destroying thirteen square miles, a highly controversial act. Images of the city would have been quite familiar to the readers of the *New York Times*.

Since 1970, massive urban revitalization has been taking place in U.S. cities, especially the twenty-five most populous. While large cities have tended to lead this effort, cities of all sizes are involved. Broadly construed, it has many forms, related to the age, local political strength, and economic vigor of the cities in which it is taking place. In general, these forms fall into four interrelated categories: (1) Catalytic, (2) Downtown, (3) Neighborhood, and (4) Project-focused. The categories are not exclusive, as cities have and are pursuing more than one approach at the same or different times.

Urban revitalization is a slow process. In the United States, what appear as hopeful signs today are the outgrowths of thirty years' experience in addressing structural – economic, demographic and political – changes played out in cities. In addition, tactics undertaken more recently have yet to yield their full results. Finally, some places are more advanced than others. Large cities have engaged in some form of renewal since 1970, smaller cities have followed suit. Places that have been involved in certain types of urban revitalization for more than three decades can measure their success in the growth, maintenance, or slower rates of loss of their populations, the increase or stabilization of their land values and new patterns of development or changed market dynamics. . . .

In exploring urban revitalization, this [selection] . . . focuses on cities, not metropolitan regions. By emphasizing cities, it underlines the primary purpose of urban revitalization: to enhance a city's chief asset, its land. U.S. cities support their municipal services and public amenities with locally raised taxes collected directly (property) or indirectly (income/wage and sales) from the productive use of their land. Cities that have more land or more highly desired land are not only more able to meet their municipal obligations than others but also have greater choices in their urban revitalization strategies.

[. . .]

CATALYTIC

Catalytic urban revitalization encompasses bold moves, affecting large amounts of land and/or coordinated, large-scale projects, engaging public leaders at local and state levels, and often involving significant economic and political resources drawn from the public and private sectors. It affects land development and land values in significant ways.

Catalytic urban revitalization includes nonconstruction and construction efforts. Nonconstruction urban revitalization approaches generally add more land or development area to a city or make their properties more desirable. Construction-based urban revitalization programs physically transform an area, are highly visible, demand sophisticated financing, and employ large public incentives and private investment. They may have innovative design elements.

Examples of nonconstruction catalytic urban revitalization include annexation, consolidation, citywide zoning code rewrites and major changes in the delivery of education and crime prevention services.

Annexation has been one of the most important tools of urban revitalization in the past decades. While annexation alone does not guarantee increases in population and land value, it provides the opportunity for these outcomes for cities with strong economies or other attractions. This technique predates the period under discussion as cities in the Northeast and Midwest employed it in the nineteenth century. However, between 1970 and the present, cities in the West and South have actively annexed territory. For example, the movement of 48 percent of the cities into and within the top twenty-five most populous ranks was accompanied by adding land area and subsequently more development and population. Notable examples are Austin (land area up 249 percent, population up 161 percent), Charlotte (land +219 percent, people +124 percent), El Paso (land +111 percent, people +75 percent), Phoenix (land +92 percent, people +127 percent), and San Antonio (land +122 percent, people +75 percent).

City–county consolidation, another powerful technique, also predates the thirty-year period under discussion. Prior to 1970, several of the nation's twenty-five most populous cities consolidated with their counties, including Philadelphia, Denver, and San Francisco, or arranged partial consolidation, including Indianapolis and Jacksonville. New York, an outlier, encompasses five coterminous counties, consolidated in the late nineteenth century. In 2003, Louisville, Kentucky, consolidated its city and county governments, resulting in a dramatic increase in its area (from 60 to 386 square miles) and population (256,000 to 694,000). The Act gave the new entity important flexibility in addressing troublesome urban issues. . . .

Many cities have modified their zoning to enhance urban development and revitalization. Again, modifying zoning does not guarantee increases in population and land values, but cities have attempted to make more land available for development through such changes. They encompass dealing with unneeded industrial land, allowing for increased densities in selected residential and commercial areas, permitting mixed-use zones, specifying special zoning around natural resources such as waterfronts or environmentally fragile places, addressing transit-oriented development, defining and protecting special districts and simplifying routine permissions while adding more scrutiny to complicated transactions. In the 1990s, nineteen of the fifty largest cities reported undertaking comprehensive zoning changes, most taking a decade or more to complete. They include Baltimore, Boston, Chicago, Cincinnati, Denver, Detroit, Milwaukee, Minneapolis, New York, Pittsburgh, San José, San Diego, Seattle, and Tucson.

Believing that poorly functioning municipal services contributed to depopulation, low bond ratings, abysmal city reputations, a dysfunctional labor force and other ills, cities undertook major reforms, especially in education and policing, starting in the late 1980s. Several large school systems had ongoing K-12 reform agendas over the years, but after 2000 Chicago (460,000 students), Los Angeles (727,000 students), New York (1 million students), Philadelphia (200,000 students) and San Diego (133,000 students) renewed these efforts. (In addition, New York also took dramatic steps to reestablish high standards in its 250,000 student City University system.) Finally, led by New York, cities that successfully employed community policing and later zero-tolerance law enforcement with the associated Compstat (GIS crime mapping) accountability process throughout the nation engaged in new crime-fighting methods.

There are at least four types of catalytic construction-based urban revitalization. They include major infrastructure investment linked to mass transit or highways, improvements in the public realm, large-scale conversion of former industrial land to mixed-use areas, linkage programs harnessing development for citywide improvements. The 1970s construction of the Washington DC Metro, Atlanta's MARTA and San Francisco's BART systems are examples. . . .

Removing or depressing freeways also have had transformative effects. Several cities have undertaken such projects, including San Francisco (Embarcadero), Milwaukee (Park East Freeway), and Portland (Harbor Drive), but Boston's "Big Dig," begun 1991 and ongoing is the most dramatic example. While it is too soon to judge the complete effects of this strategy at this point, to date, the $15 billion project has not only contributed transportation efficiencies, but also added 300 acres of open space and 16–21 million sq. ft of private development opportunities in the land-constrained city.

Cities have undertaken public realm improvements along their waterfronts and brownfields sites. Notable examples are Baltimore (Inner Harbor), Chattanooga (Tennessee Riverpark), Chicago (Chicago River Corridor Plan and Calumet restoration), Denver (Platte River Greenway), New York City (Hudson River Park), Portland (River Place), Seattle (Central Waterfront), Providence (Providence River Relocation). Future projects that have gained public approval and financial support include Atlanta's Beltline Emerald Necklace, Chattanooga's Twenty-first Century Waterfront Plan and Washington DC's Anacostia Waterfront Initiative.

Studies of the economic value of open space and thus its usefulness as an urban revitalization strategy have emerged in recent years. They judge levels of increased development or population growth, capitalization into property values and tax assessments and its spatial impact. While their results are varied, they uniformly present positive conclusions. For example, a recent assessment of the economic impact of Chicago's Millennium Park anticipates $1.4 billion in residential development and at least $1.9 billion in entertainment and shopping revenues.

Large-scale neighborhood development on former industrial land has been a mainstay of catalytic urban revitalization. Although many cities have pursued this strategy, including Memphis (Mud

Island) and Pittsburgh (Herr Island), the case of New York's Battery Park City (BPC) is slightly different from its peers due to its linkage to affordable housing. BPC is not only an important new neighborhood, one that stimulated a surge in downtown housing in Lower Manhattan, but also, through a city/state contractual arrangements, contributes payments-in-lieu-of taxes (PILOTs) pegged to its income dedicated to low-cost housing production, a program that renewed the housing stock throughout the city's poor neighborhoods.

DOWNTOWN

Recent downtown revitalization builds on a longstanding tradition to strengthen centrally located business districts but recognizes that the era of office use domination is past. It considers the area as a "super" neighborhood that requires special attention to adapt to contemporary change. This strategy supports office, adds residential use (and associated services and amenities), recruits higher education, health, arts, and culture (museums, performing arts facilities), hospitality (convention centers, hotels), entertainment (restaurants, aquariums, casinos), provides major open space amenities (waterfronts, large parks with people-drawing facilities), and improves circulation (light rail, transportation hubs, pedestrian systems). The result is today's "new downtown." Three types of downtown redevelopment exist. They are: office-focused, mixed-use environment, and town center or "faux" downtown.

Office-focused redevelopment endeavors to add or retain corporate office space and uses through zoning changes, special permits, tax incentives, tax increment financing, municipal provision of supplementary amenities (often parking) and site acquisition or write-down assistance. This approach, originating in postwar urban renewal programs continues in varying forms today. Many cities support only commercial buildings while others also include large, mixed-use office buildings. Examples are: Lower Manhattan's 1.9 million sq. ft Goldman Sachs, Philadelphia's 1.2 million sq. ft Comcast Center and Boston's 1.8 million sq. ft Ritz Carlton Towers.

In the late 1980s, faced with chronically high vacancy rates among their older offices and lofts, cities began to redefine their downtowns, adding residential and other uses. The strategy includes fostering six types of residential use (office to housing conversions, new construction-mixed use, new construction-single use, HOPE VI and/or Low Income Housing Tax Credit projects, historic districts, building on "found land"), creating special districts for art, culture, science/health and education and improving transportation, open space and other amenities. A widespread effort, it is most advanced in cities on the East and West coasts and Chicago.

Crafting town centers or "faux downtowns" in rapidly growing suburbs that have become cities is a third downtown revitalization strategy. This includes shaping town centers with retail, entertainment and open space, often employing New Urbanist design elements (short blocks, gridded streets, onstreet parking, etc.). The town center movement emerged in the 1980s with the construction of four model projects: Miami Lakes Town Center (Florida), Mashpee Commons (Massachusetts), Reston Town Center (Virginia) and Mizner Park (Florida) as buildouts in master planned communities or retrofitted shopping centers. Later versions such as Easton Town Center (Columbus, Ohio), Southlake Town Square (twenty-five miles north of Dallas) and Legacy Town Center (Plano, Texas) incorporate housing, more expansive entertainment, hotels and office space, usually under single ownership like a shopping mall.

NEIGHBORHOOD

Neighborhood revitalization whose key goal is to restore land values and smooth market operations in depreciated areas covers a large variety of initiatives but usually focuses on improving deteriorated residential districts and associated commercial corridors and assisting low-income households. While these kinds of programs date from the late nineteenth century, today's efforts focus on reducing concentrated poverty by attracting or growing middle-income households. The strategy includes neighborhood-based physical and social improvements. They include upgrading housing stock, improving streetscapes and public space, cleaning up brownfields, greening vacant lots, energizing retail corridors, providing better schools and neighborhood centers, having a number of programs involving job training, small business assistance, health service delivery, consumer education and daycare. Citywide school reform efforts benefit neighborhoods and, recently, facilitating transportation links to the surrounding region's jobs has emerged as an essential function of neighborhood redevelopment.

Cities' neighborhood revitalization programs range from the *comprehensive* approach assessing and addressing all neighborhoods according to need, as is the case with Minneapolis's Neighborhood Revitalization Program and Philadelphia's Neighborhood Transformation Initiative, to *quasi-focused efforts* covering all low- and moderate-income neighborhoods as in Cambridge, Massachusetts, Raleigh, North Carolina, or Albany, New York, or to *targeted methods* that concentrate public and private resources on a limited number of neighborhoods as exemplified by Richmond's Neighborhoods in Bloom (NiB) program. How cities choose what strategy to pursue varies from place to place. Some may do so stimulated by federal reporting requirements associated with the community development block program while others have strong political motivation, often locating their neighborhood improvement operations directly within the mayor's office.

Neighborhood revitalization involves complicated regulatory, financing and land acquisition issues that bring the public and private sector groups together. . . . The public sector, community development corporations, private sector, anchor institutions typically drive neighborhood revitalization programs and influence them according to their goals.

In public sector-led programs, local or state governments plan and implement a given project. While this was once the dominant mode of neighborhood revitalization and is exemplified by public housing construction between 1949 and 1973, it is less common today when the public sector typically assists other participants in a variety of ways, including capital investment in infrastructure or amenities, site acquisition, financing, tax abatements and technical assistance in planning and economic development. However, some cities, like Boston, have used their CDBG dollars to fund retail revitalization. Since 1995, Boston has financed a multi-site Main Street program, undertaken in conjunction with the national program sponsored by the National Trust for Historic Preservation, . . . providing . . . assistance . . . [to] nineteen commercial strips. . . .

Today, the community development corporations (CDCs) are leaders in neighborhood development. Taken as a group, CDCs, usually 501(c)(3) nonprofit organizations, focusing primarily on physical development and operating in every state, are the single largest builders of low-income housing in the United States. Since 1968, CDCs have produced 1.25 million housing units, experiencing a tenfold increase since 1988. This production level is about equal to the number of nation's public housing units. They have also built 126 million sq. ft of industrial and commercial space and created three-quarters of a million jobs . . .

Despite these performance data, CDCs operations are relatively small operations and vary greatly by region, scale and number. For example, seven years ago, their annual *median* housing production was twenty-one units and in 2005 they *averaged* eighteen units per CDC. Regionally, more CDCs exist in the North Central (29 percent of the total) and South (29 percent) than in the Northeast (22 percent) and West. . . . Among the nation's 4,600 CDCs, individual capacity varies widely and, in recent times, some have experienced financial and administrative difficulties while others have flourished. . . . Nonetheless, CDCs' accomplishments are significant and due to the support of national intermediaries, especially LISC and the Enterprise Foundation, and regional intermediaries, notably New York's Community Preservation Corporation and Philadelphia's The Reinvestment Fund. These intermediaries not only have access to capital – marketing tax credits, filtering foundation grants like the Livable Cities initiatives and other funds – but also offer technical assistance for existing work and funded innovative projects. . . .

A few specialized private developers have low-income housing provision as their central mission while others have added it to their larger portfolios. Included are McCormack Baron Salazar (St Louis), Integral Properties (Atlanta), Related Companies of California (Irvine) and Cochoran Jennison (Boston). . . . Many of these companies have engaged in other community-building projects (e.g. schools, daycare centers) in order to strengthen their own projects or in response to federal incentives like the New Markets Tax Credits.

Anchor institutions (universities, health institutions, cultural facilities, churches, and public housing authorities) provide the third type of neighborhood revitalization. As with the other groups, their improvement projects take a long time, require substantial financial support and sustained leadership. While they undertake most efforts for internal institutional advancement, namely the need for more space, several universities have sponsored broad campaigns beyond their campuses designed to preserve the safety, improve the physical appearance and upgrade the economic status of their surroundings. Among them are: the University of Chicago, Ohio State, Howard, Georgia Tech, Georgia State, Yale, Columbia, University of Pennsylvania, University of Southern California and Saint Louis University. Other examples of the participation of anchor institutions in neighborhood revitalization are hospitals and public housing authorities (Zipperer 2005). The former often incorporate housing and social services along with facilities expansion while the latter, employing the HOPE VI program, raze or partially raze obsolete housing, redesign the site and rebuild mixed-income settlements. Researchers report positive results with regard to deconcentration of poverty, increases in adjacent property values, higher performance (employment and educational attainment) of former public housing residents (Turbov and Piper 2005; Boston 2005; Popkin *et al.* 2004).

Catalytic
- Nonconstruction (annexation, consolidation, zoning, municipal services reform)
- Construction (major infrastructure linked to mass transit or highways, public realm or open space, citywide linkage programs)

Downtown
- Office (traditional, mixed-use)
- 24/7 residential emphasis (office/loft conversion; new construction: single-use, mixed-use, special purpose, HOPE VI, historic district)
- Town center/faux downtown

Neighborhood
- Types (comprehensive, semitargeted, targeted)
- Agents (public sector, community development corporation, private developer, anchor institution)

Project-focused
- Large-scale single project (convention center, stadium, performing arts venue)

Figure 36.1

PROJECT-FOCUSED

While the efforts described above result in enhancing the economic bases of their cities, they are part of a larger vision that extends beyond a single project. Some places, however, focus on a single project for the sole purpose of attracting outside dollars to their economies either through direct consumer expenditures or user taxes. These projects tend to encompass expensive, large-scale facilities ranging from convention centers to performing arts venues to sports stadiums to research parks. They have precise site location demands related to spatial requirements, highway access and parking. They require massive financing packages (bonds, TIFS, tax abatements, land deals, etc.) and often inspire considerable controversy due to their expense or location.

While many researchers question the payback on these types efforts, cities continue to invest in them. In the past decade, no area has been more contested than the construction of sports stadia. . . . In the midst of this debate, since 1998 twenty-one cities have built eleven football, eight baseball and seven basketball facilities. They ranged in cost (in 2003 dollars) from $190 million (Tampa Florida's Raymond James football stadium, opened 1998) to $587 million (Seattle's Safeco (baseball) Field, opened 1999) (Coates and Humphrey 2003, p. 19).

CONCLUSIONS: HOPEFUL SIGNS TODAY AND TOMORROW

. . . Between 1970 and the present, cities have held their own in terms of their share of the national population. Although different cities have varied success, there are more large cities today than thirty years ago and the average size of the top 100 cities has increased. The top twenty-five places are relatively stable in terms of the cities in the group but individual rankings have changed. One reason why these conditions prevail is the active roles cities have taken in addressing their growth and development or redevelopment.

Over time, attitudes toward cities and the context in which they exist have changed, and, in response, cities have pursued a wide range of urban revitalization strategies: catalytic, downtown, neighborhood and project-focused. Cities employ some or all of the approaches depending on their needs and results are reflected in population growth and household increases, . . . and in other changes affecting land values, development patterns and market dynamics. . . .

While U.S. urban revitalization is a local affair undertaken primarily at the municipal level, with the state and federal governments offering direct and indirect financial support, a complex mixture of public, nonprofit and private groups accomplish it. The United States has no stated national policy about the economic health of its cities or what it considers an appropriate distribution of its population. Nonetheless, municipalities have undertaken an astounding number of urban revitalization activities in the past thirty years. In this period, they and their partners have refined their approaches. Their reliance on local property (and to a limited extent sales and income) taxes has been a powerful incentive for these efforts. Employing techniques ranging from annexation, zoning revision and municipal service reforms that have citywide implications to small-scale, block-level reconstruction, the nation's cities are reinventing themselves as they strive to increase their income-producing abilities, ensure their survival and, beyond that basic need, promote their advancement.

Five themes frame the character of contemporary urban revitalization endeavors. Taken together they help explain the process by which today's hopeful signs will become tomorrow's permanent accomplishments. First, whether large or small, urban revitalization approaches leave physical marks on their cities. Second, they take a long time to conceive, implement and yield results. They also require focused and sustained leadership from the groups engaged in them. Third, in the past couple of decades, municipalities and their urban revitalization leaders have used urban revitalization to redefine their cities and neighborhoods and are crafting programs/approaches to support and sustain these changes. Fourth, of the approaches, the catalytic are expensive but appear to be having big payoffs in terms of city positioning, the downtown efforts are strengthening central places, the neighborhood are creating new communities and the economic-base enhancing are adding place-defining facilities. Fifth, while urban revitalization is extraordinarily complicated (in terms of assembling resources, developing sophisticated leadership, fostering a supportive political environment), it is flourishing in its many forms.

37

Sprawl and the Tyranny of Easy Development Decisions

From *Confronting Suburban Decline: Strategic Planning for Metropolitan Renewal* (2000)

William H. Lucy and David L. Phillips

Editor's introduction

"Sprawl" is a word that urban and regional planners consistently use negatively. But the word, which dates to the eighteenth century, was not always a pejorative term. For example, in 1974, when the Real Estate Research Corporation issued the now classic three-volume *The Costs of Sprawl: A Detailed Cost Analysis* (Washington, DC: U.S. Government Printing Office, 1974), commissioned by the U.S. Department of Housing and Urban Development, the Council on Environmental Quality, and the U.S. Environmental Protection Agency, "sprawl" simply referred to low-density development. However, a quarter-century later, when the authors of *Costs of Sprawl – 2000* (Washington, DC: National Academy Press, 2002) revisited the earlier study, reporting the results of a five-year effort sponsored by the Federal Transit Administration and the Transportation Research Board, they redefined sprawl as "spread-out development that consumes significant amounts of natural and man-made resources, including land and public works infrastructure of various types." By 2005, architectural historian Robert Bruegmann, in a revisionist interpretation, *Sprawl: A Compact History* (Chicago: University of Chicago Press, 2005), argued that an "objective" and "basic" definition of sprawl is "low-density, scattered urban development without systematic large-scale or regional public land use planning." That same year, the *New Oxford American Dictionary* defined it as "a group or mass of something that has spread out in an untidy or irregular way, the expansion of an urban or industrial area into the adjoining countryside in a way perceived to be disorganized and unattractive." (See Plate I.11.)

These definitional variations reflect the high level of controversy surrounding sprawl – controversy that has flared in recent decades as the nation's population has grown, with disproportionate physical expansion in suburban and exurban areas. For example, between 1970 and 1990 the Chicago area experienced a 20 percent increase in households while its urbanized land area grew 35 percent. This pattern of development, replicated throughout the nation, especially in the south and west, has drawn attention to sprawl's costs and benefits, prompting explanations of how and why it has happened, and discussions of how to handle its effects.

What causes sprawl? American governmental structure is partly to blame. Home rule – which gives the nation's 18,000 local governments jurisdiction over their land use, responsibility for financing their municipal services, and power to borrow funds for capital improvements pegged to local property values – frames development decisions. Through zoning powers, local governments determine who lives within their boundaries and what kinds of commerce and industry will take place. In most states, local governments can exclude certain classes of use. For example, a place can prohibit or limit high-density housing, set minimum lot sizes for single-family dwellings, or disallow home offices – all factors that shape development patterns. Local governments pay for their municipal services – including schools, refuse collection, parks maintenance, police and fire protection – primarily through property taxes. In suburban areas, where the largest expense is education, localities search for ways to enhance their revenues while minimizing their obligations, often leading to land use decisions that add taxes but not children to the system. For example, a town may zone for an office park but not an apartment building. Finally, jurisdictions usually pay for capital facilities (schools, roads, police stations, open space) and improvements by issuing municipal bonds or borrowing, whose cap is set by the state as a proportion of the town's property values. Two critical U.S. Supreme Court decisions related to school finance – *San Antonio Independent School District v. Rodriguez*, 411 U.S. 1 [1973] and *Milliken v. Bradley*, 418 U.S. 717 [1974], the first disqualifying equalization funding among school districts and the second disallowing inter-district busing–reinforced the fiscal zoning practiced in many localities. Although some states have since passed legislation to equalize school financing among cities and suburbs, such laws do not appear to have reduced widespread exclusionary zoning. Some observers attribute this phenomenon to residents' reluctance to allow any use perceived to lower property values, since homes are usually a household's largest asset.

Zoning is the almost universal form of regulating land use in the United States. Dating from 1916, when the first comprehensive ordinance in New York City was passed, and upheld as a legitimate use of the police power in *Village of Euclid, Ohio, v. Ambler Realty Company* 272 U.S. 365 (1926), zoning spread quickly throughout the United States. By 1927, only a year after the Supreme Court decision, 525 of the nation's localities (almost 20 percent) had adopted zoning; by 1936 the number reached 1,246; and by 2006 Houston was the only large American city that did not have it.

As zoning gained acceptance, another phenomenon contributed to changes in the U.S. landscape: the rise of personal mobility. Widening automobile ownership allowed more households to move to peripheral areas. Newly populated areas had the option of incorporating as independent jurisdictions or seeking annexation to central cities. As time passed, localities in the north and Midwest overwhelmingly chose incorporation, while those of the south and west went either way, depending on local circumstances. The resulting fragmentation of local government would soon contribute to the mixed development patterns characterizing U.S. land use. Localities had varied development thresholds that real estate developers and investors used to their advantage – if a place did not allow a certain use, the investor could go to a place that did.

A variety of national policy initiatives also fueled the growth of outlying areas. They include the decision to finance an interstate highway system and to encourage home ownership through favorable tax policies and federal mortgage insurance programs oriented to new development, not rehabilitation of older neighborhoods. These policies initially enlarged the commuter shed, allowing workers to live at greater distances from central-city employment in homes that met contemporary consumer preferences (freestanding units with their own yards). As population in outlying areas increased, employment (and retail) followed, thus opening up other suburban commuter sheds that, under the nation's fragmented land use regulatory system, resulted in sprawl.

Sprawl has raised a number of environmental and equity issues. But it has also supplied a range of advantages to a large portion of the U.S. population, thus driving widespread consumer acceptance. Among the advantages, both perceived and actual, are lowered land and housing costs, larger lot and dwelling unit size, more lifestyle choices, higher participation rates in local government, lower crime rates, and better-performing schools. While debate swirls around the quality, quantity, and extent of these features, American households have "voted with their feet," as suburban development has dramatically outpaced urban growth in the past thirty years.

The authors of this selection, William H. Lucy and David L. Phillips, are faculty members in the Department of Urban and Environmental Planning, School of Architecture, University of Virginia – Lucy is Lawrence Lewis Jr Professor, and Phillips is associate professor. Their newest book, *Tomorrow's Cities, Tomorrow's Suburbs* (Chicago: APA Planners Press, 2006), examines growth in thirty-five metropolitan regions, tracking and explaining changes since 1950.

For another definition of sprawl and discussion of its causes, see Howard Frumkin, Lawrence Frank, and Richard Jackson, *Urban Sprawl and Public Health: Designing, Planning and Building for Healthy Communities* (Washington, DC: Island Press, 2004). This book is representative of a number of monographs offering antisprawl accounts and planning advice. Other examples are: F. Kaid Benfield, Jutka Terris, and Nancy Vorsanger,

Solving Sprawl: Models of Smart Growth in Communities across America (Washington, DC: National Resource Defense Council, 2001); Peter Calthorpe and William Fulton, *The Regional City* (Washington, DC: Island Press, 2001); and Jane Silberstein and Chris Maser, *Land Use Planning for Sustainable Development* (London and New York: Lewis Publishers, 2000). James Howard Kunstler's *The Geography of Nowhere: The Rise and Decline of America's Man-made Landscape* (New York: Simon & Schuster, 1993) is a fast-paced diatribe against U.S. development patterns. Finally, Kevin M. Kruse and Thomas J. Sugrue's (eds) *The New Suburban History* (Chicago: University of Chicago Press, 2006) and Jon Teaford's *The American Suburb: The Basics* (New York: Routledge, 2007) place suburban history in its local, metropolitan, and national contexts and offer new interpretations of the roots and levels of diversity, conflict, and competition among contemporary suburbs.

Several problems with metropolitan markets contribute to the tyranny of easy development decisions. The notion of a tyranny of easy development decisions is intended to be ironic. One meaning of "tyranny" is strong, even complete, control leading to unjust results. In metropolitan markets, lax influence over development on the metropolitan fringe leads to unintended consequences over which central city and suburban public officials have little influence. Cities and suburbs are subjected to a tyranny of remote decisions. These decentralized decisions have unjust consequences over which city and inner suburban influence is minuscule. The ironic consequence is that freedom for some is equated with tyranny by others.

Tom Daniels describes such decision situations in *When City and Country Collide:* "Developers, like any businesspeople, prefer to have as much predictability and certainty as possible for their projects. Real estate development runs on borrowed money. Developers bear the risk of building a project that can't be sold, losing money if their projects are denied or delayed, or selling the project at a loss if the economy turns sour. In short, the sooner a development is built and sold, the sooner the developer can pay off real estate loans and turn a profit."

In some instances, low risk of developers in developing the fringe, by which we mean the countryside beyond the suburbs, may correspond with low influence by public sector regulators. Daniels has identified eight obstacles to "coordinated, long-term, and effective growth management in the metro fringe: (1) Fragmented and overlapping governments, authorities, and special districts. (2) The large size of fringe areas. (3) Lack of community, county, or regional vision. (4) Lack of sense of place or identity. (5) Newcomers, social conflicts, and rapid population growth. (6) The spread of scattered new development. (7) Too few planning resources. (8) Outdated planning and zoning techniques." The net effect of these obstacles usually is limited public sector influence on development patterns due to confusion about goals or insufficient means to accomplish them in those relatively rare instances when goals are clear.

In other words, land development outcomes can be a product of what is easy to accomplish rather than a product of what consumers prefer. Businesses interpret market opportunities through three lenses: (1) new inventions, products and services, (2) changes in regulations and incentives, and (3) modulation of supply to demand based on trial-and-error explorations, including trying to shape preferences. Uncertain relationships between new opportunities, ease of action, and consumer preferences yield ambiguous messages from markets. Whether producers respond to consumers or consumers respond to producers whose products are determined by multiple influences is obscure and, one suspects, in flux. Ambiguous messages from markets also complicate calculations about relationships between quantity and quality in strategic planning and between goals for jurisdictions and regions.

REGIONAL STRATEGIC PLANNING

Greater regional planning, policy, development, and management capacity is needed to cope with the tyranny of easy development decisions and its accompanying conditions – excessive income disparities among local government jurisdictions and too little reinvestment in middle-aged and older housing and neighborhoods. With what kinds of goals should regional strategic planning be concerned? Both quantity and quality goals are important. Quantity issues involve how a region can attract more jobs and how a region can grow faster. The bias in quantity planning is usually that more is better. Quantity planning interacts with quality planning through concern about being competitive. Quality of life has a vast impact on attracting and retaining highly skilled, mobile employees in the contemporary era of footloose businesses that are not tied to natural resources, rail and water transportation, or the hometown of the founder. A region's most important asset is its people. Highly skilled employees, even more than businesses, are footloose, able to live wherever they choose. The quality of life that a region offers, comprising such factors as environmental conditions, education, safety, culture, and recreation, significantly affects the quantity of business activity and its pace of expansion. . . .

Interregional and intraregional strategic planning also interact through quality-of-life issues. Because regions are composed of many jurisdictions, some of which overlap and many of which are geographically distinct, complex patterns of interjurisdictional competition and cooperation occur. Jurisdictions compete with other jurisdictions within the region to attract a larger quantity of businesses with certain characteristics – high property values, high salaries, low pollution, and not too much traffic. They may compete by offering tax incentives or other direct benefits to specific businesses, but they also compete by trying to provide a good quality of life. Each jurisdiction to some extent faces variations on the potential conflict between more quantity and quality.

[. . .]

ACTORS' DECISIONS AND METROPOLITAN HOUSING MARKETS

Three types of actors make uncoordinated decisions that produce the tyranny of easy development decisions, which leads to suburban and exurban sprawl. First, government decision makers, especially in local government, consider pursuing quantitative economic growth and qualitative improvements, reacting to interest groups, citizens, and their own beliefs. Second, developers and lenders are "satisficers" who limit risk while seeking satisfactory profits. In producing residential, commercial, and industrial buildings, they prefer easy development decisions with satisfactory profits to high-risk projects with the possibility of maximum profits but significant potential for losses. Third, consumers make housing location choices in neighborhoods and jurisdictions in "distorted markets." Decisions by government officials, developers, consumers, and citizens are influenced by jurisdictions' geographic settings within metropolitan areas. Interactions among these actors also shape prospects for reinvestment in established neighborhoods and trends in income disparities among local jurisdictions. The interplay between these actors occurs on the stage for regional strategic planning.

Strategic planning for regions should take into account the roles of individuals, businesses, and governments. Some of these roles can be generalized, and some will be specific to each region. . . .

[. . .]

Metropolitan housing market problems include the following:

- Markets that are distorted by democracy.
- Markets that lack a full range of institutions to arrive at better outcomes.
- Markets that are multiple and involve dimensions beyond the usual producing and selling of commodities and quasi-commodities, including government, neighborhood, and school dimensions.
- Markets with uncertain outcomes.

Strengths, weaknesses, dangers, and opportunities in competitive contexts in specific settings, the essence of strategic planning analyses, can be traced to these general market circumstances that affect every setting.

PARTICIPATORY DEMOCRACY

Markets are distorted by local participatory democracy. The supply of housing does not react sufficiently to demand for more development between the center and fringe, partly because current residents often oppose substantial change in their settings. Whether the change is adding condominiums on single-family dwelling sites or replacing fifty-year-old, 800 sq. ft houses with new 2,000 sq. ft single-family dwellings, neighbors are more likely to fight than welcome the proposed changes. . . . From residents' vantage point, this "market distortion" is the means by which they try to protect the quality of their neighborhoods. Using threats of votes and campaign contributions, current residents often intimidate elected officials and their appointees to maintain the status quo. In most instances the social gains are too uncertain, and the gratitude of future residents too unlikely, to persuade elected officials to oppose the preferences of intense minorities who have votes to cast and money to dispense.

[. . .]

REINVESTMENT INCENTIVES AND INSTITUTIONS

Markets lack the full range of lending, redeveloping, planning, and implementing institutions needed to arrive at better reinvestment results. The problem of democratic distortions of markets is compounded by institutions that are deficient in "patient money" – money that waits for profits. Therefore, holders for money need to be especially motivated to achieve satisfactory reinvestment. Those with sufficient motivation may be current owners; local institutions such as universities, hospitals, and churches; federal, state, or local governments; or public-spirited entrepreneurs.

Time delays are costly. Local governments can be organized to reduce delays, or they may be lackadaisical or intimidated by citizen opposition, with delay a consequence of either posture. Uncertainty about whether delays will be lengthy increases risk and decreases willingness to wait. In established neighborhoods, development delays are likely. As opportunities to move money rapidly among capital markets nationally and internationally have increased, the supply of patient money may have decreased. Since renewal and reinvestment in housing and neighborhoods has not been a major goal in the United States, institutions and incentives for reinvestment are inadequate.

[. . .]

COMPLEX HOUSING MARKETS

Market conditions make real estate commodities unusually complex. The real estate maxim that property values are a function of "location, location, and location" reveals additional complexity. Features of land and dwelling are far from irrelevant, but dwellings are sought as investments as well as for shelter, complicating the purchase of housing compared with most purchases of goods and services. Many other considerations enter location decisions. Limiting attention to land and structures leaves out too much and obscures complex real markets. Real housing markets include characteristics of governments, neighborhoods, and schools within commuting-to-work zones. Commuters vary in their tolerance for travel distance, time, cost, stress, and danger. Rarely is there only one jurisdiction, neighborhood, and school within a tolerable commuting territory. Sometimes more than one state is included within commuting zones, adding considerable policy and tax variety to location choices. Moreover, locations also have negative characteristics, such as the social behavior and personal characteristics of neighbors. Moving decisions and destination choices include pushes away from negative characteristics as well as pulls toward positive ones.

HOUSING MARKET UNCERTAINTIES

Uncertain outcomes from housing markets are to be expected. For the majority of lenders, developers, and buyers, limiting uncertainty is a goal. A common strategy to limit uncertainty is to make conventional choices, hoping that "following the crowd" into the suburbs or exurbs will limit risk.

Market uncertainties have many dimensions. Concerning a residence, it is one thing to assess its physical soundness, consider its match with potential occupants' stylistic preferences, and imagine whether living in it would equal its appeal on a twenty-minute walk-through visit. But what about transportation options – after a job change?

What about school conditions – after redistricting? What about neighborhood safety – after the income status of neighbors falls? What about time for house and children – after one parent falls ill or departs permanently? What happens to the jurisdiction if the major employer moves out, the state government cuts state aid by 25 percent in a financial crisis, and the police chief pleads guilty to collusion with drug dealers? What happens if a supposedly safe destination is chosen, perhaps in a gated subdivision, and then the school bus is hit broadside while turning from the subdivision on to a high-speed road, the developer goes bankrupt before the park is built, and the commuting distance disrupts family life on late work nights?

These possibilities are difficult for individuals to calculate or even approximate, as well as being difficult for any two members of a single household to evaluate identically. As a consequence, many people make cautious location decisions in which personal safety and investment security figure prominently.

Discussion of these four conditions – participatory democracy, reinvestment incentives, housing market uncertainties and complexities – reveals numerous obstacles to housing market participants' ability to make fully informed decisions. The potential risks to producers and buyers are substantial. Most real estate producers and consumers cope with these risks by aiming for what they consider relatively safe and secure investments – that is, they are risk averse. Rapid change adds to the difficulty of being adequately informed as well as contributes to greater risks. We argue that the net effect of these conditions falls short of meeting the social or societal aspirations of market participants. The exercise of preferences in the current market structure leads to imbalance, disequilibrium, substantial risk, and questionable satisfaction. This disequilibrium is a challenge for strategic planning. . . .

[. . .]

COMMUTING INFLUENCES DECISIONS

Local governments can be categorized as "destination" jurisdictions or as "pass-through," "balanced," or "feeder" jurisdictions relative to current employment locations and development trends. These categories influence how local public officials and citizens evaluate alternative development patterns. Destination jurisdictions have many commuters as well as local residents employed within their boundaries. Feeder jurisdictions have more residents employed and looking for work than jobs available within their boundaries. Pass-through jurisdictions are between destination and feeder jurisdictions. Balanced jurisdictions are also between destination and feeder jurisdictions but are substantial employment subcenters and may have more jobs than residents employed and seeking work. Balanced jurisdictions may be old suburbs or new edge cities.

Central cities are often destinations. The extent to which they continue to be destinations depends on whether they have lost jobs during the postindustrial era. Some inner suburbs are in the balanced category, while others are pass-through jurisdictions. Outer suburbs and exurbs, and many other suburbs in large metropolitan areas, are usually feeder jurisdictions, although there are occasional edge city exceptions to this generalization. These categories are not stagnant. Positioning evolves with changes in metropolitan scale, national and international employment trends, and the success and failure of local enterprises. Policies of local elected officials will be influenced by perceptions of where their jurisdiction is positioned, but also by where they believe it has been positioned and where they think it is headed. Often these policies concern whether their jurisdiction has enough local revenue sources, and the prospect of future resources, to meet anticipated public service costs.

COSTS OF GROWTH

Typical patterns of metropolitan evolution also influence policies of elected officials. Elected officials of feeder jurisdictions are confronted with the short-run and long-run conflicts between quantity and quality. Residential development in fringe jurisdictions typically moves ahead of attracting employment and retail sales, so these jurisdictions become feeders. Local jurisdictions usually rely mainly on real property taxes to pay for local public services, especially for public education. . . .

Elected officials react to these high real property tax rates, and the conflicting demands for more schools and lower taxes, by trying to increase business properties within their jurisdiction. This desire by elected officials contributes to the tyranny of easy development decisions. Loudoun County, Virginia, provides an example of a jurisdiction where, until the end of the 1990s, public officials had faith that business growth would produce tax benefits. Loudoun, which is west of Washington, DC, and Fairfax County, was the eighth fastest-growing county in the United States from 1990 through 1997 and the third fastest in 1998.

The belief that business properties would constitute a net gain to the public fisc was so ingrained in the assumptions of Loudoun County supervisors that until 1998 the county had never conducted an analysis of revenue and expenditure effects of new businesses. In that year, the prospect that WorldCom Inc. would locate its headquarters and 30,000 jobs in Loudoun led to a revenue–expenditure analysis. The analysis assumed that one-third of the employees would live in Loudoun, many of them with children in public schools. Little if any net revenue gain was expected after taking account of new public expenditures, especially for schools. Local officials estimated that houses would need to sell for $400,000, instead of the typical $200,000, to pay for school, infrastructure, and other service costs. While this analysis reframed the local debate about how much to encourage or resist growth, the counterargument in this unresolved controversy was stated by the chair of the Loudoun Economic Development Commission: "Let's say we slow down the economic engine. That, in my opinion, is setting the framework in motion for the residential growth to keep on growing without the commercial development." Implicit in this argument was the belief that residential growth cannot or will not be limited. It implicitly acknowledged that the tyranny of easy development decisions was in control in Loudoun County. The Loudoun Board of Supervisors was split in 1999 over how to deal with this dilemma. It united, however, in turning to the Virginia General Assembly for relief, arguing that some means of tapping residents' income growth was needed since it was so far outstripping increases in real property values.

BOUNDARY LOCATIONS

Attracting more employment has several spatial effects. Within suburban and exurban jurisdictions, local government officials often encourage employment growth near their boundaries. Boundary locations maximize tax base gains while minimizing traffic within the jurisdiction, or so it seems initially. In suburban and exurban jurisdictions, the most attractive boundary for business development often will be the boundary closest to the central city. This tendency has two effects on adjacent jurisdictions. Elected officials and residents of the next inner jurisdiction resent it. It is close to the outer boundary of the next inner jurisdiction, which often has been reserved for residences, based on a goal of separating residences from nuisances such as traffic. Thus, residents' initial security from congestion and cut-through traffic is threatened. . . .

EFFECTS ON TRANSPORTATION

A second spatial effect is that the commuting shed extends another twenty miles or so beyond each new area of major job creation. Employment opportunities in the fringe jurisdiction create commuting opportunities twenty miles farther out in other jurisdictions. These more remote commuters then drive through the jurisdiction that has received the business enhancement, either to work in that jurisdiction or to travel beyond it to other jurisdictions. By this sequence, each residential area toward the central city receives more traffic congestion and may be a potential cut-through traffic neighborhood. Avoidance of cut-through traffic is reinforced as a local neighborhood and political goal.

Each of these settlements takes on a unimodal transportation pattern. Only private automobiles can cope effectively with this arrangement of land uses. Thus, the tyranny of easy development decisions passes through more turns of the screw, leading toward a widening territory of mediocre development that threatens the cumulative quality of life within each jurisdiction and for the region as a whole.

CONTENDING POLITICAL GOALS

In each category of jurisdictions (destination, feeder, pass-through, and balanced), contending forces are inspired by different interests and goals. Some groups are interested in expansion, because they gain from it. Other groups are opposed to expansion, because they fear uncertainties that come with it. Some elected officials believe their duty is to represent constituents and their jurisdiction to enhance their position. Other elected officials are more apt to accept a larger sense of regional merit, believing that the concerns of residents and elected officials from other jurisdictions deserve consideration.

Some elected officials believe their role is to cope with projections of population and employment as though the projections are immutable forces of nature. Other officials try to imagine a better quality life and figure out how to move toward it. Each of these attitudes, values, and interests will be in play. Their frequency and distribution at a given time and space will influence the pattern and density that occur as structural imperatives unfold, based partially on jurisdictions' metropolitan locations.

[. . .]

SPRAWL, REGULATIONS, AND RISK

The key to development locations lies in the concept of the tyranny of easy development decisions. The automobile, affluence, personal preferences, infrastructure investments, and metropolitan spatial form created conditions in which vast territories can be developed. Numerous local governments with the power to control land use govern these vast territories. Typically, they adopt zoning codes that require separating residences from location of work, commerce, and even schools. Isolated pods of automobile-dependent development inevitably result. State governments limit local governments' powers. Landowners retain powerful rights to development. Land developers, once they have launched development processes, also gain development rights. The jurisdictions and subareas where they can exercise these rights are numerous. Local governments' growth management successes, therefore, may have unintended spillover effects in nearby jurisdictions. . . .

The net effect is that numerous development opportunities exist, which lending institutions have been ready to finance, and other institutions have been willing to insure against some risks. A large proportion of these relatively low development-risk situations have been in fringe locations. In this decentralized decision-making system, effectively plugging the plentiful leaks in the proverbial dike is very difficult. The tyranny of easy development decisions follows from too many development opportunities and leads to excessive sprawl, insufficient reinvestment, and increasing income disparities among local jurisdictions.

38

How to make a Town

From *Suburban Nation: The Rise of Sprawl and the Decline of the American Dream* (2000)

Andrés Duany, Elizabeth Plater-Zyberk, and Jeff Speck

Editor's introduction

In June 2007, the *International Herald Tribune* ran a feature headlined "Twenty-five Examples of Good Urban Design" that declared: "It's not necessarily the billion-euro development, star-architect-designed gallery or shiny new Ferris wheel that makes locals feel good about their town . . . the measure of a city is more about everyday wonders." The article listed street clocks in Prague; New York City's High Line; a fire station in Gelsenkirchen, Germany; London's Regent's Park; the Simone de Beauvoir footbridge in Paris; Tokyo Midtown, a 6 million sq. ft mixed-use district in Roppongi; and nineteen other "everyday wonders." These projects represent the wide span of urban design, a subfield of urban planning, landscape architecture, and architecture.

Though urban design was not given an official name/identity until the late 1950s, it has roots in the civic design of the turn of the twentieth century. Three early books framed the field: Camillo Sitte's 1889 *City Planning According to Artistic Principles*, translated by George and Christine Crasemann Collins (London: Phaidon Press, 1965); Charles Mulford Robinson's *The Improvement of Cities and Towns, or, The Practical Basis of Civic Aesthetics* (New York: Putman's, 1903); and Werner Hegemann and Elbert Peet's *The American Vitruvius: An Architect's Handbook on Civic Art* (New York: Architectural Book Publishing, 1922). The authors of these books analyzed public space, building composition, and public art. They recommended replication of historic designs – the medieval market square or the Baroque public space – and adornment of streets, parks, and squares with classical decorative art. (See "Modernism and Early Urban Planning" in this *Reader*.)

Modernism, emerging in the early twentieth century, squashed these traditional city-building ideas. Led by architect Le Corbusier, modernists designed buildings to look as if they were mass-produced – Le Corbusier's own Domino house (1914), a simple unadorned, rectilinear, concrete structure, is a prime example. For Corbusier, it was a short step from designing houses to designing cities. The bold superblocks, wide streets, and high-rise offices and residences of his *Cité contemporaine* (1922), *Plan voisin* (1922), and *Cité radieuse* electrified colleagues and critics alike. In 1928, Corbu and his followers formed the Congrès International d'Architecture Moderne (CIAM) to promote their designs. They created the CIAM grid, a giant matrix that specified the per-capita square footage for housing, work, recreation, and many other functions. Guided by the grid, designers went to work finding clients among the many public authorities engaged in large-scale urban redevelopment projects after World War II – either rebuilding war-torn areas in Europe (London's Barbican, for example) or practicing urban renewal in the United States (Boston's West End).

By the 1950s a backlash against CIAM-inspired urban design set in. When Jose Luis Sert, Dean of the Harvard Graduate School of Design, convened a two-day symposium on urban design in April 1956, the more than 200 designers, journalists, and elected officials in attendance applauded journalist Jane Jacobs as she castigated city planners for their promotion of vast barren urban renewal projects. Two years later, the University of Pennsylvania and the Rockefeller Foundation sponsored a follow-up conference, at which attendees reiterated their critiques but spoke of the need to find a new direction for the field. To this end, Rockefeller started a grant program in urban design. Kevin Lynch received funding for an observational project, which looked at how people found their way in cities, resulting in *Image of the City* (Cambridge, MA: MIT Press, 1960). Other grants were awarded to Jane Jacobs, for *The Death and Life of Great American Cities* (New York: Random House, 1961); Christopher Tunnard and Boris Pushkarev, for *Man-made America: Chaos or Control*? (New Haven, CT: Yale University Press, 1963); and Erwin Gutkind, for the eight-volume *International History of City Development* (New York: Free Press, 1964–1972).

Those 1960s authors articulated what have become the key principles for today's urban designers. (See "Dimensions of Performance" and "Downtown is for People" in this *Reader*.) They are: (1) enhancing the vitality and legibility of the public realm; (2) increasing the pleasure and comfort of city residents and visitors; (3) using public investments to stimulate private investment in property development; (4) blending the natural and built environment in sustainable ways; and (5) giving meaning to places, whether it be historic, cultural or aesthetic. Urban designers pursue these goals through work in every geographic scale from the vest-pocket park (see William H. Whyte's *Social Life of Small Urban Places*) to the region (Peter Calthorpe and John Fregonese's *Envision Utah* plan).

The Rockefeller-funded writers inspired a number of followers who codified the principles. In 1987, for example, California planners Alan Jacobs and Donald Appleyard issued a manifesto articulating four requirements for good urban design. It provides minimum density, integration of activities, pedestrian-oriented public spaces, and diverse building types ("Toward an Urban Design Manifesto," *Journal of the American Planning Association*, 53: 1, 1987). Four years later, architects Peter Calthorpe, Andrés Duany, Elizabeth Plater-Zyberk, and a handful of others translated these ideas into "New Urbanism," a call for compact, mixed-use, pedestrian-friendly places (Ahwahnee Principles; see www.lgc.org/ahwahnee/principles.html). They later asserted that this formula was best expressed with traditional designs – gridded streets with wide, tree-lined sidewalks, two-story houses with front porches, corner stores, downtowns with housing mixed with retail.

The Duany group founded the Congress for New Urbanism (CNU) to promote New Urbanism through publications (books, newsletters, scholarly articles), executed projects (it claims 210 completed by 2008), conferences, and an informative website (www.cnu.org). Articles in popular magazines like *Newsweek*, real estate advertisements, and even a movie setting (*The Truman Show*, 1998) have popularized the movement. New Urbanists, in hopes of further institutionalizing their models, are encouraging municipalities to replace traditional zoning ordinances with "form-based" codes that mandate not a certain type of land use or a certain density but rather a prescribed building type, on the grounds that the *look* of a community's built forms is key. (See Plate 1.14 and "Shaping Cities through Development Regulations" in this *Reader*.)

But new urbanism is by no means the only paradigm for urban design. There are many others scattered throughout the world, but they are somewhat less formulaic and defy easy categorization. In 2005 New York's Museum of Modern Art, in an exhibit entitled "Groundswell," showcased more than twenty urban design projects from a wide range of countries – German parks, British town centers, Japanese riverfronts, even the adaptive reuse of a sanitary landfill in the United States. Although uniquely tailored to their sites, the projects all have common characteristics, including their urban locations, dramatic transformation of the public realm, and unusual visual experiences.

This selection is a New Urbanist piece from a book criticizing contemporary suburbanization, written by the movement's entrepreneurial founders, Andrés Duany and Elizabeth Plater-Zyberk, and one of their chief lieutenants, Jeff Speck. Duany and Plater-Zyberk, who are married, helped start the edgy Miami-based firm Arquitectonica in 1977. Three years later, the couple launched their own firm, DPZ. Also based in Miami, it produces neotraditionalist projects. Their most recent book, written with Robert Alminana, is *New Civic Art: Elements of Town Planning* (New York: Rizzoli, 2003). Since 1996, Plater-Zyberk has served as Dean of the University of Miami School of Architecture. The third author, Jeff Speck, was DPZ's director of town planning before serving as Director of Design at the National Endowment for the Arts. He is now in private practice with Canopy Development, Northampton, Massachusetts.

For more reading on the evolution of urban design, see Sigfried Gideon, *Time, Space and Architecture* (Cambridge, MA: Harvard University Press, 1941); Gordon Cullen, *The Concise Townscape* (London: Elsevier, 1961); Paul D. Spreiregen, *Urban Design: The Architecture of Cities and Towns* (New York: McGraw-Hill, 1965); Edmund Bacon, *Design of Cities* (New York: Viking Press, 1967; Collin Rowe, *Collage City*

(Cambridge, MA: MIT Press, 1978); Jonathan Barnett, *Introduction to Urban Design* (New York: HarperCollins, 1982); Francis Tibbalds, *Making People-friendly Towns: Improving the Public Environment in Towns and Cities* (London: Longman, 1992); Jonathan Barnett, *Redesigning Cities: Principles, Practice, Implementation* (Chicago: APA Planners Press, 2003); and Jon Lang, *Urban Design: A Typology of Procedures and Products* (London: Elsevier, 2004).

To sample urban design writings, see Matthew Carmona and Steven Tiesdell (eds), *Urban Design Reader* (Oxford: Elsevier, 2007), and Michael Larice and Elizabeth Macdonald (eds), *The Urban Design Reader* (Abingdon: Routledge, 2007). The American Planning Association's *Planning and Urban Design Standards* (Hoboken, NJ: Wiley, 2006) is an excellent reference book offering measurements and minimum standards for streets, parking spaces, tree planting, and other elements of urban design.

Christopher B. Leinberger's *The Option of Urbanism: Investing in a New American Dream* (Washington, DC: Island Press, 2008), the most recent of the many offerings on New Urbanism, shows how to blend finance and design to achieve walkable places. Witold Rybcznski's *Last Harvest: From Cornfield to New Town* (New York: Scribner, 2007) traces a developer and his quest to convince a rural town to permit a New Urbanist subdivision. Other books focusing on New Urbanist practice are: Hank Dittmar and Gloria Ohland (eds), *The New Transit Town: Best Practices in Transit-oriented Development* (Washington, DC: Island Press, 2004); Douglas Kelbaugh, *Repairing the American Metropolis: Common Place Revisited* (Seattle, WA: University of Washington Press, 2002); and Peter Calthorpe and William B. Fulton, *The Regional City: Planning for the End of Sprawl* (Washington, DC: Island Press, 2001).

Three different offerings treat areas of broad concern for urban designers: Richard B. Peiser with Adrienne Schmitz (ed.), *Regenerating Older Suburbs* (Washington, DC: Urban Land Institute, 2007), Julie Campoli and Alex S. MacLean, *Visualizing Density* (Cambridge, MA: Lincoln Institute of Land Policy, 2007), and Mike Jenks and Nicola Dempsey, *Future Forms and Design for Sustainable Cities*, Oxford: Elsevier, 2006).

Much information about urban design can be found on a group of superb websites. The site of Project for Public Spaces (www.pps.org/) has a section on great cities. The Resource for Urban Design Information is a British site (www.rudi.net). Some cities have active design advocacy groups with websites – see the Municipal Art Society of New York (www.mas.org), San Francisco's Planning + Urban Research Association (www.spur.org) and Philadelphia's Planphilly (www.planphilly.com).

As badly as we have been shaping our built environment, we still possess the ability to do it right. The principles and techniques of true urban design may have been forgotten, but they are not lost; they can be relearned from the many wonderful older places that still exist. By emulating the past, a number of recent projects have demonstrated that designers can make new places that are as impressive as the towns which inspired them.

REASONS NOT TO, AND REASONS TO DO SO ANYWAY

[. . .]

New Towns are not always the answer. The appropriateness of a greenfield development depends on the particular characteristics of the surrounding region. Certain facts must be accepted as given. If a region is not growing statistically – in population or wealth – it should not be growing geographically. The result of such unwarranted dispersal is the draining of the inner city and the wasteful distribution of new infrastructure. Even in regions that are growing, the objectives of economic efficiency and social justice suggest that growth be focused in areas that are already at least partially developed.

Why create new places at all when existing places are underutilized? It must be clearly stated that many social and environmental ills would best be solved, at least temporarily, by a moratorium on greenfield development. There is a ready supply of vacant land available for infill projects, both in the inner city and in existing suburbs. But

. . . forces conspire to make exurban investment more attractive to developers than infill work. For the time being, as we fight these incentives for suburban growth, we must admit that it is still occurring, and in the worst possible form: automobile-based sprawl.

. . . Unless unjustified greenfield development is stopped – an unlikely prospect – designers should endeavor to ensure that what gets built on the urban fringe is environmentally sound, economically efficient, and socially just. . . . [Following are design] . . . principles that should inform any conscientious attempt at healthy suburban growth. . . .

REGIONAL CONSIDERATIONS

The most important design criteria of any new village or town – and the least often satisfied – are regional. Currently, most development occurs not according to geographical logic but according to the random disposition of resources: the first parcels to be developed are often those whose owners have the financing, rather than the ones that are the best located or the least environmentally sensitive. . . . Ideally, parcels under consideration for growth should be rationally located within a comprehensive regional plan that seeks to limit automobile dependence and preserve open space. If not immediately adjacent to existing development, the new development should be at a concentration of infrastructure and, if possible, at a likely transit stop. In the best regional plans, existing and future rail lines serve as a basis for locating new neighborhoods and town centers.

MIXED-USE DEVELOPMENT

Regardless of location, a new neighborhood can avoid unduly contributing to sprawl by being of mixed use. At the bare minimum, every residential neighborhood must include a corner store to provide its residents with their daily needs, from milk to aspirin. While it is only a start, a small corner store does wonders to limit automobile trips out of the development, and does more than a social club to build the bonds of community.

The corner store should be constructed in an early building phase. It will not, at first, be economically self-sufficient, due to the small number of houses around it. It should not be expected to turn a profit until the neighborhood matures, and for that reason the retail space should be provided rent-free by the developer as an amenity, much in the way a conventional developer would construct an elaborate entry feature or a clubhouse. Since it can be very effective in marketing real estate – if properly staffed with a gregarious busybody – the corner store is a fairly easy concept for enlightened developers to understand.

[. . .]

The corner store is, of course, only the first step toward a true mix of uses. A neighborhood-scale shopping center may be appropriate for a larger population or when adjacent to through traffic. Such a concentration of retail – around 20,000 sq. ft, including groceries, dry cleaner, video rental, and other daily needs – should be designed as part of any large development in anticipation of future demand. Any town plan with two neighborhoods or more should include such a town center, which is built when there are enough citizens to make it viable.

A mixed-use neighborhood also includes places to work, the more the better. Perhaps the smallest, aside from the home office, is the neighborhood work center, a place where residents can share the costs of a secretary, office equipment, and meeting rooms. Such neighborhood work centers are emerging in a for-profit form as the local Kinko's, a business that is flourishing as more people choose to work at home.

Ideally, every neighborhood should be designed with an even balance of residents and jobs. While this flies in the face of convention, it is not impossible, to implement. All that is needed is for the housing and commercial developers to agree to work in the same location with a coordinated plan. When there is only one developer for both, it is even easier. Riverside is a new Atlanta neighborhood recently built by Post Properties, a company large enough to develop housing and office space at the same time. Their first phase of construction included a quarter-million square feet of office space and two hundred apartments, all of which were rented immediately at rates 40 percent above the market average. . . .

A common criticism of "forcing" the workplace into residential areas is that, even though the workplace is near the homes, it is not near the homes of the people who work there. This assertion may be true at first, but not over the long run. There is no doubt that most of the workplace in new towns will be staffed initially by people who commute from some distance away, just as most of the new houses will be occupied by people with steady jobs elsewhere. But the study of older communities shows that this relationship improves within a generation. When they can, people will relocate their home or business to be near their business or home. It is the planner's imperative to offer them the opportunity to do so.

Criticism of traditional town planning can be shortsighted, as it presumes that fully integrated communities can be conjured up overnight. True towns take time; a designer can only provide the ingredients and conditions most likely to lead to a mixed-use future. Eighteen years after it was planned, Seaside just built its first school, but has yet to build its town hall. . . .

Which brings up the final component of mixed use: civic buildings. After housing, shops, and workplaces, civic buildings are a required element for any new community. Indeed, land should be reserved for them at the most prominent locations, such as a high ground, a main intersection, or the town square. Larger civic buildings – city halls, libraries, churches, and the like – require the most patience, as they are typically the last to get built, but they must be planned for if they are to exist at all. In the meantime, smaller civic buildings such as the neighborhood recreation center or the bandstand on the town green square can serve as social centers and contribute to a sense of community identity.

The most important civic building is the neighborhood elementary school, which should never be more than a fifteen-minute walk from any home. This may seem a radical proposition these days, when schools seem to be sized primarily for the efficiency of the janitorial service, but there are many arguments in its favor. It has become clear that small schools are key to effective learning. Recent studies have demonstrated that schools with fewer than 400 students have better attendance rates, fewer problem children and dropouts, and often higher test scores. . . .

True neighborhoods mix different uses within individual buildings as well. Many mixed-use buildings containing apartments, offices, and shops have been constructed in new traditional towns such as Seaside. . . .

CONNECTIVITY

If a new neighborhood is to contribute more to its region than traffic, it must do more than just mix uses. Its relationship to its neighbors is important as well. In order to avoid the inefficient hierarchical street pattern of sprawl, in which virtually every trip uses the same few collector roads, the new neighborhood must connect wherever practical to everything around it, even if its neighbors are nothing but single-use pods. One must say "wherever practical," because it is obviously not possible to connect across superhighways or river beds, nor is it advisable to connect to oil refineries or trucking depots. But all compatible land uses should be connected, especially between residential areas, the most common adjacency.

This is easier said than done. Whenever we design a new neighborhood, we make every effort to convince the adjacent subdivisions to allow us to connect to them. We'll go so far as to place the most luxurious housing directly abutting the neighbors, whatever the quality of their housing. We hand them photographs and testimonials from our other developments, and appraisals demonstrating their impressive financial performance.

[. . .]

Connectivity is also an important issue as it concerns highways and arterials. . . . [T]he concept of the highwayless town implies two basic rules: highways and arterials approaching neighborhoods should skirt them rather than split them; and when they do come into contact with a neighborhood, they should take on low-speed geometries. Unfortunately, this contradicts current conventions. We battle over these rules in almost every development we work on, thanks to public works directors who prioritize traffic volume over neighborhood viability. . . .

When faced with a major road, how should a neighborhood respond? That depends on whether the road is designed as a civic thoroughfare or as an automotive sewer. When it is properly detailed as an avenue or a main street – as is appropriate within the neighborhood – or as a parkway or boulevard at the neighborhood edge, the thoroughfare becomes a worthy setting for buildings and will benefit aesthetically from their presence. Princeton, New Jersey, has just such a main street – a delightful collection of shops fronting a major regional arterial – and in Kansas City's Country Club District expensive estates directly face a heavily trafficked roadway. Seen to best advantage across deep lawns, these houses provide motorists with a grand entry into the city.

Only when noxious, high-speed traffic is inevitable should a road face the backs of houses. If a developer resorts to this solution, he must build a wall as well, or the backyards become uninhabitable. Since most major roads are designed to create high-speed traffic, the "sound wall" is the standard solution in the new suburbs.

The ruling principle is that as long as the road is designed with low-speed geometries, traffic generally treats the neighborhood the way that the neighborhood treats it. Friendly house fronts tell drivers to slow down, while blank walls and house backs tell them to speed up. An intermediate solution, appropriate for roads of moderate speed, is to face the road with the short ends of the blocks, so that it is met with the sides of houses, with a deep lawn as an additional buffer. Sunset Boulevard in Beverly Hills is configured this way. Every block ends on the boulevard, resulting in an intersection every 300 ft or less. Such frequent spacing can raise the hackles of the traffic engineers, who tend to want much wider spacing between intersections, so that cars can travel at higher speeds. These engineers must be reminded of the difference between a boulevard and a highway, and that the latter has no place in residential areas.

MAKING THE MOST OF A SITE

Modern development is notorious for its unique approach to nature, typically: *level the site first, design it later*. This attitude has been the rule rather than the exception since the 1800s, when Jefferson laid his perfect grid across the continent. It comes as no surprise, then, that the typical American builder would rather spend $100,000 on bulldozers and artificial drainage than on a sensitive site plan.

We know better now, and there are many justifications for preserving a site's natural qualities, aside from the obvious ecological benefits. First, natural features – not just waterfront and hillsides, but wetlands and trees – can add significantly to property value. Second, the character of the landscape can help people understand and negotiate their environment. It is much easier to give directions, even in cul-de-sac suburbia, if one can say "take a left at the pond." Finally, for planners, varied and idiosyncratic sites are actually easier to design, and much more interesting. While flat and featureless land gives few hints as to where to begin, a complex site tells the designer pretty clearly what it wants to be. . . .

[. . .]

THE DISCIPLINE OF THE NEIGHBORHOOD

. . . The five-minute walk – or *pedestrian shed* – is roughly one-quarter mile in distance. It was conceptualized as a determinant of neighborhood size in the classic 1929 New York City Regional Plan, but it has existed as an informal standard since the earliest cities, from Pompeii to Greenwich Village. If one were to map the neighborhoods of most prewar cities, they would average about one-quarter mile from edge to center. While some flexibility is advisable – the West Coast designer Peter Calthorpe recommends a ten-minute walk in order to engage a larger number of households to a transit stop, and college students seem to put up with twenty minutes, . . . most new traditional town plans are designed around the five-minute measure. One-quarter mile is usually the distance from which you can actually spot your destination. More important, experience suggests that it is a distance short enough that most Americans simply feel dumb driving, making it a perfect rule of thumb for our auto-dependent times.

The first step in designing an open site is to use its natural features to locate the centers and edges of five-minute-walk neighborhoods. Neighborhood centers are typically located at the geographic center of the available land, but can be shifted in response to site conditions, such as a view or a major road at one edge. . . . For sites large enough to hold multiple neighborhoods, two organizational options are available: the neighborhoods can be distinct, separated by a green belt, in which case each remains a village, or the neighborhoods can be directly adjacent, sharing a boulevard at their seam, in which case they can coalesce into a town or even a city. In both cases, the overall structure links neighborhood centers with avenues in a fairly direct transit loop....

[...]

Once the center and the edge of a neighborhood have been located, the distribution of uses follows naturally. The areas of highest density and urbanity surround the center, which is the location of a major public space such as a plaza, square, or green, depending on the local tradition. The center is also the location for shops and a transit stop. From the center outward, housing densities fall, such that, in villages, the conditions at the edge can be downright rural. Different building types are "zoned," not by use, but by size, and changes in zoning occur at mid block rather than mid street, so that each street tends to have the same building types on both of its sides. This is quite different from the asymmetrical experience typically encountered in suburbia.

As one leaves the center and approaches the neighborhood edge, building densities decrease and there occurs a corresponding shift in the design of the street. Every single aspect of the public realm transforms from urban to rural. Closed curbs and gutters become open swales; trees stop lining up and become more varied in species; sidewalks narrow and eventually disappear; and front yards become gradually deeper. In this way, there is an authentic and gentle transition from culture to nature. This sort of detailing, which is essential to giving a neighborhood a unique sense of place, requires designers and developers to exercise a degree of care that is now rare. One can only hope that the financial success of the new places designed in this manner will eventually encourage more developers to invest in such precise design.

The gradual transition from center to edge occurs most clearly in villages, which by definition are single neighborhoods sitting free in the landscape. In towns or cities, where multiple neighborhoods meet across shared main streets, the neighborhood edges may instead be designed as areas of increased density and activity. In this case, the urban/rural transition is reserved for the outer edge of the entire collection of neighborhoods.

In addition to this radial organization, the neighborhood also possesses a Cartesian substructure. ... The larger streets that lead to the center divide the neighborhood into quadrants, each of which is sized to be the independent realm of the small child. As such, each is equipped with nothing but the slowest roads, and contains a local "pocket park" – often no bigger than a single house lot – located within a three-minute walk of every dwelling. The neighborhood thus grants freedom of motion and a certain degree of autonomy even to its youngest citizens.

MAKING TRANSIT WORK

The neighborhood structure is naturally suited for public transit, be it light rail, trolleys, buses, or jitneys. But there are also three rules that transit must follow in order to appeal to users, regardless of the urban framework:

1 *Transit must be frequent and predictable.* The challenge is not to prove this obvious principle but to create a transit system in which frequency is economically viable. This objective can be achieved only at certain densities; studies suggest that a minimum of seven units per acre is necessary if transit is to be self-supporting. For lower densities, the careful organization of neighborhood centers, to be served by smaller vehicles, can result in a successful network. This network, however, would likely require financial support.
2 *Transit must follow a route that is direct and logical.* Riders shy away from transit systems in which the path is not efficient and easy to understand conceptually. Anyone who has ever taken a shared hotel bus to the airport knows how intolerable an uncertain, zigzagging route can be. Yet bus routes often dogleg interminably. The desire for a trustworthy, unchanging route is one factor that helps explain riders' preference for light rail over buses.
3 *The transit stop must be safe, dry, and dignified.* In most suburban communities, transit passengers are made to feel like impoverished transients, waiting by the side of the road on a graffiti-covered bench or inside an ungainly plastic bubble. No wonder, then, that the only people who take the bus are those who have no choice,

creating a self-perpetuating underclass ridership. In contrast, the structure of the traditional neighborhood offers the possibility of a transit experience that is both comfortable and civilized. When the transit stop is located at the neighborhood center, next to the corner store or the café, the commuter has the opportunity to wait for the bus or trolley indoors with a cup of coffee and a newspaper, with some measure of comfort and dignity. For this condition to occur with regularity, transit routes and urban plans must be developed in concert. Ideally, transit authorities should also work directly with shop owners, who typically welcome the extra business that a transit stop can generate.

THE STREETS

. . . On well traveled streets within a neighborhood, there is no justification for travel lanes wider than 10 ft and parking lanes wider than 7 ft. If either are any wider, the cars speed. However, on less traveled residential streets, another logic should prevail, that of the "yield street." Common in almost every prewar American neighborhood – but now summarily rejected by public works departments – the yield street uses a single travel lane to handle traffic *in both directions*. When two cars approach each other, they both slow down, and one eases slightly into a parking lane while the other passes. Because traffic is necessarily slow, accidents are virtually unheard of on such streets. While inappropriate for heavy volume, yield streets cause few delays when used for minor residential streets in low-density neighborhoods.

Although this type of street is endorsed in the engineers' official manual, it is virtually impossible to get one approved. Almost everywhere we've worked, our demand for yield streets has threatened to delay our projects. It seems irrelevant that these streets exist in every older city, and that we have all driven on them regularly without incident.

[. . .]

THE BUILDINGS

A good town plan is not enough to generate a desirable public realm; individual private buildings must also behave in a manner that contributes to pedestrian life. Once again, a study of how the most valued historic neighborhoods differ from conventional sprawl uncovers the rules for a pedestrian-friendly architecture.

While conditions should vary throughout the neighborhood, houses should generally be placed close to the street in order to define its space, with fronts that are relatively simple and flat. Setbacks should range from about 10 ft near the neighborhood center to about 30 ft near the neighborhood edge. To encourage sociability, the front yard should include porches, balconies, stoops, bay windows, or other semi-private attachments. These attachments should be allowed to encroach within the setback area, so that they represent a gain in space rather than a loss for those who build them. With the proper incentives, front porches need not be mandated, although the town planner who wishes to create streets and squares of dependable character may do so in specific areas.

Attached row houses, a common urban type, should generally be placed closer to the sidewalk than freestanding houses – right up against it, with room only for the stoop. In this case, the first floor must be raised at least 2 ft off the ground for privacy. People don't seem to mind sipping tea directly adjacent to passing pedestrians if those pedestrians can't easily see over the window sill. Residential spaces within 5 ft of a sidewalk must never be located at ground level, period. If one must place a ground-level room within 10 ft of a sidewalk, it must be protected by a porch or a dense garden.

For retail buildings, the setback rule is straightforward: don't have one. Traditional retail, to be successful, must pull directly up to the sidewalk, so that people can see the merchandise in the window. Parking lots in front are of course forbidden: there is little that is more destructive to pedestrian life. All parking that cannot be handled on the street can be provided by mid-block lots that are hidden behind buildings. The connection from mid-block parking to street-shop entrances is a tricky one, and must be handled with extreme care. The most effective technique is the traditional *pedestrian passage* . . . in which a carefully detailed walkway – often articulated with trellises, fountains, stairways to second-floor apartments, and landscaping – connects the rear parking lot to the street. Experienced retailers recognize this passage as a merchandizing opportunity, and flank it with windows and indoor/outdoor displays. In Palm Beach, a series of charming "paseos" helps to make Worth Avenue one of America's most successful shopping destinations.

Whether commercial or residential, taller buildings are to be encouraged because they use land more efficiently while doing a better job defining the public space. Most houses should be a minimum of two stories tall. One-story shops and offices, the suburban standard, fail to provide for mixed use and are a waste of valuable land. They should be combined with each other or with housing whenever possible. Where no other mechanism exists to make this happen, municipalities should direct their housing subsidies to the construction of apartments above shops.

There is one last rule that much of suburbia needs to follow: traditional architectural detailing, if used at all, should be used accurately, or it results in parody. There is no specific argument or justification for this rule, except for the horrible feeling that one gets when it is broken. . . .

PARKING

When building new places, one quickly finds that the amount of dwellings, shops, and offices that one can provide depends primarily on the amount of parking that can be accommodated. As one of our clients puts it, "parking is destiny." Unfortunately, parking is often a very antiurban destiny, as most municipalities' parking requirements make higher densities impossible without multilevel parking garages, something that most developers can't afford to build. The high cost of structured parking – $12,000 per place, versus $1,500 in a surface lot – is the reason why almost every new suburban building is either less than three stories tall or more than ten stories tall; only a tower can pay for a parking garage.

When new towns are being built in the suburbs, parking requirements cannot be dismissed, as they can be in an older city. Most developers and their lenders insist on ample parking, anyway. The key to sizing parking lots properly in the suburbs is to recognize that the existing requirements are written for the purest sprawl, in which no alternatives to driving exist, and on-street parking is rarely allowed. Each of the factors that distinguishes traditional towns from sprawl – on-street parking, mixed use, transit (when present), pedestrian viability, etc. – also reduces the number of parking spaces that are needed. For example, mixed use means that a school and a cinema can share a parking lot, since they have complementary schedules; the same is true of an office building and an apartment house. Therefore, it is improper to apply the standard suburban parking requirements – often as high as five off-street parking spaces for every 1,000 sq. ft of construction – to a mixed-use neighborhood. A more appropriate requirement is three per 1,000 sq. ft, *including on-street spaces*. This is the number that acknowledges the opportunity for shared parking.

That's still a lot of parking, enough to undermine most attempts at urbanity. But it is important to remember that *where* is more significant than *how much*, and that the quality of the street space comes first. An essential rule of thumb is to provide no more offstreet parking than can be concealed behind buildings, and no more buildings than that amount of parking can support.

THE INEVITABLE QUESTION OF STYLE

Traditional neighborhood design has little or nothing to do with the issue of architectural style. This point may seem obvious to lay readers, but the question of style must be addressed for one reason: it is the architectural style of most Traditional Neighborhood Developments that causes them to be dismissed as "nostalgic" by much of the design profession. While the word style is hardly used in architectural circles – "What style is your architecture?" is a question that makes most designers cringe – the fact is that the current architectural establishment could be accurately described as violently allergic to traditional-style architecture. For many architects, it is impossible to see past the pitched roofs and wooden shutters of Seaside and Kentlands to the progressive town-planning concepts underneath.

Why the negative reaction? Because modernist architects associate it with ideology, style takes on moral overtones. In an age of technology and diversity they believe that it is morally unacceptable to build with techniques of an earlier era or in styles used by repressive societies. Now, there is no denying that the avant-garde has contributed tremendously to the vitality of our culture, from urban skyscrapers to war memorials. It has fared less well, however, in the common vernacular – the suburban building for everyday uses – where, at odds with the human need for communication and personalization, it has been thoroughly debased. . . .

As a result, there now exist essentially three different types of architecture: cutting-edge modernist, authentic traditional, and a gigantic middle ground of compromise that includes lazy historicism, halfhearted modernism, and everything in between, most of which could be called kitsch. While cutting-edge modernism has proved popular for monuments, commercial structures, some apartment buildings, and the spectacular houses of well-to-do patrons, it has not penetrated the middle-class housing market. The vast majority of home buyers are only interested in traditional architecture or, sadly, the middle ground of compromise.

It is on this ravaged battlefield that the campaign for traditional town planning is being waged. Most of our audience – the citizens and public servants who must approve our projects if they are to be built – do not appreciate or trust modernist architecture. To present the ideas of neighborhood design in an underappreciated modernist vocabulary would bring them up against insuperable skepticism, even more than they already face for being different. It is hard enough convincing suburbanites to accept mixed uses, varied-income housing, and public transit without throwing flat roofs and corrugated metal siding into the equation.

[. . .]

39

Planning Metropolitan Regions

From *Planning for a New Century: The Regional Agenda*, ed. Jonathan Barnett (2001)

Gary Hack

Editor's introduction

Recently, a new form of regional planning has emerged in response to post-World War II settlement patterns. Labeled "New Regionalism," it has many sources ranging from geography to city planning to community organization. The intellectual foundations of New Regionalism are traceable to a line of thinking that started with Jean Gottmann's *Megalopolis: The Urbanized Northeastern Seaboard of the United States* (New York: Twentieth Century Fund, 1961), which identified the 600-mile corridor from Boston to Washington as a continuous metropolitan area. This assessment contributed the idea that a region could be enormous and relatively unbounded, dominated by several older cities linked by rail and highway and surrounded by suburban development.

Two decades later, Joel Garreau's *Edge City: Life on the New Frontier* (New York: Doubleday, 1988) introduced a new discovery: "edge cities," areas having 5 million or more sq. ft of office space, 600,000 sq. ft of retail, more jobs than bedrooms, locations near highway interchanges, a unitary image, and recent rapid growth. In Garreau's view, the region is polycentric, composed of the older cities' high-density downtowns and newer less dense edge cities, with interstate highways providing the connective tissue.

Robert E. Lang's turn-of-the-century *Edgeless Cities: Exploring the Elusive Metropolis* (Washington, DC: Brookings Institution Press, 2003) offered yet another finding. Lang argued that while the edge city certainly existed, it was more an anomaly than the norm. He identified "edgeless cities," sprawling areas that accommodated two-thirds of the nation's non-downtown office space in office buildings scattered throughout the landscape. Edgeless cities perform city office functions but do not have city form, are lacking retail, and have no boundaries. In this analysis, the polycentric metropolitan region encompasses centers of varying size and location (primary and secondary downtowns and edge cities) and countless isolated low-density office buildings arrayed along the highways with single-purpose shopping malls and residential areas filling the interstices. In Lang's model, the region has some centers with skyscrapers concentrated in small geographic areas – a couple of square miles – a few edge cities, and swaths of low-slung office buildings spread over hundreds of square miles.

Adding to this descriptive literature on new regional forms are public policy studies zeroing in on regions (usually as defined by the Census–see Selection 5 for discussion). No group has done more to shape thinking on these matters than the Brookings Institution's Metropolitan Policy Program led by Bruce Katz. The Metropolitan Policy Program focuses on the nation's metropolitan areas because they house more than 80 percent of the U.S. population and contain 85 percent of U.S. jobs. The group's steady stream of research and recommendations regularly appear in monographs, reports, interactive databases, and other communications, all easily accessible online (www.brookings.edu/metro/metro.htm). Its work centers on providing data on regional contexts and pursuing strategies that promote economic growth that is robust, sustainable, and inclusive. Of note is its Living Census series (analyzing topics ranging from immigration to poverty to downtown living) and its Census Plus database (offering customized up-to-date reports on such subjects as ethnicity and commuting). Some Brookings initiatives draw attention to and monitor conditions in metro areas of national importance that have commanded popular attention but have no systematic, in-depth, and ongoing review. Exemplary are the Katrina Index, tracking key post-hurricane recovery indicators in the Gulf region, and the Great Lakes Economic Initiative, covering the twelve middle American states, significant contributors to the U.S. economy and trade, which have had a difficult transition from their industrial past to today's knowledge-based economies. Finally, in anticipation of the national elections in 2008, the program produced *A Blueprint for National Prosperity* that posited a countrywide strategy based on strengthening regions.

The Citistates Group, a network of journalists who report on city and regional issues, is also influential, especially in its use of the press to shape opinion and policy. It undertakes commissioned region-focused studies. Led by syndicated columnist Neal Pierce, Citistates Group, which Pierce founded in 1996, has produced studies on twenty-four metros, including Charleston (2007), New England (2005–06), Boston (2004), and South Florida (2000). In addition, Pierce covers a "city-region" beat in his weekly column, which appears in more than fifty papers nationwide and reports on subjects of urban interest, such as megaregions and transit-oriented developments. To read the reports and columns, and to find additional information, see www.citistates.com.

Katz and Pierce, along with an emerging group of researchers and professionals (including Catherine Ross, Georgia Tech's director of the Center for Quality Growth and Regional Development, California-based writer William Fulton, Portland-based planning consultant John Fregonese, and National Resources Defense Council president Frances Beinecke), represent a loose coalition coalescing around the "New Regionalism." This new branch of regional planning concentrates on issues related to governance (as opposed to government), open definitions of regional boundaries, coalition building, visioning, and benchmarking. Key to success is crafting strong leadership and broad agreement on common values and pursuits.

One means of achieving these goals is large-scale visioning exercises with follow-up initiatives. Usually a public regional planning agency, a nonprofit civic entity, or an industry group sponsors the visioning exercise, convening stakeholders throughout a region to focus on issues of growth and development, often framing the effort as a way to enhance the place's economic competitiveness and secure necessary public investment. Visioning may be undertaken because of legal mandates or as a voluntary campaign. For example, as a condition of federal funding, metropolitan planning organizations (MPOs) established under federal transportation law must regularly produce transportation improvement plans (TIPs), relating mobility, land use, and air quality. Volunteer efforts usually arise from coalitions of good-government, environmental, design/planning, and university groups.

"Envision Utah" is one of the earliest and best-known examples of large-scale visioning. A public/private partnership, it began as an extensive research, outreach, and engagement campaign in 1995, asking participants to identify potential growth issues in the Greater Wasatch Area, a 23,000 square mile region encompassing ten counties, ninety-one municipalities, and 157 special districts. Four years later, having involved thousands of people in the 1.6 million population, the partnership produced its report, *Quality Growth Strategy*. The report sketched six broad goals encompassing air quality, transportation, land use, natural resources, housing, and infrastructure investment to be achieved through implementation of initiatives outlined in thirty-two individual strategies. It represented agreement on items generally known as Smart Growth principles – funding public transportation, promoting walkable communities, conserving critical open space – depicting the outcomes graphically and in writing.

Envision Utah inspired visioning endeavors nationwide. Some examples are myregion.org's *How Shall We Grow: Central Florida Regional Growth Vision* (2007), Reality Check Plus: Imagine Maryland's *Today's Vision, Tomorrow's Reality*, and Joint Venture: Silicon Valley Network's *Building the Next Silicon Valley* (2003). See their websites for more information: www.envisionutah.org; myregion.org/; www.realitycheckmaryland.org; and www.jointventure.org.

Elsewhere, government-sponsored regional planning is undergoing rapid change as states attempt to support broader, more coordinated efforts. Consider Illinois and California. The Illinois state legislature merged two regional agencies (Northeastern Illinois Planning Commission and the Chicago Area Transportation Study) to create the Chicago Metropolitan Agency for Planning (CMAP), which has a steady revenue stream from the state and will be responsible for all federal and state planning activities for the seven-county region surrounding Chicago. (See www.chicagoareaplanning.org.) In California, regional blueprint planning – a voluntary, cooperative effort among metropolitan planning organizations (MPOs), councils of government (COGs), and the state transportation agency (Caltrans) – stimulated comparative analysis of the transportation and land use needs of regions across the state, one of the first steps in developing statewide budgeting priorities, required by the passage of a $43 billion infrastructure bond bill. (See calblueprint.dot.ca.gov.) Notably, nonprofit regional planning advocacy groups, Chicago Metropolis 2020 and the California Center for Regional Leadership, took the lead in lobbying for these collaborations. For more information, see www.chicagometropolis2020.org and www.calregions.org/statepolicy.

The vision and organizational structure are just the beginning of new regional planning. Armed with these tools, sponsoring organizations conduct educational and advocacy campaigns to align public and private decision making with the recommendations. Many track progress through indices or annual benchmarking reports. The idea is to guide, not mandate, the myriad actions to gain the larger vision.

This selection offers an example of the out-of-the-box thinking that should inform today's regional planning before or after a visioning exercise. Written by Gary Hack, the dean and Paley professor in the School of Design at the University of Pennsylvania, it discusses regional issues on micro and macro scales. An urban designer and an acute observer of the urban scene, Hack has written *Site Planning*, third edition (Cambridge, MA: MIT Press, 1984) with Kevin Lynch and *Global City Regions: Their Emerging Forms* (London: Taylor & Francis, 2000) with Roger Simmonds. He edited *Practice of Local Planning* (Washington, DC: International City/County Managers Association, 2008) with Eugénie L. Birch, Paul Sedway, and Mitch Silver. Hack has been thinking about regional planning since his graduate student days, when he participated in a lengthy exploration of the Northeast megalopolis, visiting and interviewing its leaders over several months. As PennDesign dean, he has supported seminal student studios that have worked on large-scale regional planning issues to help inform the work of Florida's myregion.org and the national efforts of America 2050.

For additional reading, see William Fulton, *The Reluctant Metropolis: The Politics of Urban Growth in Los Angeles* (Baltimore, MD: Johns Hopkins Press, 2001), a case study of the tensions surrounding regional planning and development in Los Angeles. For more on the Utah experience, see Peter Calthorpe and William Fulton, *The Regional City: Planning for the End of Sprawl* (Washington, DC: Island Press, 2001). For an update on California's regional blueprint planning, see California Center for Regional Leadership, *California Regional Progress Report* (San Francisco, 2007).

When we think about the future of metropolitan areas, it is not very useful to envision them as cities surrounded by a ring of suburbs. The terms "city" and "suburb" are themselves too limited to describe urban patterns . . . We have to begin thinking about cities as metropolitan regions, or city regions, as a matrix of development that extends over wide areas and includes many centers. Some centers may have a longer history than others, but the original central city will not necessarily have a dominant position in the new metropolitan region. The Phoenix metropolitan region is one prototype of an area where the historic center is not the most important location, even for commerce. In Los Angeles, downtown is just one of many centers; in Atlanta, the areas known as Midtown, Buckhead, and Perimeter Center rival the importance of downtown, and political boundaries are not obvious on the ground.

GIVING A NAME TO NEW REGIONAL REALITIES

Calling the settled area that spans Pennsylvania, Delaware, and New Jersey "Philadelphia" is no longer accurate, as the region includes other cities, like Camden and Wilmington, and counties in the three states. . . . [M]ore and more people are using the term Delaware Valley for the region. Most advertising for the Dallas–Fort Worth area refers to it as the Metroplex, although this term is also used in various places around the country for sports stadiums and convention centers. Miami is part of Metro-dade, and around San Francisco it is the Bay Area. These new names reflect the new realities.

WHAT'S WRONG WITH CURRENT DEVELOPMENT?

[. . .]

As the new metropolitan region is here to stay, we need to be clear about what is wrong with the emerging development patterns. . . . Today, as older areas are deserted for new development on the urban fringe, it may be an inefficient use of public resources to let excess infrastructure lie fallow at the centers of cities while constructing new utilities as part of developments at the perimeter. When lengthening travel times and congestion in a metropolitan area make it uncompetitive and unattractive to people, the sprawling form of the city certainly becomes a problem. Individuals and businesses locate still further out to avoid congestion until . . . congestion equalizes itself. Dispersal can also be a problem if the infrastructure of sewers, water supply, roads, and transit can't support development at the new regional scale – although, if densities are low enough, collective water and sewer systems may not be a necessity.

There is often a mismatch between the location of jobs in the dispersed metropolis and those in need of employment. There may be a second disparity between the ability of public jurisdictions to raise resources through the forms of taxation they have available and the demands for services and social assistance placed upon them. The latter is a problem of governmental organization, not particularly an issue of urban form. And the supposition that every part of a region must be closely tied to the historic center is not necessarily the only, or even the most desirable, way for regions to develop.

Because of the way that most metropolitan regions have evolved, road and transit systems are often incongruous with the new geography. There are usually good radial expressways but poor circumferential highways, although only a small fraction of trips may be between center and periphery. There are few mass transit systems running between suburbs or between older communities at the fringe. Over the next generation, planning of such roads and rail lines, as well as new mass transit systems that connect radial routes, will become very important. Many of the proposals in the recent regional plan for New York advocate this, and in the Delaware Valley metropolitan area a mass transit system is being planned linking Camden and Trenton, New Jersey, parts of the region that are outside the traditional center. We will see a lot more of these kinds of projects.

THE EVOLUTION OF "SPRAWL" INTO NEW PATTERNS

People talk about suburban sprawl as if metropolitan areas were all developing at a uniform, low density. In fact, the evolving metropolitan region has a distinct underlying structure. Employment and shopping clusters have grown up in portions of metropolitan areas that have good highway access. . . . Often these clusters include offices, services, shopping, hotels, entertainment areas – all the functions of an old city center, except that you can't get from one building to another without getting into your car. A major order of business for the next generation will be turning these clusters into functioning urban centers.

The next development opportunities in these clusters will involve the huge areas that have been set aside for parking. Decanting the parking into garages will open up land for additional businesses, housing, and facilities, and will create an opportunity for making the connections that are not there today. The difficulty is that all the highways leading to the cluster are likely to be congested already. Local governments resist approving more density unless there are new transportation systems, and a number of developers have begun to create transportation management organizations in suburban clusters. But if the new development is residential and thus has a different traffic pattern, or if the new development is for entertainment uses that encourage office workers to stay beyond the peak hour, it may not result in more congestion, and might well take advantage of otherwise vacant parking. Another way to reduce traffic congestion is by cutting the number of internal trips within the cluster. This can be accomplished by creating mixed-use areas that allow people to go to lunch and run other errands on foot. As a result, daytime traffic congestion should diminish.

The conundrum of emerging commercial clusters is that they cannot become true urban centers without mass transit and much higher densities, yet they cannot get government approval for such changes because of worries about congestion and spillover effects in adjacent areas. A breakthrough strategy may involve a combination of aggressive transportation management measures, locating new public facilities such as schools and cultural facilities in clusters, and creating new public–private arrangements, such as public parking authorities that can build garages that free up parking lots for added density.

. . . [A]nother type of suburban cluster that can be found in most metropolitan areas . . . emerge[s] in an unplanned way as communities map zones for garden apartments and town houses near arterial streets that have been zoned for strip shopping centers and franchise stores. Occasionally there are small office buildings along these commercial strips. There are often no connections among these assorted developments and frequently no sidewalks along the streets, but people create pathways across parking lots or undeveloped areas or lawns wherever they can. These clusters have all the ingredients of a traditional urban neighborhood, except, again, that it is difficult to go from one building to another without using a car. Replanning these suburban residential clusters should be another priority for the next generation, and in high-growth areas there is an opportunity to plan such areas as more urban places before they are developed. These clusters will become the key points of access for suburban transit systems.

Over the next generation, the traditional city center will no longer be the only venue for cultural and entertainment events. Increasingly, these will be dispersed to other parts of the metropolitan region, near where many of then patrons live. Performing arts centers have recently been created in a number of newer urban areas such as El Cerrito, California, and West Palm Beach, Florida, as well as on university campuses in smaller communities and outlying areas. Theaters, concert halls, and other civic spaces can become some of the glue that holds lively and interesting new centers together.

THE FUTURE OF TRADITIONAL DOWNTOWNS

As outlying centers acquire the advantages traditionally enjoyed by center cities, what can one say about the future of traditional downtowns? They will become even more specialized subcenters, certainly not the only centers or even the most dominant ones, but important nonetheless. Their comparative advantage will be found in the areas of tourism, conventions, communications, entertainment, healthcare services, higher education, governmental services, private businesses closely linked to governments (e.g., lawyers dealing with the courts), logistics-sensitive industries that value centrality, and specialized retailing. Private headquarters functions and financial services will continue to be a component, but in only a few cities will they dominate downtown. Central areas will also remain attractive to those who wish to live and work in close proximity and to groups who value convenience over space.

NEW NEIGHBORHOOD DIVERSITY

Only 25 percent of the housing market today is made up of households with two parents and school-age children. Three-quarters of households are in some other category: single parents, single people without children, married couples whose children have left home, married couples who do not expect to have children, unrelated individuals living together, and so on. While inner-city neighborhoods have always seemed quite diverse, the profile of people living in outlying communities has also become quite different from the typical image of 1950s suburbia. Although all the houses in a neighborhood may look similar, a close look at mature suburbs will reveal considerable diversity. . . .

It is also a mistake to stereotype new metropolitan development. Specialized districts are emerging in outlying parts of the metropolitan region. As examples, there are close knit but dispersed artist communities on the urban fringe (such as Bucks County in Pennsylvania); semirural communities centered on weekend farming or ranching or keeping horses on the edges of many cities; villages for people who work on the internet and commute to a more urban location or an outlying office park once or twice a week; golf and recreation communities that may have been designed for retirees but today are a lifestyle option.

NEW REGIONAL DIVERSITY

We are also witnessing lifestyle-based specialization of metropolitan areas. Orlando, for example, is largely a tourism- and entertainment-based community. But it has also attracted a large number of retirees who perhaps locate there in the hope that their grandchildren will visit them. The area around Rochester, Minnesota, also pulls in a large number of retirees who want to be near the medical care offered by the Mayo Clinic. University communities are attractive to those who are footloose because of the range of cultural opportunities they offer, and there has been a growth in retirees around many former military base communities.

There are high-tech communities built around research and education that define new metropolitan forms. The Research Triangle in North Carolina is a classic case of a location planned to take advantage of the three major universities located in three nearby communities, and by spanning the area between them a new form of settlement has emerged. Another obvious example is Silicon Valley in California, which owes its original impetus to Stanford University, but now sprawls across two counties. Then there are locations for special kinds of manufacturing, such as the Route 202 corridor west of Philadelphia from Wilmington, Delaware, to New Brunswick, New Jersey, where almost all the major pharmaceutical companies in the United States are located. There are regional manufacturing centers as well, such as Smyrna, Tennessee, in the Nashville metropolitan region, where a whole series of automotive parts manufacturers have located around the Nissan plant because of the just-in-time manufacturing practices.

[. . .]

The daily and weekly geographic range of people in these linked settlements also appears to be expanding. Today many people live in more than one part of the region. East Hampton, Long Island, used to be a summer resort but many of its residents currently spend three days of each week there and four days in New York year-round. As the need to be near a fixed base of employment declines, people can make different lifestyle choices, and areas previously too distant from business centers are witnessing a great deal of growth – Portland, Maine; Northampton, Massachusetts; Bucks County, Pennsylvania; Charlottesville, Virginia, among many others.

People can commute outbound as well as inbound, and increased numbers of people who seek urbanity are choosing to locate in central cities even though their jobs may be elsewhere. The parts of the older cities that are most successful today in attracting residents are the highest-density, mixed-use districts and the low-density garden neighborhoods. Notably passed by are many medium-density neighborhoods two or three miles from the center that have few of the virtues of urban life and none of the advantages of green countryside. There is a growing interest in living in city centers that have cultural activities, shopping, restaurants, and entertainment within pedestrian range, and generous if often rundown housing. Childfree households, including many active elderly, seek out such areas. For the 25 percent of households with school-age children, the single most important deterrent to moving to central areas has been the poor quality of schools. Charter schools may begin to address this issue.

Outside city centers, new housing being constructed is almost always at much lower densities than in previous years. In the Hough neighborhood of Cleveland, an area that burned during the riots of the 1960s, the city is making land available at low prices and providing tax abatements to people who wish to build new homes. Dozens of housing units, including many mansions, are under construction, and the neighborhood is becoming a near downtown garden neighborhood. It has many locational advantages; it is next door to major hospitals and cultural institutions and close to downtown employment. Clearly the tax advantages are an important incentive, but the transformation of a tough area like Hough would not have happened without the desire of many to live close to urban amenities. And Hough is not alone; in cities across the country centrally located garden neighborhoods are growing in popularity (e.g., Brookline in Massachusetts, Chestnut Hill in Philadelphia, Highland Park in Dallas, Oak Park in

Chicago, Roland Park in Baltimore, Georgetown in Washington, River Oaks in Houston, among others). Densities are increasing in many of these areas, and they are becoming more diverse ethnically and economically than when originally settled. They provide an important model for the revival of the rim of declining areas that surround many historic centers.

In the United States, there is a long history of clustering of economic activities in traditional urban centers – steel in Pittsburgh, autos in Detroit, finance in New York, and so on. But it is important to note that the new economic clusters are almost entirely at the fringes of metropolitan areas, not at their centers. And there is significant dispute over the issue of whether a healthy center is essential to continued prosperity of suburban clusters, or the reverse, whether suburban growth and development guarantee revitalization of historic centers. The patterns of interdependence appear to be subtle.

Some people might move to an outlying area because it offers the richness of opportunities once found only in traditional city centers, others may be attracted to living on farms, and still others may be drawn by educational opportunities in the region. Planning policies for metropolitan areas have to recognize both the diversity of regions and the diversity within particular regions. There is no single prescription that fits each case; policies must emerge from each area's comparative advantages.

REGIONAL TRENDS IN OTHER COUNTRIES

In looking at what future metropolitan policies should be, it is useful to compare the initiatives being taken in other countries. . . . Between 1960 and 1990, in all of the cities studied there was rapid urbanization accompanied by rapid motorization. Metropolitan regions spread, often beyond the control of local governments of the central cities. . . . [T]he settlements became much less dense – with population densities in European urban areas declining by 50 percent or more. . . .

What are cities around the world trying to do today to structure this growth? The prevailing idea of planners remains Ebenezer Howard's century-old notion of creating New Towns, aimed at intercepting development destined for what is regarded as oversized urban centers. Howard's solution advocated decentralization, not to suburban areas, but to planned communities 50 km or 100 km away from the old urban cores, or to intermediate-sized cities even more distant. Green belts or urban limit lines are used to enforce the perimeter of urbanization. The London and Paris regions are textbook examples of this approach, and cities like Singapore and Hong Kong have pursued similar strategies.

New Towns require huge capital resources and consistent political will, exercised over fifteen- to twenty-year periods, which few countries, much less cities, can muster. In the rapidly developing world, there are many more examples of failed than successful attempts to pursue such strategies – including the Jakarta and Bangkok regions, where political instability and corruption in land markets and infrastructure decisions confounded plans for decentralization.

[. . .]

[A] number of European and [Asian] cities have restructured their forms of government to grant powers to regional entities to control infrastructure and the form and timing of development. As an example, the regional government of Madrid – whose powers were recently enlarged by the national government – has prepared an aggressive plan for restructuring the city in a polynucleated pattern, with a grid of infrastructure that equalizes access and service across the urbanizing area. A public development entity has been created to take the lead in carrying out key projects that promote this agenda. . . .

. . . In Tokyo . . . investment in infrastructure plus extraordinarily tight controls on the urbanization of farmlands has limited the spread of development and promoted infilling and much greater densities. . . . [A] series of new subcenters has been built in Tokyo over the past two decades. They are located in two rings: a set of new "downtowns" in the inner city, connected by the Yamanote JR line, built around the terminals of commuter rail lines, and forming a ring about five miles around the Imperial Palace; and an outer ring, twenty to thirty miles from central Tokyo. Each subcenter is served by mass transit, commuter rail, and expressways, and infrastructure continues to be developed to expand access between the centers. Each of the outlying centers has a distinct character and mix of functions. . . . With these, and many other efforts aimed at improving the quality of life, Tokyo is today a multicentered and highly livable regional city covering several hundred square miles.

PLANNING FOR THE NEW REALITIES IN THE UNITED STATES

American cities are unlikely to muster the will to emulate Tokyo, but one of the important lessons to be learned from Tokyo is the necessity of having a regional development strategy. Following from this is the need for some entity to promote the regional agenda; it cannot be left to local municipalities. Part of the difficulty of creating and maintaining a regional agenda is the common perception that central cities and their surrounding areas are in competition with each other. Each time an expanding business relocates from a central area to the perimeter, the sense of competition is reinforced. Certainly central cities have had to bear a disproportionate burden of the costs of metropolitan development and are faced with large populations in need of public assistance and services. Absent significant state aid, they have been forced to tax at rates that discourage new businesses from locating there, further exacerbating their problems. . . . Put differently, recognizing interdependence allows urban regions to organize more effectively to be competitive, functioning regional areas.

EVERY METROPOLITAN REGION NEEDS A REGIONAL DEVELOPMENT STRATEGY AND A PERMANENT MECHANISM FOR PROMOTING IT

Books and articles advising every metropolitan region to emulate Portland, Oregon, have worn out their welcome. Certainly this is a good course of action if local politics permit. However, there is a need for alternatives.

One alternative is a private advisory group. The Regional Plan Association (RPA) of New York and New Jersey has published three influential plans over the past seventy-five years and has had a surprising degree of influence by simply filling the vacuum and framing the discourse about the New York-centered region. . . . RPA has no formal authority; it must rely on public persuasion and its influential board of directors to see its intentions realized.

Another example of an advisory regional plan is Chicago Metropolis 2020, recently prepared by the Commercial Club of Chicago in association with the American Academy of Arts and Sciences. EcoCity Cleveland has rethought its regions in terms of the preservation of natural systems. The advantage of regional plans generated by private organizations is that they are relatively easy to start and nobody has to agree to their creation. The disadvantage is that there is no enforcement mechanism.

Some citizen-based initiatives such as 1000 Friends of Oregon and Save the Bay have achieved results that transcend local boundaries by focusing on a narrow set of purposes – stopping land filling of San Francisco Bay and opening its edges to the public, and promoting compact, transit-centered development at the edges of Portland, in these two cases. But other such interest groups have also made it extraordinarily difficult and time-consuming to construct region-shaping infrastructure, such as new airports, expressways, and transit lines, and there is a need for coalition of interests for such projects.

Efforts in the United States to adjust governmental boundaries to match regional realities have been largely unsuccessful, except in those areas where annexation allows city governments to expand their boundaries outward. . . . However, there is one regional governmental institution, the metropolitan planning organization (MPO), in every metropolitan area. Under federal law the transportation priorities for a region have to be set by the MPO. Many of these organizations are merely forums for negotiation among the local governments that are represented. But a few, such as the Association of Bay Area Governments around San Francisco and the San Diego Association of Governments, have produced regional plans as a context for their regional transportation priorities. Outside of transportation, however, such a plan is, again, purely advisory. . . .

It is often easier to create new governmental entities for special purposes than to modify those that exist. Regional authorities have been established to create, finance, and operate metropolitan infrastructure, such as transit, ports, airports, expressways, and sewer and water systems, but rarely have these been coupled with effective regional planning efforts. The new Georgia Regional Transportation Authority may prove to be an exception.

One possibility for effective regional policies would be the creation of a regional development district, with responsibility for growth centers and major infrastructure projects, coupled with the capacity to lead a process of visioning and debating regional opportunities. Much of the extraordinary costs of infrastructure in growth areas might be recouped through taxes on increased land values. . . . Examples of effective mechanisms for regional action exist in many places in the United States – regional service districts in Boston and Atlanta; urban development corporations in New York State; business improvement districts with bonding capacity, as in New York and Philadelphia; and tax increment districts found in many states. It remains to put these mechanisms together into an effective and coordinated capacity to pursue regional development. A further complexity is that many metropolitan regions cross state lines. But there are a number of models of bistate authorities that have been effective, including the Port Authority of New York and New Jersey, which has served as a capable advocate for economic development of the region as well as an operator of key transportation facilities.

MORE POWER AND RESPONSIBILITY NEED TO DEVOLVE TO LOCAL GOVERNMENTS

At the same time that greater regional capacity is needed, many of the functions operated at the scale of counties or big municipalities could benefit by devolution to smaller entities. Housing improvement programs could be more effective if carried out by neighborhood-based entities. Local zoning decisions are most effectively administered by bodies elected on a neighborhood level, who are sensitive to local needs and aspirations. Here the differentiation between areas of the city that are of regional importance (large commercial and office districts, industrial areas, ports, and other regional facilities, and so on) from local areas is critical. Current charter proposals in Los Angeles advocate devolution of many zoning functions to local councils. Many municipal services are also better administered on a small area basis, as New York's creation of community service districts – which correspond to community planning boards – has demonstrated. Charter schools offer an important alternative to the monopoly of school boards in public education. . . .

PLANNING IS A NECESSITY FOR SUCCESSFUL CITIES AND REGIONS

If the past century of development leaves any lesson, it is that the detailed relationships, even design, of urban areas ultimately affects their desirability and attractiveness. Suburban clusters fall far short of their potentials precisely because they lack the planning for ways that pedestrians move from place to place, have a poor mix of facilities, waste land for underused parking areas, and lack access alternatives. It is not accidental that many of the most desirable urban areas are cities with active planning efforts. Improving the parts of the city that have declined or have evolved in haphazard ways will require more, not less, planning. Locality and region are the dual necessities of every urban resident. The next century will require a policy framework that recognizes these as critical aspects of modern life and that creates an accommodation between them.

[. . .]

PART SEVEN

Emerging Issues in Urban and Regional Planning

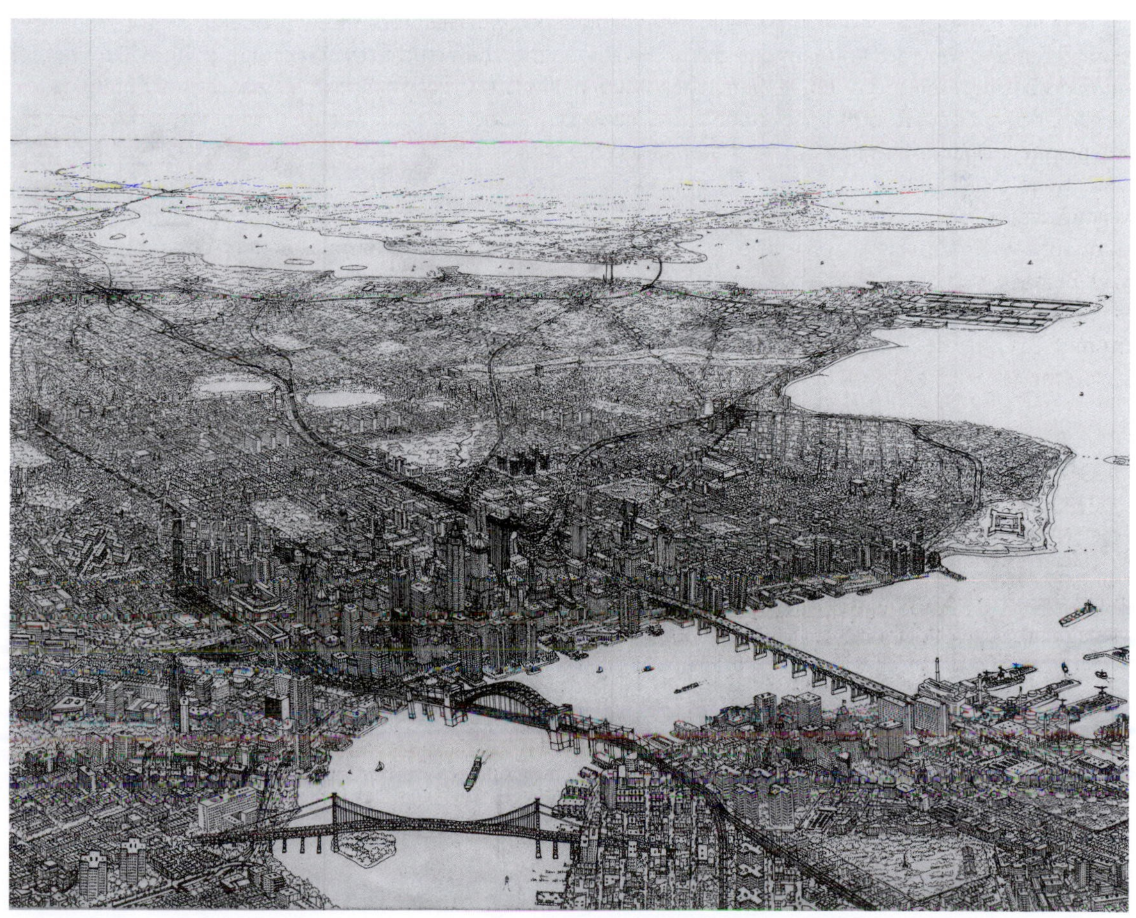

Growth of a typical Northeastern city: 2000. (Drawing by Andrew Whittemore)

INTRODUCTION TO PART SEVEN

Looking ahead to the future is something urban planners must do. Although focused on the immediate issues of neighborhood, city, and region, planners also pay attention to the broad global and domestic projections that provide context for their work.

Planners glean ideas about the future from many sources. For international trends, they look to non-governmental organizations like the Millennium Project of the World Federation of UN Associations, UN-HABITAT, Cities Alliance and the World Bank. Through its *State of the Future* report, issued annually since 1992, the Millennium Project provides a worldwide outlook – a futures index that monitors general well-being and lists ever-changing global challenges. In 2007 the report led with this observation: "People around the world are becoming healthier, wealthier, better educated, more peaceful, and increasingly connected and they are living longer, but at the same time, the world is more corrupt, congested, warmer and increasingly dangerous." It then highlighted rising urbanization and the associated demands for shelter, education, water, transportation, and security in urban places. It also underlined today's green-house gas threat, noting that "China's surpassing the United States in CO_2 has put global climate change among the top issues of the world." (Jerome C. Glenn and Theodore J. Gordon, *2007 State of the Future*, www.millennium-project.org/millennium/sof2007.html.) The 2008 list of global challenges includes a call to support sustainable development, balance population growth with natural resource consumption, deal with disease and crime, and accelerate scientific and technological breakthroughs to improve the human condition. Published in print and online, circulated through youtube.com (www.youtube.com/view_play_list?p=2C7D2B78000F1C2D), and illustrated here, the Millennium Project outlines world conditions broadly.

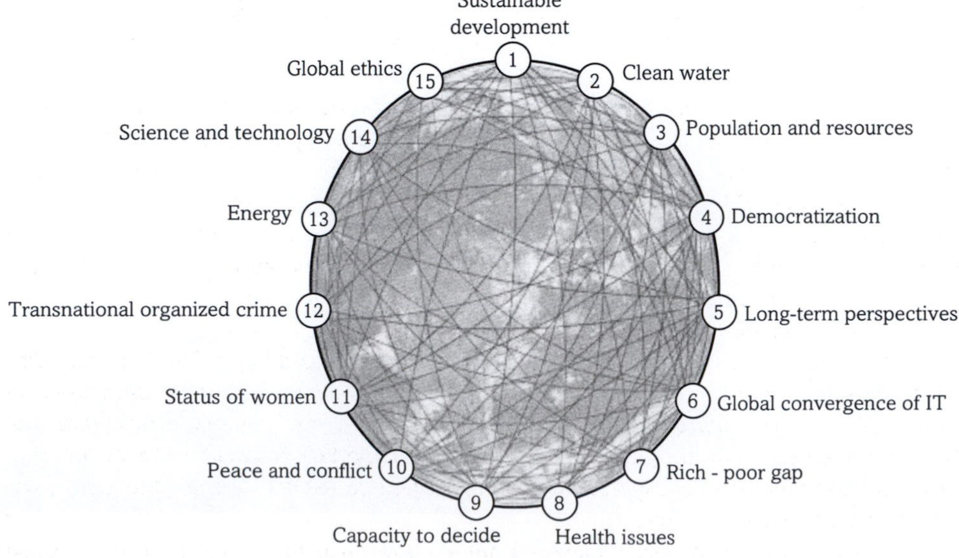

Fifteen challenges facing humanity. (The Millennium Project of WFUNA, www.millennium-project.org)

For domestic perspectives, planners turn to thinktanks like the Brookings Institution's Metropolitan Policy Program, America 2050, and the Urban Institute. Both the Brookings Institution (through its *Blueprint for American Prosperity*, www.brookings.edu/PROJECTS/BLUEPRINT.ASPX) and America 2050 (through its megaregion efforts, www.america2050.org/) have embarked on campaigns to educate the public (and urban planners) about U.S. demographic and economic growth projections. And, closer to home, professional organizations' research, publications, and practice assessments put meat on the bones of all of these groups' predictions.

For planning-specific information, they rely on professional associations like the American Planning Association (APA) and the Urban Land Institute. For example, in 2007 the American Planning Association's publications (e.g. *Planning Magazine*, *Journal of the American Planning Association*, *Planning Advisory Service Reports*) covered congestion pricing, walkable neighborhoods, innovative GIS and zoning, climate change, infrastructure needs, public–private partnerships, and anticipated future land consumption.

What does all this information mean to planners today? Part Seven, dedicated to emerging issues in urban and regional planning, focuses on three areas: (1) differential urban growth in the global South and global North; (2) climate change and urban adaptation; (3) new angles on traditional issues, including revitalizing older industrial cities, enhancing disaster preparedness, managing large infrastructure projects, and understanding the drivers of the internal spatial organization and shape of cities and suburbs.

The first selection, "The Promise of Urban Growth," comes from *Unleashing the Potential of Urban Growth* (2007), the annual state-of-the-world-population report issued by the United Nations Population Fund (UNPFA). Highlighting an important benchmark achieved in 2008 – half of the earth's inhabitants now live in cities – it argues that urbanization is on a rapid and unstoppable trajectory and that the majority of future growth will occur in the global South, especially in Asia and Africa. Do not waste time trying to halt urbanization, the report advises. Focus instead on selecting, preparing, and servicing suitable land for the newcomers.

Robert E. Lang and Dawn Dhavale, in "Beyond Megalopolis: Exploring America's New Metropolitan Geography," are among the scholars who are studying the location of population growth in the United States, one of the few countries in the global North that anticipates increases. The authors, as well as other researchers, believe it will occur in a handful of existing regions, places they variously label "megapolitans" or "megaregions." Lang and Dhavale further assert that the U.S. Census Bureau should recognize this phenomenon with a new designation, Megapolitan Area. Whether or not such a designation comes to pass, recognizing the importance of these larger regions and the land use arrangements peculiar to each – many low-density megas in the South are nearly built-out, while those of the Midwest and Northeast still have room to grow – can contribute to more meaningful regional planning.

In dealing with future growth, urban planners will need to take into account global warming and rising oil prices and their effect on settlement patterns. In "Land Use and Urban Form: Planning Compact Cities", Timothy Beatley offers examples of greenhouse gas-reducing designs from Holland, Finland, Denmark, Ireland, and Germany. The designs feature compact development, mixed land uses, street connectedness, increased density, and new forms of energy (e.g. solar, wind) – all qualities that, according to Beatley, are adaptable to the United States.

To move the Beatley argument forward, John R. Nolon provides a legal roadmap in "Shifting Ground: the Land Use Law Solution." It is not rocket science, he asserts. Just apply and adapt traditional planning tools: zoning ordinances, building codes, comprehensive plans, urban renewal rules, urban design guidelines, and municipal policy regarding public buildings.

Planning for resilience is a related concern. Richard A. Matthew and Bryan McDonald's "Cities under Siege, Urban Planning, and the Threat of Infectious Diseases" offers disaster management principles for urban places. Their proposals – which apply to both natural and manmade disasters (hurricanes, floods, terrorist attacks) – call on planners to provide transportation schemes to accommodate fast evacuation and comprehensive, easily accessible information infrastructure systems (pipes, wires, etc.) to enhance preparedness and recovery.

Turning to urban revitalization policy, planners are already trying to repurpose former industrial sites and will continue to do so as the economy continues to globalize/go global, thus making more places obsolete. Miguel Barcelo, in "22@Barcelona: A New District for the Creative Economy," outlines Barcelona's use of new activities and innovation to stimulate the local economy. First came a vision plan, a call to transform a well located but low-performing manufacturing district into a live work community for knowledge workers. Next came implementation. Guided by a special urban development agency, the city offered financial incentives to attract businesses in the technology, biomedical research, and media sectors; zoning code revision to accommodate a necessary mix of dense land uses; and city capital budget allocations to provide amenities and services.

Paying for infrastructure to support new development demands resourceful planning, according to Alan A. Altshuler and David F. Luberoff, authors of "The Changing Politics of Urban Mega-projects." They argue that today's planners, like doctors, have to engage in "do no harm" planning, a process that requires strong political support, risk taking, entrepreneurial public leaders, neighborhood mitigation plans, and creative financing packages. To ensure success, the authors also recommend high levels of government cost sharing, strong environmental regulations, and voter referenda to garner local approval.

Finally, Alain Bertaud in "The Spatial Organization of Cities: Deliberate Outcome or Unforeseen Consequence?" suggests that political leaders, working with planners, can manage the conditions discussed above (rapid urbanization, climate change, urban revitalization, execution of large infrastructure projects) if they set clear objectives, are patient, and blend market forces and zoning, investment in transportation, and other planning tools. He argues that knowledge about how daily travel, land consumption, and density choices shape urban places is fundamental to such an approach.

40

The Promise of Urban Growth

From *State of World Population, 2007*

United Nations Population Fund

Editor's introduction

Syndicated columnist Neal Pierce has labeled the twenty-first century "the century of the city," and indeed it is. Today more than half the world's population is urban (3.2 billion people), and within two decades 60 percent will be (5 billion). Put simply, cities are increasing in population by about a million people per week. And the global South, or developing countries, already having the largest number of urbanites, will supply the majority of this growth (73 percent), with their share of the world's urban population rising from 71 percent in 2008 to 79 percent by 2030. Notably, the global South, with its much lower proportion of urbanites (48 percent), has a much higher growth potential than the global North because the developed countries are already highly urbanized (76 percent). Furthermore, by mid century, more than three-quarters of the world's cities will be in Asia, Africa, and Latin America; Asia alone will have 51 percent of the total.

Overall, different-sized cities have varied growth rates, raising planning issues of differing scales and types. For example, within the next decade, small cities (under a half million) will capture more than half (52 percent) of the urban population, but their number will grow slowly (9 percent), going from 455 to 494 in 2015. Larger cities (1 million to 10 million) will proliferate faster, jumping from 394 to 499, a 25 percent increase. Providing municipal services in any city is complicated, especially in a high-growth scenario, but servicing a city of under a half million is far different from servicing one of 10 million. All in all, planning for cities with populations of a half million to 10 million is of critical importance because they will constitute more than 90 percent of all cities. Interestingly, the massive agglomerations (cities of 10 million and up) that have attracted substantial attention in the past currently represent only 2 percent of the total number of cities and house a small proportion (4 percent) of the world's population, figures that will remain unchanged in the near future.

It's important to understand not only the rapid pace and the locales of urban growth, but also the source of growth. Natural increase (more births, fewer deaths), rather than rural-to-urban migration, has become the most important driver of urban growth everywhere except China. It follows that a disproportionate share (60 percent) of urbanites will be under age eighteen by 2030, putting extreme pressure on educational systems and, later, employment. (Low levels of educational attainment and slow job creation have contributed to today's 20 percent poverty rates in the global South.) In addition, as time passes, the share of elderly will also rise, bringing additional stress on cities' services. Finally, an anticipated increase in households will affect housing markets, especially in the global South. Due to current inefficiencies, about a billion urbanites in developing countries live in squatter or informal settlements today.

The global South and global North have many urban problems in common, though variations in wealth, urbanization rates, and settlement patterns affect the extent and scale of these problems. Both regions are faced with questions of unequal distribution of income, affordable housing, and infrastructure. They have environmental tribulations, including pollution of various sorts, especially land degradation. They are confronting the wider threat of global warming, seeking ways to foster ecologically sound development, searching for sustainable or less energy-consuming practices for urban living.

At the most fundamental level, the differences between the countries of the global South and those of the global North derive from their urbanization histories. With a shorter record of urbanization and a smaller proportion of total population urbanized, countries of the global South potentially have more options in shaping their settlement patterns in the future. Yet they must deal with major problems in managing growth, often due to unresolved questions of land tenure, land registry systems, and weak or nonexistent regulatory and financial tools. At the same time, countries of the global North, well advanced in urbanization, face their own challenges. Their shifting economic bases and demographic profiles have placed new demands on their cities and metropolitan areas. While these countries tend to have more resources than those of the global South to deal with these conditions, they often lack the political will to address them. In many instances, their government structures are too fragmented to cope with modern urban issues, resulting in conflicting social, political, environmental, and economic needs among older cities/suburbs and outlying areas.

Further, the world's four large geographic areas – Asia, Africa, Latin America/Caribbean, and Europe/North America – have variations in growth rates and population distribution, issues to be considered in urban planning. A look at Asia and Africa, and their current and projected urban population distributions, illustrates key variations in the source, scale, pace, and direction of metropolitan development.

Asia's urban population is concentrated in four countries – China, India, Indonesia, and Pakistan – that together contain more than two-thirds of the total. Add Japan and the percentage rises to three-quarters of the total Asian urban population. In the next twenty-five years these proportions will hold steady as Pakistan, India, Indonesia, and China experience urban population growth rates ranging from 64 percent to 122 percent. Approximately 42 percent of Asian urbanites live in cities of a million and above, and in the coming decade Asia will contain three of the world's five largest urban agglomerations (Mumbai, Dhaka, Delhi) and will have thirteen cities of 10 million or more, twenty-three of 5 million to 10 million, and 244 of 1 million to 5 million.

The spatial arrangements of Asia's urban places often relate to economic development. In industrialized countries, urban development is characterized by spread-out growth outside of formal city boundaries, along infrastructure routes at increasing distances from the core. In its most exaggerated form, this pattern becomes a mega-urban region, exemplified by metropolitan agglomerations around Tokyo, Seoul, Bangkok, Jakarta, Hong Kong, and Manila. In less industrialized cities in Asia – including Delhi, Dhaka, and Mumbai – urban growth, fueled by poor, rural inmigrants and immigrants, is also spread out but less directed by the formal economy. In all cases the public and private sectors have been unable to meet the demand for affordable housing and basic municipal services. As a result, pockets of affluence exist beside sections of abject poverty.

In contrast, Africa's three most urbanized countries – Nigeria, Egypt, and the Democratic Republic of the Congo – currently account for one-third of the continent's total urban population, and with projected growth rates ranging from 81 percent to 222 percent they will continue to do so. However, Africa's urban population is far more dispersed than Asia's. Africa has far fewer large cities than Asia does. This pattern will persist, as demographers estimate that by 2020 Africa will have only two cities of 10 million and up, two of 5 million to 10 million, and fifty-eight of 1 million to 5 million. Nonetheless, these large cities, of which the largest is Lagos, suffer from many of the same problems as those of the less industrialized countries of Asia.

In the "century of the city" planners from the global North and global South have a clear agenda to deal with growth, address poverty, and promote sustainable development. They, of course, are incorporating the Millennium Development Goals, discussed in Selection 1, into their work. They are sharing transferable best practices among fast-urbanizing places, developing strong research capabilities to supply missing knowledge and data helpful to new cities, and training urban planners and other professionals to deal with future urban challenges. Above all, they are beginning to craft innovative solutions appropriate to the level of development and governance of a given place.

In the global South the primary concerns are securing appropriate land that is not vulnerable to natural and human-made disaster; serving settlements with basic infrastructure, including water, sanitation, and streets;

providing adequate shelter, education, healthcare, and transportation. In contrast, in the global North, while the land development issues are similar with regard to avoiding environmentally vulnerable places, the infrastructure program focuses not on basic provision but on upgrading and repairing existing facilities, and shelter and service questions revolve around equitable delivery systems.

This selection, drawn from the United Nations Population Fund's *State of World Population 2007* report, calls for a major attitude shift regarding cities. It argues that cities are positive, not negative, forces in economic development. It also refutes the notion that cities take up too much land. (Today they occupy 2–3 percent of ice-free land, and with anticipated urban growth the number will rise to only about 6 percent). Finally, it provides an important conceptual framework to tie policy to regional development patterns.

The report's lead author, George Martine, is a consultant to the United Nations Population Fund (UNFPA); president of the Brazilian Association of Population Studies; and scientific director, Population, Development and Environment Research Project, Committee for International Cooperation in National Research in Demography (CICRED). The UNFPA, founded in 1969, has a strong focus on the health of women, adolescents, and HIV/AIDs patients. With a budget of $605 million in 2006, the agency had programs in 154 countries promoting reproductive health, gender equality, and poverty alleviation. Its *State of World Population* report, published annually, is authoritative in demographic matters.

For a general introduction to twenty-first-century population trends, see Joel E. Cohen's prize-winning book *How Many People can the Earth Support?* (New York: Norton, 1995). For an assessment of broad policy issues related to world metropolitan development, see Peter Hall and Ulrich Pfeiffer, *Urban Future 21: A Global Agenda for Twenty-first Century Cities* (London: Taylor & Francis, 2000) and Allen J. Scott (ed.), *Global City Regions: Trends, Theory Policy* (New York: Oxford University Press, 2001). Robert Neuwirth in *Shadow Cities: A Billion Squatters, a New Urban World* (New York: Routledge, 2005) portrays the human side of squatter settlements. Shlomo Angel, Stephen C. Sheppard, and David L. Civco's *The Dynamics of Global Urban Expansion* (Washington, DC: World Bank, 2005) employs remote sensing to track current and future land consumption patterns in ninety cities. Aprodicio A. Laguian, Vinod Tewari, and Lisa M. Hanley (eds), *The Inclusive City: Infrastructure and Public Services for the Urban Poor in Asia* (Baltimore, MD: Johns Hopkins University Press, 2007) and Douglas Webster, Jianming Cai, and Chuthatip Maneepong, *Metropolitan Governance in China: Priorities for Action in the Context of Chinese Urban Dynamics and International Experience* (Washington, DC: World Bank, 2006) review service provision and governance issues in Asia. Editors Brian Roberts and Trevor Kanaley's *Urbanization and Sustainability in Asia: Case Studies of Good Practice* (Philippines: Asian Development Bank, 2006) covers other important countries, including Indonesia, Bangladesh, Pakistan, and the Philippines. Christine Kessides, *The Urban Transition in sub-Saharan Africa: Implications for Economic Growth and Poverty Reduction* (Washington, DC: Cities Alliance, 2006) reviews urbanization in the forty-two sub-Saharan countries, including Nigeria, Democratic Republic of the Congo, and Kenya.

THIS ICEBERG IS GROWING

"The growth of cities will be the single largest influence on development in the twenty-first century." These were the opening words of the UNFPA's [United Nations Fund for Population Analysis] 1996 *State of the World Population* Report. This statement is proving more accurate today. Until now humankind has lived and worked primarily in rural areas. But the world is about to leave its rural past behind: By 2008, for the first time, more than half of the globe's population, 3.3 billion people, will be living in towns and cities. The number and proportion of urban dwellers will continue to rise quickly. Urban population will grow to 4.9 billion by 2030. In comparison, the world's rural population is expected to *decrease* by some 28 million between 2005 and 2030. At the global level, *all* future population growth will thus be in towns and cities. Most of this growth will be in developing countries. The urban population of Africa and Asia is expected to double between 2000 and 2030. It will also continue to expand, but more slowly, in Latin America and the Caribbean. Meanwhile, the urban population of the developed world is expected to grow relatively little: from 870 million to 1.01 billion.

This vast urban expansion in developing countries has global implications. Cities are already the locus of nearly all major economic, social, demographic and environmental transformations. What happens in the cities of the less developed world in the coming years will shape prospects for global economic growth, poverty alleviation, population stabilization, environmental sustainability and, ultimately, the exercise of human rights.

[. . .]

An outstanding feature of urban population growth in the twenty-first century is that it will be composed, to a large extent, on *poor* people. Poor people often fall through the cracks of urban planning; migrants are rejected or simply ignored in the vain hope of deterring further migration. Realistic planning for future urban growth calls for explicit consideration of the needs of the poor. It also requires gender analysis: The particular needs and capabilities of poor men and boys. And, as population structures change, attention to youth and the needs of the elderly will become ever more important.

[. . .]

BASING POLICIES ON FACTS, NOT BIASES

Policy makers have understandably been much concerned with the speed and magnitude of urban growth. Many would prefer slower growth or none at all; slower growth would theoretically give them more flexibility to deal with urban problems. Generally, they attempt to slow growth by restricting migration, but . . . this rarely works.

Moreover, such efforts reflect a poor understanding of the demographic roots of urban growth. Most people think that the migration is the dominant factor; in fact, the main cause today is generally natural increase. Reclassification of formerly "rural" areas and residents as "urban" also contributes to urban growth.

In developing countries, city growth during the "second wave" . . . is being driven by higher rates of natural increase than in Europe and North America at the height of their urbanization processes.

The latest comprehensive research efforts to separate natural increase from the other components of urban growth put the contribution of natural increase at about 60 percent in the median country. The remaining part of urban growth – roughly 40 percent – is a combination of migration and reclassification.

As time passes and countries become more urban, the proportion of urban growth attributable to natural increase invariably rises. That is, the higher the level of urbanization in a country, the smaller the pool of urbanites contributing to natural increase.

Of course, country experiences vary a good deal. In India, recent assessments of components of urban growth 1961–2001 found that the share of growth attributable to urban natural increase ranged from 51 percent to about 65 percent over the period. Some 65 percent of current growth in Latin America stems from natural increase, despite steep declines in fertility rates, especially in urban areas. China, where migration has recently predominated, is unusual.

Given the importance of natural increase and the failure of antimigration policies, it seems obvious that fertility decline is much more likely than migration controls to reduce the rate of urban growth. Since high fertility in rural areas often underlies rural–urban migration, lower fertility in both rural and urban areas can decelerate urban growth. Such a reduction would give policy makers more time to prepare for the expansion of the urban population.

Policies that aim to slow urban growth should therefore shift their attention to the positive factors that affect fertility decline – social development, investments in health and education, the empowerment of women and better access to reproductive health services. On reflection, it is surprising how rarely this agenda has influenced policy decisions, as opposed to an antimigration approach. . . .

[. . .]

WRONG WAY STREETS AND NEW AVENUES

To meet the needs of burgeoning urban populations, stimulate both urban and rural development and achieve the Millennium Development Goals

(MDGs), planners and policy makers should reconsider their bias against urban growth. It is ineffective and counterproductive. Moreover, it stands in the way of initiatives to reduce poverty.

There is clear evidence that urbanization can play a positive role in social and economic development. Historically, the statistical association between urbanization and economic growth has been strong. . . . Cities are the main site of economic growth in most countries and account for a disproportionately high share of national economic production. . . . Proximity and concentration give cities the advantage in the production of goods and services by reducing costs, supporting innovation and fostering synergies among different economic sectors. For instance, cities can provide much cheaper access to basic infrastructure and services to their entire populations. As a result, urban poverty rates are, overall, lower than those in rural areas; the transfer of population from rural to urban areas actually helps to reduce national poverty rates. . . .

TRYING TO KEEP THE MASSES OUT: A FAILED STRATEGY

National governments have tried two strategies to restrict the rapid expansion of urban settlements of the poor: (a) ambitious schemes to retain people in rural areas or to colonize new agricultural areas and (b) regulating urban land use, backed up either by evictions or, more frequently, by denying essential services such as water and sanitation. . . .

ADDRESSING THE SHELTER NEEDS OF THE POOR

Once policy makers accept the inevitability of urban growth, they are in position to meet the needs of the poor. One of the most critical issues is shelter. As UN-Habitat has made eminently clear over the years, the main difficulties faced by the urban poor are linked, to a greater or lesser extent, to the quality, location and security of housing. Overcrowding, inadequate infrastructure and services, insecurity of tenure, risks from natural and human-made hazards, exclusion from the exercise of citizenship and distance from employment and income-earning opportunities are all linked together. Shelter is at the core of urban poverty: Much can be done to improve the lives of people through better policies in this area. . . .

A QUANTUM LEAP: MEETING A NEW SCENARIO FOR SHELTER

How can national and international institution help to create a livable urban future for the masses of urban poor . . . ? Here it is necessary to distinguish between approaches aimed at meeting the needs of the urban poor *currently* living in cities and those aimed at relieving the pressures caused by large *future* growth. . . .

Planning for *future* rapid expansion of shelter needs in towns and cities, while at the same time addressing the accumulated demand of the past, calls for a critical change in the approach of municipal and national governments. They will have to mobilize their technical and political resources for, rather than against, the land, housing and service requirements of the urban poor. . . .

One strategy would be to focus on providing *access to serviced land* for the growing millions. Hard realism must permeate this vision. Governments of rapidly urbanizing countries are simply unable to provide housing and desirable urban services for most of their current urban poor. They will hardly be able to cater to the needs of a rapidly growing number of additional urbanites. It is even more unrealistic to imagine that these New Urbanites will be able to compete successfully in what are sure to be aggressive real estate markets.

Under these conditions, providing minimally serviced land goes to the heart of the matter. The object would be to offer poor people a piece of land accessible by wheeled transport (from buses to bicycles) with easily made connections to, at least, water, sanitation, waste disposal and electricity. The first lodging will often be a simple shack, made of whatever scraps are available. But it will probably improve: The history of informal settlements reaches us that, if poor people feel secure about their tenure, and have reasonable access to livelihoods and services, they will improve their dwellings over time. . . .

Providing poor people with minimally serviced land is not an easy solution: Given the voracity of the economic interests involved, the murkiness of titles in many developing cities and the uncanny ability of informal land markets to turn a profit by exploiting the poor, dealing in land use is always fraught with difficulties. . . .

REGULATING URBAN LAND MARKETS: MISSION IMPOSSIBLE?

The main *technical* difficulties involved in providing land for the urban poor concern (a) locating and acquiring enough buildable land, (b) devising sustainable ways of financing its transfer to the poor and (c) regulating the functioning of land markets.

An alleged shortage of land has been a main obstacle to more effective housing for the poor. The need to safeguard environmental and agricultural land from chaotic urban expansion is a genuine concern. However, most cities still have buildable land in good locations, but it is owned or controlled by private interests or by state agencies with no interest in socially directed uses of the land. The real shortage is thus not of land, but of serviced land at affordable prices.

Meeting the land needs of the poor is easier in the context of well regulated land and housing markets. Not only do effective markets make more land available to the poor but they also favour economic growth.

Lack of good regulation actually increases poverty: metre for metre, people in informal settlements pay more for land and services than people in wealthier residential areas. Unregulated markets also make it difficult for governmental bodies to collect property taxes or reduce land speculation and to build up resources on this basis for socially oriented planning of land use.

Financing socially oriented housing has always been difficult, but there is no shortage of innovative proposals, once past the hurdle of antiurban bias. Given regularized land markets, the support of local governments, NGOs and international funding agencies could be marshaled towards a more proactive approach.

International and multilateral agencies could make a difference. New rules for the UN system, promulgated by the Secretary General in August 2006, will enable the United Nations to address this structural shortcoming and provide more effective support on affordable housing finance. That support will include pro-poor mortgage financing systems, now being tested in the field as an alternative to conventional social housing policies. Particular attention will have to be given to the gender constraints that exist in formal credit channels precluding women from tapping into this market. Access to microfinance has proven to buttress women's empowerment and helps to reduce urban poverty.

ADVOCACY, VOTES AND ACTION: THE NEED FOR LEADERSHIP

These initiatives call for a new awareness and an unprecedented level of political support at the local and national level. Most politicians are, at the best of times, unwilling to confront the power of the urban real estate market. The added complexity of attending to the land needs of the poor as described above, is even less enticing. A critical initiative, without which most efforts will fail, is to regulate increments in land value. In other words, it is necessary to introduce fiscal measure that prevent speculators and developers from hiking up the price of land and services unreasonably as soon as socially motivated land allocation is proposed. . . .

[. . .]

Planning for the land needs of the poor is only one aspect of the broader issue of land use, which will become more urgent as the urban population grows. The aim should be to minimize the urban footprint by regulating and orienting expansion before it happens. The interactions between urban growth and sustainability will be particularly critical for humankind's future. Cities influence global environmental change and will be increasingly affected by it. This calls for a proactive approach, aimed at preventing environmental degradation and reducing the environmental vulnerability of the poor. It is particularly critical in developing countries, whose urban population will soon double, and in low-elevation coastal zones.

Given the prospects of urban growth, only proactive approaches to inevitable urban growth are likely to be effective. Minimizing the negative and enhancing the positive in urbanization requires both vision and permanent concern for poverty reduction, gender equality and equity and environmental sustainability. . . . Policy steps to help reduce the social and environmental costs of urban expansion include:

- Orienting future expansion . . .
- Generating early-warning indicators . . .
- Planning infrastructure and housing policies . . .
- Identifying populations at risk . . .

PREPARING FOR URBAN TRANSITION: A LAST WORD

The anti-urban policies common in the developing world during the last quarter-century misapprehended both the challenges and opportunities of urban growth. Urban poverty is unquestionably an important and growing problem in many developing countries. Environmental problems are increasingly clustered in urban sites. Yet to blame cities for poverty and environmental problems is to miss the point. Dispersing or deconcentrating population and economic activities would not bring relief – even if it were possible.

For humankind to benefit from the urban transition, its leaders must first accept it as both inevitable and important for development. They must recognize the right of the poor to what the city has to offer and the city's potential to benefit from what the poor have to bring.

Rather than attempting in vain to prevent urban expansion, planners must objectively examine the available policy options for addressing it and building on its possibilities. Urban improvement and slum upgrading draw a lot of attention from city governments and urban planners. Such action is necessary, but it is not enough: Cities must look urgently to the future.

[. . .]

The space taken up by urban localities is increasing faster than the urban population itself. Between 2000 and 2030, the world's urban population is expected to increase by 72 percent, while the built-up areas of cities of 100,000 people or more could increase by 175 percent. The land area occupied by cities is not in itself large, considering that it contains half of the world's population. Recent estimates, based on satellite imagery, indicate that all urban sites (including green as well as built-up areas) cover only 2.8 percent of the Earth's land area. However, most urban sites are critical parcels of land. Their increased rate of expansion, and where and how additional land is incorporated into the urban makeup, has significant social and environmental implications for future populations. From a social standpoint, as shown [earlier], providing for the land and shelter needs of poor men and women promotes human rights. It is critical for poverty alleviation, sustainable livelihoods and the reduction of gender inequalities. . . .

The territorial expansion of cities will also effect environmental outcomes. The conventional wisdom has been that the expansion of urban space is detrimental in itself. Since cities are situated at the heart of rich agricultural areas or other lands rich in biodiversity, the extension of the urban perimeter evidently cuts further into available productive land and encroaches upon important ecosystems. At the same time, however, there is increasing realization that urban settlements are actually necessary for sustainability. The size of the land area appropriated for urban use is less important than the way cities expand: Global urban expansion takes up much less land than activities that produce resources for consumption such as food, building materials or mining. . . .

The conclusion that using land for cities is potentially more efficient only heightens the need for careful and forward-looking policies. . . . It calls for a vision based on solid analysis, and encompassing a broader notion of "space" than the one imposed by political and administrative city limits. It also demands a longer time horizon than the terms of politicians and administrators.

DENSITY, URBAN SPRAWL AND THE USE OF LAND

. . . The built-up areas of cities with 100,000 or more presently occupy a total of about 400,000 sq. km, half of it in the developing world. Cities in the developing countries have many more people but occupy less space per inhabitant. In both the developing and industrialized countries, the average density of cities has been declining quickly: at an annual rate of 1.7 percent over the last decade in developing countries and 2.2 percent in the industrialized countries. In developing countries cities of 100,000 or more are expected to *triple* their built-up land area to 600,000 sq. km in the first three decades of this century. Cities in the developed world expand at an ever faster rate per resident despite their smaller population size and lower rate of population growth. They will increase their built-up land area by two and a half times between 2000 and 2030. At that point, they will occupy some 500,000 sq. km. . . .

Recent trends to lower densities may accelerate as globalization has its effect on lifestyles and production processes. Whatever the case, the data show that developing countries now share the trend to urban sprawl. Urban sprawl results from the combination of different types of pressures on territorial expansion. For purposes of simplicity, these can be classified into two groups: residential suburbanization and peri-urbanization.

THE DISCREET CHARM OF SUBURBIA

The modern trend to urban sprawl began in North America after World War II, where . . . the intensive use of the automobile for daily commuting was both a case and consequence of urban sprawl. This patter of settlement spawned new locations for trade and services and this, in turn, further promoted automobile use and outward city growth. . . . Suburbanization appears to be more complex in developing countries. Given their pervasive poverty and inequality, the culture of the automobile and its far-reaching impact on urban civilization arrived later and continue to be restricted to a minority. At the same time, the relative precariousness of public transportation and infrastructure has prevented wealthier people from moving to the suburbs in large numbers and commuting easily from there. . . . Some extension of the American pattern of settlement to outlying areas of cities has been observed recently in most low- and middle-income countries. More affluent suburbs are increasingly found in most cities. In short, the globalization of markets and consumption patterns is leading to the reproduction of urban settlement patterns in the mould of the American dream. Nevertheless, the suburbanization of the affluent is insufficient to explain the growing trend to urban sprawl, especially in developing countries, We must look for additional explanations.

SPRAWL AND PERI-URBANIZATION

The growth of cities in the developing world is dynamic, diverse and disordered – and increasingly space-intensive. The process of urban growth, largely in noncontiguous transitional zones between the countryside and city, is increasingly being referred to as "peri-urbanization." Peri-urban areas often lack clear regulations and administrative authority over land use. They suffer some of the worst consequences of urban growth, including pollution, rapid social change, poverty, land use changes and degradation of natural resources. But, as opposed to suburbia, they are home to a variety of economic activities.

Peri-urbanization is fuelled, in part, by land speculation, nurtured by the prospect of rapid urban growth. Speculators hold on to land in and around the city, expecting land values to increase. They do not bother renting, especially if they fear that users might gain some right to continued use or controlled rents. People who need land for residential or productive purposes must therefore find land further from the centre. Changes in the structure and location of economic activity contribute greatly to peri-urban growth. Better communications and transportation networks make outlying areas increasingly accessible. Globalization encourages economies of scale in production and distribution, which, in turn, encourage large facilities occupying large tracts of land. This deconcentration and decentralization of production is often found on the outskirts of the more dynamic cities, where growing workplaces and workforces can no longer find space in city centres, making spillover growth inevitable. In turn, the periphery offers cheaper infrastructure, land and labour, which encourage further peri-urbanization.

Peri-urban areas often provide more accessible housing for poor residents and migrants in informal and scattered settlements. Poor settlements in such areas tend to be more insecure and subject to removal, while their residents generally lack services and infrastructure. They compete with agriculture for space and both can be displaced by other economic uses. Land conversion, market opportunities and rapid flows of labour, goods, capital and wastes force land prices up. Peri-urbanization also increases the cost of living for the original rural population.

Peri-urban areas encompass a wide range of activities, including farming, husbandry and cottage industries, together with industrial expansion, land speculation, residential suburbanization and waste disposal. They fulfil other key functions for urban areas, from the supply of food, energy, water, building materials and other essential to the provision of ecological services such as wildlife corridors, microclimates and buffer areas against flooding. This involves a complex readjustment of social and ecological systems as they become absorbed into the urban economy.

Since peri-urban areas are generally beyond or between legal and administrative boundaries of central cities, the capacity of government authorities to regulate occupation is particularly weak. As a result, the process of urbanization can be, to a great extent, unplanned, informal and illegal, with frequent struggles over land use. Environmental degradation is also an issue in peri-urban areas. Specific health hazards arise when agricultural and industrial activities are mingled with residential use. Some peri-urban areas become sinks for urban liquid, solid and, sometimes, airborne wastes. The type, impact and gravity of such problems vary considerably. . . .

To sprawl or not to sprawl?

There is much debate among experts over the advantages of compact versus decentralized cities, but no consensus. Disagreement arises over the varied sources of sprawl, methodological issues and conflicts in values. . . . Few planners defend sprawl, but some question whether intensifying use can deliver a more sustainable urban future. They also question whether dense occupation is acceptable to the general public. . . . Much of the discussion, whether it accepts or bemoans urban sprawl, assumes that the dispersed city is how people want to live – but this may simply reflect the bias of the discussants, who are mostly from developed countries. Dispersed suburban settlement seems simply unrealistic for urban masses in developing countries. . . . Whatever the . . . difficulties, the "green dimension" should have full consideration in this debate . . . The discussion often neglects to notice that sprawl is increasingly attributable to peri-urbanization and to the mobility of economic activity, especially in developing countries. . . .

[. . .]

REALISTIC POLICIES FOR URBAN EXPANSION

[. . .]

The inevitable growth of developing-country cities and their peri-urban surroundings demands a coordinated and proactive approach. Within the overall framework, there must be a new set of realistic, equitable and enforceable regulatory regimes. In this process, care should be taken not to disturb sensitive lands and watersheds, Provisions for land, infrastructure and services for the poor should be a key concern. The local population should be involved in any discussion of future growth in order to guarantee people's rights while increasing the success rate of planning efforts. . . .

Sorting out the land issues in future urban growth is but one aspect of the question, though an important one. A broader political and spatial approach, within a longer time frame, is also required to deal with other sustainability and organizational issues. Sprawl and peri-urbanization tend to fragment urban space in unpredictable ways, producing nuclei of different sizes and densities, with a variety of common or unique problems. The solution lies not so much in prescribing relative density of urban area as in good local governance that can guide urban development, and yield appropriate densities. In the current situation, fragmentation of the urban territory brings both administrative inefficiency and environmental setbacks. The boundaries of the city's administration rarely coincide with its actual area of influence. In the case of larger cities, this area usually extends over neighbouring subregions, which may include smaller cities as well as peri-urban and rural areas. . . .

The key question, therefore, is who will take the initiative in an urban world marked by these growth processes? The suggestion made here is to approach the organization and regulation of spatial process that affect social and environmental well-being from a regional rather than a strictly urban perspective. This concept of "city-regions" is useful in this new social, economic and political order. It provides an easily understandable starting point in advocating for a more coordinated and effective approach to dealing with the growing problems of sprawling urban and peri-urban areas, and on behalf of the poor as an essential and dynamic element in urban development.

41

Beyond Megalopolis: Exploring America's New "Megapolitan" Geography

Metropolitan Institute Census Report Series

Robert E. Lang and Dawn Dhavale

Editor's introduction

In the United States and throughout the world, regions are the fundamental units in national and global economies. Regional planners usually focus on improving large systems in the environmental (e.g. water, open space), mobility (highways, rail), and economic (labor flows, concentrated poverty) arenas.

Regions result from a complex interplay of the forces of concentration and dispersal that shape development patterns. This is especially true in countries with market systems, like the United States, where land transactions are primarily lightly regulated private sector activities. A region forms as individual firms and households follow their needs in deciding where to locate. For them, political boundaries are less critical in determining where they are than the ability to access the labor, land and capital, and markets to conduct business or earn a living and raise families here.

Advances in communications and transportation have allowed some firms and households, taking advantage of advances in communication and transportation, can disperse, while other firms require close proximity to special facilities or social networks. Businesses that have routine and repetitive operations have different locational requirements than those that demand fast-turnaround production or advice from highly skilled professionals or face-to-face contacts. For example, a 200 person tax preparation firm that services only wealthy customers does not need a large office in a high-rent district of a major city – it could set up shop in less expensive suburbs or even in a foreign country, as long as it has a central meeting space for its top salesmen that is proximate to clients. A highly trained computer specialist, in contrast, will want to live and work in a place like Silicon Valley that values his or her skills.

While measuring the agglomeration, or clustering, of firms and households has led some economists to delineate regions loosely, others have more precise definitions. In the United States, for example, the U.S. Bureau of the Census defines a number of regional statistical units. It divides the United States into four large geographic areas (Northeast, Midwest, South, and West) and further breaks those areas into twelve subregions. The Northeast has two divisions: New England (with the six states from Maine to Connecticut) and Middle Atlantic (consisting of New York, New Jersey, and Pennsylvania). So does the West, with its Pacific (five states from Alaska to California) and Mountain (eight states from Arizona to Montana). The Bureau also has much smaller regional designations – the Metropolitan Statistical Area, the Micropolitan Statistical Area, and the Combined Statistical Area – that encompass a principal city and its surrounding counties.

Many state and federal government agencies use these designations as the basis of their operations. For example, the nation's 341 Metropolitan Planning Organizations (MPOs), groups who develop twenty-year regional transportation plans as a condition of receiving federal highway and transit funds, employ Census definitions for their service areas. (A typical MPO encompasses a principal city and contiguous counties.) While the U.S. Department of Environmental Protection Agency has ten administrative regions that follow state lines, it also supervises the nation's watersheds, whose ecological boundaries provide another way to define a region.

Other regional organizations designate areas to serve their institutional needs. For example, the Regional Plan Association (RPA), a private, non profit planning entity, attends to a tri-state region involving parts of New York, New Jersey, and Connecticut that fall into a large commuter shed surrounding New York City. The Twin Cities Metropolitan Council's urban services area is the geography for which it supplies sewers. Private corporations have their own regional definitions, based on their market areas.

Area definitions like the RPA district have roots in early twentieth-century theories that conceived of regions as monocentric entities focused on a central city and its environs. Today the regional concept has evolved into a polycentric one, reflecting the current decentralization of population and employment and incorporating multiple centers within a region. Even the RPA, which once centered its work on New York City, now has plans for subcenters in Connecticut, New Jersey, and New York State.

Regionalism is not unique to the United States, though; it pervades contemporary urban policy and thinking globally. For example, Europeans have identified their strongest economic unit as the "Blue Banana" (extending from London, through Germany, to Milan), and are outlining others (the Sunbelt, from Milan to Valencia; and the Yellow Banana, from Paris to Warsaw) for planning purposes. The European Community (EC) recognizes the importance of regions through its Committee on Regions, consulted by the central authority on issues of employment, education, health, and social and economic cohesion. The British national government has had longstanding regional planning strategies commencing with its postwar New Towns programs and extending to the present with its efforts in the South East region (SERPLAN) and the Thames Gateway plan, both designed to relieve pressure on London.

According to the United Nations, about 300 global city-regions exist worldwide, twenty of which have populations of 10 million or more. British urban planner Sir Peter Hall first identified special global city-regions (then called "world cities") in the mid twentieth century and completed a major research project for the European Union that has added more detail to the concept. The Globalization and World Cities Research Network (GaWc), an Anglo-American group, has energized recent research in this area, describing the complex and intricate interrelationships of such regions, demonstrating that their boundaries are not strictly drawn but determined by time/distance practicalities. The growing consensus about the form of global city-regions is that they are large, polycentric areas that include a traditional downtown, a new center to accommodate modern office buildings, a variety of economic clusters at their edges for back office and other activities not needing central locations, and specialized subcenters for such functions as education, entertainment, sports, and conventions.

American planning scholars and practitioners are building on this knowledge, applying the analysis to the United States. Inspired also by French geographer Jean Gottmann's seminal study of the Northeast, *Megalopolis: The Urbanized Northeastern Seaboard of the United States* (Cambridge, MA: MIT Press, 1964), they have identified ten large regions anticipated to capture the majority of the nation's population growth in the next five decades. Because the field is still evolving, such regions have been referred to with various names.

The authors of this selection use the term megapolitan region, or megaregion. And they advocate for the inclusion of this designation in the U.S. Bureau of the Census metropolitan definitions. In their essay they explore the implications of employing megaregions in crafting public policy around two critical themes: transportation and sprawl control.

The authors, Robert E. Lang and Dawn Dhavale, are associated with the Metropolitan Institute of Virginia Tech, where Lang is Director and Dhavale is a research assistant. Lang has written extensively on demographic matters in *Edgeless Cities: Exploring the Elusive Metropolis* (Washington, DC: Brookings Institution Press, 2003) and *Boomburbs: The Rise of America's Accidental Cities* (Washington, DC: Brookings Institution Press, 2007) with Jennifer B. LeFurgy. He has coedited three volumes of the Brookings Institution's *Redefining Urban and Suburban America: Evidence from Census 2000* (Washington, DC: Brookings Institution Press, 2003, 2005, 2006) and recently updated *Edgeless Cities* in "Beyond Edgeless Cities: Office

Geography in the New Metropolis" (Washington, DC: National Association of Realtors, 2006) with Thomas Sanchez and Jennifer LeFurgy.

To read more about spatial economics and the formation of regions, see Masahisa Fujita and Jacques-François Thisse, *Economics of Agglomeration: Cities, Industrial Location, and Regional Growth* (Cambridge: Cambridge University Press, 2002), especially Part I, "Agglomeration and Economic Theory," which provides an overview of the most current thinking.

Recent publications on global city-regions include Roger Simmonds and Gary Hack (eds), *Global City Regions: Their Emerging Forms* (London: Spon, 2000); Michael Storper, *The Regional World: Territorial Development in a Global Economy* (New York: Guilford Press, 1997); Allen J. Scott (ed.), *Global City-Regions: Trends, Theory, Policy* (Oxford: Oxford University Press, 2001), with an excellent essay by Sir Peter Hall, "Global City-Regions in the Twenty-first Century"; Andreas Faludi (ed.), *European Spatial Planning* (Cambridge, MA: Lincoln Institute of Land Policy, 2002); and Peter J. Taylor, *World City Network: A Global Urban Analysis* (London and New York: Routledge, 2003).

To learn more about U.S. efforts in advancing megapolitan research and planning, see the America 2050 website, www.america2050.org, which has posted several studies and articles on the topic, including Armando Carbonell and Robert D. Yaro's "American Spatial Development and the New Megalopolis," in *Land Lines*, April 2005, and Armando Carbonell, Pria Hidisyan, Mark Pisano, and Robert Yaro, *Global Gateway Regions*, September 2005. Note that America 2050 defines large-scale regions in a different way than the authors of this selection do.

A significant majority of Americans live in [ten U.S. Megapolitan Areas], which by 2003 accounted for more than two in three U.S. residents. The top fifteen most populous metropolitan areas are found in Megas. In addition, thirty-nine of the nation's fifty most populous metros lie within the ten Megapolitan Areas, as do seventy-two of its 120 CSAs. . . . [S]ix Megapolitan Areas appear in the east while four more are found in the west. Despite capturing over two-thirds of the populations, US Megas make up less than 20 percent of the land area in the lower forty-eight states.

Megapolitan Areas extend into thirty-five states, including every one east of the Mississippi River except Vermont. Four states – Connecticut, Delaware, New Jersey, and Rhode Island – are completely Megapolitan. Eight states have parts of two Megapolitan Areas within their borders: Alabama, California, Florida, Nevada, Pennsylvania, Texas, Virginia, and West Virginia. Only two Megapolitan Areas lie completely in one state: Florida's Peninsula and Arizona's Valley of the Sun. California and Nevada link up via two Megapolitan Areas – NorCal (from San Francisco, CA, to Reno, NV), and the Southland (from Los Angeles, CA, to Las Vegas, NV). Oregon and Washington share the Cascadia Megapolitan Area. The I-35 Corridor reaches into four states, while the Gulf Coast, Piedmont, and Midwest Megas include parts of five, six, and seven states respectively. The Northeast Megapolitan Area spreads from the Mid-Atlantic to New England, touching a dozen states and the District of Columbia. The region is thus represented by almost a quarter of the U.S. Senate – or twenty-four Senators.

. . . [K]ey interstate highways link major metros within Megapolitan Areas. Interstate 95 plays a big role in Megapolitan mobility from Maine to Florida. Because of the big population centers in the Northeast and Peninsula, the number of people living within fifty miles on either side of this interstate exceeds all others in the nation. The west's bookend to I-95 is I-5, which runs through three separate Megapolitan Areas. . . . [I]n 2000, over 64 million people lived within just fifty miles of I-95, and over 37 million lived within the same distance of I-5. Most of this population is found in the two Megapolitans along I-95 and the three straddling I-5. Interstate 10 also links three Megas – the Southland, Valley of the Sun, and Gulf Coast. Other places where key interstates help define Megapolitan growth include the I-35 Corridor, which goes from Kansas City, MO, to San Antonio, TX, and I-85 in the Piedmont linking Atlanta, GA, to Raleigh, NC.

CURRENT POPULATION AND GROWTH RATES

. . . As a group, Megapolitans outpaced the national growth rate for the first three years of the decade. The United States gained 3.3 percent more people from 2000 to 2003, while the Megas added 3.9 percent. The Megapolitan Areas gained 7.5 million new residents over the period. Just two Megapolitan Areas trailed the nation as a whole in growth. The Gulf Coast grew faster than NorCal since 2000, and by 2003 surpassed it by about 40,000 people.

The Northeast and Midwest are by far the most populous Megas, with more than 50 [million] and 40 million residents by 2003. Together, at 90.5 million people, they surpass Germany, the biggest European Union nation, with 82.5 million residents in 2004. And unlike Germany, the Northeast and Midwest are still growing, albeit slower than the other Megapolitan areas. The Midwest and Northeast form the old core industrial heart of the nation . . . and still represent the largest-scale transmetropolitan development in the United States, even with their relative population decline compared to the Sunbelt.

The fastest growing Megapolitan Areas lie in the Sunbelt. Several of them experienced gains above 5 percent for the period 2000 to 2003. The fast-growth Megas, ranked by their development pace, are: Valley of the Sun, Peninsula, I-35 Corridor, Southland, and Piedmont. Two Megapolitans fall below the 10 million resident mark, but, based on an extrapolation of current growth rates, Cascadia will pass this size in 2025, while the booming Valley of the Sun gets there by 2029.

Just thirty-five years ago, the Valley of the Sun was a modestly settled region, but rapid growth has moved the area within striking distance of Megapolitan status. A similarly fast expanding region is the Front Range of the Rockies in Colorado. The Front Range extends from Fort Collins, CO, in the north to Colorado Springs, CO, in the south. Denver's sprawling metropolitan area dominates this region. Front Range cities are linked by I-25, which helps shape it into a corridor. The Front Range was not included in this analysis because it is projected to have just 7 million residents by 2040. However, feedback on this study may prompt a reconsideration of methods and result in the Front Range being added as the eleventh Megapolitan Area.

LAND AREA

Megapolitan Areas vary by size. . . . The Midwest is the largest with 119,822 square miles, or an area slightly smaller than the state of New Mexico. The Piedmont is almost equally expansive with 91,093 square miles. The more populous Northeast, by contrast, comes in at just 70,062 square miles. By this calculation, the Northeast would appear to be the densest Megapolitan Area. However, the square mileage figure for the Southland is significantly distorted by the inclusion of Riverside and San Bernardino Counties in California. . . . Big western counties may feature vast open space. This is certainly true in the case of Riverside and San Bernardino, the latter of which is the physically largest county in the United States.

[. . .]

POLICY IMPLICATIONS AND IMPACT

Any new geographic category can reshape public policy. Given that the Megapolitan Areas label as proposed here redefines the space where two in three Americans reside, its impact could prove significant. There are countless ways that Megas may alter the policy landscape. . . . [T]wo of these issues [are] urban sprawl and transportation planning. . . .

MEGAPOLITAN SPRAWL

. . . The emergence of Megapolitan Areas comes not just from rapid population growth over the past several decades – it also reflects how the nation is developing. Since 1950, the most significant urban pattern has been decentralization. Even by the time Gottmann first observed the megalopolis, the emergence of the "spread city" (a term coined by the Regional Plan Association of New York in the 1960s) was apparent. Suburbs from Boston to Washington were racing toward one another. When they joined, which many have, they would make the Northeast a single extended Megapolitan space.

What began with the original Northeast megalopolis now extends to nine other places. The combination of fast growth and massive decentralization transformed once distant cities into galaxies and corridors of linked urban space. Ever-expanding exurbs will continue to strengthen and add to these linkages. Thus the physical connectivity that provides a basis for Megapolitan Areas should strengthen over the next few decades.

Not only are Megapolitans one consequence of sprawl, the different ways these regions develop also provide insight into how urban decentralization varies around the nation. More importantly, this knowledge can improve the way regions respond to the consequences of sprawl. . . . [U]rbanization patterns vary considerably and produce distinct regional metropolitan built forms. . . . [S]prawl, as measured by built density, differs in character by region. . . . [A] "dense sprawl" in places such as Los Angeles, where even the edge of the region may have subdivisions with small lots[,] . . . contrast[s] [with] the edges of southern metropolitan areas [that] feature lower density development and constitute a quasi-rural environment. . . . The edge of Megapolitan development in the Southland is sharp and well defined, as indicated by the very small share of people living in the nonurbanized fringe. Conversely, the Piedmont edge is amorphous given that one in three people live outside its urbanized areas.

Even the intensely built Northeast . . . has over 5.2 million residents living in places with less than 1,000 people per square mile. The Piedmont has just over 6 million in these same places, while the Midwest has almost 6.7 million. Nationally, nearly 25.8 million Megapolitan residents live in low-density non-urbanized areas, with the bulk coming from the Northeast, Piedmont, and Midwest – or east of the Mississippi.

From [this] analysis, it appears that there is a Southland versus Piedmont style of sprawl. Knowing this, the Megapolitan Areas could develop regionwide strategies for addressing sprawl. For example, given that the region is already densely built, altering the pattern of Southland-style sprawl could mean better mixing land uses to facilitate pedestrian or transit-oriented development. But the same strategy would not work in the Piedmont where densities are low.

Some Megapolitans also face the prospect of "build-out," or the point at which large-scale greenfield opportunities disappear. Large sections of the Southland and peninsula are near build-out. The Los Angeles and Miami metropolitan areas in particular face this prospect. For example, less than 10 percent of developable land remains in Broward County, Florida (home to Fort Lauderdale) because of environmental concern for the Everglades. In much of the Southland and the Peninsula, sprawl has simply nowhere to go. These two Megapolitan Areas will have to shift their growth models if they are to accommodate even a portion of their projected population gains.

THE NEW MPOS: MEGAPOLITAN PLANNING ORGANIZATIONS

If officially designated, Megapolitan Areas would be the U.S. Census Bureau's largest geographic unit. Their rise could spark a discussion of what type of planning needs to be done on this scale. In Europe, Megapolitan-like spatial planning now guides new infrastructure investment such as high-speed trains between networked cities. The United States should do the same. The interstate highways that run through Megapolitan Areas, such as I-95 from Boston, MA, to Washington, DC; I-35 from San Antonio, TX, to Kansas City, MO; and I-85 from Raleigh, NC, to Atlanta, GA, would greatly benefit from unified planning. A new Census Bureau Megapolitan definition would legitimize large-scale transportation planning and trigger similar efforts in such areas as economic development and environmental impact.

Federal transportation aid could be tied to Megapolitan planning much the way it has recently been linked to metropolitan areas. The Intermodal Surface Transit Efficiency Act of 1991 required regions to form Metropolitan Planning Organizations (MPOs) in order to receive federal money for transportation projects. . . . In a similar vein, new super-MPOs could result from future legislation that directs Megapolitan Areas to plan on a vast scale. It is then that tangible impacts would begin to occur. For example, an analysis of traffic flow along the Northeast Megapolitan Area from Boston to Washington would show that a key pressure point is the Delaware tolls on the Delaware Turnpike. The state uses these tolls to raise revenue, but the resulting traffic on busy weekends and holidays creates a major inefficiency in the Northeast Megapolitan Area's transportation system. Also, the resulting pollution lowers the air quality in places such as southern New Jersey. The super-MPO that would be charged with transportation planning throughout the Northeast Megapolitan Area may be in a position to alter how these tolls are collected, including having the toll booths positioned in just one direction.

At the moment, there is no guiding vision of how to invest the nation's transportation funds. We are now only keepers of past visions, most notably the Interstate Highway System. The Interstates – good or bad – at least demonstrated national will for investment. They also completed a national project, begun in the nineteenth century with canals and rail, to provide equal access and capacity across a continental nation. The investment paid off, as witnessed by the emergence of Sunbelt boomtowns such as Phoenix. But the next stage of American spatial evolution is at hand. The United States has moved beyond the simple filling in of its land and is now witnessing Metropolitan growth. Likewise infrastructure investment must move beyond the basic links between the entire United States to focus on significantly improving the capacity within its Megapolitan Areas.

[. . .]

42

Land Use and Urban Form: Planning Compact Cities

From *Green Urbanism: Learning from European Cities* (2000)

Timothy Beatley

Editor's introduction

Three trends are pushing issues of sustainable development to the forefront of American planning. The trends are: global warming, the United States' overwhelmingly urban population (nearly 80 percent), and land consumption rates in this country that are higher than population growth. As a result, planners are developing practices that promote sustainability in metropolitan areas.

In the context of planning, sustainability encompasses many activities. They include supporting regional ecosystems and designing multipurpose, varied-sized open spaces meeting needs for respite, recreation, aquifer protection, stormwater management, flood control, and urban agriculture. (See "Environmental Planning" in this *Reader*.) Other activities that promote sustainability are improving the functioning of municipal infrastructure; quantifying the financial value of greening in public capital investment decisions; and instituting zoning, subdivision, and building codes that mandate energy-efficient and resource-conserving land uses and structures. Planners who engage in such practices aim to create the green city.

What is a green city? First, it is an ideal – yet to be attained by any urban place in the world but certainly achievable in the twenty-first century. According to economist Matthew E. Kahn, a green city is a healthy place that has "clean air and water, pleasant streets and parks." It is "resilient in the face of natural disasters and faces little risk of infectious disease. Its residents have strong, green behavioral habits, like taking public transit, practicing recycling and water conservation, using renewable energy." But defining the green city also requires distinguishing between places that already exist and those yet to be built. Existing places have their basic form while new ones, obviously, do not. Existing places will have to adapt and mitigate non-green conditions, while new ones can start fresh. Older industrial cities have extensive park systems, but they also may have massive brownfields, or obsolete combined sanitary and storm sewers. They also may encompass acres of vacant and abandoned land. Newer, rapidly growing, spread-out cities have little parkland, rely heavily on automobiles, and accommodate growth through greenfield conversion under zoning ordinances that legislate low-density, large-lot sites. Increasingly, both types of cities are implementing infill (or fill-in) strategies to increase densities and use infrastructure more efficiently. Some, for example, are investing in mass transit and encouraging transit-oriented development (TOD). All together, they are inching towards creating compact urban forms. (See Plate II-8.)

Whether old or new, green cities have a common feature: they take full advantage of their natural environments. They use the sun and wind for power, have markets for locally grown food, recycle their garbage, support water conservation and mass transit. They reduce or minimize their imprint on the land. They employ planning expertise, modern technology, and common sense to meet these ends. Some places – including

Portland, Oregon, Denver, Seattle, Oakland, and Minneapolis – are already more advanced than others in their march towards being green.

To some extent, these leading cities have benefited from their lower levels of industrialization and relatively late development histories. But national environmental legislation has played an even bigger part. A generation ago, the federal government launched initiatives to clean up polluted air, water, and lands. Through environmental legislation, the nation established minimum standards for air and water. Subsequent laws focused on transportation, sewage treatment, and brownfields, offering substantial funding for specified initiatives. Driven by concerns about environmental quality (not broader greening issues), this legislation has achieved major milestones in cleaning up industrial pollution and thus has contributed to growing greener cities.

Regrettably, though, while the federal government has tinkered with the laws over time, it has never conceived of them holistically. For example, federal transportation legislation, in calling for compliance with federal air quality minima, has successfully forced localities to consider the environmental implications of their mobility investments. However, the monitored pollutants include six emissions (nitrogen dioxide, sulfur dioxide, lead, carbon monoxide, particulates, and ozone) but not carbon dioxide (CO_2), a major contributor to global warming. The twentieth-century clean air vision was one that dealt with polluting agents of the past but not those of concern today, as planners deal with climate change threats.

With its piecemeal environmental strategy and its support of other population-dispersing, sprawl-inducing policies (including highway construction in outlying areas and income tax relief for home ownership), the federal government has had mixed effects on metropolitan areas. The air and water are definitively cleaner, but populations are consuming more land than ever. Finally, the nation's failure to endorse the Kyoto Protocol (1997) has left a gap that might have been filled with more enlightened policies.

In the absence of comprehensive federal policy on greening, cities (and states) have begun to take it up, largely driven by concerns about global warming. Since 2005, under the leadership of Seattle mayor Greg Nickels and working through the U.S. Conference of Mayors, more than 700 mayors have signed the U.S. Mayors' Climate Protection Agreement. This pact sets a 7 percent reduction in 1990 CO_2 levels by 2012, a higher target than that of the Kyoto Protocol. To reach the goal, mayors are focusing on transportation, land use, building codes, and municipal energy consumption. Chicago, the self-declared greenest city in America, has promoted many measures, including green roofs, installing plants on the top of buildings (it has approximately 200 encompassing about 2.5 million sq. ft, some supported by a specially legislated tax increment financing district in the Loop and others through city grants), tree planting (400,000 saplings in the past decade, with approximately 5,000 in sixty miles of new, vegetated street medians), and bike lanes (about 250 miles). New York City issued a greening plan called PlaNYC 2030, pledging a 30 percent CO_2 reduction by 2030, but failed to secure the state legislature's approval for a key component, congestion pricing (charging a fee for driving in a designated high-traffic areas of the city). Nonetheless, it is undertaking major rezoning to concentrate new development proximate to transit.

Among the more notable efforts in urban greening has been the institution of green building standards under the Leadership in Energy and Environmental Design (LEED) Green Building rating system initiated in 1998 by the U.S. Green Building Council, the Congress for the New Urbanism, and the Natural Resources Defense Council. To date, the U.S. Green Building Council lists approximately 6,000 projects worldwide as having registered for the LEED ratings. The group has added a new program, LEED-ND, to its rating system. Piloted in 2007–08, LEED-ND rewards dense, compact development, awarding points for location (near existing development and infrastructure or infill), walkability (wide sidewalks with shade trees and other interesting features), mixed use (proximate location of housing, stores, and workplaces), multimodalism (access to public transit, bike trails), and water and energy conservation (use of landscape features to minimize runoff, collect rainwater, and recycle gray water; installation of solar panels or other renewables). The LEED measures are important tools for achieving greener cities in the United States because an anticipated 100 million more people in the next thirty years will stimulate an enormous building boom. By 2025, analysts predict, half of the built environment will be new and will include the construction or reconstruction of 50 million housing units and 100 billion sq. ft of nonresidential building.

The United States has much to learn about urban greening from other parts of the world. Singapore is a model of greening, with an integrated land use and transportation strategy starting with its Long Range Concept Plan (1972) that ringed the central business district with New Towns linked by mass transit. Successive revisions (1991 and 2001) concentrated development in four high-density regional centers served by public transit and aimed at transforming large swaths of heavy industrial land into mixed-use, high-density communities while doubling and upgrading open space and addressing the jobs–housing balance.

In Europe, exemplary cities are compact, have multimodal transportation systems as well as extensive open space, and practice energy conservation. England, for example, has a tradition dating from the end of World War II of containing its largest cities with green belts. Its Docklands reclamation project transformed an eight-and-a-half-square-mile brownfield area into a large mixed-use green district. More recently, London has mandated global warming prevention measures in the mayor's Climate Change Action Plan (2007) addendum to its comprehensive plan. Copenhagen and other European cities have encouraged biking and walking by providing an extensive network of routes reaching from downtown to the outskirts and instituting traffic calming devices such as narrowed streets to guarantee pedestrian and biker safety. The Danes issued the Copenhagen Agenda for Sustainable Cities to reinforce best practices. This selection reviews several European examples of compact development.

Its author, Timothy Beatley, is Theresa Heinz Professor of Sustainable Communities, Department of Urban and Environmental Planning, University of Virginia. Beatley has written extensively on environmental issues, most recently *Native to Nowhere: Sustaining Home and Community in a Global Age* (Washington, DC: Island Press, 2004). His other works on environmental planning include *The Sustainable Urban Development Reader* (New York and London: Routledge, 2004) with Stephen M. Wheeler and *The Ecology of Place: Planning for Environment, Economy, and Community* (Washington, DC: Island Press, 1997) with Kristy Manning. He helped develop the Copenhagen Agenda for Sustainable Cities.

For more on the physical issues of urban greening, see Donna Erickson's *Metro Green: Connecting Open Space in North American Cities* (Washington, DC: Island Press, 2006), which provides several case studies, maps, and plans about the development of open space systems in the United States and Canada. *Visualizing Density* (Cambridge, MA: Lincoln Institute of Land Policy, 2007) by Julie Campoli and Alex S. MacLean is an illustrated compendium of development at various densities, showing with great detail different forms of compact and spread-out urban patterns. Economist Matthew E. Kahn in *Green Cities: Urban Growth and the Environment* (Washington, DC: Brookings Institution, 2006) explores urban form, economic development, and environmentalism, asking how growth affects cities' prospects for becoming more–or less–green. Kent E. Portney delves into the relationship of economic development and greening in *Taking Sustainable Cities Seriously: Economic Development, the Environment, and Quality of Life in American Cities* (Cambridge, MA: MIT Press, 2003). For a look at sustainable design, see Nicholas Low, Brendan Gleeson, Ray Green and Darko Radovic's *The Green City: Sustainable Homes, Sustainable Suburbs* (Abingdon: Routledge and Sydney: University of New South Wales Press, 2005), LEED-ND initiative chair Douglas Farr's *Sustainable Urbanism: Urban Design with Nature* (Hoboken, NJ: Wiley, 2007), and landscape architect Diana Balmori's *Land and Natural Development (LAND) Code: Guidelines for Sustainable Land Development* (Hoboken, NJ: Wiley, 2006).

To even the most casual observers visiting Europe from the United States, there are clear and immediate differences between the basic land use patterns and spatial form of American and European cities. Historically, European cities have been fundamentally more compact, with a distinct separation between urban and rural. The cities are walkable, have good public transit, and are generally much less reliant on the automobile. For these reasons these cities are important sources of both practical planning and policy guidance, and sheer inspiration, for American cities and regions struggling to control urban sprawl.

[. . .]

While American cities consume land and growth spatially at a much faster rate than population growth, European cities . . . have generally tended to grow more compactly, at higher densities, and with greater emphasis on the redevelopment and reuse of land within existing urbanized areas. And, even in small countries with relatively large populations such as the Netherlands, the percentage of land devoted to cities and developed areas is a modest 13 percent – this, in a country with one of the highest population densities in the world. In a less dense country such as Sweden, only about 2 percent of that nation's land base is reported to be in built-up or urban uses. In Sweden, expansion of urban areas has closely tracked urban population growth. Between 1960 and 1990, for example, population in Swedish cities grew by 31 percent, and the extent of urban or built-up areas grew by a larger but still modest 47 percent. Despite some degree of urban spread, European cities have been largely able to maintain their compactness and density.

[. . .]

European cities also illustrate that density and compactness are not antithetical to economic productivity but, on the contrary, may actually enhance it. Some of the densest, most compact cities are the wealthiest and most economically productive in the world. [Researchers] have shown this by calculating per capita regional domestic product for the cities in a worldwide database. The results indicate that per capita Gross Regional Product certainly need not be hindered by density, and may actually be enhanced by it.

[. . .]

STRATEGIES FOR COMPACT URBAN DEVELOPMENT

Although compact urban form has been a priority for centuries, few countries have embraced the virtues of urban compactness in current planning as much as the Dutch. The Netherlands has implemented an explicit national compact cities policy since the mid 1980s. The vast majority of new development in the country occurs in designated development areas (so-called VINEX sites, an abbreviation for Vierde Nota over de Ruimtelijke Ordening Extra, or the Fourth Memorandum on Town Planning Extra), which are generally within or contiguous to existing cities and urban areas. These development areas must meet a minimum overall density target of thirty-three dwelling units per hectare. This system has been reasonably effective at promoting compact development patterns, although problems have been experienced in recent years (e.g., land speculation) and some believe that the density levels must be much higher. Compact growth, especially in the Randstad, is further supported by national transportation and locational policies. . . .

The compact city is achieved through careful use of each piece of land. Densities within cities are several times those of United States and Canadian cities. Residential areas are usually planned with multifamily units or modest townhouses. Private conspicuous consumption of space is rare. Despite recent policies to bring more market forces into the Dutch housing system, and also more upscale housing, the national planning authorities are careful to maintain the policies for higher density. Most notably, thanks to a successful battle by Dutch planners, there are few land-gobbling shopping centers or office projects on the suburban fringe, and efficient parking is planned. Visitors today enjoy the picture-book Green Heart area, with its spick-and-span stewardship of every corner, complete with grazing cows, meticulous canals, and dikes. Not the little fingers of the legendary boy, but a concerted planning assault has held back the streams of urban sprawl.

[. . .]

In many . . . cities . . . , the promotion of compact development within each city structure or comprehensive plan is given a clear priority. Stockholm's new structure plan, for example, takes sustainability as a main theme and endorses a compact regional spatial pattern. A number of former industrial sites are identified as the primary areas for future growth in the city. . . . Generally, this planned development pattern locates new growth in close proximity to existing transportation modes. These sites are to be developed not as new suburbs, but as mixed urban villages, with shopping, light industry, and residences, all within relatively close proximity to Stockholm's inner city. Similarly, the city of Helsinki has prepared a set of "Sustainable Development Principles for City Planning," and the city's 1992 Master Plan also takes a strong sustainability orientation. Among other things, the plan identifies areas of future development that are "almost without exception, located within the urban structure, supplementing, consolidating, or renovating existing building sites for new uses."

[. . .]

Many . . . cities . . . have followed a strategy of urban form that allows large blocks of open space or green wedges to come very close to urban neighborhoods. In Helsinki, large tentacles of green space penetrate into the very center of the city, providing ecological corridors and connections with surrounding countryside. One of the largest of these green tentacles is Keskuspuisto central park – an 11 km-long unbroken green wedge that ends at the periphery of the city in old growth. In Copenhagen, high-density development is clustered along transit lines (like "pearls along a necklace"), with large wedges of green space between them.

Amsterdam has taken a similar form, beginning with its General Extension Plan, approved in 1935. A number of large public parks date to this period, most notably the Amsterdamse Bos, which constitutes one of the most important of the green wedges. These green wedges, separating the major new *lobes* of urban development in the city, also connect with and lead into the Randstad's Green Heart, a large area of farms and open space in the middle of this urban agglomeration.

[. . .]

European cities also generally exhibit a much higher level of mixing and integration of functions. While separate Euclidean-style residential districts do exist, it is much more common to see an interspersing of retail shopping, grocery stores, and even industrial activities in close proximity to where people live. Urban residents are frequently within easy walking distance of a wide variety of services, shops, and community functions. These characteristics, along with the compact nature of the cities, generally means that residents are able to get to many more places and do more things by foot. In the Netherlands, some 35 percent of all trips nationally under 2.5 km are made on foot, with another 40 percent by bicycle. Compact urban form, from the regional level to the design of new districts, makes walkable lifestyles possible.

COMPACT GROWTH DISTRICTS, WITH A GREEN EMPHASIS

There are many examples of local efforts to design and build new housing areas that continue these patterns. Indeed, the major new growth areas in almost every city studied are situated in locations adjacent to existing developed areas and are generally designed at relatively high densities. Good examples of this can be seen in the new growth areas planned in Utrecht (Leidsche Rijn), Freiburg (Rieselfeld), Amsterdam (Ijburg), Copenhagen (Orestad), Helsinki (Viikki), and Stockholm (Hammarby Sjostad). Ijburg, in Amsterdam, for example, is being planned to accommodate a housing density of about 100 dwelling units per hectare.

. . . There are few better examples of Dutch attempts to promote compact development than the new growth district Leidsche Rijn, in Utrecht. An overall master plan and detailed building plans have been prepared for the area, and construction of the first phase of the project has just begun. The plans are ambitious, and this extension of Utrecht will accommodate some 75 percent of the growth in the region for the foreseeable future. In total, the district will accommodate 30,000 houses, as well as 30,000 jobs. The district covers an area of about 2,500 ha, with an overall development density of about thirty-seven units per hectare.

The most impressive aspect of the new district involves the efforts to connect it to the existing city. The site is immediately to the west of existing Utrecht and takes in the preexisting villages of Vleuten and DeMeern. The plan for a compact, contiguous extension of the city has not been easy, and major physical barriers have had to be overcome, most notably the A-2 highway and a large shipping canal (the Amsterdam–Rhine Canal). The plan involves a creative roofing of the highway in several places for a total distance of about 300 m. An extensive area of greenhouses has had to be relocated as well.

The commitment to connectivity and building on to the existing structure and fabric of the city is truly impressive. In this pursuit, three new bridges are to be constructed, again with the goal of strengthening the connections between the old centrum and the new district. One of the bridges (the middle one) will be for bicycles only, while another will be reserved for public transit (buses, at least in the beginning) and bicycles. A third bridge will accommodate these modes, as well as automobiles.

This project also exemplifies an effort to promote a mixture and diversity of people and activities. Approximately 70 percent of the housing will be private sector, and 30 percent public sector; a range of housing types will be provided, including multifamily flats, single-family attached, and single-family detached. While the Utrecht centrum will remain the major shopping area for residents, grocery and other daily shopping will be available close to home. The inclusion of commercial and industrial uses also provides for local employment, and some of the homes will be provided with additional space for inhome offices.

Several innovative environmental features will be included. All homes will be connected to a district heating system (hot water piped to the district from the residual heat of a nearby power plant). Extensive use is to be made of natural stormwater collection, through the use of *wadis*, or green drainage swales to which water is directed. All homes will be provided with two different lines of water – one for clean water (from the city's water treatment facility) for drinking, and a second line supplying less clean water (from a major water transfer pipe running through the area) to be used for car washing, garden watering, and other non-drinking purposes.

A goal from the beginning is to discourage the use of automobiles. There will be three new stations for the train, as well as new bus service. One transferium (parking deck) will be provided to give automobile drivers the chance to shift to the bus. Higher-density housing and much of the employment will be clustered near the train stations. (It is not clear, however, that the automobile is sufficiently restricted here: the plan still calls for 1.2 spaces per unit.) The area is also designed to facilitate bicycle use. For example, the network includes extensive bike-only routes. The project designers have managed to keep most of the new development within 5 km of the centrum, and they believe this is the distance most people can be expected to ride their bicycles ("autonomic cycle lanes with a mesh-work from 500 m"). The project also contains extensive green infrastructure. The plan calls for a "central park" of about 300 ha, in the center of the district, with ecological connections between the district and surrounding areas. (The Randstad's Green Heart is immediately to the west.) The central park includes, among other things, sporting facilities and allotment gardens for residents.

URBAN VILLAGES AND THE IMPORTANCE OF CITY CENTERS

Although there are certainly examples of communities where the city center is declining in population and intensity of use, European cities on the whole . . . have been able to maintain and strengthen the vitality of these centers. In no small part, this is a function of density and compactness, but it is also the result of numerous efforts to maintain and enhance the quality and attractiveness of the city center.

In most . . . cities, the center area has remained a mixed-use zone, with a significant residential population. A common pattern is the location of residential units above retail shops and offices (as in most Dutch cities and towns, for example). Consequently, significant numbers of people still live there. In the city of Munster, some 12,000 residents live in the old center, a fairly sizable portion of the city's population. In Amsterdam, about 80,000 people live in the historic inner city, and the center remains "the economic heart of the entire agglomeration." Sixty-five thousand people live in the historic core of Copenhagen, and some 55,000 residents live in the historic center of Bologna. Odense, Denmark, has a healthy city core, and some 70,000 residents live within its city-center statistical zone. An extensive system of pedestrian streets and the presence of many amenities have made the center an attractive place to live near.

Many European cities are investing major resources in supporting new development in city-center locations. Examples of such projects on quite a massive scale can be seen in Berlin. . . . In Berlin, the largest new development site is the Potsdamer Platz, where 1.1 million sq. m of office and residential space are under construction. This new area will be served by major new transit investments – with a planned modal split of 20 percent private car and 80 percent public transit – and will be served with central heating and cooling.

SEVEN

[. . .]

Other cities have undertaken aggressive programs for improving the look, feel, and functioning of their city centers. The city of Groningen, for instance, has undertaken a host of actions under its *binnenstad beter* ("a better inner city") program. Described as a "cohesive set of measures to improve the center," the measures have included the creation of new pedestrian-only shopping areas (creating a system of two linked circles of pedestrian areas), the installation of yellow brick surfaces in walking areas, the installation of new street furniture, and the adoption of a new public transit concept. The public transit concept includes making the center car-free, providing free peripheral parking and direct bus service to the center, replacing street parking with a series of underground car parks on the edge of the center, and increasing the number of cycling paths. Together these measures are designed to improve the access to and attractiveness of the center.

These cities also present excellent, and generally successful, examples of redevelopment and adaptive reuse of older, deteriorated areas within the city center. Exemplary cases include Amsterdam's eastern port area (8,000 new homes in an abandoned port area) and Bologna's urban plan, which identifies a number of development sites, all within the existing urban area and including the reuse of a former cigarette factory site. In Dublin, the Temple Bar district has emerged as a major new cultural and commercial district in a once derelict part of the city. Through a combination of tax incentives and creative design ideas (e.g., the creation of a new public square, street resurfacing, and other pedestrian improvements), the district has become a lively and desirable location.

Some cities are attempting to further encourage the injection of residential units in city-center and commercial areas, as a way to both efficiently provide housing and enhance the vitality and livability of these city-center areas. Traditionally, British and Irish center cities have had fewer residents than continental European cities. The cities of Dublin and Leicester have operated (only moderately successful) programs that provide financial incentives to encourage people to relocate to city centers. One such effort is called the Living Over The Shop program, which seeks to take advantage of the traditional housing opportunities that exist above shops and retail establishments. While this is a common housing option in most Dutch and German cities, much less can be seen in cities such as Dublin. Several major obstacles include the cost of bringing buildings up to the building and fire codes, the fact that shop owners may not be able to effectively use the tax credits provided, and the perception by shop owners of the liabilities of having new residents living above them. Nevertheless, the living-over-the-shops notion is increasingly viewed as a promising strategy for promoting sustainable urban form as well as for providing affordable housing.

Compact urban form and efforts at infill and urban redevelopment go hand in hand in many of these cities. The city of Amsterdam has pursued its own compact cities policy since 1978, a clear reaction to the siphoning off of the city's population by the outlying growth center communities. As a result, the city has vigorously pursued developing and redeveloping close-in areas. One of the centerpieces of this policy has been the redevelopment of the eastern docklands area: formerly vibrant docks during the nineteenth century that were made obsolete by the shifting of activities to the city's western docks. The city's physical planning department prepared a master plan for the area that envisions development at about 100 units per hectare, a mixture of uses and activities (including a major new shopping area), and direct (bicycle, pedestrian, auto) connection to the city center. Eventually, the area will accommodate some 8,500 new dwelling units, about 1,500 of which have already been completed.

[. . .]

NEW TOWNS AND GROWTH CENTERS

One possible strategy for accommodating population growth in a compact form is the creation of New Towns and growth centers. Europe offers many positive examples of successfully designed and implemented New Towns, many of which approximate the ideal characteristic of sustainable communities. Sweden and the Netherlands offer the most successful guides.

The Dutch growth centers policy began in the mid 1960s and lasted until the mid 1980s. It was a realistic recognition of the trend of people moving out of city centers, and it reflected a desire to direct this outward growth to compact nodes. Under a strategy known as *concentrated deconcentration* (or clustered deconcentration) fifteen new growth centers were designated, most in the Randstad, and significant national government subsidies were provided to build them. Each growth center had a target population of between 55,000 and 80,000, and each involved sites ranging from small villages, to medium-sized cities, to entirely new towns. Despite certain perceived negative aspects (e.g., a higher per-unit cost), they have generally been successful and were built as intended. They can be generally characterized as compact and transit-oriented, and of mixed housing types. The most dramatic examples are Almere and Lelystad, which are New Towns shaped literally from the sea, built on a reclaimed polder. Their main limitations have been in failing to achieve economic self-sufficiency (there is high out-commuting in Almere, for instance, much of which is by car) and the effect that some of the growth centers have had in siphoning off population from older cities such as Amsterdam.

Nevertheless, Almere is an impressive design creating an urban form that largely approximates our image of an ideal sustainable place. The town's spatial form consists of a polynuclear structure, with three main centers: the first was Almere Haven, then Almere Stad and Almere Buiten. The central train station is at Almere Stad, which represents the main city center, where most regional entertainment, cultural, and shopping functions are located. A walking boulevard extends immediately from the central train station, with shops and restaurants on the ground level and apartments and multifamily housing above. The city's *stadhuis* (city hall) is located at the other end of the boulevard. Accessibility is a key feature of Almere. Perhaps the most interesting transportation mode is the city's fixed-route bus system. City buses run along bus-only routes and consequently provide very rapid service between different parts of Almere. A separate system of bike lanes, which generally pass under roads and the bus-only routes, makes it easy and comfortable to travel by bicycle. Few living areas are more than a few hundred meters away from a bus stop or a train station, and there are plans to increase the number of train stations from three to seven.

Overall density of the city is about thirty-five dwelling units per hectare, although density is higher in the center of Almere Stad. There are now new plans to intensify activities in the center of Almere Stad, and a new Stadscentrum 2005 plan was recently proposed by Rem Koolhaus and his Office of Metropolitan Architecture. Even now, Almere's urban form is fairly compact, and each neighborhood has services and small-scale shopping as well as schools, all accessible by a short walk or bicycle ride. One of the most impressive features about Almere is its green network. As a rule, residents are no farther than 500 m from green areas, and there are extensive forested and nature areas just a few minutes' bike ride away. Other sustainability features include a district heating system fueled by a natural gas cogeneration plant. . . .

[. . .]

LESSONS FOR AMERICAN CITIES: PROSPECTS FOR MORE COMPACT URBAN FORM

[. . .]

The experiences of the European cities profiled here represent in many important ways more compelling models of future American urban growth: strategies that reinforce and strengthen existing urban patterns and texture, including new developments that are of much higher density but that (while not excluding the automobile) include by design from the beginning public transit, bicycles, and walking, and that perhaps (most importantly) are tightly connected with existing cities. A high premium is clearly placed on connections to, and new growth that links to and is interwoven with, the prevailing urban form.

The European cities profiled here also effectively demonstrate that it is possible to achieve compact urban form at the same time that green spaces in and around cities are protected. Indeed, in many Scandinavian cities, such as Helsinki, large wedges of green space and nature extend into the very center. In many respects, it is the very compactness that allows these networks of green space to exist in such proximity to large populations. Good public transit systems that make outlying green spaces easy to reach, even from the very core of the city, is another important ingredient.

There are many lessons to be learned from these European cities in terms of both design and building of new residential districts. . . . In the new growth districts . . . a strong emphasis is placed on connecting with and building on to the existing city and its fabric, building at densities that make walking and other alternatives to the automobile possible (not to mention the more efficient use of land), and designing new communities with town centers, diverse housing types, and mixtures of uses and activities. Even the best new American communities typically lack these qualities.

[. . .]

In thinking about the American translation of compact urban form, it is also important to recognize that substantial additional density is possible and desirable even while maintaining an emphasis on single-family housing units. Single-family detached, or attached, units can be organized in ways that enhance walkability, connectedness, and higher density, while also recognizing the important American home values of privacy and private yard space. Indeed, many of the European development projects profiled here (e.g., many new Dutch projects) also appear to respect these amenity values. It is quite common to include, for example, extensive low-rise, single-family new housing with private backgrounds and gardens. There clearly are ways to build more densely, but with the American sensibilities and vision of what a desirable home includes kept firmly in mind.

One of the clear lessons from research on the visual preferences of Americans (especially the work of Anton Nelessen) is the importance of aesthetics and design in determining acceptability of density. Incorporation of trees, sidewalks, on-street parking, varied rooflines, and so on would substantially improve the attractiveness of higher-density forms of housing. Objections to density are often founded in a fear about what the visual implications or ramifications will be (and a sort of sterile concrete, higher-bulk image of what multi-family and higher-density housing would entail). Careful design and the incorporation of desired amenities would do much to improve acceptability of density in American communities.

[. . .]

In large part, the different spatial and physical results in European cities are the result of a fundamentally different approach to planning and development, one that American cities should consider adopting in whole or in part. In most American cities, real public planning has arguably been abdicated to developers and the private sector. Planners, and the democratic planning bodies they work for, exercise but a modest control – usually setting broad land use parameters through largely reactive policy instruments, typically zoning and subdivision ordinances. The European model of serious public control and guidance of future growth, the integration of the different spatial levels of planning, and an aggressive and strong public role in shaping the design of new development areas is one that American planning must eventually learn to emulate. Playing a much more active (and forward-looking) role in acquiring land (and both influencing growth patterns and reaping speculative gains) is also needed. At the very least, planning in American cities must do a better job of laying down a sustainable template of connected streets, transportation and other investments, ecological infrastructure, and the spatial outlines of community.

43

Shifting Ground to Address Climate Change: the Land Use Law Solution

From *Government Law and Policy Journal* (2008)

John R. Nolon

Editor's introduction

When former Vice-president Al Gore, star of *An Inconvenient Truth*, the hit documentary on global warming, attempted to put solar panels on the roof of his Belle Meade, Tennessee, home in 2007, the town's building inspector halted the operation, citing a zoning code violation: all electricity-producing elements had to be placed on the ground. After Gore's contractors forcefully protested the ruling – in shady residential neighborhoods, solar panels only work up high – the city quickly amended its ordinance to allow rooftop solar panel installations. The zoning ordinance in this municipality with a population of under 3,000 – only thirty-three pages long and uncomplicated – was easily modified. Around the country, many local governments are in the same situation. They have the regulatory framework to deal with climate change but they need to bring their rules up to date. They will call on urban planners for help in adapting their codes because the escalation of oil prices and carbon dioxide is making energy issues urgent.

In 2008, U.S. residents constituted 5 percent of the world's population, but they consumed 25 percent of the world's energy and emitted 20 percent of earth's carbon dioxide, the largest greenhouse gas (GHG) component. And according to the U.S. Environmental Protection Agency (EPA), Americans burned profligate amounts of CO_2-producing fossil fuels – oil, gas, and coal – for transportation (33 percent of the total carbon dioxide, with 60 percent of it coming from personal cars); industry (28 percent, with 50 percent emanating from generating steam or heat for industrial production); and residential (20 percent) and commercial (18 percent) buildings, mainly for heating, cooling, lighting, and appliances/equipment.

But in addressing energy issues related to climate change, America's policies are dispersed, uncoordinated, and uneven, involving federal, state, and local governments. Overseeing all federal efforts is the cabinet-level Committee on Climate Change Science and Technology Integration, co-chaired by the secretaries of Energy and Commerce. Its members include the secretaries of Transportation, Defense, Interior, Health and Human Services, State, and Agriculture as well as the EPA administrator and chairman of the Council on Environmental Quality.

President George W. Bush tied his administration's initiatives to reducing "greenhouse gas intensity." This measure pegs GHG emissions to the gross domestic product (GDP), and Bush pledged an 18 percent GHG intensity reduction from 2002 levels by 2012. (This figure can be deceiving, however, as intensity can decline if GDP rises – in these circumstances *even if* actual GHG emissions increase.)

Congress responded with the Energy Policy Act of 2005 and the Energy Independence Security Act of 2007, focused on energy conservation through standard setting. Included are automobiles (expressed in the

Corporate Average Fuel Economy standards – CAFE – set to go to thirty-five miles per gallon for cars and light trucks by 2020); appliances and equipment (managed by the EPA under the Energy Star program for such household items as refrigerators and air conditioners); and light bulbs (incandescent lights will be eliminated by 2014). The legislation also offers some financial incentives (primarily loans) for developing wind, solar, nuclear, biofuel, and other renewable energy sources and funds research on climate-change science and technology.

Other federal legislation can potentially contribute to energy savings indirectly. For example, the Intermodal Transportation Efficiency Act and its successors have supported the construction of mass transit – especially in such auto-dependent places as Phoenix and Houston – and improvements in freight transport, including the $2.4 billion Alameda Corridor project that dramatically reduced truck traffic by smoothing connections between Long Beach and Los Angeles ports and transcontinental railways. Recently, the U.S. Supreme Court mandated the EPA to regulate GHG emissions under the Clean Air Act (*Massachusetts v. EPA* 127 S. Court 1438 [2007]), a decision that affects setting the National Ambient Air Quality standards (NAAQ) standards.

Congress is currently considering America's Climate Security Act (S.2191), sponsored by senators Joseph Lieberman (D. CT) and John Warner (R. VA). It proposes a cap-and-trade system that would establish state GHG emission limits and allow auctions of excess allowances (resulting from GHG reductions), using the proceeds to support low-carbon research and implementation programs. In December 2007 the Lieberman–Warner Act passed out of the Senate's Environment and Public Works Committee, and in June 2008 was debated on the Senate floor. (To track this Bill, see lieberman.senate.gov/rwb/track_a_bill.cfm.)

While the federal government has been reluctant to impose stringent GHG emission controls, some states have stepped up to the plate. California's AB 32 Global Warming Solutions Act of 2006 calls for a return to 1990 levels by 2020. Another sixteen states have instituted various GHG reduction initiatives. Also emerging are multi-state coalitions like the Regional Greenhouse Gas Initiative (RGGI: nine Northeast and Mid-Atlantic states), Midwestern Greenhouse Gas Accord (MGGA: ten states), and Western Climate Initiative (WCI: seven states and three Canadian provinces). RGGI, the most advanced, is pledging major power-plant CO_2 emission reductions.

In the area of public utilities, many states have employed their regulatory powers to attack global warming, a reasonable approach considering that electricity generation is a major GHG source. Techniques include the institution of renewable portfolio standards (requiring electric companies to supply a set percentage of their energy from renewable sources according to a timetable), public benefit funds (adding a charge to individual electricity bills to support energy-efficiency programs such as weatherization grants for low-income households), and variable rate tariffs, which involve charging less for off-peak use. These broad strategies will engage planners locally. For example, if electric companies use wind, solar, or nuclear sources, urban planners will help with siting requirements and managing related citizen engagement exercises.

In this policy environment, local governments are key players. They promote mitigation (reducing human-caused factors, like sprawl, that contribute to climate change) and adaptation (introducing natural or human systems programs like LEED-ND, a sustainability rating code or transit-oriented development). Since states delegate to municipalities the power to regulate construction and land use, cities can retool their building codes, subdivision rules, zoning ordinances, and comprehensive plans to encourage energy-saving settlement patterns and development. In addition, cities can use their capital budgets for improvements to infrastructure (streets, water, sewer) and the public realm (parks, streetscapes, and forestry).

To supplement green regulations and spending programs, cities can encourage climate change adaptations by expediting permitting or reducing permitting fees for sustainable construction, instituting a carbon tax (based on electricity use), or floating revenue bonds to support energy-saving projects. One example is San Francisco's bonding the development of renewable energy for city agency use.

In 2005, the U.S. Conference of Mayors, under the leadership of Seattle's Mayor Greg Nickels, created the U.S. Mayors' Climate Protection Agreement that pledged to reduce GHG emissions to 1990 levels by 2012. In November 2007 the group sponsored the first Mayors' Climate Protection Summit to share best practices in land use, mass transit, green buildings, and public finance and to host a hearing of the U.S. House of Representatives' Select Committee on Energy Independence and Global Warming. By spring 2008, 852 mayors – representing more than 80 million people – had signed on to the GHG reduction initiative.

In this selection John R. Nolon, Professor of Land Use Law at the Pace University School of Law, and Visiting Professor of Environmental Law, Yale University School of Forestry, outlines how municipalities can modify their existing regulations and practices to address climate change. An expert in land use law and comprehensive planning, Nolon edited *Losing Ground: A Nation on Edge* (Washington, DC: Environmental Law Institute, 2007) (with Daniel B. Rodriguez), which deals with land use issues in disaster-prone areas; *Land Use Law for Sustainable Development* (Cambridge: Cambridge University Press, 2006) (with Nathalie J. Chalifor, Patricia Kameri-Mbote and Lin Heng Lye), which compares local approaches to global warming; and *Land Use: Cases and Materials*, sixth edition (Minneapolis, MN: Thompson West, 2003) (with Morton Gitelman, Patricia Salkin and Robert R. Wright), the definitive textbook on this subject. In addition, he served as a consultant to the Council on Sustainable Development under President Clinton and to New York State's Quality Communities Advisory Board.

To find out more about municipal approaches to energy saving and climate change, see Daniel Lerch's *Post-carbon Cities: Planning for Energy and Climate Uncertainty: A Guidebook on Peak Oil and Global Warming for Local Governments* (Sebastopol, CA: Post Carbon Press, 2007) and the U.S. Conference of Mayors' *U.S. Mayors' Climate Action Handbook* at www.iclei.org. For the most current writing on legal issues, see the *Harvard Environmental Review* (www.law.harvard.edu/students/org/elr), New York University's *Environmental Law Review* (www.law.nyu.edu/journals/envtllaw/), and the Environmental Law Institute's *Environmental Law Reporter* (www.eli.org).

Advocacy and professional groups are excellent sources of information in this fast-moving area. See, for example, the websites of the following organizations: Environmental Defense Fund (www.edf.org/) for cap-and-trade arrangements; the Natural Resources Defense Council (www.nrdc.org) for information on LEED programs as well as general sustainability issues; and the Post Carbon Institute (www.postcarbon.org) for locally based greening efforts. The American Planning Association (www.planning.org/) is a pioneer in Smart Growth initiatives that are by definition green; the Urban Land Institute (www.uli.org/) follows urban-focused global warming concerns; and the Association of Landscape Architects is a partner in a sustainable sites initiative (www.sustainablesites.org). All provide up-to-date information on the professional front.

STRATEGIES FOR MITIGATING CLIMATE CHANGE

Robert Socolow, a professor of engineering at Princeton, set an action agenda for mitigating climate change by identifying fifteen strategic "stabilization wedges," each one capable of preventing the emission of at least a billion metric tons of carbon annually using existing technologies. The genius of Socolow's strategy is that it divides the daunting and discouraging task of climate change mitigation into categories that enable us to order our response efficiently. It makes a formidable challenge seem more doable and allows us to identify the actors who are capable of effective adaptation within each wedge and to formulate strategies that enable and empower those actors to succeed. One of Socolow's wedges focuses on reduced use of vehicles (vehicle miles traveled), which lowers the use of fossil fuels consumed by vehicles. A second aims at creating energy-efficient buildings and appliances. A third fosters wind energy and a fourth energy produced through solar power. A fifth aims at preserving forests and vegetated soils to capture and sequester carbon.

This article conceives and describes a Land Use Stabilization Wedge: a strategy that aggregates these five wedges and further organizes strategic energies. [See Figure 43.1.] This builds on Socolow's optimistic assertion that "an excuse for inaction based on the world's lack of technological readiness does not exist." I assert that the existing legal authority of state and local governments to regulate and guide land use and building is a powerful "technology already deployed somewhere in the world." The Land Use Stabilization Wedge aggregates sev-

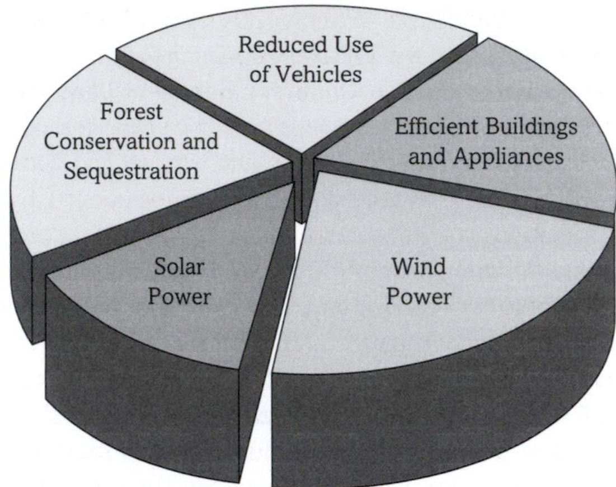

Figure 43.1 The land use stabilization wedge: climate change mitigation

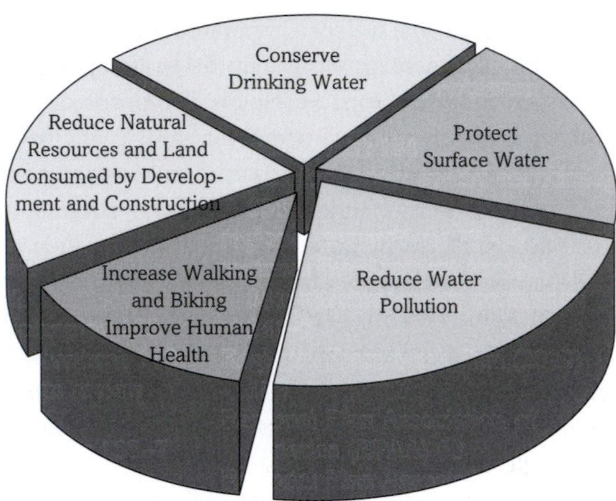

Figure 43.2 The land use stabilization wedge: corollary benefits

eral of Socolow's initiatives and employs multiple mitigation techniques available to citizens in every locality in the country.

The Land Use Stabilization Wedge comprises all the ways the device of land use control can reduce CO_2 and other greenhouse gas emissions. These include:

1 Shifting development patterns so that less driving occurs.
2 Reducing the size of housing units.
3 Creating more compact and thermally efficient buildings.
4 Reducing the materials consumed in building construction.
5 Creating more energy-efficient buildings.
6 Utilizing more efficient equipment and appliances.
7 Permitting and encouraging the use of wind energy generation facilities.
8 Permitting and encouraging the use of solar energy generation facilities.
9 Preserving undisturbed vegetated areas that sequester carbon.
10 Retaining agricultural lands and the production of farm products close to urban centers, further reducing transportation costs.

This article touches on corollary benefits that result from the implementation of the Land Use Stabilization Wedge. These include reduced use of drinking water, reduced impervious coverage and flooding, prevention of water pollution, and others. [See Figure 43.2.] These objectives can be achieved by local governments in most states through the legal authority already delegated to them to regulate land use and building construction. The Land Use Stabilization Wedge targets local governments as key actors in climate change mitigation, understanding that considerable support and assistance from state and federal agencies and the cooperation and guidance of the private sector are essential to their success.

POTENTIAL EFFECTS OF MITIGATION THROUGH LAND USE AND BUILDING CONTROL – SHIFTING GROUND

The U.S. Census Bureau projects that the nation's population will grow by 100 million by the year 2043. With a projected household size of 2.6 persons, this yields 40 million new households. This new population, and the need to replace aging homes and buildings will cause the private sector to build 70 million new homes and 100 billion sq. ft of nonresidential space. About two-thirds of the development on the ground by 2050 will be built between now and then. How that growth is placed on the landscape in human settlement patterns is critically important.

In the past decade approximately 60 percent of households have chosen to live in single-family homes on individual lots. For a variety of reasons, the projected 40 million new households will be more urban-oriented and willing to live in dynamic, walkable neighborhoods in cities and urban suburbs. Market projections indicate that urban housing located in compact developments will increase in price more rapidly than single-family, suburban homes. It is quite possible that the market demand will support "shifting ground", so that the historical numbers are reversed. If the rates change and 60 percent of these new households (24 million) chose instead to live in more compact, mixed-use environments and 40 percent (16 million) choose the single-family pattern, this will shift fully 8 million households (over 20 million people) from one human settlement pattern to the other.

The new paradigm for development, one consistent with the Land Use Stabilization Wedge strategic approach, is a more compact, dense, and mixed-use human settlement pattern, one capable of being implemented through coordinated local land use law. This envisions a shift in the dominant pattern of development from single-family, single-use neighborhoods to neighborhoods characterized by smaller homes, clustered and stacked, mixed with service and retail uses reachable by foot or on bicycle, with nearby schools and recreation, served by transit stops, now or in the future.

The movement of vehicles is responsible for about one-third of U.S. CO_2 emission and that number is growing. "Single family homes use more energy per person that multifamily homes. Large homes use more energy than smaller homes. The farther new homes are from existing population centers, from work and shopping, the greater the additional energy use in transportation per home and per person." A little over one-third of the increase in driving is associated with demographic change; the rest is attributed to "land use patterns that have led to increases in average trip distances (38 percent) and in the number of trips made (25 percent)."

Portland, Oregon, is one city likely to achieve significant greenhouse gas emission goals, owing to the urban growth boundaries adopted in 1974 that were designed to protect farmland and contain sprawl. Climate change mitigation, in this case, is an unintended benefit that is due to the increased density, reduced vehicle trips and vehicle miles traveled, and increased transit ridership that land use law reform achieved.

According to the Urban Land Institute's *Growing Cooler* report, "much of the rise in vehicle emissions can be curbed simply by growing in a way that will make it easier for Americans to drive less. In fact, the weight of the evidence shows that, with more compact development, people drive 20 to 40 percent less, at minimum or reduced cost, while reaping other fiscal and health benefits."

Compact development, as defined in the *Growing Cooler* report "does not imply high-rise or even uniformly high density development . . . that will result in the 'Manhattanization' of America." It refers to development at about twelve to fourteen dwelling units per acre, which is 75 percent above the 2003 national average density for all housing development. The report concludes that "shifting 60 percent of new growth to compact patterns would save 85 million metric tons of CO_2 annually by 2030." This is aimed at abating the alarming increase in driving caused by the dominant single-family growth pattern, which will increase driving by 59 percent by 2030 while the population increases by 23 percent, according to the U.S. Department of Energy's forecasts.

If it were possible to shift half of these 8 million households from single-family settlements to higher-density urban development – the type associated with transit-oriented development – the positive effect on the environment and climate change would be dramatic.

1 In higher density urban developments, the amount of CO_2 emitted per capita can be fifteen metric tons less annually, when compared with single-family living. Multiplied by 10 million people shifted into higher-density urban developments, the potential CO_2 reduction equals 150 million metric tons annually.
2 Residences in higher-density urban and compact developments are smaller than the national average for single-family homes. Using an estimate of 1,500 sq. ft for these developments, compared with the single-family average of 2,600, yields a savings of 1,100 sq. ft. This space does not need to be heated and cooled. Less space to construct reduces the fossil fuel consumed in manufacturing and assembling building materials.
3 Additional CO_2 stabilization occurs when local governments zone to encourage wind and solar generation, preserve undisturbed landscapes,

and preserve farm land close to urban market demand.

The corollary benefits of the compact development pattern are equally dramatic. The Hudson Park project in Yonkers, New York, discussed in the next section, is a representative example of a higher-density, transit-oriented development in an urban neighborhood. Its first phase contains 118 dwelling units per acre: four or five times more dense than the average compact development project. If half of the 8 million new households were shifted from single-family settlements to this type of development, the results would include:

1 74 billion fewer cu. ft of storm water annually.
2 33 billion sq. ft less impervious coverage.
3 100 billion gallons of potable water per year.

REDUCING USE OF VEHICLES

How can the Land Use Stabilization Wedge reduce the number of trips taken and the vehicles miles traveled in the United States? Comprehensive plans and zoning laws adopted by local governments, when aggregated, create the blueprint for the development of land and buildings for their region. Through changes in plans and zoning laws, communities can create transit oriented development and transportation efficient development that shift development patterns from a single-family dominant pattern to one that fosters compact, mixed-use development. This new pattern greatly reduces automobile dependency, vehicle trips, and vehicle miles traveled: a method of implementing Socolow's vehicle travel reduction wedge.

Central cities and their older and developing suburbs constitute the relevant region for transportation planning purposes. In these regions, Metropolitan Planning Organizations (MPOs) prepare capital plans for all types of transportation infrastructure, including transit services. Developing mechanisms to coordinate state and MPO transportation planning with local land use planning is key to the success of connecting higher-density urban developments and compact developments to transit services now or in the future and is arguably required under federal law.

Whether legally mandated or not, for practical reasons land use planning among localities in a transportation region must be coordinated with transportation infrastructure planning and development. Local land use plans and zoning determine how much population can increase over time, and this, in turn, determines demand for various types of transportation services. Transit lines for rail and Bus Rapid Transit (BRT) services cannot be planned in isolation, station by station. The economics of transit station development and rail and bus lines are dependent upon land use densities; there must be a sufficient number of commuters in a relevant group of adjacent communities to provide a minimal level of ridership throughout the area served by the transit system. Where transit service is not feasible because of insufficient land uses and densities, other modes of transportation must be planned.

TRANSPORTATION-EFFICIENT DEVELOPMENT (COMPACT DEVELOPMENT)

Compact developments may not be intense enough to support ridership at various locations in a transportation region. In the near term, they may have to be developed as "transportation-efficient" communities that are ready to receive transit services in the future as the region grows. Compact developments not near existing transit services can incorporate a variety of land use and transportation features that reduce vehicle miles and trips. Land use plans can allow for mixed uses, a variety of housing types and sizes, parking and bicycle facilities, and transportation-related improvements. These can be coordinated with planned capital improvements such as interconnected sidewalks and trails, bike paths, and jitney service from moderate-density hamlets to regional transit stations. Together these initiatives can reduce congestion, car dependency, and provide for transit stops in the future.

The town of Malta, just outside Albany, New York, used an innovative land use technique that can be employed by communities to manage and define future growth in a way that creates more livable places that are transportation-efficient and transit-ready. It adopted a central business district overlay zone that is transit-ready. The Malta zoning law provides densities at the compact development level and contains a number of standards that will create a typical mixed-use and walkable neighborhood. Currently, the town is not served by transit, but the Capital District Transportation Plan calls for BRT service in the future. In anticipation, the overlay zone provides mass transit. It states that "to promote pedestrian activity and multimodal transportation, developments should be located within 1,320 ft of an existing or future transit stop as approved by the Planning Board."

The Town of LaGrange, in Dutchess County, New York, adopted a mixed-use Priority Growth District, or PGD, that directs development to a specific location and contains design and amenity standards that provide an alternative to the large-lot single-family zoning prevalent in suburban areas that are distant from the metropolitan center and transit services. The PGD concept is particularly well suited for outlying suburban communities, where the rate of growth is significant but where there is still a rural character and significant natural resources to be preserved. The pressure to provide new homes in these suburban growth areas can be addressed through the identification of Priority Growth Districts where roadways and other infrastructure either exist or can be accommodated in ways that reduce the length and number of automobile trips and create the possibility for some type of transit service in the future.

LaGrange worked with Dutchess County to create a PGD zone where there was an existing suburban transportation corridor and intersection. The zone in effect creates a new hamlet, serving new and existing residential development and providing some retail services. It combines mixed-use development, a variety of housing types, including affordable units, and trails and sidewalks. The zone encompasses 616 acres, and provides for up to 220,000 sq. ft of commercial space, including up to 160,000 sq. ft of retail, a supermarket and restaurants, a 50,000 sq. ft government center with a library, and between 560 to 680 housing units of several types: senior housing and assisted living units, apartments, townhouses, and single-family residences. It will be served by central water and sewer with potential to serve additional adjacent growth, and is located along a state highway.

TRANSIT-ORIENTED DEVELOPMENT (HIGHER-DENSITY URBAN DEVELOPMENT)

In many urban areas served by transit stations, densities of housing at fifteen to forty dwelling units can be achieved. Around transit stops, particularly, higher urban density development can be planned for and supported by zoning and infrastructure planning. These types of developments, as demonstrated above, significantly reduce per capita carbon emissions and yield numerous other climate change and environmental benefits.

The Bloomington, Minnesota, City Code provides for an "HX-R" zoning district (high-intensity mixed use with residential) that is aimed at getting people out of their cars. It attempts to reduce vehicle trips and vehicle miles traveled by maximizing high-intensity development in close proximity to transit. The ordinance prohibits drive-through uses that obstruct sidewalks and discourage walking. It provides a minimum density of thirty dwelling units per acre for residential development. It also provides a minimum floor area ratio of 1.5 and a maximum of 2. This maximum may be increased through density bonuses to encourage retail and service businesses, below-grade parking, development of plazas or parks, affordable housing, public art, and sustainable design. Parking is restricted in the ordinance in order to promote walking, biking, and transit use. Parking must be located below grade, within structured ramps, or in individual on-street spaces parallel with and adjacent to low-volume streets. Bicycle parking must be provided near building entrances. Development directly adjacent to transit stations must provide sidewalk and bikeway connections to the transit station as well as to adjacent sites. The Bloomington zoning strategy evinces a commitment to development that is truly transit-oriented. It restricts parking, connects to nearby transit, locates retail and service uses within short walks of residences, and thereby reduces vehicle trips and vehicle miles traveled.

The City of Yonkers, New York, struggled for years to jump-start its downtown and adjacent industrial waterfront on the Hudson River, an area that is served by three commuter train stations, less than a half-hour trip from New York City's Grand Central Terminal. During the past two decades, the city amended its waterfront urban renewal plan over a dozen times before the private market began to respond. Governmental commitments to provide urban recreational and design amenities, build an impressive central library, renovate historic buildings, clear deteriorated buildings, remediate brownfields – all within walking distance of the renovated central rail station on the river – began a process that has led to considerable success.

The zoning and land use techniques that the City of Yonkers used were numerous and are instructive. It adopted a highly detailed master plan for the waterfront area that contained certain specifications regarding the types of development the city wanted on available vacant land in the area. An innovative zoning technique – called the Master Plan Zone – was adopted that provided as-of-right status for developments that conform to the design standards contained in the master plan. Compliance with New York State's extensive environmental review requirements was waived for such projects, since the impacts of development contemplated by the master plan had already been studied in detail and mitigation provided.

Early in this process, a developer was selected through a request-for-proposals process to plan the redevelopment of two centrally located sites, immediately adjacent to the train station. As the city developed its plan and conducted its environmental impact review, the private redeveloper began site planning and provided economic and market input. Information provided by citizens, environmental consultants, other professionals, and the developer were integrated as the process progressed and the master plan and designs for the two sites were adjusted.

The result is the development of Hudson Park, a two-phase project that contains nearly 500 middle-income rental residential units, public pedestrian access to a renovated waterfront, restaurants, office and retail space, and immediate access to the train station through carefully designed walkways and entrances that provide security to riders. Hudson Park is a dramatic transit-oriented development where parking provided is approximately 50 percent less than the amount required by traditional urban zoning. This is possible because the buildings and area appeal to commuters who travel to work by train and the developer's marketing was designed to attract them. The developer saved $25,000 in development costs for each parking space not constructed, and residents save $6,000 annually for owning one car instead of two. Three high-quality restaurants and a number of retail stores catering to the middle-income population of these buildings have appeared since the first 250 residents moved into phase one of the Hudson Park development. This project and the public amenities provided by the government are credited with sparking considerable additional private sector interest in the area.

EFFICIENT BUILDING LOCATION, CONSTRUCTION, AND OPERATION

Suburban and urban communities can mitigate carbon emissions and promote energy efficiency by adopting building design and location standards, such as those promoted by the Leadership in Energy and Environmental Design (LEED) criteria promulgated by the U.S. Green Building Council. They can do this in at least three ways: by committing themselves to meeting LEED and other energy standards in newly built or renovated municipal buildings, or in those funded by the municipality; by requiring new privately built or renovated buildings to meet such standards; and by adopting zoning standards for appropriate districts similar to those contained in the Council's evolving Neighborhood Development Rating System.

There are four levels of LEED certification for individual buildings which can be attained by accumulating points for implementing design standards in the categories of sustainable site development, water savings, energy efficiency, materials selected, and indoor environmental quality. The LEED standards can serve as a model for incorporating energy-efficient design standards into local building codes and requirements. LEED standards also contain design features normally associated with land use planning and zoning. For example, in a LEED for Homes Certification, a new home receives ten points, one-third of the required number of points for certification, just for being smaller than the national average. A project can also earn points towards certification by developing at higher densities, by being located near public transportation, or by using energy-efficient appliances.

BUILDING CODE ADAPTATION

New York is one of twenty-two states that have adopted a set of building codes that must be enforced at the local level but that allow local legislatures to add more restrictive standards. These codes create the standards that local building inspectors must enforce when asked for a building permit by a private contractor or developer prior to undertaking a building project. Under Section 379 of the New York Executive Law, the legislative body of a local government may adopt local ordinances imposing more restrictive standards for construction to ensure energy efficiency and minimize carbon loading.

The Town of Greenburgh, New York, amended its code to add new energy conservation requirements more restrictive than the adopted statewide mandatory energy code. Greenburgh's local law requires that all new homes constructed in the town comply with Energy Star guidelines introduced by EPA in 1992. The program provides several methods of making a home at least 15 percent more energy efficient through such mechanisms as effective insulation, high-performance windows, efficient heating and cooling equipment, and various energy efficiency products. The law applies to one and two-family dwellings and multi-family buildings of three stories or less. In 2006, the Town of Babylon, New York, adopted a law requiring all newly constructed commercial buildings, office buildings, industrial buildings, multiple residences, and some senior citizen residences to comply with LEED standards.

ZONING LAW REFORM

The Boston Zoning Code Green Building Amendments were adopted in 2007 to "ensure that major building projects – buildings over 50,000 sq. ft – are planned, designed, constructed, and managed to minimize adverse environmental impacts; to conserve natural resources; to promote sustainable development; and to enhance the quality of life in Boston." The Boston legislation incorporates by reference the US Green Building Council's LEED rating system. The LEED building certification standards do not impose requirements but rather allow developers to choose among a variety of criteria to obtain sufficient points for the project to become a certified LEED building. Compliance with the local law is required but developers are allowed to choose voluntarily which LEED standards to meet.

The U.S. Green Building Council is providing additional guidance to municipalities interested in promoting energy efficiency at the neighborhood development level. Under its LEED for Neighborhood Development Rating System, it integrates Smart Growth, New Urbanism, and green building standards into a system for designing and rating neighborhood development. Under this system, both the location and the design of buildings can be certified as meeting the Council's standards for environmentally responsible and sustainable development.

The U.S. Green Building Council adopted the LEED-ND program as a pilot. At the end of 2008, the early results will be evaluated and a revised rating system will be instituted. Among the standards contained at the pilot stage are reduced automobile dependence, creation of a bicycle network, compact development, diversity of uses and housing types, affordability of housing, the proximity of housing and job sites, reduction of parking footprint, proximity to transit facilities, and transportation demand management. These are matters that go to the heart of traditional local land use regulation and are at the forefront of integrating transportation and land use planning. Communities can incorporate the lessons of the LEED-ND program in their land use plans, regulatory standards, and development approval processes.

REGULATION AND USE OF PUBLIC BUILDINGS AND PROPERTY

The City Council of Scottsdale, Arizona, adopted a formal Green Building Policy for municipal buildings in March 2005. The mandatory policy for municipal buildings requires that "all . . . city buildings of any size will be designed, contracted and built to LEED Gold Certification levels or higher." The Township of Cranford, New Jersey, passed a local ordinance in 2005 adopting a policy that township-owned and funded projects will meet LEED® Silver ratings.

There are 40,000 localities in the United States. They are recycling solid waste, planting trees, greening public buildings, using biodiesel fuel in vehicles and machinery, developing methane recovery systems in landfills, using solar energy to power municipal buildings, installing geothermal pump systems to heat and cool public facilities, replacing incandescent traffic signals with light-emitting diode signals, mounting police on bicycles, adopting anti-idling protocols for municipal vehicles, and exhibiting extraordinary creativity along the way.

WIND POWER

Although wind-generated power constitutes a small fraction of the nation's power needs (around 1 percent), it is growing quickly and could eventually meet over 20 percent of the nation's demand for energy. General Electric, whose Renewable Energy Global Headquarters are in Schenectady, is in the process of building nearly 900 1.5 mW wind turbines, many in upstate New York. A 1.5 mW turbine can supply the power needs of over 400 single-family homes. This trend is encouraged by New York State's adoption of a state policy establishing a goal that 25 percent of energy consumed by 2013 will be produced by renewable sources such as wind, solar, biofuels, tidal energy, and other mechanisms.

One way that municipalities may encourage wind power use is to purchase electricity from wind farms to run locally owned utilities or to heat and cool town buildings. A village in Illinois purchases 4,500 mWh a year of electricity from a nearby wind farm to provide power to its water utility, saving nearly 5 million lb of carbon dioxide emissions annually.

Localities may also amend zoning laws to permit and encourage homeowners to install individual wind generation systems. Individuals are beginning to install backyard wind turbines on towers 50–70 ft high that generate enough power for their household use. In some cases, excess power is created that can be directed back to the local power company grid, sometimes for credit or cash. Some claim that a single wind turbine of this size can produce enough electricity for two average-sized homes in an area with moderate wind speeds, raising a host of regulatory and real estate law issues. These types of "distributed generation systems" are supported by the American Planning Association's Energy Policy Guide. Under the New York State Real Property Tax Law, local tax assessors are permitted to offer property owners who construct small wind energy systems an exemption or partial exemption from local real property taxes for the increased value of the property due to the addition of the facility to the land.

Local governments are adopting comprehensive plan components that contain local energy goals and policies, moratoriums that prevent the construction of wind generation facilities until they can be properly regulated, and a number of zoning, subdivision, site plan, special use, and environmental review mechanisms to balance the benefits of wind-generated power and the detrimental effects such facilities can have on the community. While these laws can be used to limit and discourage wind generation facilities, they can also become part of the Land Use Stabilization Wedge by encouraging the construction and use of wind generation projects both large and small through zoning and site plan provisions, tax abatement, and other initiatives.

SOLAR POWER

Local governments can mitigate climate change in at least two ways that employ solar energy generation: equip public building with solar facilities and adopt land use regulations that encourage their use by homeowners and businesses.

The New York State Comptroller reports that Albany County, the Ulster County towns of Woodstock and Rosendale, the Ulster County village of New Paltz, the Nassau County town of Hempstead, and the Tompkins County town of Lansing received financial and technical assistance from the New York State Energy and Research Development Authority (NYSERDA) for their public building initiatives. The audit, conducted for the period January 2003 to July 2007, determined that by installing solar panel electrical systems these municipalities could save roughly a million dollars and reduce the release of greenhouse gases carbon dioxide, nitrous oxide, and sulfur dioxide by over 6.6 million lb during the life of the panels, which should exceed forty years. With state assistance these municipalities paid roughly a quarter of the total project costs. An impressive number of state and federal initiatives are available to local governments as well as private property owners that lower the capital costs of solar installations.

In 1979, the state legislature granted express power to local governments to add provisions to their zoning regulations to permit and encourage solar energy systems and equipment, including access to sunlight. The legislature declared that access to solar energy is a valid public purpose and left it to each local government to adopt regulations suitable to its local environment and circumstances. This authority, which probably existed as an implied power prior to the Act, makes local power to permit solar power facilities explicit. Local governments may amend their zoning to permit solar energy systems in all zoning districts, to provide waivers of any height, area, or bulk requirements that obstruct solar facilities, or to create zoning overlay districts within which solar access is particularly appropriate.

CARBON CAPTURE THROUGH SEQUESTRATION

In developing suburban areas, there are often significant land areas that have been undeveloped for some time that contain undisturbed vegetated areas. As noted earlier, suburban communities can mitigate change by zoning to accommodate the bulk of population growth in compact developments as the towns of Malta and LaGrange are doing. By so doing, they may find it easier politically to adopt strong environmental protection ordinances applicable to the land outside these higher density zones. Density bonuses can be provided to developers of compact developments and cash contributions received in exchange, which can be used to purchase the development rights of valuable open space areas that contain critical natural resources.

The preservation of such resources will provide valuable environmental benefits such as carbon sequestration, food production, wetlands and habitat preservation, stormwater management and flood prevention, watershed protection, and the prevention of erosion and sedimentation. Soil organic carbon accumulates in undisturbed naturally vegetated areas. Further carbon stabilization occurs when developing communities preserve existing farmland where food products can be produced closer to population centers, thereby reducing transportation costs. Wetlands preservation, seen though the lens of climate change mitigation, offers the additional benefit of carbon sequestration since most wetlands have been undisturbed by previous development.

In local zoning and subdivision regulations, standards that prevent the disturbance of soils and vegetation on development sites have similar effects. The emerging field of "low impact development" experiments with pervious alleys and green roofs in urban projects and, in compact developments, vegetated swales that replace curbs and gutters for stormwater control, cluster development, tree retention, and retaining permeable topsoil on site during and after construction.

CONCLUSION

Climate change has altered the federal and state agenda and will reshape funding programs and priorities for programs and projects that promise to reduce fossil fuel consumption, dependency on foreign oil, and greenhouse gas emissions. There are relatively few local initiatives in the nation that utilize the Land Use Stabilization Wedge techniques described in this article. Localities that do move in this direction should enjoy considerable success in soliciting state and federal funding for land use and transportation planning, environmental studies, workforce housing, transportation and urban amenity capital projects, and other support needed to create successful transportation and land use demonstration projects.

Local governments, with their power to plan and regulate land use, are a critical ally of state and federal governments in the race to mitigate climate change. They have always been laboratories for experimentation – crucibles of change – from the time that New York City invented the comprehensive zoning ordinance through a host of celebrated land use movements: post-Euclidean zoning, growth management, the advent of local environmental law, and Smart Growth. Now we have the Land Use Stabilization Wedge: the climate change mitigation movement. While models exist for greening public and private buildings, reducing vehicular travel, preserving undisturbed lands, and fostering wind and solar power exist, much needs to be done.

44

Cities under Siege: Urban Planning and the Threat of Infectious Disease

From *Journal of the American Planning Association* (2006)

Richard A. Matthew and Bryan McDonald

Editor's introduction

Two characteristics – density and complexity – make urban places vulnerable in the face of natural and man-made disasters. Density, the presence of a large population living in limited space, not only heightens the human impact of a disaster but also complicates rescue and recovery. Complexity, the interdependence of economic, social, political, and physical systems, unlocks multiple chain reactions after a catastrophe, and necessitates crafting cross-cutting interventions to restore security and well-being. These two qualities underlie thinking about disaster management in highly urbanized places where urban planners play important roles in devising emergency preparedness and rebuilding programs.

Between 2001 and 2005, Americans viewed the interplay of urban density and complexity close-up. They witnessed the terrorist attack on New York City on September 11, 2001, and the landfall of Hurricane Katrina on August 29, 2005. They saw 2,600 New Yorkers die in three hours, and in New Orleans, 1,300 deaths, plus 25,000 left-behind residents who suffered in deplorable conditions for days. They watched as events unfolded – the economic hardships of New York's Chinatown residents whose businesses suffered because of their proximity to the attacks and New Orleans's two-day crime wave that required National Guard intervention. During recovery they reviewed first-responder failures, ineffective local government, faltering social welfare systems, embattled economies, and crumbling infrastructure. Finally, they have seen that rebuilding is prolonged, messy, and fraught with deeply divisive views.

As of 2008, neither New York nor New Orleans is back to where it was pre-disaster. The sixteen-acre World Trade Center site lies nearly vacant. Seven World Trade Center, adjacent to the site, opened in 2006, but the Freedom Tower, the centerpiece of the area's redevelopment plan, will not be ready for occupancy until at least 2012. None of the other on-site replacement buildings is even under construction, though ground preparations are nearly complete. Lower Manhattan's central business district, once the nation's third largest after Midtown New York and Chicago, has slipped to fourth place, below Washington, DC. Within days of the attack the area's transportation infrastructure was running with a temporary fix, yet permanent reconstruction is years away due to government indecision, inadequate financing, and local political bickering. (For example, cost overruns on the $2.2 billion Santiago Calatrava-designed World Trade Center Transit Hub and the $888 million Fulton Street Transit Center by Grimshaw Architects and Arup Engineers are requiring time-consuming design changes.) Moreover, as time passes, market forces are transforming this downtown. Since 2001, the area has experienced an 18 percent loss of commercial space and a 160 percent increase in housing units gained through office conversions (67 percent of the total) and new construction. Its residential population is growing, from 22,000 in 2001 to 55,000 in 2008, but its jobs are shrinking. (Private sector employment is down 20 percent since 2001.)

Meanwhile, New Orleans, having evacuated most of its population of 485,000 after the storm, now has recovered about 50 percent, according to the *New Orleans Index*, published by the Brookings Institution's Metropolitan Policy Program, but its recovery is painfully slow. Residents have returned to areas that did not experience flooding – 48 percent of the city's current 144,000 households live in these places – ones that held only a third of the total before the hurricane – while more severely impacted neighborhoods are still quite deserted. Funding needed to rebuild is dribbling in – by spring 2008, New Orleans had received only 35 percent of its FEMA infrastructure allocation, and the State of Louisiana had closed on only 57 percent of its Road Home housing assistance loans. Seventeen "target recovery areas" designated in 2007 for redevelopment are still being planned a year later. At the peak of the emergency shelter program in July 2006, the city had 73,000 emergency trailers; in spring 2008, 23,000 were still housing the displaced. Selected indicators show how weak the economy is – jobs are down nearly 20 percent, the number of firms is down 22 percent, and university enrollment is down 27 percent.

Take heart, say Lawrence J. Vale and Thomas J. Campanella, editors of *The Resilient City: How Modern Cities Recover from Disaster* (New York: Oxford University Press, 2005). Cities are tough; they do bounce back. Citing the recoveries of Chicago (after the 1871 fire), San Francisco (1906 earthquake), Warsaw (World War II population depletion), and Tangsham, China (1976 earthquake), they explore what makes places resilient, concluding that national pride, remembrance, and opportunism are key to rebuilding. Most important, they advise, "any full measure of urban resilience must take account of all such efforts to mitigate a disaster *a priori*."

This last observation echoes the subject of this selection, preparedness, an area that calls on urban planners' skills. Three principles guide their work. First, neither terrorism nor natural disaster will halt the march of urbanization that likely will occur in highly vulnerable places. UN-Habitat reports that since 1975 the world's urban population has grown 120 percent – from 1.5 billion to 3.3 billion – even while disasters have increased fourfold. The disasters have damaged ecological systems, economies, and urban land markets. They have wreaked havoc on urban slums, which are often informal settlements in marginal locations (steep hills, flood-prone land), with flimsily constructed housing, and poor services. UN-Habitat anticipates that the majority of future urban growth will occur in disaster-prone informal settlements of Africa and Asia, setting the scene for more disasters.

Second, mitigating disaster requires sound preparedness plans that gauge risk and formulate appropriate strategies. For example, addressing potential floods or earthquakes calls for assessing land suitability and building codes; infectious disease demands offering vaccinations; preparing hospital wards and quarantining affected districts and terrorism requires higher surveillance through cameras, body or package inspections and other techniques.

Third, thoughtful urban plans may help reduce risks or assist with recovery but only if there is the political will to enact them. Planners can establish early warning systems and plot evacuation routes; map vulnerable areas; organize sustainable land uses; propose the types, locations, and costs of protective capital investments (e.g., dams, retention basins, emergency shelter); design disaster-resistant infrastructure and buildings; and organize participatory processes and public education programs relating to the above-listed elements. They cannot, however, force politicians or people to follow their directions.

While this selection focuses on one type of disaster, infectious disease, it applies to a range of other security issues. It begins with key questions, ones that, refashioned, can address other kinds of disasters: What should cities do to prepare for a major disease event? Do urban healthcare systems have adequate surveillance and surge capacity? Have cities stockpiled appropriate medications and worked out effective quarantine, evacuation, distribution, and risk communication strategies? Are regional cooperation protocols in place? Is decision-making authority clearly established at the most desirable level? Are linkages between the private and public sectors, and across government agencies, secure and have they been tested? And are lines of communication established to ensure access to information, including classified information, and other forms of assistance that may be required? As the reader will see, the answers are equally generic.

The selection's authors are Richard A. Matthew, Associate Professor of Planning, Policy and Design at the School of Social Ecology, University of California, Irvine. Matthew also serves as Director of Irvine's Center for Unconventional Security Affairs (CUSA), with Bryan L. McDonald, CUSA Assistant Director. In 2004, they edited *Landmines and Human Security: War's Hidden Legacy* (Albany, NY: SUNY Press) with Ken Rutherford.

In recent years, a huge literature on urban security issues has appeared, prompted by highly publicized disasters and terrorist attacks that have had major impacts on urban places. This literature comes from the world organizations like the United Nations, university research centers dedicated to the topic, and from individual scholars. It ranges from broad assessments like the UN-HABITAT 2007 annual report covering issues from disasters to urban crime (United Nations Human Settlements Program, *Enhancing Urban Safety and Security: Global Report on Human Settlements, 2007*, London: Earthscan, 2007) to more focused collections like the two volumes published by the University of Pennsylvania Press, appearing within months of Hurricane Katrina. (See Ronald J. Daniels, Donald F. Kettl, and Howard Kunreuther's *On Risk and Disaster: Lessons from Hurricane Katrina*, 2006, that discusses how to assess risk and assign responsibility in modern society. It articulates ten principles for shaping hazard management, first among them that "in responding to disasters, all Americans should be treated equally," p. 10. See also Eugénie L. Birch and Susan Wachter's *Rebuilding Urban Places after Disaster: Lessons from Katrina*, 2006, which reveals the complexity of rebuilding. Geologists, economists, historic preservationists, landscape architects, city planners, musicians, educators, social workers, health providers, sociologists, architects, businesspeople, and others offer their views, which are sometimes diametrically opposed.)

There is also a stream that looks at government responses to natural disasters. One of the most recent examples is Thomas A. Birkland's *Lessons of Disaster: Policy Change after Catastrophic Events* (Washington, DC: Georgetown University Press, 2006). It joins two classic books – David R. Godschalk, Timothy Beatley, Phillip Berke, David J. Brower, and Edward J. Kaiser's *Natural Hazard Mitigation: Recasting Disaster Policy and Planning* (Washington, DC: Island Press, 1999) and Rutherford H. Platt's *Disasters and Democracy: The Politics of Natural Events* (Washington, DC: Island Press, 1999).

Finally, H. V. Savitch's *Cities in a Time of Terror: Space, Territory and Local Resilience* (Armonk, NY: Sharpe, 2008) is representative of the literature on terrorism. Savitch exposes contradictory themes evoked by urban terrorism: cities are vulnerable to it because of their density and the capacity of terrorists to evoke mass fear; at the same time, cities are resilient because they fight back with formal and informal surveillance. Savitch's discussion of antiterrorist mechanisms shows examples of what various countries have used, but is not an exhaustive evaluation of the techniques because his focus is on the political implications of terrorism in democratic societies.

In the twentieth century, improvements in water and sanitation systems, vaccination programs, and antibiotics provided great hope that the dangers posed by infectious disease could be minimized, and perhaps even eradicated. But despite great advances in public health, microbes have proven resilient and persistent. Diseases such as malaria continue to take a major toll across the developing world. . . . The main obstacles to the effective control of disease [in the developing world] are migration, urbanization, socioeconomic conditions, and a lack of adequate communication between researchers, health workers and the population. Emerging diseases such as HIV/AIDS and SARS demonstrate how quickly infectious diseases can spread, and how vulnerable every community is.

Centralized food production and the dramatic growth in fast-food consumption have increased social vulnerability to food-borne disease, especially in Western countries. Hospitals have become reservoirs of dangerous antibiotic-resistant diseases such as *Staphylococcus aureus*. . . .

Beyond threats from naturally occurring diseases, concerns about bioterrorism have increased in the past decade. . . . Whether the result of careful planning by an Al-Qaeda cell or the impulsive behavior of a disgruntled worker in a food processing facility, the threat of a serious intentional disease event cannot be ignored.

In this context, it is important to ask what cities should do to prepare for a major disease event. Do urban healthcare systems have adequate surveillance and surge capacity? Have cities stockpiled appropriate medications and worked out effective quarantine, evacuation, distribution, and risk communication strategies? Are regional cooperation protocols in place? Is decision-making authority clearly established at the most desirable level? Are linkages between the private and public sectors, and across government agencies secure and have they been tested? And are lines of communication established to ensure access to information, including classified information, and other forms of assistance that may be required? . . .

ASSESSING THE THREAT

Natural and accidental infectious disease threats

Naturally occurring infectious diseases emerge and spread primarily through interactions among people, animals, and their shared environments. Throughout history, such diseases have had a significant impact on human societies. The 1918 Spanish influenza pandemic killed more than 500,000 people in the United States and another 20 [million] to 30 million globally. The 2003 outbreak of Severe Acute Respiratory Syndrome (SARS) provides a more recent example of the speed with which diseases can emerge and spread. . . .

. . . The great fear of public health officials is that H5N1 [Avian Flu] will become easily transmissible among humans, something that is typical of influenza arising from avian reservoirs, and that could happen very quickly. Should this occur, the CDC estimates that it would cause between 89,000 and 207,000 deaths, with an economic impact ranging from $71.3 to $166.5 billion. . . .

Naturally occurring infectious disease extracts a toll beyond human lives. Direct economic costs include workers' compensation, lost productivity, decreases in tourism and travel, and reduced markets for goods. . . .

NEFARIOUS INFECTIOUS DISEASE THREATS

During the Cold War, many countries such as the United States, Canada, France, Russia, and South Africa investigated pathogens as potential weapons that they might use or to which they might need to respond. While biological weapons were outlawed in 1975 by the Biological and Toxin Weapons Convention, significant clandestine research programs continued in South Africa and the Soviet Union after the treaty was signed. Today the expertise and technologies developed during the Cold War are surfacing on the black market around the world. The development and proliferation of advanced biotechnologies in recent years may also aid efforts to develop biological weapons, and raise the possibility of creating more virulent or previously unknown types of biological weapons.

Tampering with the food supply is widely regarded as a plausible and potentially devastating scenario. . . . The vulnerability of the milk supply holds a special significance, as there have been several cases of accidental contamination of milk products that have affected large numbers of people, including an outbreak of *Salmonella typhimurium* in 1985 linked to the post-pasteurization contamination of milk from a U.S. dairy that sickened 170,000 people, and an ice cream premix that was contaminated with *Salmonella enteritidis*, causing illness in over 220,000 people in forty-one states in 1994. . . .

Many analysts believe that the conditions for a natural epidemic, a successful act of biological terrorism, or a large-scale biological accident are in place. Since cities function as hubs in economic, transportation, information, and social networks, they are likely accident sites, logical points of concentration for naturally occurring diseases, and ideal targets for the nefarious use of biological agents. Indeed, urbanization amplifies many infectious disease threats. Hence the capacity of cities to prevent and manage infectious disease deserves careful consideration. . . .

NATIONAL SECURITY AND URBAN VULNERABILITY

A well funded cluster of government entities, including the Departments of Defense, Homeland Security, and State; various intelligence agencies; the National Security Agency; and the Centers for Disease Control, work together to provide national security from external threats. While the capacity of the United Sates to deter a foreign army is effective, other threats can move into the U.S. virtually unimpeded by integrating themselves, deliberately or unintentionally, into legitimate transboundary flows. For example, drugs can be smuggled in shipments of coffee, money earned through criminal means (also called "dirty money") can be laundered through complex and seemingly legitimate banking transactions, and an infectious disease may be carried on to an airplane by an innocent and unsuspecting carrier, infecting people entering several other countries long before manifesting any symptoms. This last example is precisely how SARS emerged and spread during 2003. An innocent carrier might even be intentionally infected with an agent such as smallpox. . . .

The United States Department of Homeland Security's National Response Plan (NRP) calls for developing a multidisciplinary, all-hazards approach to addressing domestic incidents. The NRP recommends that prevention, preparedness, response to, and recovery from natural, accidental, and nefarious hazards be broad and inclusive. It advises involving all necessary stakeholders in developing robust response systems that are both mindful of local needs and conditions and integrated into state and federal response networks. This ongoing process provides an opportunity for planners to consider such threats and work to integrate local and regional planning expertise into the prevention and preparedness steps of the response process.

ELEMENTS OF A NEW URBAN SECURITY PARADIGM

In the past several years, governments carried out several exercises designed to assess urban preparedness for mass casualty crises involving biological, chemical, or nuclear weapons. These exercises were local and regional, and even international. While these exercises simulated intentional attacks, reports analyzing them underscore weaknesses in two areas that are generally relevant to the topic of urban preparedness for a major disease event: command and decision making, and medical capacity and procedures.

According to these reports, responsibility for command and decision making is poorly defined at all levels, and those likely to be in charge generally have given little thought to how they would respond to a situation involving a biological weapon. It is important to remember that those first on the scene will be in charge, at least initially. In the case of a biological threat, it is during this crucial initial phase of response, before dedicated resources and experts can be brought to the scene, that decisions will determine the scale of human and other costs. Even when the experts arrive, who is in charge will matter greatly to local residents. If no one who is credible with local residents reassures them that the situation is under control and provides clear answers to simple questions about medical assistance, water, and the need for evacuation or relocation, they may become antagonistic, excessively alarmed, or distrustful of official communications.

In addition to managing the public, on-the-spot decisions will be required concerning quarantines and evacuations. Throughout this initial response, the competing goals of saving lives, preventing further injury, reassuring the public, limiting liability, and gathering evidence, usually invested in different agencies, can lead to confrontation and failure unless there is an effective decision-making center. In short, if cities have not thought about response and trained the appropriate personnel, the consequences will inevitably be higher. . . .

Second, the various exercises strongly suggest that procedures to establish medical priorities and distribute medical supplies to affected populations remain problematic. Moreover, in the event of a major incident involving large numbers of casualities, the medical and healthcare systems of large urban areas would be overloaded in a matter of hours. Few concrete steps have been taken to plan, prepare, and educate the population about what to do in the even of a major crisis. . . .

Safer cities: big picture, small steps

The findings from the exercise reports, . . . recent research on disaster management, . . . and extensive interaction with first responders and public health experts help . . . identify six . . . steps to reduce urban vulnerability to infectious disease. As a general rule, planners should be integrated into existing disaster management systems so that they can bring their expertise to bear on immediate challenges such as urban evacuation and food security, and so that they become sensitive to the ways in which planning decisions might affect the public's vulnerability to infectious disease threats.

Early warning systems

The infectious disease threats discussed in this article fall outside the routine experiences of urban public health officials and doctors. But as in many areas of national and human security, the speed and accuracy with which a threat is detected will greatly affect its overall cost to society. . . . Thus it is important that urban and health planners work with first responders to assess the early warning needs of urban areas and design appropriate information systems that can support local decision making and harmonize with state and federal systems.

Resource management

Ironically, some of the resources a city needs in order to survive and flourish, such as a steady supply of food, can also function as effective disease vectors. . . . One partial solution is to encourage higher levels of reliance on local agriculture and markets, including through land use regulation and infrastructure planning. If planners encourage a greater degree of self-sufficiency in food production and consumption, they also enhance security and public health.

Logistics

Preparing for a major disease event should include securing, testing, and communicating to decision makers the location and adequacy of medical supply stockpiles, triage practices, and surge capacity in hospitals and other facilities. This will greatly facilitate access to critical resources such as vaccines and medical care should such an event occur.

Adequate supplies of medication, water, and food are of little value if they cannot be distributed in a timely fashion. [F]irst responders suggest that they are frustrated by the limited access to many strategically important and densely populated urban areas. For example, the economic value of the Long Beach–Los Angeles Port system is enormous, and it has often been identified as a potential target of terrorist activity. A nefarious or accidental biological release into the port could also pose a serious rescue challenge. How should the disease be contained? How should people be evacuated quickly? Transportation planners should work with public health officials and first responders to develop systems that will channel information to decision makers during a crisis, and also assist in the design of evacuation routes and lockdown procedures. Transportation and infrastructure planners should also systematically consider how their day-to-day recommendations might affect the city's ability to respond to an infectious disease incident in the future.

The efficient distribution of a vaccine or other medication could also be a serious problem for many cities. . . . In many areas, arenas, concert halls, and sports stadiums have been designated for use in emergency situations, even though they may not be well suited to such roles. City planners can help both in identifying the right locations for such surge deliveries, and in thinking about how to design and prepare such spaces for their roles in times of crisis. Planners can also help identify those in the public, private, and nonprofit sectors who could be trained to assist with logistical issues, such as handling confidential patient information in the case of a major disease event.

Cooperation

It will also be key to ensure smooth coordination between local private and public sector entities, across agencies, and between local and outside sources. People want to help during an emergency. Anyone who has directed a major event scene knows that a cascade of offers arrives in a very short time. Coordinating these offers is an enormous challenge. . . . Planning directors might begin by securing invitations to make presentations of their information resources and staff capabilities to private sector groups. . . .

Command

As noted earlier, exercises found disaster command and decision making to be poorly defined. It should be clear who will be in charge during a major disease event, what intelligence will be essential to support decision making, what resources will be available to support decisions, how decisions will be communicated, and how their implementation will be monitored, with new information cycling immediately back to the command center. . . . In the case of a major disease event, someone may have to order an evacuation or a lockdown. While studies demonstrate that citizens respond well to sensible plans implemented by confident authority figures, a lack of confidence can result in panic. First responders in Southern California worry that if many people attempt to flee an incident by car, congestion could transform the area into a giant parking lot. Land use and transportation planners have the information to support humane, realistic strategies that would effectively move or contain large numbers of people.

Communication

Finally, those in command must manage public reaction to the crisis by communicating with the public and others, including state and federal officials. While historically radio and television have been the preferred ways of reaching the public, the internet and cellphones are displacing these media. . . . Self-organizing networks based on a mixture of cellphone, internet, television, and radio will surely be the ad hoc communication systems of future disasters. . . . One goal of current information technology research is to find ways to channel information from self-organizing networks to decision makers. Transportation, land, and infrastructure planning information systems, such as websites or phone hot lines that provide information about current levels of traffic congestion, may be useful to individuals as well as official responders in these situations. Planners should consider this when designing such systems, and seek input from others in the community with disaster planning responsibility.

CONCLUSIONS

Many analysts believe that today's large urban areas are more vulnerable to a range of transnational infectious disease threats than has been the case for almost a century. It is therefore important to prepare for natural, accidental, and nefarious infectious disease events in urban areas. We have identified six components of emergency preparedness and response needed for a satisfactory urban defense. . . . Many of these steps relate to access and movement, areas in which planners have information, expertise, and practical influence. Overall, we recommend that planners seek out and work with both public and private sector groups with roles in disaster planning: design land and transportation planning information systems to aid and support decision makers during crises; encourage greater self-sufficiency in food production and consumption; assist in the design of humane, realistic evacuation strategies and routes; and consider the effects of their day-to-day recommendations on disease risk and response.

22@Barcelona: a New District for the Creative Economy

From *Making Spaces for the Creative Economy*, ed. Waikeen Ng and Judith Ryser (2005)

Miguel Barceló Rota

Editor's introduction

In 2002, Carnegie Mellon professor Richard Florida turned urban revitalization and economic development practice in a new direction. He prompted this shift with his new book, *The Rise of the Creative Class: And How it's Transforming Work, Leisure, Community and Everyday Life* (New York: Basic Books, 2002), which argued that, to remain competitive in the twenty-first century, cities had to attract not companies but a special class of highly educated workers that he labeled "the creative class." The lawyers, architects, writers, artists, engineers, financiers, musicians, educators, and doctors who make up this class, Florida wrote, have replaced other types of workers and already constitute 30 percent of the nation's labor force. But, Florida warned, these workers are footloose, in search of places that accord with their lifestyles. They prize neither job security nor company loyalty, the values of their parents. Instead, they have a "work hard, play hard" mentality. They seek places with high concentrations of people like themselves and easily accessible entertainment, recreational, and cultural amenities.

Florida's message was clear. Cities that were attractive to the creative class would thrive. Older industrial cities needed to reinvent themselves, and new cities had to craft the desired amenities. For example, cities with interesting work sites – like reconditioned factories or green buildings or joint live–work arrangements – could be winners. Cities that fostered downtown housing (loft conversions, intown historic districts) or unique locations near waterfronts or anchor institutions like universities, stadiums, and museums would rise in popularity. If they added jogging or bike trails, encouraged "funky" restaurant districts or modern entertainment sections centered on a stadium or arena, had a lively arts and culture scene, or helped institutions expand in central places, they could become "hot," ranked as the best places to live or do business by any number of magazines. Successes occurred in older places in America's Northeast and Midwest (including Lower Manhattan, Philadelphia, Boston, Minneapolis, and Chicago) and in the rising cities of the South and West (including Atlanta, Austin, Portland, Seattle, San Diego, and San Francisco). (See Plate II, 10.) European countries had their winners, as well.

Some cities already possessed the required accoutrements, but most had to work hard to provide them. They changed their land use regulations (allowing mixed-use districts, converting industrial to residential, or permitting live–work arrangements and residential-oriented retail in central business districts); invested in parks, streetscapes (lighting, sidewalks, plantings), and light rail lines or trolleys. They also built stadiums, arenas, concert halls, jazz centers, and museums. They preserved or upgraded "gaslight" or historic districts, adding period lighting and helping with storefront improvements. They opened farmers' markets. They sponsored special cultural and sports events – like "First Friday" gallery hopping, and marathons threaded through city streets.

They addressed key service needs like crime prevention and sanitation. They allowed business improvement districts to flourish. In sum, they made urban living attractive.

Seeking the creative class is a worldwide movement with its greatest manifestation in North America and Europe. Manchester, England; Bilbao, Spain; Pittsburgh, Pennsylvania; Singapore, and Amsterdam – all have had varying degrees of success with a variety of methods. Bilbao, for example, constructed a critically acclaimed museum of art designed by Frank Gehry, now a magnet for tourism and a potent symbol of successful city reinvention. Pittsburgh relied on biosciences and synergies with its many universities. Manchester is the regeneration-for-the-creative-class poster child. Once Britain's leading industrial city, it fell into great decline in the last quarter of the twentieth century. In 1996, a terrorist bomb destroyed its downtown. The national government assisted with its reconstruction. A prize-winning design offered a new image and modern amenities–open space, transportation, and snazzy stores and offices. Other contributions to the city's revival were investments in sports facilities (due to the awarding of the XVII Commonwealth Games in 2002), entertainment venues, arts and culture programming, and loft conversions by many firms but led by the aptly named "Urban Splash." By 2006 the press declared Manchester the best place in England to do business, dubbing it the second city (after London) in the country.

Focusing urban revitalization on the creative class is controversial, however. Some analysts criticize this strategy for neglecting undereducated, chronically unemployed residents. Critics also argue that creative-class approaches benefit already established arts and entertainment institutions, or strengthen new elites, with few spillover effects. (See Plate II, 9.) Among these critics are CUNY sociologist Sharon Zukin, author of *Loft Living: Culture and Capital in Urban Change* (New Brunswick, NJ: Rutgers University Press, 1989) and *Landscapes of Power: From Detroit to Disney World* (Berkeley, CA: University of California Press, 1991), and Cleveland State University researcher Mark Rosentraub, author of *Major League Losers: The Real Cost of Sports and Who's Paying for It* (New York: Basic Books, 1999) and *Guns and Butter of Redevelopment: Big Ticket Items, Neighborhoods and a New Convention Center* (Cleveland, OH: Maxine Goodman Levin College of Urban Affairs, 2003).

Other scholars have more favorable interpretations of the creative-class and arts-and-culture redevelopment strategies. For example, Philadelphia researchers Mark J. Stern and Susan C. Seifert report their discovery of "natural cultural districts" ("geographically defined networks created by the presence of a density of cultural assets in particular neighborhoods"). In their report *Cultivating "Natural" Cultural Districts* (Philadelphia: The Reinvestment Fund, TRF, and Social Impact of the Arts Project, SIAP, 2007), they argue that these clusters support social capital formation and are focal points of grassroots economic development. Jeremy Nowak, president of The Reinvestment Fund (TRF), a major community development intermediary, further explores this subject in *Creative Society and Neighborhood Revitalization: A Culturally Driven Investment Framework* (Philadelphia: TRF and SIAP, 2007), which discusses how to build a business model based on these findings. In *Cultural Planning and the Creative City* (Minneapolis, MN: Humphrey Institute of Public Affairs, 2006), University of Minnesota's Ann Markusen calls for an integrated cultural planning approach, noting the possibilities of linking neighborhood and city-based initiatives that build on arts communities of varied scale and capacity. Finally, Williams College economist Steven C. Sheppard and his team of student researchers quantified the positive effects of cultural clusters in deteriorated communities in *Culture and Revitalization: The Economic Effects of MASS MoCA on its Community* (North Adams, MA: Center for Creative Community Development, 2006).

This selection focuses on the efforts of municipal leaders in Barcelona, Spain, to reinvent their city through a creative-class strategy. Stricken with the full panoply of postindustrial problems – declining population, high rates of unemployment, crime, poverty, brownfields – Barcelona also had a magnificent but underused waterfront highly contaminated by industrial pollution and cut off from the rest of the city by rail. And the city lacked many amenities, especially public open space.

In the post-Franco era, municipal leadership decided to compete for the 1992 Olympic Games and, in preparation, undertook critical public investments in sports facilities. After securing the bid, they seized the opportunity to disperse other improvements through many neighborhoods. The most spectacular accomplishment was the transformation of a dilapidated industrial district, including three miles of waterfront, as the Olympic

Village. The city rerouted the rail, added a seaside boulevard, new apartments, hotels, and associated retail amenities. In addition, it revamped public transportation and the airport and built more than 200 local parks, plazas, schools, and markets. It also refurbished the center-city promenade known as the Ramblas, rid the adjacent neighborhood of crime, and built a museum and cultural center to attract new activities and tourists.

The work continued well after the Olympics. Buoyed by the galvanizing effects of hosting an international event, the city initiated another one, the 2004 Forum. Dedicated to culture, sustainable development, and peace – themes explored through seminars, performances, and exhibitions – the Forum took place in a redeveloped district, the Diagonal, a short distance from the Olympic Village. To accommodate the event, Barcelona extended mass transit, added parks, built housing, offices, and retail (for use post-Forum), and mitigated the environmental effects of nearby power and sewage treatment plants. In parallel efforts, the city relocated its port facilities, releasing additional land for redevelopment. All told, the city spent about €27 billion for the Olympics and 2004 Forum-related facilities. While these public works did not arrest Barcelona's population decline–the city, now 1.5 million people, lost 250,000 inhabitants since 1980–they succeeded in slowing the rate of decline in residents. And they did transform the city's image, as well as jump-start new economic functions including tourism and business formation.

But municipal leaders weren't done yet. They next turned their attention to Poblenou, a 500 acre, 117 block area bracketed by the Olympic Village and 2004 Forum. They created a municipal development corporation, 22@Barcelona, to oversee work that includes rezoning and reparcelization, investment in modern communications infrastructure (fiber-optic grids, Wi-Fi connections, upgraded electrical systems), a pneumatic solid-waste disposal system, new streets with bicycle lanes, open space, and incubator facilities for media, telecommunications, biosciences, and energy enterprises. The agency has forged partnerships with universities to relocate academic and research activities to the area. Several international and national corporations have moved in, including Microsoft, General Electric, American Express, and Telefónica Moviles (headquarters for the Spanish telephone company's cellular business). The development group is also supporting selective historic preservation and sponsoring job training and other programs to attract the necessary labor force. In short, 22@Barcelona is remaking the Poblenou district as a model of twenty-first-century sustainability and creativity, hoping it will one day accommodate 4–15 percent of the city's economic activities. This project, which will take decades to complete, bears monitoring. See www.22barcelona.com.

Miguel Barcelo Rota, author of this selection, is president of 22@Barcelona, the municipal development agency charged with overseeing the Poblenou redevelopment. He has written extensively on the use of economic clusters in urban revitalization. Underlying his work and that of other urban planners concerned with this topic are two streams of inquiry. The first is urban economics and its attention to agglomeration and industry cluster theory first described by Princeton economist Edward S. Mills and now discussed in basic urban economics textbooks such as Arthur O'Sullivan's *Urban Economics*, sixth edition (New York: McGraw-Hill, 2007). The second is urban sociology, especially network theories first introduced by Saskia Sassen in her classic *The Global City: New York, London, Tokyo* (Princeton, NJ: Princeton University Press, second edition 2001) and in an edited work, *Global Networks, Linked Cities* (New York and London: Routledge, 2002). Sassen established the concept of urban places being critical for the social interchange that contributes to economic growth. City planner Elizabeth Currid applies these ideas to the New York arts scene in *The Warhol Economy: How Fashion, Art and Music Drive New York City* (Princeton, NJ: Princeton University Press, 2007).

For additional reading on creative-class approaches to urban redevelopment, see Waikeen Ng and Judith Ryser (eds), *Making Spaces for the Creative Economy* (Madrid: International Society of City and Regional Planning, 2005), a worldwide survey of cities' knowledge economy-focused planning programs. For discussion of specialized sectors within the creative economy, see Ann Markusen, *The Artistic Dividend Revisited* (Minneapolis, MN: Humphrey Institute of Public Affairs, University of Minnesota, 2004); Ira Harkavy and Harmon Zuckerman, *Eds and Meds: Cities' Hidden Assets* (Washington, DC: Brookings Institution Center for Urban and Metropolitan Policy, 1999); and Richard Florida, *The University and the Creative Economy* (Pittsburgh, PA: Heinz School of Public Policy, Carnegie Mellon University, 2006).

BARCELONA AND ITS METROPOLITAN AREA

The Metropolitan Area of Barcelona (MAB) houses one of the biggest industrial and demographic concentrations in Europe. The MAB covers more than 3,200 sq. km, and is inhabited by 4.5 million people. The MAB also acts as a service centre for an extensive hinterland populated by 17 million inhabitants, including Valencia, Saragossa, the Balearic Islands and southeast France.

Thanks to its strategic geographical location and an extensive and fully interconnected transport network, Barcelona is a natural link between Europe, the Mediterranean and Latin America, and provides access to a potential market of more than 40 million people. This is why, throughout history, the city has been a natural point of exchange and attraction of ideas and business.

Barcelona's economic structure is composed of a polycentric network of cities with high levels of industrialisation, strong external connections and highly diversified production. These circumstances have created an economically stable structure, and a dynamic environment that employs synergies to drive the growth of new companies and projects. As a result, Barcelona has experienced sustained economic growth in recent years (above the European average), and has become one of the EU regions with a highest growth potential for the near future.

In addition, Barcelona has an excellent climate, along with rich cultural, entertainment and educational offerings. These assets have helped Barcelona become a leading European city in terms of quality of life, and one of the preferred locations in Europe for international businesses.

Regarding knowledge creation and transfer, Barcelona is home to one of the main university communities in Europe, including various highly prestigious business schools at the international level. It also avails of an extensive network of research centres, innovation centres and technology transfer centres which have placed the city at the forefront of research in Spain.

Nevertheless, despite these key assets, international indicators show that Barcelona is currently missing from the list of the most interesting cities for innovation and development of clusters. This is primarily due to the fact that the principal knowledge creation and distribution systems are still often based on a linear conception, which belong more to the old industrial production model than to the new demands of the creative society in terms of organisation, functionality and connections.

For this reason, the main public and private entities in Barcelona decided to jointly undertake an ambitious strategy called "Barcelona, the City of Knowledge," which is aimed at turning Barcelona into an innovation metropolis at the global level. The 22@Barcelona project constitutes a paradigm in terms of the overall regeneration plan, given that it is transforming the old economic heart of the city. The project area was formerly based on the industrial production model, but today its development is being guided by a new concept of "knowledge space" based on networks and corporate cooperation.

RENEWAL OF POBLENOU

For over 100 years, Poblenou was the industrial area *par excellence* in Barcelona, and Catalonia's principal economic engine. The 22@Barcelona project is aimed at recovering Poblenou's productive economic vocation by transforming its obsolete industrial areas into an innovative productive district, equipped with modern infrastructure and high levels of urban quality to attract and develop highly skilled activities.

Known throughout its time as the "Manchester of Catalonia", Poblenou was one of the most important centres of industrial innovation in Southern Europe up until the 1960s. Throughout the latter half of the twentieth century, the development of transport infrastructure and the building of specialized industrial facilities on the periphery of Barcelona led to factories relocating from the city centre out to the metropolitan ring. This trend gained momentum during the oil crises and resulted in the loss of more than 1,300 industries in Pobelnou between 1963 and 1990. Transport-related activities ended up occupying the major share of the space, and subsequently the space was gradually abandoned by other activities. These factors contributed to the deterioration of the urban surroundings.

The regeneration of Poblenou began with the interventions that took place as part of the

Olympic Games efforts, and involved important infrastructure and urban renewal projects. Two significant operations performed during this period are particularly relevant to the regeneration of Poblenou. The first project was the construction of the Barcelona ring roads, which removed the tramlines that used to cover the entire industrial heartland and greatly improved its connections with the rest of the metropolitan region, the port and the airport. The second project was the construction of the Olympic Village in this part of the city, which signified the first modern residential neighbourhood in Barcelona's coastal district. This project initiated the revival of Poblenou's seafront through the creation of new beaches and parks for civic use. The positive outcome of the Olympic Village project led to new improvements in Poblenou that connected it to the rest of the city, and continued the coastal transformation by creating new residential areas. As a result of these operations, Catalonia's traditional industrial heartland was gradually transformed into an accessible, central area, equipped with modern dwellings and quality public space.

CITIZENS' DEBATE

This new perspective ignited an interesting citizens' debate over the future of Poblenou's industrial areas, which, by the late 1990s, had still not been renewed. On the one hand, a section of public opinion supported the transformation of the industrial areas into residential space, as had been taking place up until that point. The market exercised major pressure in terms of maintaining this trend towards transforming these strategically located seafront areas. On the other hand, some sections of the public opposed transforming Barcelona into a predominantly residential city and, instead, were in favour of reviving the area's historic vocation as a production centre by adapting the urban and economic fabric to create new forms of wealth.

It was in this context that Barcelona's Economic and Social Strategy Plan, in its 1999–2005 programme, proposed the creation of a new urban model to encourage the development of emerging sectors of the knowledge economy. During this time, the Pacto Industrial de la Región Metropolitana de Barcelona (a public agency established to promote regional competitiveness) commissioned the Catalan Institute of Technology to study the development of skill-intensive activities and their impact on the transformation of the area. The study, entitled "The Digital City," showed that the most innovation-intensive activities are naturally concentrated in major urban nuclei all over the world, where there are high standards of infrastructure, flexible production space, a varied cultural, educational and entertainment offering, and a high quality of life. The study concluded that on account of Poblenou's privileged location and the extensive space available, regenerating the old production areas was a unique opportunity in Europe to create an important platform for innovation that would fulfill all the requirements of skill-intensive activities.

Barcelona's City Council decided the debate on the future of the industrial areas in Poblenou in favour of the regeneration of its historic production activities. Following this decision, a group of experts from various disciplines were given the task of drawing up new urban regulations that would control the transformation of this privately owned space and direct it towards the objectives established by the public sector. In July 2000, the modification of the General Metropolitan Plan concerning the renovation of Poblenou's old industrial areas, popularly known as 22@Barcelona project, was definitively and unanimously approved by the City Council.

Urban and economic strategy

Given its dimensions and implications, the 22@Barcelona project represents both a necessity and an opportunity. On one hand, the project satisfies the need to promote the renewal of degenerated areas of the city in order to revive their traditional economic and social vitality. On the other hand, the transformation of these extensive, centre city production areas present a unique opportunity to return Poblenou to its original status as the economic heart of the city by creating an important scientific, technological and cultural node aimed to consolidate Barcelona's position as one of the principal international innovation platforms.

As an urban regeneration project 22@Barcelona reinterprets the function of the old industrial fabric of Poblenou in a contemporary way, applying a New Town planning model based on sustainability, efficient infrastructure and quality of life. Under this premise, the project establishes a new model for compact, combined urban space by promoting the coexistence of production activities with residential areas and cultural facilities. In this way, the project seeks to create a more diverse, sustainable, balanced city, characterised by enhanced economic strength and cohesion.

As an economic regeneration strategy, the 22@Barcelona project propels the dynamic "Triple Helix" innovation model, based on the interaction between science and technology, and government and business, by concentrating the main institutions of the innovation system and creating new networks of cooperation. As such, the project creates areas of excellence to the overall competitiveness of the production fabric, and consolidates it in terms of international projection.

22@CODE: A NEW URBAN SPACE MODEL

The transformation brought on by project 22@Barcelona rests on substituting the former "22-A" zoning code, which designated Poblenou's productive space exclusively for industrial purposes, with a new urban classification called "22@". This new classification creates an innovative productive district that meets the requirements of the knowledge economy. Through an effective balance of incentives and duties, the new urban regulations allow the public sector to lead the transformation of private space towards the following objectives:

Mixed uses. The coexistence of complementary activities encourages innovation in production processes and creates an urban environment that is more balanced and sustainable. This ensures the vitality of the public areas throughout the day and allows people to live close to their places of work. Thus, the 22@ zoning code breaks with the traditional zoning model of the manufacturing economy, and clearly focuses on a model of a mixed and diverse city where production activities and research centres coexist with continuous training and technology transfer, commerce, housing units and green areas.

Urban density. In order to generate the critical mass necessary to develop agglomeration economies (and to promote a more rational use of space), the 22@Barcelona project avoids the low density that characterised industrial areas and advocates a dense and complex urban space. The new 22@ zoning code ensures the feasibility of this transformation through a system of incentives that requires every development project to contribute to financing the re-urbanisation of each street. This system helps to create new public spaces for green areas, facilities and housing through compulsory and gratuitous land transfers from developers to the community. Thus, the gradual renovation of industrial space promotes the revival of the social and commercial energy that characterised Poblenou in the past, and creates a space with a high standard of living in the centre of Barcelona.

Focus on knowledge-intensive activities. The 22@Barcelona project also aims to concentrate the most innovative activities of the creative society (the so-called "@ activities") within its area. These activities are characterised by their intensive use of information and communication technologies, space and highly qualified human capital. In order to ensure that these strategic activities are present in the new production spaces, the 22 zoning code creates a link between the building ratio of projects and the type of activity to be performed in the buildings. Thus, renewal plans that contain a specific percentage of the "@ activities" in their functional programs are entitled to a higher building ratio than projects that do not involve skill-intensive activities.

Through this system of incentives, urban regulations encourage promoters to reach agreements with the most dynamic and competitive companies to construct spaces that are tailor-made to their requirements.

Innovation Support Centres. In order to promote knowledge transfer in the production system, it is important that creation, transmission and knowledge distribution centres be located close to companies. This is why 22@Barcelona seeks to create a new system of facilities that provide support to the production system, instead of being linked exclusively to housing, as has traditionally been the case.

Therefore 10 percent of the converted space is obtained from the compulsory land transfers, and are to be used in relation to the so-called "7@

facilities" to house universities' centres of scientific and technological innovation, laboratories, design and R&D departments and on-going training centres. With this measure, the project encourages proximity and exchange between the principal innovation partners and ensures the on-going availability of highly qualified human capital.

Modern infrastructure. The most dynamic companies of the creative economy require a high level of services so that they can develop their activities in the most effective way possible. In this regard, 22@Barcelona has undertaken the overall reurbanisation of the sector through a new infrastructure network that is adapted to today's urban, social and environmental requirements. As a consequence, free competition is promoted among urban service operators, and priority is given to energy efficiency, acoustic comfort and responsible management of natural resources.

This new standard of urban services includes modernized networks for electricity, telecommunications, centralized air conditioning and waste collection and improves mobility of the sector, along with the quality and sustainability of the public spaces.

Flexibility. Given the complexity and extent of the old industrial areas in Poblenou, gradual and flexible renewal is envisaged. This means that each regeneration plan adapts to the environment's urban, economic and social characteristics without causing traumatic changes from the current land use. In order to accelerate the transformation of the district, the 22@ zoning code is also flexible in terms of the agents involved, given that the envisaged transformation will be a combination of both public and private initiatives.

Barcelona's City Council originally defined planning for six strategic sectors to act as focal points for urban revitalization, in addition to promoting the renovation of the remaining areas on behalf of the private initiative.

Finally, the project is also flexible in terms of its transformation mechanisms. In contrast to traditional urban planning, 22@Barcelona does not define the detailed precise planning of the district. Instead, it allows for different kinds of initiatives that vary in magnitude and building typologies, while respecting the typological and morphological diversity of previous industrial designs. In this way, the design of the district meets the quality, functionality and presentation requirements for a diverse range of final users.

TRANSFORMATION MANAGEMENT

22@Barcelona is a municipal corporation created by the Barcelona City Council to manage and promote the regeneration of the old industrial areas in Poblenou, which now have the new 22-A zoning code.

As an agency for urban development, the Council-run company 22@Barcelona is aimed at promoting and managing the creation of more than 4 million sq. m of new GFA, the re-urbanisation of 35 km of streets and approximately 240,000 sq. m of new public land for facilities, green spaces and subsidised housing in the former inner-city industrial areas.

As an agency for economic development, 22@Barcelona's mission is to promote the introduction and development of strategic content in these new areas and to favour the international projection of the new economic, scientific, training and cultural activities in the district.

STATUS REPORT

In the project's final phase, the transformation has been focused on developing facilities for knowledge-based activities as the vital step towards enabling economic regeneration in the area.

Consequently, since the project was approved, urban transformation of the 22@Barcelona district has been very intense and sustained. Over the past four years, refurbishment has begun in over 50 percent of Poblenou's industrial areas, guided by a total of forty-two approved plans for urban amelioration. These plans have brought approximately 1.3 million sq. m of potential productive space to the market, and have concentrated 60 percent of the supply scheduled for Barcelona in the coming years. The plans also designate more than 100,000 sq. m of land dedicated to new facilities, open spaces and approximately 2,300 subsidised housing units, of which a minimum of 25 percent must be used as rented accommodation.

In the business sphere, 22@Barcelona has also received a very warm welcome. More than ninety companies, leaders in their respective sectors, are already installed in the 22@Barcelona district, or are in the process of having their corporate headquarters built. In the new locations alone, productive activity in the district has increased by almost 280,000 sq. m. As a consequence, since approval of plan 22@, Poblenou has seen a significant increase in its productive structure, which is clearly evolving towards knowledge-intensive urban activities. Approximately 90 percent of the new companies and institutions setting up in the 22@Barcelona district are intensive users of ICT, space and qualified staff. In other words, they respond to the prototype for "@ activities".

After this initial phase that boosted the creation of new spaces and infrastructures, 22@Barcelona has begun a new, highly intensive phase of economic transformation of the area. A new set of projects, known as the "Seven Motors" for the 22@Barcelona district, have been designed to attract and create high value-added activities.

ATTRACTING TALENT: SEVEN MOTORS FOR DISTRICT 22@BARCELONA

The Council-run company 22@Barcelona promotes a set of initiatives, based on the dynamic "Triple Helix" innovation model, with the objective of structuring the area's economic transformation, and promoting the creation of areas of excellence in certain spheres of knowledge, to boost Barcelona's leadership in the knowledge economy. These initiatives, known as the "Seven Motors" for the 22@Barcelona district, reflect strategic objectives in terms of their present and future growth potential in Barcelona, and involve the main public and private agents in the processes of innovation. The "Seven Motors" are:

22@Media

Directed at encouraging the culture of excellence in the audiovisual sector, the 22@Media initiative promotes the creation of a large multidisciplinary centre to bring together currently dispersed and atomized major public and private agents in the sector. The objective of concentrating agents is to multiply their competitiveness and international projection.

The first phase of the 22@Media engine is the "Barcelona Media Park", a 60,000 sq. m facility that will be operative in mid 2007, and will bring together technical areas and offices related to the audiovisual sector. It will include university and continuing education, research and technology transfer centers, areas and services for incubating audiovisual companies, as well as temporary residences for students, teachers and entrepreneurs, and exhibition and interaction areas.

Barcelona Media Park is the result of collaboration of a public university (Universidad Pompeu Fabra), a leading private company in the Catalonian audiovisual sector (Mediapro), several entities from Barcelona City Hall (Department for Culture, Local Development Agency, 22@Barcelona) and the regional government (Centre for Corporate Innovation and Development). This is a clear example of an innovation medium in which public–private collaboration has made it possible to create an emerging system by bringing together leading players in the audiovisual sector.

22@TIC

Aimed at the Information and Communication Technologies (TIC in Spanish) sector, 22@TIC covers a set of initiatives to generate the critical mass required to position the 22@Barcelona district as a European leader for the ICT sector. To its advantage, Catalonia already has several highly renowned groups in the field that are able to attract investment and R&D.

The creation of this critical mass is based firstly on recruiting leading companies and institutions, and in creating spaces for small and medium-sized enterprises. The recent relocation of some important public institutions and private companies to Poblenou is a good example of this. These include the Department for Universities, Research and the Information Society of the Catalonian Government, as well as companies such as T-systems, Indra and Auna. As a result of an agreement between 22@Barcelona, the property developers Castellvi and the Dutch firm of consultants Zernik, the 22@Barcelona district will also begin construction of a landmark building for software and telecommunication SMEs, with high value-added exclusive services.

Secondly, the productive fabric need requires infrastructures for knowledge generation and transfer, such as the ICT Technological Centre promoted by the Catalonian government. Areas for diffusion and interaction, such as the future ICT and Productivity House promoted by Barcelona City Council, are designed to help spread the impact of ICT into society and business innovation.

Finally, mention should also be made of other initiatives directed at creating specialized business environments, such as the E-learning cluster promoted by the Open University of Catalonia and the Multilingual cluster led by the Digital Barcelona Foundation. 22@Barcelona will also include projects that promote the development of innovative services, such as GeriaTIC, a new welfare centre model which puts ICT developments to work for the well-being of the elderly.

22@Bioempresa

Catalonia has a strong clinical and biological base and great biotechnical potential, and 22@Bioempresa aims to support new companies (*empresas* in Spanish) in this sector that are in the growth and maturity phase by creating new services and areas adapted to their specific post-incubation needs.

Towards this end, 22@Barcelona has concluded agreements with both the Barcelona Scientific Park and the Biomedical Research Park to promote the creation of post-incubation areas for new companies. In conjunction with the Catalonian Investment Agency, 22@Barcelona has also undertaken a programme for attracting biotechnical companies to help create a business environment that encourages competition and exchange of experience among both consolidated and newly created companies.

22@Campus

Located in the area which housed the Universal Forum for Cultures in 2004, 22@Campus is designed for excellence in themes related to sustainability, which was one of the central themes developed during the cultural event. The new inter-university Campus for Technology and Business is specialised in mobility, energy and water technologies, as well as architecture, town planning and construction. 22@Campus is aimed at concentrating teaching, research, innovation and production activities related to each of these areas of knowledge.

The objective of this initiative is to become a clear exponent of the dynamic "Triple Helix" innovation model. 22@Campus itself is the result of an agreement between the Department for Universities, Research and the Information Society of the Catalonian Government, the City Halls of Barcelona and Sant Andria del Besos and the Barcelona Provincial Council. Construction of the first elements in this new space will begin shortly, thanks to the relocation of the Barcelona Industrial School and the recent agreement with the Consorci de la Zona Franca (Duty-free Zone Consortium) to create a business incubator. The construction will include a building for the temporary residence of teachers, students, entrepreneurs, professionals and researchers.

22@Emprendedores

The aim of 22@Emprendedores is to develop suitable conditions for district 22@Barcelona to become a magnet for entrepreneurs (*empredores* in Spanish), and to consolidate its position as an international platform for business creation and development. 22@Emprendedores has an integral and transversal programme, which includes various initiatives directed at creating a complete network of support infrastructures for national and international entrepreneurs, such as specialised incubators, temporary residential areas and finance programmes.

Among the programme's main assets is the presence in the district of the "Barcelona Activa", a Local Development Agency, which is the largest business incubator in Europe. 22@Emprendedores signed an agreement with the Consorci de la Zona Franca to construct a new building for entrepreneurs, which will be dedicated to the incubation and growth of technology-based spinoff companies developed in the area's research centres, such as the Polytechnic University of Catalonia and the Ramon Llull University.

22@Tecnologico

The technology centres amass in-depth knowledge of the optimum environment in which new technologies are generated in order to transfer this to the productive system. Therefore, these centres are one of the main intermediation structures between research and business. 22@Tecnologico is transversal in nature, and includes several initiatives to attract technology centres linked to those sectors with the greatest potential for future development in Catalonia. 22@Tecnologico then encourages centres to locate within the district's vicinity in order to promote the advantages of proximity and interrelation.

Mention should also be made here of the agreement to locate Alstom's Centre for Research, Development and Innovation in the 22@Barcelona district, along with leading technological centres such as the Centre for Technological Innovation "Barcelona Media", promoted by the Catalonian Regional Government's Department of Industry.

22@Poblenou

To foster quality of life and employment in Poblenou, 22@Poblenou encourages interaction between different urban agents through the creation of new links and cooperation channels. It consists of several initiatives, including:

The Association of 22@Barcelona Companies and Institutions, whose objective is to foster environments that facilitate cooperation between "@" companies or institutions located in Poblenou, and to become actively involved in configuring district 22@Barcelona as a European platform of innovation. . . .

The Educational Project, the result of a joint agreement between 22@Barcelona, Barcelona City Hall and the educational institutions in Poblenou, is designed to encourage the use of new technologies with students and promote work internship experience at companies in district 22@Barcelona.

VISION 2010

In summary, the 22@Barcelona initiative is a paradigmatic example of ambitious strategic planning and close public–private collaboration which has characterised Barcelona's transformation over the last twenty-five years, and which some have termed the "Barcelona Model."

With the support and involvement of the key urban institutions, 22@Barcelona has done well in the first phase of implementation; far exceeding initial expectations. However, the second phase of development will be decisive for the project's success, given that the competitiveness of other emerging regions will be decided over the next few years as well.

Barcelona firmly believes that, by maintaining a creative culture, the city can make the leap forward to obtain a leading position in the global knowledge society. The vision is that in the next five years 22@ would become more than a local town-planning zoning code, and would become a symbol of innovation. To this end, collaboration networks will be woven through leadership, perseverance and collective enthusiasm to turn the 22@Barcelona district into the geographical expression of the culture of excellence, and consolidate it as an international reference for Tolerance, Technology and Talent.

46

The Changing Politics of Urban Mega-projects

From *Land Lines* (2003)

Alan A. Altshuler and David E. Luberoff

Editor's introduction

Cities and regions in the United States have been undertaking large projects since the early nineteenth century. These efforts first revolved around infrastructure such as canals, water supply systems, and flood control or open space reserves located at the periphery of cities. In the New Deal era, the construction of civic buildings, early public housing on multi-acre sites, and bridges and their ramps took place in cities but did not widely disrupt residential neighborhoods. After passage of the 1949 Housing and Slum Clearance Act and the 1956 Federal-Aid Highway Act, however, the federal government underwrote projects in densely populated areas, causing significant population dislocation.

Postwar slum clearance and highway programs had substantial government funding as well as eminent domain power to condemn large swaths of land. Between 1949 and 1968, Congress provided $10 billion to almost 1,000 cities for about 2,000 urban renewal projects and an initial $41 billion to build a 47,000 mile interstate highway system. Under the Fifth Amendment to the U.S. constitution, local officials exercised eminent domain. Highways had long been eligible as a public use, but the Supreme Court qualified slum clearance only in the 1950s, in the landmark case *Berman v. Parker* 348 U.S. 26 (1954). (An extension of eminent domain for economic development received a positive but split Supreme Court decision, *Kelo v. City of New London* 125 S.Ct. 2655 (June 23, 2005), and is now causing a great deal of controversy.)

These programs had enormous effects on the practice of urban and regional planning, especially after the federal government mandated planning in 1954 for urban renewal and in 1962 for highways. In the former case, cities had to produce a plan (nicknamed a "701 Plan" after the section of the legislative mandate) or evidence that they were engaging in planning, that involved citizen participation as part of their applications for urban renewal funding. In the latter, places having populations of 50,000 or more had to create metropolitan planning organizations (MPOs) to develop regional transportation plans. Naturally, these requirements dramatically increased the demand for planners. In the next thirty years, their numbers, measured by membership in the American Institute of Planners, the professional association, grew exponentially from 240 in 1945 to 2,900 in 1960 and more than 11,000 in 1976.

As government officials embarked on large-scale projects they displaced thousands of people, often in poor, minority communities. According to historian Raymond A. Mohl, writing in John F. Bauman *et al.* (eds), *From Tenements to the Taylor Homes: In Search of an Urban Housing Policy in Twentieth Century America* (University Park, PA: Penn State Press, 2000), the highway program destroyed at least 330,000 houses and, at its height in the 1960s, uprooted 32,400 families a year, while urban renewal earned the epithet "Negro Removal." These programs, especially urban renewal, enraged the victims, released a stream of critical literature from the right (Martin Anderson, *The Federal Bulldozer*, Cambridge, MA: MIT Press, 1964) and the left

(Chester Hartman, "The Housing of Relocated Families," in James Q. Wilson (ed.), *Urban Renewal: The Record and the Controversy*, Cambridge, MA: MIT Press, 1966), and launched the career of architectural critic Jane Jacobs, whose *The Death and Life of Great American Cities* (New York: Random House, 1961) railed against urban renewal and urban and regional planners.

Despite the negative reactions, these programs also produced positive results. While large-scale projects were not successful everywhere, there were sufficient achievements to justify the continuing interest of urban leaders in pursuing them. The achievements include reconditioned decaying central business districts (Pittsburgh's Golden Triangle and Baltimore's Charles Center) and an efficient national highway network. Indeed, some of the projects will be sources of local economic strength in the twenty-first century: universities (Cleveland State), hospitals (Massachusetts General), cultural precincts (Lincoln Center), and downtown residential districts (Bunker Hill, Los Angeles). Today large-scale projects continue to unfold in many U.S. cities – the most famous is Boston's "Big Dig," the $14 billion reorganization of the city's highway network and the associated creation of 360 acres of open space, including forty acres downtown.

Whether regarded as negative or positive, large-scale projects have transformed the environment for urban and regional planning. The federal government gave responsibility for the planning and financing of large projects to cities, and citizens became more active in the projects from their inception. This resulted in what the authors of this selection label "Do no harm" planning. In this mode, planners needed new skills – they had to know how to mitigate negative effects on project surrounds, devise effective public–private partnerships, and manage citizen participation.

The authors, Alan A. Altshuler and David E. Luberoff, are at Harvard University, where Altshuler is Ruth and Frank Stanton Professor in Urban Policy and Planning, Graduate School of Design and the Kennedy School of Government, and Luberoff is the Executive Director of the Rappaport Institute for Greater Boston, John F. Kennedy School of Government. Altshuler's books include *The City Planning Process: A Political Analysis* (Ithaca, NY: Cornell University Press, 1965), *Regulation for Revenue: The Political Economy of Land Use Exactions* (Washington, DC: Brookings Institution Press, 1993) with José A. Gómez-Ibáñez, and *Governance and Opportunity in Metropolitan America* (Washington, DC: National Academies Press, 1999). Altshuler also served as secretary of transportation for the Commonwealth of Massachusetts from 1971 to 1975. Luberoff has been a columnist on infrastructure issues for *Governing* magazine. The Urban Politics section of the American Political Science Association named their monograph *Mega-projects: The Changing Politics of Urban Public Investments* (Washington, DC: Brookings Institution Press, 2003), the best book of the year in 2003.

For background relating to the construction of large-scale projects in the early nineteenth century, see Robert Fishman (ed.), *The American Planning Tradition: Culture and Policy* (Washington, DC: Woodrow Wilson Center Press, 1999). For more on postwar large-scale projects in Europe and the United States, see Bent Flyvbjerg *et al. Megaprojects and Risk: An Anatomy of Ambition* (Cambridge: Cambridge University Press, 2003); Susan S. Fainstein, *The City Builders: Property Development in New York and London, 1980–2000* (Lawrence, KS: University Press of Kansas, 2001); and a pioneering work by Sir Peter Hall, *Great Planning Disasters* (Berkeley, CA: University of California Press, 1982).

For case studies of the American experience, see Bernard J. Frieden and Lynne B. Sagalyn, *Downtown Inc.: How America Rebuilds Cities* (Cambridge, MA: MIT Press, 1989); David L. A. Gordon, *Battery Park City: Politics and Planning on the New York Waterfront* (New York: Routledge, 1997); and Lynne B. Sagalyn, *Times Square Roulette* (Cambridge, MA: MIT Press, 2001). Lynne B. Sagalyn's essay "Public/Private Development: Lessons from History, Research, and Practice," *Journal of the American Planning Association* (73: 1, March 2007) traces the growth of creative financing techniques for urban revitalization from the 1970s to the present.

From the earliest days of the Republic, civic boosters have prodded American governments to develop large-scale physical facilities – mega-projects, we label them – ranging from canals and railroads in the nineteenth century to rail transit systems and convention centers today. Until the mid twentieth century, such projects tended to involve modest public expenditures by contemporary standards and they rarely caused significant disruption of the existing urban fabric.

This pattern altered abruptly in the 1950s and early 1960s. Central city economies had, with rare exceptions, stagnated through the Great Depression and World War II, and they continued to do so in the early postwar years. Local business and political leaders concluded that if central cities, particularly those developed prior to the auto age, were ever to thrive again, they would require major surgery. Specifically, they needed to clear slums to provide large downtown sites for redeveloped office districts; to facilitate high-speed automotive movement between suburban and central city locations; and to provide larger airfields with attractive terminals for the nascent commercial aviation industry.

Recognizing that they could not finance these expensive projects with locally generated funds, urban leaders campaigned aggressively for federal assistance, and they were successful in obtaining considerable amounts of funding. We attribute their success mainly to the following factors: (1) public confidence in government was unusually high in the postwar period; (2) business leaders generally accepted the need for government activism to sustain prosperity; and (3) although cities lacked the political clout to secure expensive programs on their own, they were able to participate in much broader coalitions, most notably, those focused on housing (which expanded to include urban renewal) and highways. Urban aviation advocates were less successful, but as aviation traffic boomed they were able to fund new airports and expand old ones by relying primarily on revenues from landing fees and terminal leases.

During the late 1950s and the 1960s these efforts combined to produce an unprecedented wave of urban public investment. While often successful on their own terms, these projects tended to be highly disruptive as well, destroying in particular vast amounts of low-income housing and urban parkland. Project advocates maintained that the public should accept such impacts to advance the greater good. Robert Moses, New York's famed master builder, never tired of citing a French proverb: "You can't make an omelet without breaking eggs."

During the late 1960s and early 1970s, however, neighborhood activists allied with those involved in the emerging environmental movement against the full panoply of mega-project programs that had come into being during the 1950s. They succeeded not just in blocking large numbers of planned expressways, renewal schemes and airport projects, but also in securing the adoption of numerous statutes, regulations and judicial doctrines, thus strengthening the hand of critics in urban development controversies. For a time it seemed to most observers that the era of mega-project investment in cities was over.

"DO NO HARM" PLANNING

The forces committed to mega-projects have proven highly resilient and adaptive, however. While the character of such investment has changed dramatically since the 1970s, its volume has remained high. Nevertheless, mega-project advocates have had to work within new constraints; they have had to learn the art of making omelets without leaving a residue of broken eggs. We label this art, as exercised in the domain of urban land use, "do no harm" planning. Its essential components are the selection, siting and design of projects to minimize disruptive side effects, and the aggressive mitigation of any harmful impacts that cannot be avoided entirely. Most obviously, governments have ceased clearing slums and building expressways through developed neighborhoods, and only one major new passenger airport, in Denver, has been constructed since the early 1970s.

Public investment in facilities such as rail transit systems, festival retail markets, sports stadiums and arenas, and convention centers has surged. Within the transportation sector, moreover, investment priorities have shifted toward the reconstruction of existing highways, new construction on suburban fringes and airport terminals rather than runway improvements. The great advantage of such projects is that they are relatively easy to site either at some distance from existing development or in older commercial districts that have few preservationist defenders.

Where cities and states have gone forward with major highway and airport projects, they have taken extraordinary steps to minimize social and environmental impacts. The new Denver airport, for example, is on a previously rural fifty-three-square-mile site twenty-five miles east of downtown. Its location and scale were determined primarily by two considerations: land assembly without the disruption of existing residential enclaves; and future airport operation without significant noise impacts overflowing the airport boundary. Boston's $14.6 billion Central Artery/Tunnel project, known colloquially as the "Big Dig," appears very different, in that it is located in the heart of downtown, but it is virtually identical in its do no harm planning orientation. It is almost entirely underground as it passes close to built-up areas (replacing a previous elevated roadway); it has been threaded into the urban fabric without the taking of a single home; and it will add significantly to the city's parkland.

COMMON THEMES

In addition to do no harm planning, our review of mega-projects built over the past two decades identified the following themes as particularly salient.

Business support

While insufficient by itself, strong business support has generally been an indispensable condition for mega-project development. Within the business community, leadership has almost invariably come from enterprises with deep local roots, particularly in real estate ownership, development and finance. The strongest supporters of Denver's new airport, for example, were those who owned property with commercial development potential near the new site; downtown businesses concerned that the city's existing airport was too small to allow for the region's continued development; and the banks and financial service firms that had lent money to many of the city's property owners and developers. Similarly, the most active and effective support group for Boston's Big Dig has been the Artery Business Committee, a coalition of those who own major buildings adjacent to the artery's corridor and several major employers with historic roots in downtown Boston.

Public entrepreneurs

In addition to well mobilized constituencies, aggressive, deft government officials have been indispensable to the success of recent mega-project proposals. Indeed, it was frequently they who originated project ideas and first sparked the formation of supportive coalitions. Even when others initiated, they commonly took the lead in crafting strategies, tactics and plans; in lobbying for state and federal aid; in securing other types of needed legislation and regulatory approvals; and in dealing with project critics.

Though business groups initiated some projects, they seemed more frequently to invest in proposals originated by public entrepreneurs. The business constituents were by no means easy marks, of course. Like venture capitalists in the private sector, they considered a great many ideas brought to them by public entrepreneurs (and others), but invested only in those few that looked particularly good for their enterprises, were to be carried out mainly or entirely at public expense, and had a reasonable chance of securing the myriad approvals required.

Illustratively, Boston's Big Dig was conceived by Fred Salvucci, a transportation engineer who had become active in battles against planned highway and airport projects during the 1960s and then served as transportation secretary for twelve years under Governor Michael Dukakis. During the first Dukakis administration (1975–79) the main constituencies for a new harbor tunnel (business) and for depressing the central artery (neighborhood and environmental groups) were at loggerheads. While temporarily out of office from 1979 to 1983, however, Salvucci concluded that the politically feasible strategy might be to marry these projects, while also relocating the tunnel to an alignment far from a neighborhood that it had historically threatened. This strategy in fact resolved the local controversy, and prepared the way for a successful campaign for massive federal aid, led again by Salvucci with critical business support.

Denver Mayor Federico Pena broke a similar type of logjam that had persisted for years over whether to expand Denver's existing Stapleton Airport or build a new facility on a large site outside the city's borders. Concluding that the obstacles, both political and environmental, to expanding Stapleton were insuperable, but that city ownership and operation of any new airport remained a critical objective, he negotiated successfully with adjacent Adams County for a massive land annexation. To achieve this objective, he accepted conditions protecting county residents from significant airport noise and guaranteeing Adams County most of the tax benefits that would flow from economic development around the new airport. With local agreements in hand he, like Salvucci, then led a successful campaign for special federal assistance.

Mitigation

Do no harm plans avoid substantial neighborhood and environmental disruption but it is impossible to build a mega-project with no negative side effects. The commitment of do no harm planning is to ameliorate such impacts as much as possible, and to offset them with compensatory benefits when full direct mitigation cannot be achieved. The boundary between mitigating harm and providing net benefits to protesting groups is often indistinct, however, so the norm of mitigation provides leverage as well for skilled activists whose demands are at times tangential to the mega-projects whose budgets they seek to tap. Mega-project champions in turn reflected on the fate of such projects as New York City's proposed Westway, which failed because of what seemed at first a minor legal challenge. They were deathly afraid of litigation and were frequently willing to make very expensive concessions in return for agreements by critics not to sue.

During permitting for the Big Dig, for example, Boston's Conservation Law Foundation (CLF), a group whose signature strategy was litigation for environmental purposes, threatened to sue unless the state committed to accompany the highway project with a multi-billion dollar set of rail transit investments, mainly for expansion. CLF's rationale was that the transit projects would prevent the new road from filling up with traffic, which in turn would generate more air pollution. Modeling done for the project (as well as data from other regions) showed that the Big Dig would not in fact have significant air pollution effects, and that investing in rail transit extensions would be a particularly inefficient way to offset pollution effects if they did occur. Nonetheless, both Democratic and Republican state administrations acquiesced to CLF's demands because they did not want to risk litigation, which at the very least threatened project delays and might also have imperiled the breadth of local consensus in support of the Big Dig.

Bottom-up federalism

A naive observer of American politics might assume that the federal government distributes grants to achieve national goals. In fact, however, the grantor–grantee relationship is usually much more complicated than that. Recipient jurisdictions are typically active participants in the coalitions that bring new programs into being and provide them with critical support each budget season. The programs of aid for mega-project investment that we examined were all distinguished more by their openness to local initiative than their sharp definition of national purpose. If grantee jurisdictions had a great deal of influence collectively on program structure, moreover, they had even more when it came to projects, and they were able to exercise it individually.

Every project we studied was initiated by subnational officials and interest groups, and it was they who took the lead at every stage in the decision process. While limited in their discretion by federal program rules, they were alert as well to opportunities for securing waivers, statutory amendments and add-on funds, with the assistance of their congressional delegations. Stated another way, when federal aims are diffuse and weakly defended, principal–agent theory (as applied to the intergovernmental system) needs to be read bottom-up rather than top-down.

High and rising costs

Do no harm designs and related mitigation agreements have tended to produce projects that are vastly more expensive than their historic predecessors. According to Brian Taylor (1995), the average cost per centerline mile of urban freeways rose by more than 600 percent in real terms from the 1960s to the 1980s, and costs were even more extreme in some of the mega-projects we examined. Whereas Taylor found that urban freeways cost on average about $54 million per centerline mile (in 2002 dollars) in the 1980s, for example, the Big Dig cost $1.9 billion per centerline mile. Judith Grant Long (2002) reports in a similar vein that the average cost of new stadiums and arenas more than quadrupled in real terms from the 1950s to the 1990s, and we have calculated that light rail development costs increased by nearly two-fifths from the 1980s to the 1990s.

Both older and more recent projects have been marked by a consistent pattern of substantial cost increases between authorization and completion. The projected cost of Boston's Big Dig, for example, has roughly tripled in real terms since its approval by Congress as an interstate highway project in 1987. The cost of Denver International Airport more than doubled from the late 1980s, when it received voter approval and its federal funding commitments, to its completion six years later.

While a full study of this issue was beyond the scope of our work, we judge that the consistent pattern of underestimation has two primary causes. First, project advocates have very strong incentives to estimate optimistically as they seek political commitments of support. Second, mega-projects are often so complex – both technically and in terms of the mitigation agreements that will often prove necessary to keep them on track – that early cost estimates are typically little more than guesses within very broad ranges.

Locally painless project funding

The hallmark of successful mega-project financing is that projects should appear costless, or nearly so, to the great majority of local voters. The easiest way to achieve this result is to rely on funding from higher-level governments. Where such aid is unavailable or insufficient, the challenge is to identify other sources of revenue to which local voters are generally insensitive, which means, above all, avoiding local property and income taxes and spreading the burden beyond host city residents.

This challenge became increasingly salient after 1970 with rising antitax sentiment, the end of federal renewal aid, and the surge in capital spending for such facilities as stadiums, arenas and convention centers, for which federal aid was only rarely available. In the growing domain of mass transit, moreover, federal matching ratios have tended to decline since 1980.

The revenue strategies adopted to deal with these challenges have been varied and ingenious. New terminals and runways at major airports have been funded largely by increased landing fees, lease payments, and (since the early 1990s) ticket surcharges authorized by the federal government but imposed locally. Stadiums, arenas and convention centers are commonly funded by taxes that fall mainly on nonresidents, such as taxes on hotel rooms, car rentals and restaurant bills. Where broad-based taxes have been unavoidable, the preferred method has been incremental add-ons to sales taxes, which typically require voter approval. Voters have often said no, but sales tax increases provide large amounts of revenue when they are adopted, and when they are not, project advocates routinely come back with revised plans. In Los Angeles and Seattle, for example, transit advocates responded to referendum defeats by scaling back their rail plans and allocating some of the projected revenue to bus service and local road improvements.

LOOKING TO THE FUTURE

Almost two decades ago, when New York City's ambitious Westway project died even though its backers had helped pioneer the do no harm planning and design paradigm, then-Senator Daniel Patrick Moynihan wondered whether it had become so difficult to build public projects that "Central Park could not conceivably be built today." Recent history suggests, however, that the mega-project impulse remains strong. The pertinent question is not whether the U.S. political system can still generate mega-projects but whether the projects that go forward are typically worth their costs to taxpayers.

In general, economists are skeptical about the cost-effectiveness of the most prominent mega-projects, from the Big Dig to the scores of rail transit systems, professional sports facilities and convention centers, built over the past twenty-five years. Project advocates invariably retort that the economists miss intangible project benefits such as fostering community pride and (in the case of transit, particularly) strengthening the likelihood of smart growth practices in new development. The national coalitions in support of highway and airport improvements, which economists tend to rate more favorably than other types of projects, have argued vociferously that current environmental rules and opportunities for critics to litigate are too onerous and should be relaxed.

There is no easy resolution of these issues because they involve tradeoffs between important, deeply held values. However, our review of a half-century of public works projects in urban areas has left us with three clear impressions. First, states and localities should be required to bear half or more of the cost of projects they undertake, because great windfalls of earmarked money from higher levels of government tend to overwhelm serious local deliberation. Second, strong environmental regulation helps ensure that local pro-growth coalitions do not leave fouled environments or devastated neighborhoods in their wake. Finally, while referenda are in general a flawed instrument of policy making, the evidence seems to suggest that the requirement of voter approval for major local projects tends to have a salutary effect on the bargaining between business groups that stand to benefit financially from the proposed investments and the more general interests of local taxpayers and residents.

The Spatial Organization of Cities: Deliberate Outcome or Unforeseen Consequence?

From *Background Paper, World Development Report* (2004)

Alain Bertaud

Editor's introduction

One of the most enduring issues of urban planning is density, defined as the amount of development in a given area. Planners have long been concerned with the topic, figuring out what determines density, how dense places should be, how to design for differing densities that are efficient, equitable, and sustainable.

From this work, three big ideas underlie current thinking. First, transportation is a determining factor in density. Second, ideas about optimal levels of density vary from place to place. Third, U.S. policy makers at all levels of government, but particularly at the local and regional levels, are faced with making new choices about density in their jurisdictions. (See Thomas L. and Katherine Daniels, "Environmental Planning," and John R. Nolon, "Shifting Ground to Address Climate Change: The Land Use Law Solution," in this *Reader*). Urban planners will be closely involved in helping shape these changes but they will be working with already established settlement patterns.

These settlement patterns are the result of national transportation policy and the presence of inexpensive fuel and plentiful land. To be sure, in the twentieth century, planners encouraged the switch from fixed routes (subway, trolley, bus, train) to flexible lines (roads, highways, freeways, and interstates) and celebrated the rise of mobility gained by proliferation of cheap cars and trucks and national investment in roads. Over time however, the results of this approach became clear: monocentric (a dominant center) regions gave way to polycentric (many centers) arrangements and, ultimately, uncontrolled sprawl – a development reinforced by new work patterns that mean that proximity to urban centers is less important. According to the Lincoln Institute of Land Policy, thirty years ago only ten metropolitan areas had more than half of their populations living ten miles or more from the center; today the number is forty-six (Julie Campoli and Alex S. MacClean, *Visualizing Density*, Cambridge, MA: Lincoln Institute of Land Policy, 2007, p. 4).

Two countertrends – the increase in downtown living and the rise of transit-oriented development – are emerging, albeit slowly and in a limited fashion – but they offer a hint of the future. Between 1970 and 2000 many of the nation's most populous cities welcomed new residents to converted office buildings and new construction in their central business districts – households were up 8 percent and population 10 percent, with households increasing 13 percent in the 1990s alone (Birch, *Who Lives Downtown?* Washington, DC: Brookings Institution, 2005). Many more places are willing to experiment with laying out higher-density nodes around critical transportation junctions. This is particularly true in places that have invested in rapid transit systems such as Portland and Washington, DC, as shown by Robert Cervero, who inventoried transit-oriented developments, collecting data on more than 100 (Cervero *et al.*, *Transit-oriented Development in the United States: Experiences, Challenges and Prospects*, Washington, DC: Transportation Research Board, 2004).

Yet these positive countertrends are, to date, miniscule relative to overall development. Land consumption in this country continues to outpace population growth. In the 1990s, U.S. land consumption increased 47 percent, while the population grew 17 percent. Meanwhile, the average-sized home rose 35 percent (from 1,375 sq. ft in 1970 to 2,057 in 2000).

Furthermore, the acceptability about density levels varies across the country. And Cervero observes Midwestern ideas of high density would make New Yorkers or Washingtonians laugh. The Lincoln Institute of Land Policy portrayed some of these differences, noting some surprising regional variations. On average, metropolitan Los Angeles is far denser than metropolitan New York. LA's suburban density level – 82 percent that of the central core – is far greater than New York's – only 12 percent. This translates to overall densities of 4,000–6,000 people per acre (p.p.a.) in the western metro, and 1,000–2,000 p.p.a. in the eastern one (Campoli and MacLean, *Visualizing Density*, p. 24). And the higher-density pattern repeats in fast-growing cities like Tampa or Tucson, characterized as experiencing dense sprawl – eighteen to twenty single-family dwellings per acre but without retail amenities. Exurbs in the East and Midwest that have isolated development – subdivisions with ten or more units to the acre laid out as islands in a sea of agricultural land. Witold Rybczynki's *Last Harvest: From Cornfield to New Town* (New York: Scribner, 2007) is a riveting account of how the developer of one such settlement in Pennsylvania convinced the town to allow denser construction than customary but still built an isolated subdivision. At the same time, many older cities with higher densities – Philadelphia, Baltimore, and Detroit, for example – continue to hemorrhage population while their suburban areas bleed out. They, too, are struggling to identify appropriate densities and find complementary uses for vacant and abandoned properties while taking advantage of preexisting infrastructure (Richard M. McGahey and Jennifer S. Vey, eds, *Retooling for Growth: Building a Twenty-first Century Economy in Older Industrial Cities*, Washington, DC: Brookings Institution Press, 2008). Smart Growth solutions, meaning compact, transit-oriented development, have worked reasonably well in a few states, but are not widespread. See Oliver Gillham, "Regionalism," and Gary Hack, "Planning Metropolitan Regions," in this *Reader*.

And in the next twenty-five years, as the population explodes, these choices about density will be critical. According to Chris Nelson, Presidential Professor at the University of Utah, in the next twenty-five years the nation will be on a $30 trillion construction binge to accommodate anticipated demographic growth – of an estimated 70 million people, 32 million households and 40 million jobs. Nelson predicts that 38 million dwelling units will be built (40 percent of the current stock) and 78 billion sq. ft of nonresidential structures (90 percent of current stock) will be added ("Planning for a New Era," *Journal of the American Planning Association*, fall 2006, pp. 393–409).

With the nation on the brink of an energy crisis due to increasing costs and the threat of global warming, making existing places more dense is a logical solution. But convincing the public that places that are more dense are not ugly, congested, and unpleasant will be a challenging task. Planners will need to educate themselves and the public about how to make higher densities work, eliminating the auto dependency that afflicts many newly built, high-density communities today. Planners and citizens alike will have to balance sustainability, equity, and efficiency as they devise answers.

In this selection, urban planner Alain Bertaud offers an analytical framework to assist with understanding density and its relationship to transportation. Combined with visual images such as those provided by Campoli and MacClean in the already cited *Visualizing Density*, this selection supplements design-based recommendations.

Bertaud, a consultant to the World Bank, was formerly the World Bank's Principal Urban Planner in the Urban Development Division, working on large-scale infrastructure and housing projects in Russia, China, Eastern and Central Europe. He is working on a monograph analyzing spatial arrangements in forty-nine metropolitan areas around the world. He has written extensively on urban spatial structures and city planning and land use and financial models. See alain-bertaud.com/ for a full list of his publications and reports.

For more reading on density and the drivers of metropolitan development, see Terry Moore and Paul Thorsnes, with Bruce Appelyard, *The Transportation/Land Use Connection* (Chicago: APA Planners Press, 2007).

Three books treat practical matters related to density. They are: the American Planning Association's *Planning and Urban Design Standards* (Hoboken, NJ: Wiley, 2006), Donald Watson (editor-in-chief), Alan Plattus and Robert Shibley (eds), *Time-Saver Standards for Urban Design* (New York: McGraw-Hill, 2003), and Sidney O. Dewberry (editor-in-chief), *Land Development Handbook: Planning, Engineering and Surveying* (New York: McGraw-Hill, 2002).

Helping policy makers and citizens come around to supporting higher density is challenging because urban planners must overcome preconceived ideas about it. Michael Kwartler and Gianni Longo's *Visioning and Visualization: People, Pixels and Plans* (Cambridge, MA: Lincoln Institute of Land Policy, 2008) and David Walters, *Designing Community: Charettes, Masterplans and Form-based Codes* (Oxford: Elsevier and Architectural Press, 2007) are valuable guidebooks.

Many planning agencies are exceptional resources on issues of density. Of particular note is the Smart Growth Office of the Delaware Valley Regional Planning Commission, whose website (www.dvrpc.org/planning/regional/SmartGrowth.htm) is a fount of information on local and regional planning studies.

Other sources of information on density include Smart Growth America (www.smartgrowthamerica.org), an advocacy group, and Planetizen, the online planning and development network. See, for example, the thoughtful essay by Terry Holzheimer, director, Economic Development, Arlington, VA, at www.planetizen.com/node/23176.

INTRODUCTION

The spatial structure of large cities evolves very slowly and grows in only a few directions. It is shaped by market forces interacting with regulations, infrastructure investments and taxes. It also results from unforeseen consequences of policies that are often implemented without any particular spatial concerns. However, once land is developed it is nearly impossible to bring it back to its natural state.

Defining an optimum city shape is also impossible because urban development objectives change with time. Typically, city mayors pursue policies that meet multiple objectives simultaneously. The choice of the appropriate tradeoffs between several often conflicting objectives is a political, not technical, decision, one best left to elected officials.

However, identifying the type of city shape that would be consistent with a specific objective is possible. For instance, a sprawling urban shape is unfavorable to the development of public transport while a compact form tends to be favorable. Urban planners should constantly monitor the impact of specific policies on city shape. They should be aware of the effects of the most common planning tools – land use regulations, infrastructure investments and taxation – on cities' spatial organization. They should make sure that the urban shape resulting from their actions will be consistent with the objectives set by elected officials. To accomplish this, planners need a good understanding of the dynamics and patterns of urban growth.

As Figure 47.1 illustrates, the number of large cities (1 million or more population) has increased since 1975. UN analysts believe this trend will continue into the future. What are the economic reasons behind this surge? One explanation is that the presence of a large, unified labor market promotes urban population growth. When a city's spatial organization allows complete mobility, households, whatever their location within the metropolitan area, can reach all job locations within a reasonable time (say less than one hour).

Planners can evaluate how various urban spatial structures support unified labor markets. They can employ three easily obtained measures: the average daily trip pattern (the length of trips in a city), average land consumption (average built-up area of a city) and the density profile (spatial location of population). These data come from census data, land use plans and satellite imagery. Planners can use these metrics also to test policy interventions such as transportation improvements or zoning changes.

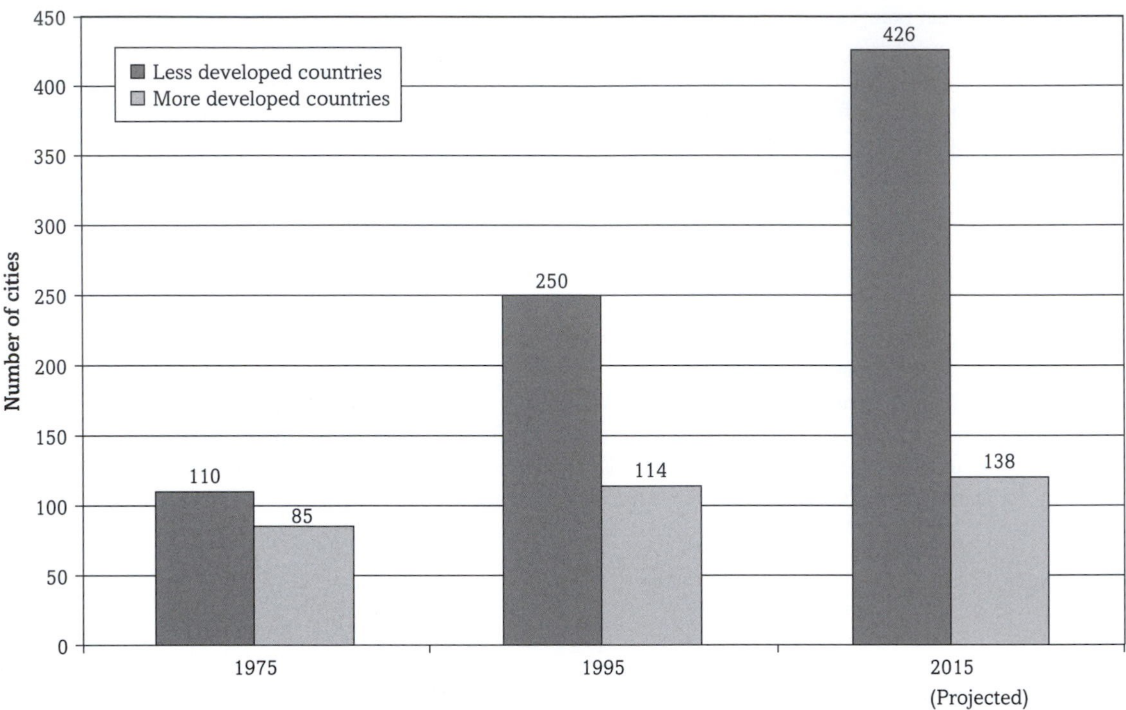

Figure 47.1 Growth of cities larger than a million people between 1975 and 1995. (Derived from United Nations, *World Urbanization Prospects, 1999 Revision*, 2000)

AVERAGE DAILY TRIP PATTERNS

The location of jobs (or trip-generating activities) determine the average daily trip pattern measurements. To capture them, analysts calculate the average distance to the center (ADC) and average distance between random points (ADR) using GIS and satellite data.[1]

In a monocentric city jobs are concentrated in the central business district (CBD). However, over the years, many jobs clustered outside the traditional CBD, prompting a polycentric structure. No city is ever 100 percent monocentric, and few are 100 percent polycentric (i.e. with no discernible downtown). Some cities are dominantly monocentric, others are dominantly polycentric and many are in between. Some circumstances tend to accelerate the mutation toward polycentricity – historical business center with low level of amenities, high private car ownership, cheap land, flat topography, grid street design – others tend to retard it – historical center with high level of amenities, rail-based public transport, radial primary road network, and difficult topography preventing communication between suburbs.

Figure 47.2 offers a conceptual view of the relationship between urban spatial structure and average daily trip patterns. In Figure 47.2(a), a monocentric city supports a unified labor market by providing the possibility of moving easily along radial roads or rails from the periphery to the center.

Polycentric cities can also support a unified labor market. It functions very much in the same way as a monocentric city: jobs, wherever they are, attract people from all over the city. The pattern of trips is different, however. In a polycentric city each subcenter generates trips from all over the built-up area of the city (see Figure 47.2(c)). Trips tend to show a wide dispersion of origin and destination, appearing almost random. Trips in a polycentric city will tend to be longer than in a monocentric city.

A cautionary note: some urban planners idealize polycentric cities, thinking that a self-sufficient community will develop around each employment cluster and that a number of self-sufficient "urban villages" would then aggregate to form a large

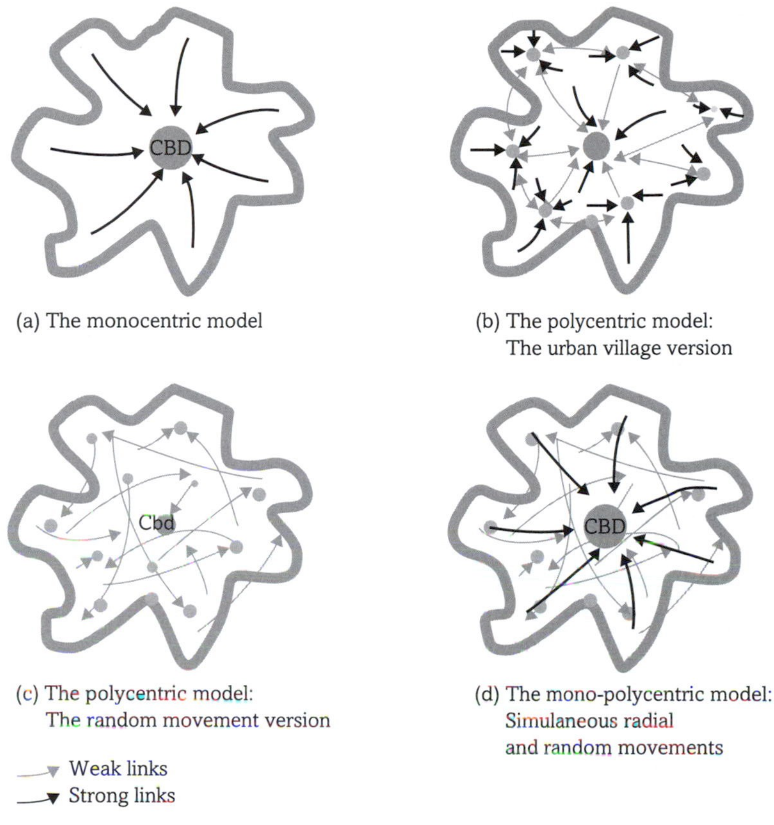

Figure 47.2 Schematic representation of trip patterns within a metropolitan area

polycentric metropolis (Figure 47.2(b)). In such a large city, they hold, trips would be very short and everybody would walk or bicycle to work. To date, nobody has ever observed this phenomenon in any large city. A metropolis constituted by self-sufficient "urban villages" contradicts the only valid explanation for the existence and continuous growth of large metropolitan areas: the increasing returns obtained by having large integrated labor markets. The urban village concept is the ultimate labor market fragmentation.

In spite of not being encountered in the real world, the utopian concept of a polycentric city as a cluster of urban villages persists. For instance, in some suburbs of Stockholm urban regulations allow developers to build new dwelling units only to the extent than they can prove that there is a corresponding number of jobs in the neighborhood. The satellite towns built around Seoul and Shanghai are another example of the urban village conceit; surveys are showing that most people living in the new satellite towns commute to work to the main city, while most jobs in the satellite towns are taken by people living in the main city.

AVERAGE LAND CONSUMPTION

The average land consumption of a city is an important parameter in defining an urban structure. Expressed as people per hectare (p/ha), Analysts derive the measure by dividing population by the built-up area (land coverage). Census has the population and satellite images provide land coverage data. Built-up area is defined as including all uses with the exception of contiguous open space larger than 4 ha, agricultural land, forests, bodies of water and any unused land. In addition, land used by airports and by roads and highways not

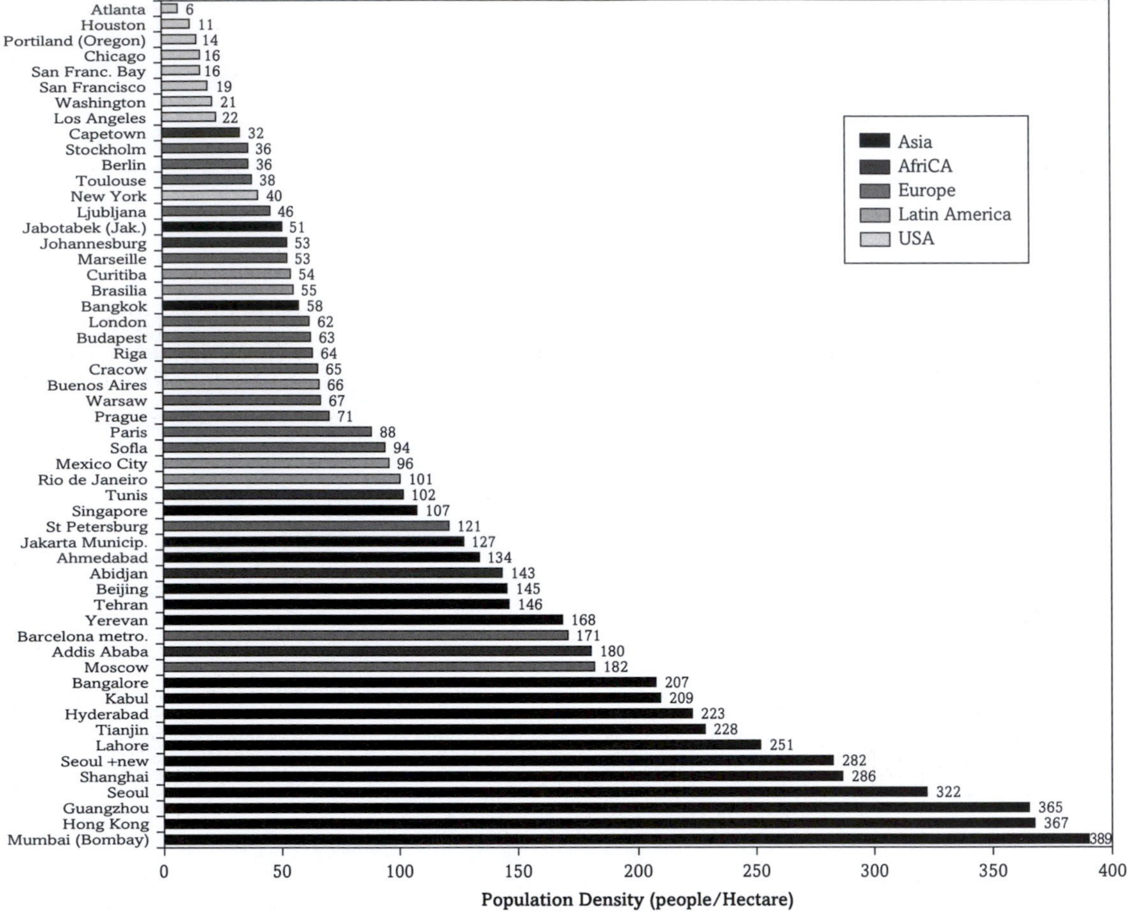

Figure 47.3 Average land consumption of fifty-four cities. (Alain Bertaud, *Order without Design*, 2005)

adjacent to urban used land is not included in the area defined as built-up area.

As seen in Figure 47.3, a comparison of average land consumption among fifty-four cities around the world shows differences of several orders of magnitude. Although other studies, not presented here, show that there is no clear correlation between density and income or population size, there is a correlation between the density of a city and its location on a continent. U.S. cities have the lowest densities; African, European, and Latin American cities have medium-range densities and Asian cities have high densities.

The cities listed in Figure 47.3 are all reasonably successful; some might be better managed than others, but the great majority of them constitute the prime economic engines of their countries. This suggests that, given the wide range of densities encountered, there is no "right," "correct," "manageable" or "acceptable" range of density per se.

DENSITY PROFILES

The density profile shows the internal distribution of population within a city's built-up area usually measured from a central point, the CBD. In a majority of cities, it is expressed as a negatively sloped exponential curve. Figure 47.4 illustrates the large difference in density profiles among U.S. and Asian and European cities. These variations can be related to the average daily trip pattern data. Dominantly monocentric cities tend to have much higher densities close to the CBD than cities that

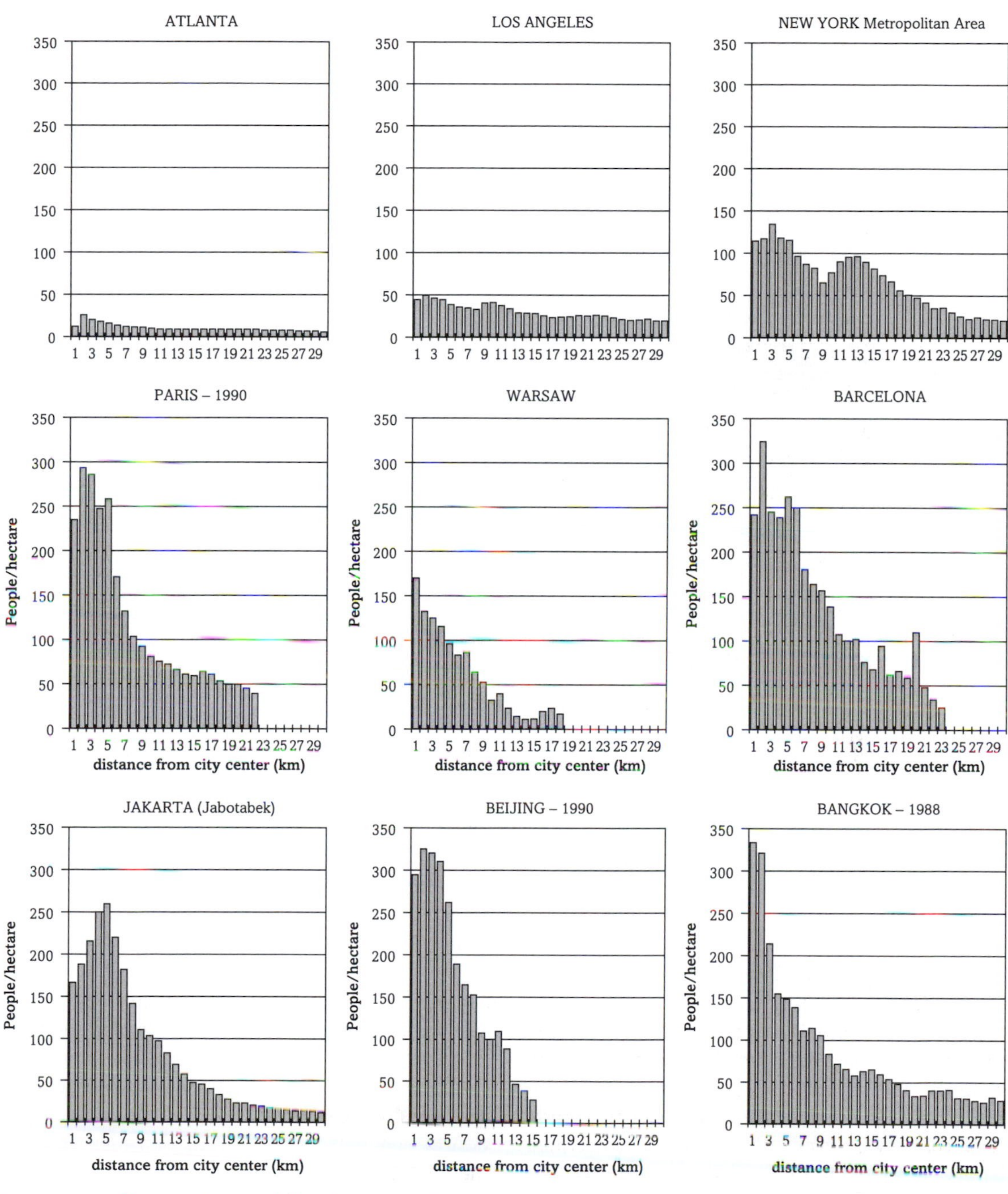

Figure 47.4 Comparative population densities in the built up areas of selected metropolitan areas. (Alain Bertaud, *Order without Design*, 2005)

are dominantly polycentric. The six non-U.S. cities shown on Figure 47.4 have densities within 4 km of the CBD ranging from 170 to 320 (p/ha) compared to a range between 20 p/ha (Atlanta) to 120 p/ha (New York).

DAILY TRIP PATTERNS, LAND CONSUMPTION AND MONOCENTRISM

For a given population, the higher the density, the smaller the built-up area. Providing the built-up

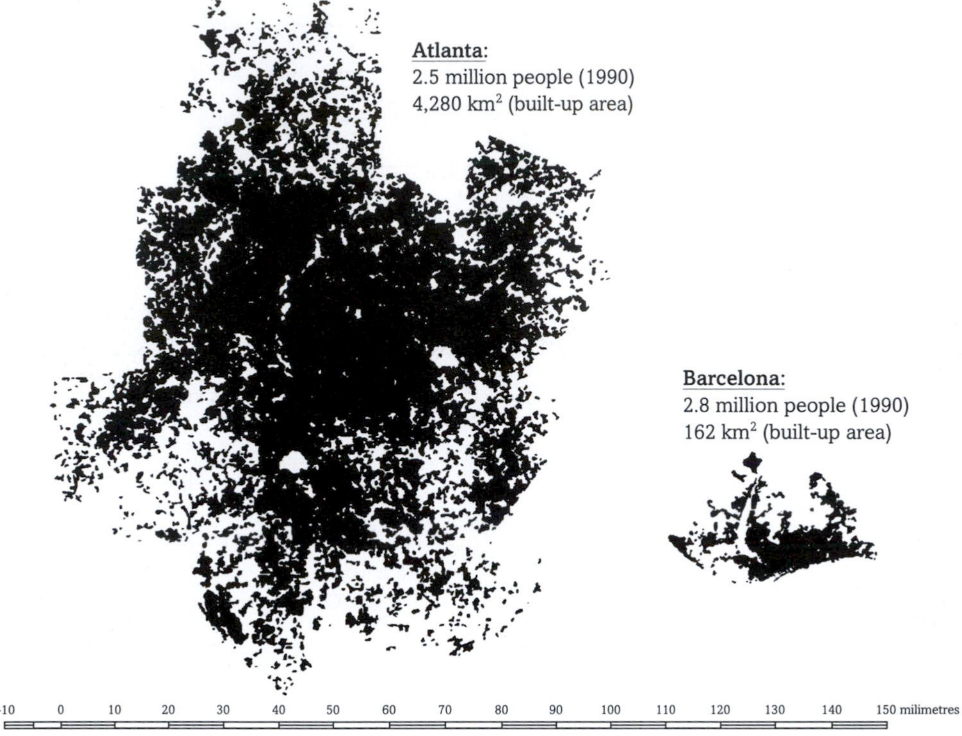

Figure 47.5 The built-up areas of Atlanta and Barcelona represented at the same scale

area is roughly contiguous – i.e. not formed of large isolated areas like satellite towns – trips in these areas will be shorter in length than in others with low density. The comparison of the built-up area of two cities like Atlanta and Barcelona with similar population (about 2.5 million in 1990) but very different density profiles illustrates this point (Figure 47.5). In Atlanta the longest possible distance between two points within the built-up area is 137 km, in Barcelona it is only 37 km. The short trip distance due to high density in Barcelona makes it possible for a significant number of trips to be done by foot or bicycle. Within the Barcelona municipality, 20 percent of trips are made by walking. In Atlanta, the number of walking trips is so insignificant that it is not even recorded.

But density is not the only factor which influences trip length. In a dominantly monocentric city, trips are usually shorter as the majority of trips come directly from the peripheral areas to the downtown where the center of gravity of the population coincides with the CBD. This is the case in New York, London, Paris, Moscow, Shanghai.

The effect of the spatial distribution of density on trip length is often underestimated. Figure 47.6 shows the variation in trip length produced by different spatial arrangements. It illustrates the ADC (average distance to the center) and ADR (average distance between random points) in an imaginary 12 km × 1 km city of 1 million people, having an average density of 100 p/ha in a built-up area of 100 sq. km. Keeping the average density, population and built-up area constant, it shows the ADC and ADR for twenty variations of spatial structures. The variables are the density of subareas, the location of subareas with different densities and the shape of the built-up area within the limit of a 12 km × 12 km square – all factors potentially shaped by market and/or public policy initiatives.

Figure 47.6 presents the spatial organization types ordered by decreasing performance for average distance to the CBD. This exercise yields three important observations:

1 The variation of performance among types is large. For example, between layout 1 and

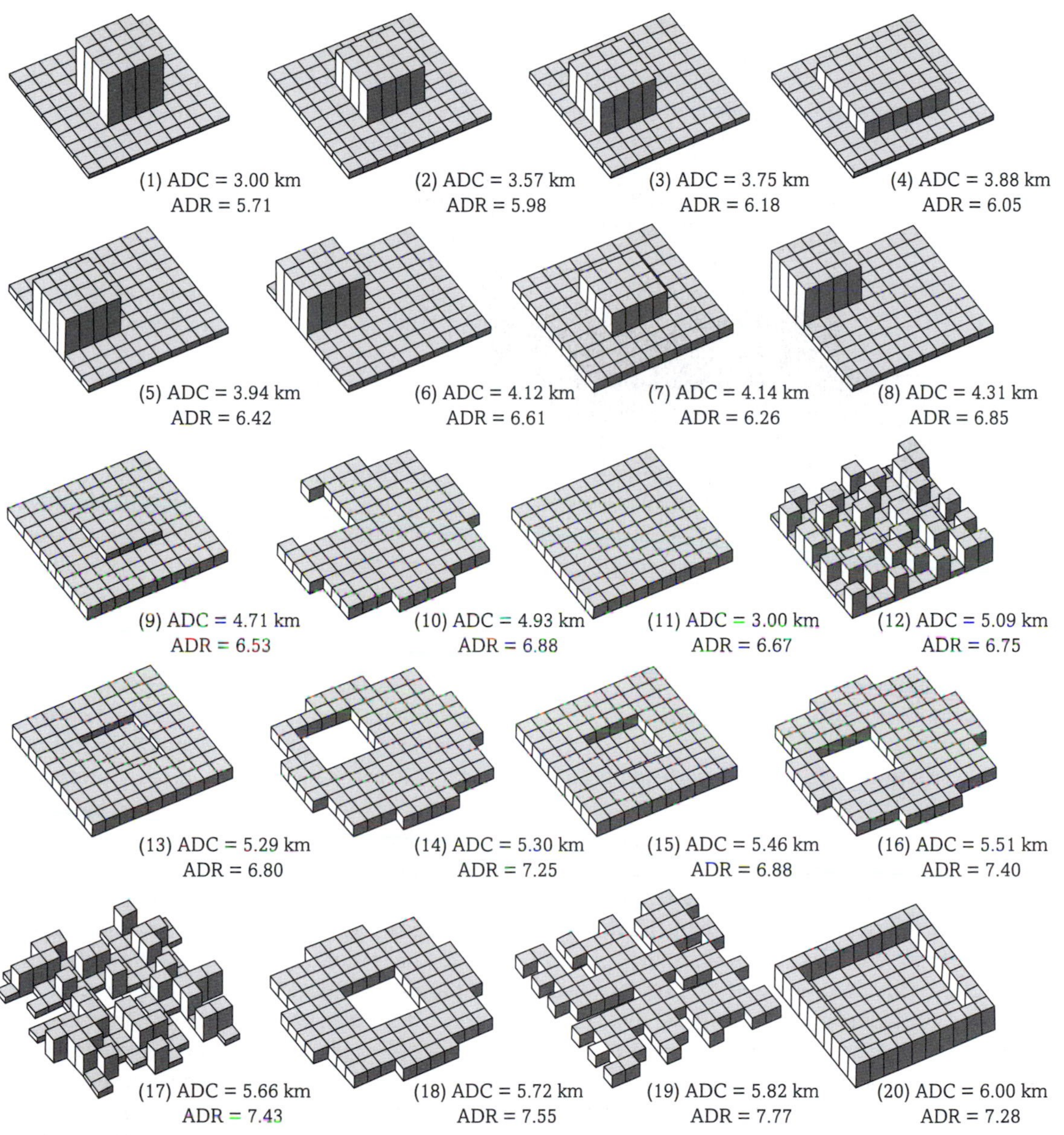

Figure 47.6 Schematic representation of different distribution of density in a city with constant average density and built-up area. *Scale*: Each cube represents 10,000 people. *ADC* Average distance per person to center of gravity. *ADR* Average distance per person between random points.

layout 20 the ADC is double, ranging from 3 km to 6 km. Among cities of identical average density, the distribution of local densities is therefore a very important factor in determining the length (and by implication the costs) of trips and transport networks.

2 The variation in ADC is much larger among different spatial arrangements than it is between the ADC and the ADR for a given shape. Shape itself is more important in city performance than whether a city is monocentric or polycentric.

3 While a poor performer ADC will generally be a poor performer for ADR, the correspondence is not linear. Some types of spatial arrangements, which are favorable to monocentric movements, are not favorable to random movements. For instance, layout ranked 13 for ADC performs better for ADR than the layout ranked 8.

URBAN SPATIAL STRUCTURE AND PUBLIC POLICY QUESTIONS

This exercise has many implications for thinking about public policy issues related to planning. Following are discussions relating urban spatial organization to managing private cars and mass transit, to alleviating transport-generated pollution and dealing with jobs access and residential mobility or housing choice, topics of particular importance in supporting a large unified labor market and the associated issue of addressing the needs of the poor.

Managing cars and mass transit

So far, densities are expressed in people per hectare of land (p/ha), but they can also be expressed in square meter of land per person (m^2/p). For instance the average density of Atlanta, 6 p/ha, corresponds to a consumption of land of 1,666 m^2 per person, and Barcelona's density of 171 p/ha corresponds to a consumption of 58 m^2 per person. In the CBD of the European and Asian cities whose density profile is shown on Figure 47.4, the density of around 250 p/ha corresponds to a land consumption of 40 m^2 per person. A private car also needs about 40 m^2 for movement and parking. In Atlanta, a car will occupy only a small fraction (0.4 percent) of the land available per person while in the center of an Asian or European city a private car would require about the same space as a person. The more cars are introduced in the CBDs of dense cities, the more they compete for space with people, commerce, open space and all sorts of amenities.

There is definitely a threshold of density beyond which private car access should be severely restricted or even banned. Unfortunately, most cities do not do this but encourage widespread use of cars by offering unlimited access to streets and subsidized or free parking. Some cities have developed exemplary programs such as creating pedestrian-only streets, as in Copenhagen or Santa Monica, CA, or restricting access to historic centers as in Rome, Cracow and Riga. London is controlling vehicles through its congestion pricing program that imposes a tax for the vehicular use of streets in specified areas. Similarly, Singapore that by Asian standards is not particularly dense, uses technology to fine-tune the pricing of downtown access while at the same time keeping the costs low.

While high densities are incompatible with the use of private car, the reverse is true for transit. Transit is incompatible with low densities and with spatial structures that are dominantly polycentric.

There is a purely geometric explanation to why low densities are incompatible with transit. Transit stations or bus stops have to be accessible by people walking from or to their residences, their jobs or whatever other activities require the trip. The walking speed of human beings in a city is limited by physiology to about 4.5 km an hour. Acceptable walking time to a transit station varies with culture and income but surveys show that most people will not walk more than ten minutes to a transit station or to a bus stop. (Notably, the acceptable walking time is usually higher for a metro station than for a bus stop.) A ten-minute walking distance at 4.5 km/hour speed corresponds to an accessibility radius of about 800 m to a bus stop or a metro station. A radius of 800 m in a street grid pattern will correspond to a catchment area varying between about 110 ha and 128 ha, depending on the arrangement between transit stop interval and transit line distances. A rule-of-thumb maximum catchment area is 120 ha per transit stop. When the number of people living, or working or shopping, within this 120 ha fall below a certain threshold, transit becomes unpractical to the user and financially unfeasible to the operator. There seems to be a consensus among various researchers and operators that the density threshold for transit is around 30 p/ha (see Table 47.1).

A comparison between Atlanta and Barcelona illustrates how densities affect mass transit operation. With nearly the same number of inhabitants, Barcelona's metro network is 99 km long and 60 percent of the population live less than 600 m from one of its 136 metro stations. Atlanta's metro

Table 47.1 Recomended built-up and residential densities for various level of transit services

	Density (p/ha)	
	Built-up	Residential
Boris Pushkarev and Jeffrey Zupan (1982)		
1 Bus: intermed serv, 1/2 mi between routes, 40 buses/day 7 du/res ac	29	42
2 Bus: freq serv, 1/2 mi between routes, 120 buses/day 15 du/res ac	62	89
3 Light rail: 5 min peak headways, 9 du/res ac, 25–100 sq mi corridor	37	53
4 Rapid tr: 5 min peak headways, 12 du/res ac, 100–150 sq mi corridor	50	71
The Institute of Transportation Engineers (1989) recommends the following minimums:		
1 1 bus/hour, 4 to 6 du/res ac, 5 to 8 msf of commercial/office	21	30
2 1 bus/30 min, 7 to 8 du/res ac, 8 to 20 msf of commercial/office	31	44
3 Lt rail and feeder buses, 9 du/res ac, 35 to 50 msf of commercial/office	37	53
Peter Newman and Jeffrey Kenworthy (1989)		
"public transit oriented urban lifestyle"	35	50

Extracted from John Holtzclaw, "Using Residential Patterns and Transit to Decrease Auto Dependence and Costs"; Natural Resources Defense Council, June 1994

Note
one acre = 0.405 ha. Persons/dw = 2.4. % residential 70.

network is 74 km long – not so different from Barcelona – but only 4 percent of the population live within 800 from a metro station. Predictably, in Atlanta only 4.5 percent of trips are made by transit versus 30 percent in Barcelona. If Atlanta, with its current spatial structure, wanted to provide its population with the same metro accessibility as Barcelona, it would have to build 3,400 km of metro tracks and about 2,800 new metro stations. An analysis of bus rapid transit yields the same results. With its low density (6 p/ha) – compared to Barcelona's 171 p/ha – Atlanta would have difficulties developing a viable bus system.

If Atlanta's low density precludes developing transit then encouraging the needed higher density, as many reports are fond of recommending, might be a solution. However, to reach a critical mass (30 p/ha threshold) over a period of twenty years, and assuming that its historical population growth rate of 2.7 percent per year continues uninterrupted, the city would have to shrink by 67 percent and move its population and jobs into the remaining area – not a feasible solution.

The Atlanta example shows how a low-density spatial structure limits the number of alternative transportation policies. And, in this case, density is not the only constraining spatial factor. A dominantly polycentric structure is also a hindrance. In monocentric cities most trips have multiple origins (the periphery) but a single "clustered" destination (the city center) which allows simple radial routes. In polycentric cities most trips have multiple origins and multiple destinations. Consequently, a dominantly polycentric structure creates a multiplicity of routes with few riders.

Reducing transport-induced air pollution

The amount of air pollution generated by urban transport depends on the length, speed and number of motorized trips and the type of vehicles. For a given urban population, the length and number of motorized daily trips are closely correlated with the average population density in built-up areas, and the spatial distribution of trip destinations and origins. Compared to high-density monocentric cities, the spatial structure of low-density polycentric cities has a double effect on transportation-generated pollution. First, it increases trip length. Second, it increases the number of motorized trips as the proportion of transit trips.

word of caution: engine technology and fuel also play an important role in the amount of vehicular pollution and can counteract or attenuate the effect of unfavorable spatial structure. Again an Atlanta–Barcelona comparison is useful. In 1999 air pollution was higher in Barcelona than Atlanta. As measured by the average yearly level of nitrogen oxides (NOx), Barcelona had 55 mg/m^3 and Atlanta, 47 mg/m^3 despite the fact that in Barcelona 30 percent of all trips use transit and 10 percent are walking trips – compared to 4 percent for Atlanta. Barcelona's laxer emission standards for vehicles, especially for cars fueled by diesel – about 55 percent of private cars use diesel in Barcelona. In addition, vehicles tend to be older in Spain than they are in the United States. And older vehicles emit more pollutants. Finally, the level of pollution exposure in dense monocentric center city areas might be higher because they may have more intense and slower traffic. Thus public policy must be calibrated to a city's situation. Higher emissions control standards, mandatory vehicular inspections and traffic management techniques (e.g. strict ban of on-street parking to increase the speed of traffic flows) can counteract high pollution exposure in central city areas.

Access to jobs and residential mobility

Urban spatial structures affect access to jobs and residential mobility, issues of particular importance to the poor. When it comes to access to jobs, the poor are better off in a dense monocentric city. However, in this type of city the poor are more likely to consume less land and floor space than in low-density polycentric cities and the quality of their environment in a number of ways may be worse. In countries where the poor cannot afford private cars or where the large size of the city precludes their walking to work, dense monocentric cities that reduce distances and allow development of public transport give the poor have a better chance to have a good job access.

However, there are caveats to the wholesale endorsement of the high-density, monocentric city as serving the poor best. In these places, land is usually expensive. Consequently, the poor can afford only minimal amounts of land and floor space for housing, often forcing them to live at high densities in settlements that require sophisticated and expensive infrastructure and services. For instance, a leaky sewer in a high-density settlement (800 p/ha, a common density in Asian cities) causes a much more severe health hazard than the same leaky sewer in a low-density area (50 p/ha). Similarly, a deficient solid waste collection system is less damaging to the environment in low-density settlements than in high-density ones.

Residential mobility – defined as being able to change residential location to maximize its own welfare – is another important general welfare issue that is important for the poor. However, many well intentioned land use laws and subsidized housing programs tend to severely limit the residential mobility of the poor.

In a free market system, urban land is always affordable to most income groups. For a given price of land, each income group will adjust its consumption of land (and therefore density) and makes its own tradeoff between distance to work and land and floor space consumption. However, land use regulations by establishing minimum plot sizes and floor area ratios restrict choices for some households who can afford only the quantity of land below the arbitrary legal minimum. Many land use regulations have the effect of segregating the poor.

In a city where the land market follows the model as shown on Figure 47.7 prices decrease with distance from the center. Assume that two income groups, A and B, are able to pay respectively $5,000 and $20,000 for land. The affordable density for each group will vary by distance to the center as represented in the figure. The two density curves for each group represent only the density – and therefore the area of land – which can be purchased for the amount of money each group is willing to pay for land. For instance, the group A can afford to live at 3 km from the center at a minimum density of about 220 p/ha (45 m^2/person), while at the same distance group B can afford a minimum density of 60 p/ha (166 m^2/person).

Now assume that a well intentioned urban planner draws a citywide zoning plan containing – as zoning plans always do – restrictions on minimum plot size, floor area ratio and setbacks. Their cumulative effect will be to set an upper limit on densities within the boundaries of each zone, as represented by the red dotted line on Figure 47.9. The result is to exclude group A from most areas of the city except between a distance of 3 km and 4 km and beyond 14 km from the center. In this

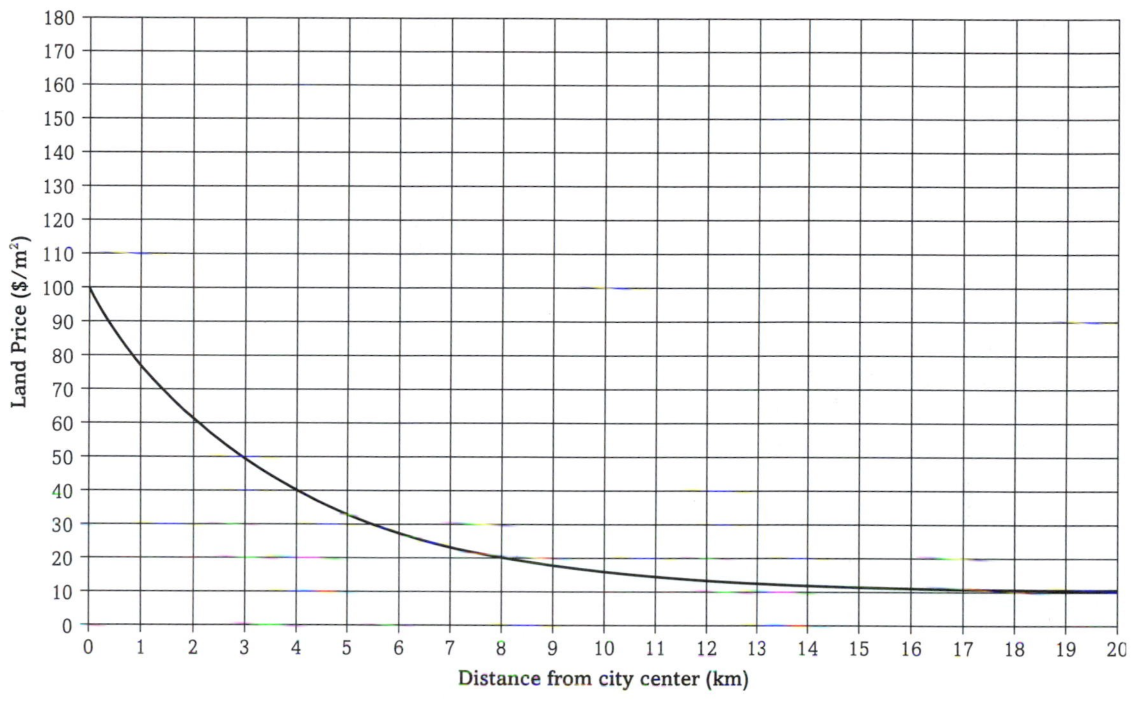

Figure 47.7 Variation of the price of land with distance to the city center in a monocentric city

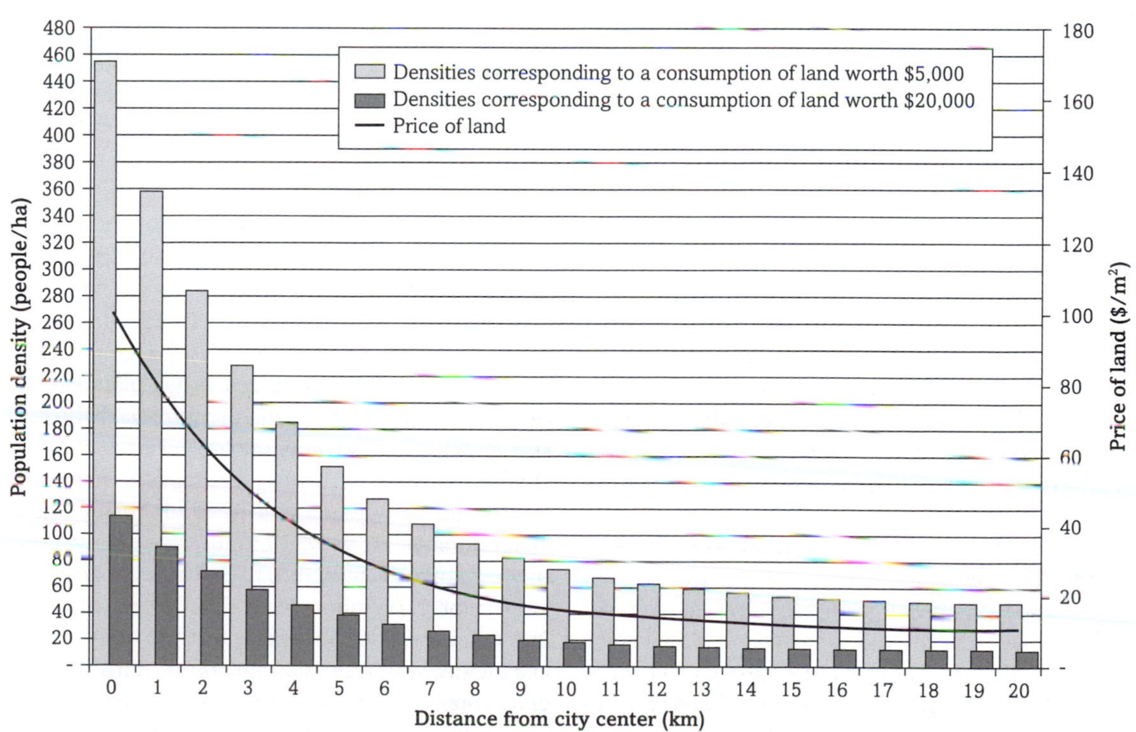

Figure 47.8 For two income groups, densities corresponding to a fixed expenditure for land when distance from the city center varies

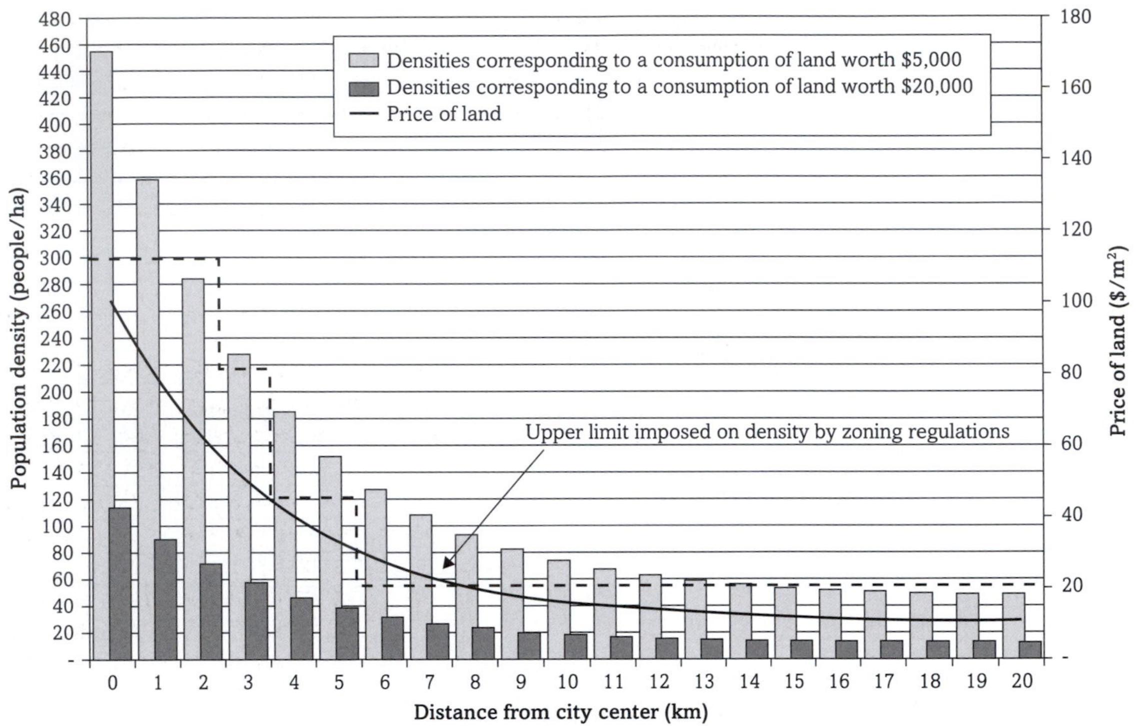

Figure 47.9 Zoning restrictions and affordable densities corresponding to a fixed expenditure for land

particular case, group A would be practically relegated to the periphery of the city. An alternative for group A would be to bypass the effect of regulations by switching to the informal sector, but in doing so group A will lose a part of its property rights.

TWO REMAINING QUESTIONS

Employing the metrics (daily trip patterns, average land consumption and density profiles) for evaluating different urban spatial structures leads to two remaining questions. How can planners affect their cities' spatial structure? Is there a global trend in the evolution of urban spatial structures?

How can planners affect their cities' spatial structure?

Varied spatial structures have different advantages. Some cities may attempt to change an existing urban shape in order to meet some specific objectives, as for instance a reduction of motorized trips. The following principles should guide their thinking. First, there is no optimal spatial structure among many types of spatial organization. However, places that have a positive density profile and dispersed noncontiguous urban development are costly to service, have many negative environmental side effects and should be avoided. Second, urban spatial structures are very resilient. It is easier to decrease density than to increase it, and it is easier for a monocentric city to become polycentric than the opposite.

In the face of market forces, urban planners can influence urban spatial structure indirectly through regulation, taxation and public infrastructure investment. Figure 47.10 offers a schematic view of the interaction of markets and government action in shaping cities. In order to influence the evolution of a city's spatial structure these three tools should be carefully coordinated and be internally consistent to meet a common spatial objective. Consistency is very rare, as regulations, infrastructure investments and taxes are often designed at different levels of government and for very different purposes, often having nothing to do city spatial structure.

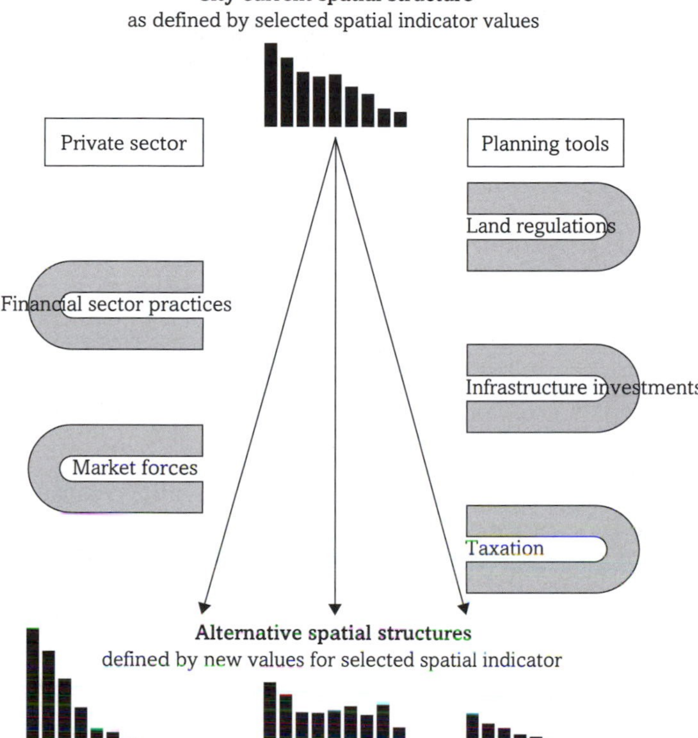

Figure 47.10 Interaction between market forces and government action. Spatial outcomes, market forces, and planning tools

While it is possible to influence spatial trends in a limited way, a more ambitious outcome requires strong, long-term and continuous political determination combined with a dedicated and well integrated team of urban planners, municipal and transport engineers and financial planners. Curitiba, Brazil, is an example of a city where these conditions were met and where without doubt the current city structure is the result of a concerted long-range effort. It is, however, not clear whether the resulting spatial structure has resulted in a welfare increase for the majority of the inhabitants of Curitiba, compared to what it would have been if the structure had been more market-driven.

Is there a global trend in the evolution of urban spatial structures?

With the increase of income and mobility in the world's large cities, dominantly monocentric cities tend to become less monocentric and the densities in of their central business districts also tend lessen. The existence of this trend, however, does not mean that every city is now tending toward a low density extremely polycentric model.

As cities become larger, the CBD also grows and in the process loses the proximity which made it attractive in the first place. It is therefore inevitable that subcenters will emerge as a city becomes larger and the degree of monocentricity decreases with size. However, some very large cities like New York, London, Buenos Aires or Shanghai retain very strong centers, which while containing an ever dwindling ratio of total jobs remain a very strong attractor for prestige retail, entertainment and culture. By contrast, some very successful cities manage to grow without any prestigious center, Atlanta or Phoenix are good examples of this type of city.

In the future, possibly two types of city will emerge. Both will be polycentric in terms of jobs distribution. The first type will retain a strong, prestigious center with a high level of amenities surrounded by a high-density residential areas inhabited by mostly high- and middle-income households. The second type will be a pure labor market without any centrally located amenities; jobs and whatever amenities are provided will be evenly distributed throughout the metropolitan area without any prestigious center.

NOTE

1 To capture them, analysts calculate the average distance per person to the center (ADC) and average distance per person between random points (ADR) using the following formulae.

The built-up area of a city of population P is divided into n cells (each cell could correspond to the portion of a census tract within the built-up area or to an arbitrary square within a grid overlapping the built-up area and for which the population within each square has been calculated.

Average distance per person to the city center Dc is:

$$\overline{D}c = \frac{\sum_{i=1}^{n} d_i \cdot p_i}{P} \qquad \text{(formula 1)}$$

where there are n cells, d_i is the distance from cell i to the center and p_i is the population of cell i. P is the total population of the city.

Average distance per person to random points Dr is given by the formula:

for cell i the average distance d_i to the n other cells is:

$$\overline{d}_i = \frac{\sum_{j=1}^{n} d_{i,j}}{n} \qquad \text{(formula 2)}$$

where d_{ij} is the distance from cell i to cell j.

The average distance per person between random points Dr is:

$$\overline{Dr} = \frac{\sum_{i=1}^{n} \overline{d}_i p_i}{P} \qquad \text{(formula 3)}$$

where p_i is the population of cell i and P is the total population of the city.

In practical terms I use a GIS to first create a set of polygons (set 1) corresponding to census tracts. I create a second set of polygons (set 2) corresponding to the built-up area (using satellite images). I calculate then the built-up density of each census tract by dividing the population of the tract by the portion of the tract that is built-up. I enter the built-up density as a new field in each census tract. I intersect the two sets to create a third set of polygons (set 3) that corresponds to the part of the census tract that are built. I create a fourth set of polygons (set 4) consisting of concentric circles at 1 km intervals centered on the CBD. The intersection of sets 4 and 3 give me a fifth set of polygons (set 5) that contains population and distance to center. For the random distance I conduct the same operation, except that I use a square grid instead of circles. The distances are calculated between the centroids of each cell.

Permissions

ILLUSTRATION CREDITS

Every effort has been made to contact copyright holders for their permission to reprint plates in this book. The publishers would be grateful to hear from any copyright holder who is not here acknowledged and will undertake to rectify any errors or omissions in future editions. Following is copyright information for the plates that appear in this book.

Cover photograph

"View of the foot of Market Street" Credit: PennPraxis from "A Civic Vision for the Central Delaware" 2007.

Part titles

The illustrations with which each part of the book opens are the work of Andrew Whittemore, © 2007 by Andrew Whittemore, and are reproduced by permission of the artist.

Plates

Plates 1–16 Planning today and yesterday (between Parts Two and Three)

1 World Urbanization © 2007 from National Geographic Society. Reproduced by permission of the National Geographic Society.
2 U.S. Large Cities 1970 to 2000. © by Eugénie Birch.
3 The 1791 L'Enfant Plan for Washington 1791 and the 1997 Monumental Core Framework. Public domain and © 1998 from *Extending the Legacy, Planning the Nation's Capital for the 21st Century* (Washington, DC: National Capital Planning Commission, 1998) by National Capital Planning Commission. Reproduced by permission of the National Capital Planning Commission.
4 Philadelphia Greenways Over Time. William Penn Plan: source: Library of Congress; Philadelphia Green Network. © 2007 from Center City District, *State of Center City, 2007* (Philadelphia, PA: Center City District, 2007) by Center City District. Reproduced by permission of Center City District.
5 New York City: Carver Houses, Before and After: East 102nd Street, 5/8/50; East 103rd Street, 9/7/51; Children Posing in Carver Houses, 6/6/57; Carver Houses Amphitheater, March, 1972 from the LaGuardia and Wagner Archives, LaGuardia Community College © New York Housing Authority. Reproduced by permission of the LaGuardia and Wagner Archives, LaGuardia Community College.

6 Chicago: Daniel Burnham's 1909 Plan for Chicago and Chicago 2020 Metropolis Plan from . . . Reproduced by permission of Chicago Metropolis 2020.
7 New York Building Zone Resolution maps in two eras: 1916 and 2000. Public domain from City of New York, *New York City Zoning Resolution* (New York: Department of City Planning, 2007 and 1916).
8 Frank Lloyd Wright's Broadacre City and Le Corbusier's Ville Contemporaine. Reproduced by permission of the Fondation Le Corbusier Artists' Rights Society.
9 Radburn. © 1929 from Regional Plan Association, *Plan for New York and Its Environs* (New York: Committee on Regional Plan of New York and Its Environs, 1929). Reproduced by permission of the Regional Plan Association.
10 The Atlanta Metro Region 2.1 Community Classifications. © 2002 from Myron Orfield, *American Metropolitics, The New Suburban Reality* (Washington, DC.: The Brookings Institution, 2002) by the Brookings Institution. Reproduced by permission of The Brookings Institution.
11 Megaregions Map and Northeast Ecostructure Map. © 2006 from Regional Plan Associaton, *America 2050: A Prospectus* (New York: Regional Plan Association, 2006) and University of Pennsylvania, *Uniting People Places and Systems, Megalopolis Unbound* (Department of City and Regional Planning, 2006). Reproduced by permission of the Regional Plan Association of New York and the Department of City and Regional Planning, University of Pennsylvania.
12 Sprawl c 2006. © 2006 by Emily Yuhas. Reproduced by permission of Emily Yuhas.
13 Planning Agents from Eugénie L. Birch, "U.S. Planning Culture Under Pressure, Major Elements Endure and Flourish in the Face of Crisis," in Bishwapriya Sanyal, editor, *Comparative Planning Cultures*, London: Routledge (2005). © 2005 from *Comparative Planning Cultures* edited by Bishwapriya Sanyal. Reproduced by permission of Routledge/Taylor & Francis Group, LLC.
14 The Transect. © 2007 from Duany Plater-Zyberk and Company. Reproduced by permission of Duany Plater-Zyberk and Company.
15 PlanNYC 2030. © 2007 from City of New York, *PLANYC2030: A Greener Greater New York* by New York City Mayors Office of Strategic Planning.
16 Metropolitan and Micropolitan Statistical Areas of the United States and Puerto Rico. Public Domain from United States Census Bureau, 2004. Reproduced by permission of the US Census Bureau Geography Division.

Plates 17–32 Representative plans, techniques and results (between Parts Four and Five)

17 Philadelphia's Society Hill 1950s and 2007. © Eugénie L Birch.
18 Growth Scenarios in Florida. © 2006 from Planning Studio by Master of City Planning Students, University of Pennsylvania. Reproduced by permission of the Department of City and Regional Planning, University of Pennsylvania.
19 State Plans for Maryland and New Jersey: State Plan for Maryland. © 2007 from *Maryland Priority Funding Areas and Protected Lands* by Maryland Department of Planning. Reproduced by permission of Maryland Department of Planning; State Plan for New Jersey. © 2001 from *Policy Map of the New Jersey State Development and Redevelopment Plan* by New Jersey State Planning Commission. Reproduced by permission of New Jersey Department of Community Affairs.
20 Listening to the City by Jaqueline Hemerdinger. Reproduced by permission of Jaqueline Hemerdinger.
21 Delaware Ohio Comprehensive Plans Over Time. © 2004 from *City of Delaware, Ohio, Comprehensive Plan* by Department of Planning and Community Development, Delaware, Ohio. Reproduced by permission of Department of Planning and Community Development, City of Delaware, Ohio.
22 GIS and Visualization. Public Domain from *Mapping for Congress: Supporting Public Policy with GIS, 2006* by ESRI and Library of Congress, Congressional Cartography Program edited by Nancy Sappington.

23 Atlanta's Emerald Necklace Beltway. © 2004 from *The Beltline Emerald Necklace: Atlanta's New Public Realm* by Alexander Garvin & Associates. Reproduced by permission of Alexander Garvin.
24 Compact Development. Public Domain from Northeastern Illinois Planning Commission, *Realizing the Vision, Regional Framework Plan* (Chicago: Northeastern Illinois Planning Commission, 2005).
25 Chattanooga Waterfront. © Chattanooga Pulse/Culture Systems. Reproduced by permission of Chattanooga Pulse/Culture Systems.
26 San Francisco's Rincon Hill and Transbay Center. © 2005 from *The Redevelopment Plan for the Transbay Redevelopment Project* (San Francisco: San Francisco Redevelopment Agency, 2005) by San Francisco Redevelopment Agency. Reproduced by permission of San Francisco Redevelopment Agency.
27 Pittsburgh's Herr's Island. Public Domain from Urban Redevelopment Authority of Pittsburgh. Reproduced by permission of Urban Redevelopment Authority of Pittsburgh.
28 Richmond: Neighborhoods in Bloom. © 2005 from *The Impacts of Targeted Public and Non-Profit Investment on Neighborhood Development* (Richmond: Community Affairs Office, July 2005) by Federal Reserve Bank of Richmond. Reproduced by permission of the Federal Reserve Bank of Richmond.
29 New Jersey Highlands Plan. © 2007 from *Highlands Regional Master Plan, Final Draft* by New Jersey Highlands Water Protection and Planning Council. Reproduced by permission of New Jersey Highlands Council.
30 Anchor Institutions, University of Pennsylvania. © University of Pennsylvania: Rendering of East Campus of the University of Pennsylvania by Sasaki and Associates. Reproduced by permission of University of Pennsylvania.
31 Priority regions and Corridors in California. © 2007 from *Goods Movement Action Plan* by Business, Transportation and Housing Agency, and California Environmental Protection Agency. Reproduced by permission of the State of Califonia, Business, Transportation and Housing Agency, Office of the Deputy Secretary of Goods Movement.
32 Fort Worth, Texas Urban Villages. © 2008 from City of Fort Worth, Texas Department of Planning and Development. Reproduced by permission of City of Fort Worth, Texas, Department of Planning and Development.

SELECTION CREDITS: COPYRIGHT INFORMATION

Every effort has been made to contact copyright holders for their permission to reprint selections in this book. The publishers would be grateful to hear from any copyright holder who is not here acknowledged and will undertake to rectify any errors or omissions in future editions of this book. Following is copyright information.

Part One The World of Urban and Regional Planning

1 Jay H. Moor and Rasna Warah, "The State of the World's Cities," *Global Outlook: International Urban Research Monitor*, a joint publication of the U.S. Department of Housing and Urban Development's Office of International Affairs and the Woodrow Wilson International Center for Scholars, 2001. Public domain.
2 U.S. Department of Housing and Urban Development, "Megaforces Shaping the Future of the Nation's Cities" *State of the Cities,* 2000. Public domain.
3 Catherine L. Ross and Nancey Green Leigh, "Planning, Urban Revitalization and Inner City: An Exploration of Structural Racism," *Journal of Planning Literature* 14, 3, pp. 367–380, Copyright © 2000 by Sage Publications, Inc. Reprinted by Permission of Sage Publications, Inc.

4 Mohammed, A. Qadeer, "Pluralistic Planning for Multicultural Cities," *Journal of the American Planning Association* 63, 4, pp. 481–94 (Fall 1997). Reprinted with permission from the *Journal of the American Planning Association*. Copyright © Fall 1997 by the American Planning Association.

5 Joel Kotkin, "Suburbia: Homeland of the American Future," *The Next American City,* Issue 11, pp. 19–22 (Summer, 2006). Copyright © Summer 2006 by Joel Kotkin. Reprinted with permission from JGPR Inc.

6 Myron Orfield, "The New Suburban Typology," from *American Metropolitics: The New Suburban Reality* (Washington D.C.: The Brookings Institution Press, 2002). Copyright © 2002 by The Brookings Institution. Reprinted by permission of The Brookings Institution Press.

Part Two History and Theory of Urban and Regional Planning

7 Richard T. LeGates and Frederic Stout, "Modernism and Early Urban Planning, 1870–1940" from Richard T. LeGates and Frederick Stout (eds) *Early Urban Planning, 1870–1940* (London: Routledge/Thoemmes Press, 1998). Copyright © 1998 by Richard T. LeGates and Frederic Stout. Reprinted by permission of Richard T. LeGates and Frederick Stout.

8 Michael B. Teitz, "Reflections and Research on the U.S. Experience," reprinted by permission from Lloyd Rodwin and Bishwapriya Sanyal, eds., *The Profession of City Planning, Changes, Images and Challenges: 1950–2000.* Copyright © 2000 Rutgers University, Center for Urban Policy Research. All rights reserved. Originally published as "American Planning in the 1990s: Part II, The Dilemma of the Cities" (1997) *Urban Studies* 34, 5–6, pp. 775–796. Reprinted with permission from Taylor and Francis. http://www.tandf.co.uk.

9 Eugénie L. Birch, "Practitioners and the Art of Planning" *Journal of Planning Education and Research* 20, pp. 407–422. Copyright © 2001 by Sage Publications, Inc. Reprinted by Permission of Sage Publications, Inc.

10 Dolores Hayden, "The Shapes of Suburbia," from *Building Suburbia: Green Fields and Urban Growth, 1820–2000* (New York: Pantheon Books, 2003). Copyright © 2003 by Dolores Hayden. Reprinted by permission of Ellen Levine Literary Agency / Trident Media Group.

11 Edward J. Sullivan and Matthew J. Michel, "*Ramapo* plus Thirty: The Changing Role of the Plan in Land Use Regulation," by Edward J. Sullivan & Matthew J. Michel, published in *Urban Lawyer*, Volume 35, No. 1, Winter 2003. Copyright © 2003 by the American Bar Association. Reproduced with permission. All rights reserved. This information or any portion thereof may not be copied or disseminated in any form or by any means or downloaded or stored in any electronic database or retrieval system without the express written consent of the American Bar Association.

12 Nigel Taylor, "Anglo-American Town Planning Theory since 1945: Three Significant Developments but no Paradigm Shifts," *Planning Perspectives* 14, 4, pp. 327–345 (1999). Reprinted with permission from Taylor and Francis. http://www.tandf.co.uk.

13 Susan Fainstein, "New Directions in Planning Theory," *Urban Affairs Review* 34, 4, pp. 451–478, Copyright © 2000 by Sage Publications. Reprinted by Permission of Sage Publications, Inc.

Part Three Classics in Urban and Regional Planning

14 Kevin Lynch, "Dimensions of Performance" from *A Theory of Good City Form* (MIT Press: Cambridge, MA, 1984). Copyright © 1984 by MIT Press. Reprinted by Permission of MIT Press, Cambridge, MA. "A Theory of Good City Form," copyright © 1981 by Kevin Lynch, published by MIT Press, 1981.

15 Jane Jacobs, "Downtown is for People," from Editors of Fortune, *The Exploding Metropolis* (New York: Doubleday Anchor Books, 1958). © 1958 Time Inc. Reprinted by Permission of Fortune Copyright.

16 Lewis Mumford, "Home Remedies for Urban Cancer," *The New Yorker* XXXVIII, December 1, 1962 pp. 148+. Copyright © 1962 by Elizabeth M. Morss and James G. Morrs, renewed in 1990. Reprinted by Permission of Gina Maccoby Literary Agency.

17 Charles M. Haar, "The Master Plan, An Impermanent Constitution," *Law and Contemporary Problems* 20, 3, pp. 353–418 (Summer1955). Reprinted with permission from *Law and Contemporary Problems.* © 1955 by Duke Law School.

18 Paul Davidoff, "Advocacy and Pluralism in Planning," *Journal of the American Institute of Planners* 31, 4, pp. 331–38 (November 1965). Reprinted with permission from the *Journal of the American Planning Association*. Copyright © November 1965 by the American Planning Association.

19 Ian McHarg, "The Metropolitan Region," from *Design with Nature* (Garden City, NY: Natural History Press, 1969). Copyright © 1992 by John Wiley & Sons, Inc. This material is used by permission of John Wiley & Sons, Inc.

20 Robert D. Yaro and Tony Hiss, "Paying for the Plan," from *A Region At Risk: The Third Regional Plan for the New York-New Jersey-Connecticut Metropolitan Area*by Robert D. Yaro and Tony Hiss. Copyright © 1996 by Regional Plan Association. Reproduced with permission of Island Press, Washington D.C.

Part Four The Plan: its Origins and Contemporary Uses

21 Edward J. Kaiser and David R. Godshalk, "Twentieth Century Land Use Planning: a Stalwart Family Tree," *Journal of the American Planning Association* 61, 3 (1995), pp. 365–385. Reprinted with permission from the *Journal of the American Planning Association.* Copyright © summer 1995 by the American Planning Association.

22 Oliver Gillham, "Regionalism," from *The Limitless City, A Primer on the Urban Sprawl Debate* by Oliver Gillham. Copyright © 2002 by Oliver Gillham. Reproduced with permission of Island Press, Washington, D.C.

23 Michael Neuman, "Does Planning Need the Plan?" *Journal of the American Planning Association* 64, 2, pp. 208–216 (Spring 1998). Reprinted with permission from the *Journal of the American Planning Association*. Copyright © spring 1998 by the American Planning Association.

24 Lewis D. Hopkins, "How Plans Work," from *Urban Development, The Logic of Making Plans* by Lewis D. Hopkins. Copyright © 2001 by Lewis D. Hopkins. Reproduced with permission of Island Press, Washington D.C.

25 Edward J. Kaiser, David R. Godschalk and F. Stuart Chapin, "The Land Planning Arena," from *Urban Land Use Planning Fourth Edition*. Copyright © 1995 by Board of Trustees. Used with permission of the University of Illinois Press.

26 Jonathan Barnett, "Shaping Cities through Development Regulations," from *Redesigning Cities: Principles, Practice, Implementation* (Chicago: APA Planners Press, 2003). Reprinted with permission from *Redesigning Cities*. Copyright © 2003 by the American Planning Association.

Part Five Planning Practice and Methods

27 Charles J. Hoch, Linda C. Dalton and Frank S. So, "Introduction: Planning for People and Places" from Charles J. Hoch, Linda C. Dalton, and Frank S. So (eds), *The Practice of Local Government Planning, 3rd Edition*, 2000. Reprinted with permission of the International City/County Management Association, 777 North Capitol Street NE, Suite 500, Washington, DC 20002. All rights reserved.

28 Frank So and Judith Getzels, "Planning Environments" from Frank S. So and Judith Getzels (eds), *The Practice of Local Government Planning, 2nd Edition*, 1988. Reprinted with permission of the

International City/County Management Association, 777 North Capitol Street NE, Suite 500, Washington, DC 2002. All rights reserved.

29 Alexander Garvin, "A Realistic Approach to City and Suburban Planning . . . Ingredients of Success," from *The American City: What Works, What Doesn't, 2nd Edition* (New York: McGraw Hill, 2002). Reprinted with permission of McGraw Hill.

30 Dowell Myers, "Demographic Futures as a Guide to Planning: California Latinos and the Compact City," *Journal of the American Planning Association* 67, 4, pp. 383–97 (Autumn 2001). Reprinted with permission from the *Journal of the American Planning Association*. Copyright © Autumn 2001 by the American Planning Association.

31 Ann-Marie Esnard, Nancy Sappington and Milton R. Ospina, "Geographic Information Systems," from American Planning Association, *Planning and Urban Deign Standards* (Hoboken, NJ: John Wiley & Sons, 2006). Copyright © 2006 John Wiley and Sons, Inc. Reprinted with permission of John Wiley & Sons, Inc.

32 Barbara Faga, "The Future of Public Participation," from *Public Consensus: The Civic Theater of Community Participation for Architects, Landscape Architects, Planners and Urban Designers* (Hoboken, NJ: John Wiley & Sons, 2006). Copyright © 2006 by John Wiley & Sons, Inc. Reprinted by permission of John Wiley & Sons, Inc.

33 Carol D. Barrett, "Introduction," from *Everyday Ethics for Practicing Planners* (Washington D.C.: American Planning Association Press, 2001). Reprinted with permission from *Everyday Ethics for Practicing Planners*. Copyright © 2001 by the American Planning Association.

Part Six Key Topics in Urban and Regional Planning

34 Vukan R. Vuchic, "Implementing the Solutions: Measures for Achieving Intermodal Balance." Reprinted by permission from Vukan R. Vuchic, *Transportation for Livable Cities*. Copyright © 1999 Rutgers University, Center for Urban Policy Research. All rights reserved.

35 Thomas L. Daniels and Katherine Daniels, "Environmental Planning," reprinted with permission from *The Environmental Planning Handbook for Sustainable Communities and Regions*. Copyright © 2003 by the American Planning Association.

36 Eugénie L. Birch, "Hopeful Signs: U.S. Urban Revitalization in the 21st Century," from Gregory K. Ingram and Yu-Hung Hong, editors, *Land Policies and Their Outcomes* (Cambridge, MA: Lincoln Institute of Land Policy, 2007). Copyright © 2007 by the Lincoln Institute of Land Policy. Reprinted with permission from Lincoln Institute of Land Policy.

37 William H. Lucy and David L. Phillips, "Sprawl and the Tyranny of Easy Development Decisions," from *Confronting Suburban Decline: Strategic Planning for Metropolitan Renewal* by William H. Lucy and David L. Phillips. Copyright © 2000 by Island Press. Reproduced by permission of Island Press, Washington, D.C.

38 Andres Duany, Elizabeth Plater-Zybek and Jeff Speck, "How to Make a Town," from *Suburban Nation: The Rise of Sprawl and the Decline of the American Dream* by Andres Duany, Elizabeth Plater-Zybek and Jeff Speck. Copyright © 2000. Reprinted by permission of North Point Press, a division of Farrar, Straus and Giroux, LLC.

39 Gary Hack, "Planning Metropolitan Regions," from Jonathan Barnett (ed) *Planning for a New Century* by Gary Hack. Copyright © 2001 by Island Press. Reproduced with permission of Island Press, Washington, D.C.

Part Seven Emerging Issues in Urban and Regional Planning

40 United Nations Population Fund, "The Promise of Urban Growth," from *State of World Population, 2007*. Public domain.

41 Robert E. Lang and Dawn Dhavale, "Beyond Metropolis: Exploring America's New 'Megapolitan' Geography" (Alexandria, VA: Virginia Polytechnic Institute and State University, 2005). Copyright © 2005 Metropolitan Institute at Virginia Tech. Reprinted by permission of the Metropolitan Institute at Virginia Tech.

42 Timothy Beatley, "Land Use and Urban Form: Planning Compact Cities," from *Green Urbanism, Learning from European Cities* by Timothy Beatley. Copyright © 2000 by Island Press. Reproduced with permission of Island Press, Washington, D.C.

43 John R. Nolon, "Shifting Ground to Address Climate Change: the Land Use Law Solution," from *Government Law and Policy Journal* (2008).

44 Richard A. Matthew and Bryan McDonald, "Cities under Siege: Urban Planning and the Threat of Infectious Disease," *Journal of the American Planning Association* 72, 1, pp. 109–117 (Winter 2006). Reprinted with permission from the *Journal of the American Planning Association*. Copyright © Winter 2006 by the American Planning Association.

45 Miguel Barceló Rota, "22@Barcelona: a New District for the Creative Economy," from Waikeen Ng and Judith Ryser, editors, *Making Spaces for the Creative Economy* (Madrid: International Society of City and Regional Planners, 2005). Copyright © 2005 International Society of City and Regional Planners (ISoCaRP). Reprinted by permission of ISoCaRP and the author.

46 Alan A. Altshuler and David E. Luberoff, "The Changing Politics of Urban Mega-projects," *Land Lines* 15, 4 (October 2003, pp. 1–4). Copyright © Lincoln Institute of Land Policy. Reprinted by permission of *Land Lines* and the Lincoln Institute of Land Policy.

47 Alain Bertaud, "The Spatial Organization of Cities: Deliberate Outcome or Unforeseen Consequence?" Background Paper, *World Development Report. The World Bank* (Washington, DC, 2004). Copyright © Alain Bertaud. Reprinted by permission of Alain Bertaud.

Index

Page numbers in *Italics* represent Tables and page numbers in **Bold** represent Figures